797,885 Books

are available to read at

www.ForgottenBooks.com

Forgotten Books' App
Available for mobile, tablet & eReader

ISBN 978-1-330-10881-9
PIBN 10027891

This book is a reproduction of an important historical work. Forgotten Books uses
state-of-the-art technology to digitally reconstruct the work, preserving the original format
whilst repairing imperfections present in the aged copy. In rare cases, an imperfection in
the original, such as a blemish or missing page, may be replicated in our edition. We do,
however, repair the vast majority of imperfections successfully; any imperfections that
remain are intentionally left to preserve the state of such historical works.

Forgotten Books is a registered trademark of FB &c Ltd.
Copyright © 2017 FB &c Ltd.
FB &c Ltd, Dalton House, 60 Windsor Avenue, London, SW19 2RR.
Company number 08720141. Registered in England and Wales.

For support please visit www.forgottenbooks.com

1 MONTH OF
FREE
READING

at

www.ForgottenBooks.com

By purchasing this book you are eligible for one month membership to ForgottenBooks.com, giving you unlimited access to our entire collection of over 700,000 titles via our web site and mobile apps.

To claim your free month visit: www.forgottenbooks.com/free27891

* Offer is valid for 45 days from date of purchase. Terms and conditions apply.

English
Français
Deutsche
Italiano
Español
Português

www.forgottenbooks.com

Mythology Photography **Fiction**
Fishing Christianity **Art** Cooking
Essays Buddhism Freemasonry
Medicine **Biology** Music **Ancient
Egypt** Evolution Carpentry Physics
Dance Geology **Mathematics** Fitness
Shakespeare **Folklore** Yoga Marketing
Confidence Immortality Biographies
Poetry **Psychology** Witchcraft
Electronics Chemistry History **Law**
Accounting **Philosophy** Anthropology
Alchemy Drama Quantum Mechanics
Atheism Sexual Health **Ancient History**
Entrepreneurship Languages Sport
Paleontology Needlework Islam
Metaphysics Investment Archaeology
Parenting Statistics Criminology
Motivational

MONTHLY NOTICES

OF THE

ROYAL ASTRONOMICAL SOCIETY,

CONTAINING

PAPERS, ABSTRACTS OF PAPERS, AND

REPORTS OF THE PROCEEDINGS

OF THE SOCIETY

FROM NOVEMBER 1881 TO NOVEMBER 1882.

VOL. XLII.

STANFORD LIBRARY

𝔓rinteð by

SPOTTISWOODE & CO., NEW-STREET SQUARE, LONDON.

1882.

273666

STANFORD LIBRARY

MONTHLY NOTICES

OF THE

ROYAL ASTRONOMICAL SOCIETY.

| VOL. XLII. | NOVEMBER 11, 1881. | No. 1. |

E. J. Stone, Esq., F.R.S., Vice-President, in the Chair.

On a simple and practicable Method of Measuring the Relative Apparent Brightnesses or Magnitudes of the Stars with considerable Accuracy. By Professor Pritchard.

Circumstances not generally interesting here to detail have recently diverted my attention from other astronomical work to what is conventionally meant by the magnitudes of the stars. It may be sufficient here to say that, having just completed the reduction of the relative coordinates of 250 stars in the cluster 39 Messier in the constellation *Cygnus*, for comparison by other astronomers at some remotely future period, I felt that perhaps only one-half of the work was done, unless some reliable measures of the relative magnitudes of the stars could be secured. Prof. Pickering's most interesting researches into the results obtainable from accuracy in star magnitudes also weighed on my mind. On consulting the various authorities on the subject, I found the question, whether of result or of method, somewhat in a chaos, so far as any considerable degree of accuracy was concerned. Even Struve, for instance, with all his care, gives on some occasions not less than three different magnitudes for the same star.

In this perplexity, I applied myself to the attempt of devising some method, which might land me in greater certainty, and after some time I hit upon a method which, after applying it to sixty stars, promises certainty, to the extent of one-tenth of a magnitude, in all the stars yet examined, from *a Lyræ* to the stars of least brightness estimated by Argelander in the " Durch-

B

mustering." I offer this contribution, in an important and hitherto obscure branch of astronomy, to the scrutiny of persons competent to form an opinion.

So soon, however, as I had devised the method, it at once occurred to me that it was so simple that it must have been employed by other astronomers besides myself. Accordingly, I found that our old associate Mr. Dawes, whose memory is so greatly respected among us, did employ each of the means separately, by the combination of which I believe I have succeeded. My method, however, of applying and discussing these means will be found to be widely and generically different in principle from his, and especially because he has assumed as the basis of all his results, that a star observed to be only just visible in a telescope of which the diameter of the object glass is 0·15 inch, is of a certain definite and uniform magnitude—viz. the sixth; whereas, I have found, and I am sure other astronomers will find, that there is no constancy in the aperture at which any single star is just visible. The result of observations recently made in this Observatory, as will be seen in the results recorded in the present communication, is that on two different nights, when no differences of meteorological circumstances of the atmosphere were apparent, the same star shone with brilliancies differing by nearly a whole magnitude. Moreover, the effect of a difference of Zenith Distances from 52° to 74° on the same night, was found to affect the apparent brightness by six-tenths of a magnitude.

In referring thus to the methods employed by Mr. Dawes, I cannot pass over the still more ingenious device of Mr. Knobel (see *Monthly Notices*, Dec. 1874, and June 1875), who, by means of a glass mirror, in its two conditions of silvered and unsilvered, effected the same sort of results as those derived from my use of a glass wedge. Had Mr. Knobel seen fit to persevere to the end in forming a complete catalogue of star magnitudes, my own present efforts would have been assisted or modified. Still, it would have been interesting and valuable, if he had more fully recorded the resulting differences of the several individual determinations of the brightness of the same star, in order that the unavoidable errors arising from this (or any other) method might be distinctly seen. In what follows, I have for the most part endeavoured to give these details, my object being to discuss one method of observation, without forming any judgment as to other methods possibly quite as good.

Similar remarks apply to the memorable efforts detailed by Sir John Herschel in the Cape Observations (chap. iii. p. 305). It is perhaps needless to say that they bear the marks of his genius and sagacious perseverance, and cannot properly be disregarded in any stellar photometric researches.

The very successful efforts also of Messrs. Knott and Baxendell, printed so far back as 1863, show what can be done by eye-observations; but as they were not instrumentally photometric, they scarcely fall within the species of methods discussed in this communication.

I will now detail my method of proceeding, and give the results of its application. A wedge of the purest and most homogeneous neutral-tinted glass that could be procured was carefully constructed for me by Mr. Grubb, of Dublin. It is about six inches long, and its greatest thickness one-tenth of an inch. To it is cemented another similar wedge of clear glass reversed, the whole forming a rectangular parallelopipedon. This wedge was provisionally divided, in its cell, to tenths of inches, and was made to slide close to the achromatic lens which formed the eyepiece of an excellent telescope of four inches aperture, attached to the tube of the large Equatoreal of this Observatory. Close on to this wedge, and immediately over the centre of the eye lens, projects a diaphragm, circularly perforated to direct the eye. Of course there is nothing new in this contrivance; its value depends on the mode of using it. A star, convenient for the purpose, was selected, presumed to be stable in its brightness: in this instance it was κ *Lyræ*, for which I shall assume the brightness or magnitude assigned to it by Argelander, 4·7; any other magnitude would serve the purpose perhaps equally well. This star was viewed on a fair night through the four-inch telescope, and the position of the wedge was noticed when the star was just extinguished. The mean of five readings consecutively taken was used in the further discussion. The aperture of the telescope was then reduced to two inches, and the reading of the wedge was taken as before, when the star became just invisible. It was contended that the effect of the thickness of wedge between the two readings produced the same effect as diminishing the aperture from four to two inches. This sort of observation repeated many times on different nights gives, as will be seen, an important constant, indicating the absorbing capacity of the wedge, of which constant great use is made in the following discussion. Thus if τ represent the thickness of the wedge through which the light passes, if L be the amount of the light incident, and L_1 the amount emergent, then

$$\log \frac{L}{L_{\prime}} = c\tau \text{ or } = \kappa I \qquad (1),$$

where I is the length of the interval between the two readings of the wedge, and $I = \tau \times \tan$ angle of the wedge : κ is the constant to be determined.

Inasmuch as the star (κ *Lyræ*) is observed with apertures whose diameters are as 2 : 1, it follows that here

$$\log \frac{L}{L_{\prime}} = \log 4 = \kappa I ;$$

therefore

$$\kappa = \frac{\cdot 6021}{I} \qquad (2).$$

This gives (κ) as a number expressed in terms of the number of intervals marked on the scale of the wedge, between the two observations. In the case of the wedge used in these researches (see Table I.)

$$\kappa = 0.053 \pm 0.0010 \qquad (3),$$

and therefore in general, from (1),

$$\log \frac{L}{L_{,}} = 0.053 \, I \qquad (4).$$

Again, if L_n and L_{n+x} represent the relative amounts of the light in two stars reputed to be of the nth and of the $(n+x)^{\text{th}}$ magnitudes—that is, differing from each other by (x) magnitudes—then by the recognised convention, proposed by Mr. Pogson, and generally accepted, we have

$$\frac{L_n}{L_{n+x}} = (2.512)^x$$

and therefore

$$\log \frac{L_n}{L_{n+x}} = \cdot 4 \begin{array}{l} \text{(diff. of magnitude} \\ \text{of the two stars)} \end{array} \qquad (5).$$

Now, suppose that I_x be the number of wedge intervals between the two readings of the wedge, at which the lights of these two stars are respectively extinguished when viewed successively through the wedge in the eyepiece of the four-inch telescope (or any other telescope), then by (4) we have

$$\log \frac{L_n}{L_{n+x}} = \cdot 053 \, I_x,$$

i.e. by (5) $0 \cdot 4$ (diff. of magnitude) $= 0 \cdot 053 \, I_{,}$,

    or   difference of magnitude $= 0 \cdot 133 \, I_x$ (6).

This expression is the fundamental theorem on which the results in this research are based. Hence it becomes apparent, that, for the particular wedge used in these researches, one interval of the wedge (*i.e.* one-tenth of an inch) indicates a difference of about one-eighth of a magnitude; and as in practice, and in the average of five readings, less than one-half of an interval is practically cognisable in effect, it follows that the method may be relied upon to about ·06 of a magnitude. It appears also that if the original division of the wedge were provisional only, it would be practicable to make the divisions of such a length that each shall be the indication of one-tenth of a magnitude.

The following is an example of the manner in which one

value of κ is obtained, and also the difference of the magnitude between that star and another (Lal. 33612) :—

1881, *October* 17.

Name of star ...	κ Lyræ		κ Lyræ		Lal. 33612	
Aperture of telescope	4 inches.		2 inches.		4 inches.	
	Wedge Reading	Deviation from Mean.	Wedge Reading.	Deviation from Mean.	Wedge Reading.	Deviation from Mean.
	div.		div.		div.	
	19·9	0·34	32·1	0·54	31·1	0·46
	20·2	0·04	31·3	0·26	30·3	0·34
	20·4	0·16	31·4	0·16	30·4	0·24
	21·0	0·76	31·9	0·34	31·4	0·76
	19·7	0·54	31·1	0·46	30·0	0·64
Mean Reading	20·24	0·37	31·56	0·35	30·64	0·49

For κ, the Difference of Mean Readings $= \overset{\text{div.}}{31\cdot56} - \overset{\text{div.}}{20\cdot24} = \overset{\text{div.}}{11\cdot32}$;

therefore

$$\text{by (2)} \quad \dots \quad \dots \quad \dots \quad \kappa = \frac{0\cdot6021}{11\cdot32} = 0\cdot053.$$

For Magnitude of Lalande 33612, the $\Big\}$ Difference of Mean Readings ... $= 30\cdot64 - 20\cdot24 = 10\cdot40$;

therefore

by (6) ... Diff. of Magnitude $= 10\cdot40 \times \cdot133 = 1\cdot38$ magnitude.

The numerical quantities given in equation (3) have been derived from the individual determinations exhibited in the following table.

TABLE I.

Date.	κ	Date.	κ
1881, Oct. 17	0·053	1881, Oct. 26	0·053
	·050	27	·050
18	·060		·041
	·056	29	·050
19	·060	Nov. 2	·058
	·061		·049
25	·051		

of which tabular values the mean is $\kappa = 0\cdot053$ with a probable error of 0·001.

From the observations made in the course of this investigation, it became apparent that on different nights κ *Lyræ*, and consequently other stars, shone with very considerable variations in brightness. For instance, on Oct. 17, the mean reading of the wedge when κ *Lyræ* was just extinguished was 20·24 divisions; on Oct. 18, 22·30 divisions; and on Oct. 26 it was 29·04 visions—indicating an extreme difference of brightness, equiïlent to one magnitude. At the same time, it is observable that le individual determinations of κ do not exhibit any such inïtability. This is very important. Independently of the general ïfect of meteorological circumstances, there was also perceptible ın effect produced on the brightness of stars at different distances from the zenith, owing, no doubt, to the different thicknesses of the atmosphere through which the light necessarily passed. In order to ascertain whether it was possible to connect numerically this alteration of brightness with corresponding alterations of Zenith Distances, I constructed a table from which the ratio of the thickness of the atmosphere at any Zenith Distance (Z) to the thickness at the zenith itself could be obtained at sight. I assumed that the effect was approximately the same as that which would be produced by a homogeneous atmosphere of the height of $5\frac{1}{2}$ miles; but when I had completed it, I found that a similar table had already been constructed, and published in the "Annuaire de l'Observatoire de Montsouris," and reprinted in the "Edinburgh Observations," vol. xiv. The latter table is here subjoined; it seems to assume a height of atmosphere of 50 miles. It is perhaps curious to observe that for moderate altitudes either table may be used with equal advantage for approximate purposes.

TABLE II.

Zenith Distance.	Ratio of Thickness.	Zenith Distance.	Ratio of Thickness.	Zenith Distance.	Ratio of Thickness.
1°	1·00	31°	1·17	61°	2·02
˄	1·00	32	1·18	62	2·09
	1·00	33	1·19	63	2·15
	1·00	34	1·20	64	2·23
5	1·01	35	1·22	65	2·30
6	1·01	36	1·23	66	2·39
7	1·01	37	1·25	67	2·48
8	1·01	38	1·26	68	2·58
9	1·02	39	1·28	69	2·68
10	1·02	40	1·30	70	2·80
11	1·02	41	1·32	71	2·93
12	1·03	42	1·34	72	3·07
13	1·03	43	1·36	73	3·22
14	1·03	44	1·38	74	3·39
15	1·04	45	1·41	75	3·58
16	1·04	46	1·43	76	3·79
17	1·05	47	1·46	77	4·02
18	1·05	48	1·49	78	4·2

Zenith Distance.	Ratio of Thickness.	Zenith Distance.	Ratio of Thickness.	Zenith Distance.	Ratio of Thickness.
19°	1·06	49°	1·51	79°	4·59
20	1·06	50	1·54	80	4·92
21	1·07	51	1·58	81	5·31
22	1·08	52	1·61	82	5·75
23	1·09	53	1·65	83	6 25
24	1·09	54	1·68	84	6·83
25	1·10	55	1·72	85	7·51
26	1·11	56	1·77	86	8·28
27	1·12	57	1·81	87	9·18
28	1·13	58	1·86	88	10·20
29	1·14	59	1·91	89	11·37
30	1·15	60	1·96	90	12·69

In order to obtain a formula for the atmospheric absorption of light at different Zenith Distances, I observe that by the notation of (1), we have, for the ratio of light incident in, and emergent from, the atmosphere

$$\log \frac{L}{L_{,}} = K \times \text{(thickness of the atmosphere traversed)}; \qquad (7)$$

consequently, if L_n and L_{n+r} represent the measures of the light of κ *Lyræ* at the Zenith Distances (Z_1 and Z_2), and τ the difference of the tabular thickness at these Zenith Distances, then

$$\log \frac{L_n}{L_{n+x}} = K \tau, \qquad (8)$$

or from (5)

$$K \tau = \cdot 4 \text{ difference of magnitude.} \qquad (9)$$

Now, on Oct. 17 κ *Lyræ* was examined with the 4-inch telescope and wedge, at the Zenith Distance 36° 52′, and the light was extinguished at a wedge-reading of 20·14. It was again observed at the Zenith Distance 70° 58′, the corresponding wedge-reading being 22·64. This indicates a difference of magnitude of 0·33, also the tabular length of the path for a Zenith Distance 36° 52′ is 1·25, and for Zenith Distance 70° 58′ it is 2·93; consequently, the tabular difference of path is 1·68. Hence from (9)

$$\cdot 4 \times \cdot 33 = K \ 1·68$$
$$K = 0·079,$$

and

$$\log \frac{L_n}{L_{n+r}} = 0·079 \tau, \qquad (10)$$

and finally by (5)

$$\left. \begin{array}{l} \text{Difference of magnitude} \\ \text{caused by atmosphere} \end{array} \right\} = 0\cdot198 \left\{ \begin{array}{l} \text{Tabular difference} \\ \text{of thickness} \end{array} \right. \quad (11).$$

This again is the fundamental expression for the atmospheric effects on the light of a star arising from mere alteration in its Zenith Distance, *provided the meteorological elements remain the same.* For instance, κ *Lyræ* was on the same night observed again at the Zenith Distances 48° 37' and 60° 22'. On applying the proper quantities taken from Table II., it appears that the theoretical alterations in magnitude due to the observation at these successive Zenith Distances were respectively 0·05, 0·09, and 0·19. The actual observed differences of magnitude were 0·05, 0·18, and 0·10 respectively; the error in the latter cases amounting to about one-tenth of a magnitude—*i.e.* if the effect of alterations of Zenith Distance were not considered, there would be an error in the stars of least altitude of 0·33 magnitude; but if the correction in (11) be applied, the resulting error will be at the worst one-tenth of a magnitude; and it will be observed that in this instance K is obtained from one observation only; and moreover what is here stated is to be regarded rather as the mere indication of a method than as its completion.

In Table III. are given the values of K which have been derived from the observations made at the greatest variation of Zenith Distance, and the comparison between the observed and computed magnitudes at intermediate altitudes.

TABLE III.

Date.	Star.	Z.D.	Thickness of Air.	Observed Diff. of Mag.	Value of K	Computed Diff. of Mag.	C-0
Oct. 17	κ Lyræ	36° 52'	1·25				
		48 37	1·50	0·05	0·079	0·05	0·00
		60 22	1·99	0·18		0·09	−0·09
		70 58	2·93	0·10		0·19	+0·09
18	κ Lyræ	52 12	1·61				
		61 48	2·08	0·20	0·151	0·17	−0·03
		71 59	3·07	0·35		0·38	−0·03
19	κ Lyræ	43 4	1·36				
		55 58	1·77	0·25	0·213	0·22	−0·03
		60 0	1·96	0·07		0·10	+0·03
25	κ Lyræ	52 38	1·64				
		58 13	1·88	0·15	0·139	0·09	−0·06
		73 43	3·34	0·44		0·50	+0·06
26	κ Lyræ	48 34	1·50				
		54 17	1·69	0·07	0·180	0·09	+0·02
		62 9	2·10	0·20		0·18	−0·02

From all these remarks combined it follows that it is not admissible to connect aperture with magnitude without taking into the account the general meteorological circumstances at the time of observation, and also any considerable variations of the Zenith Distance of the stars compared; but that if the wedge-reading of extinction of the standard star be observed from time to time—say, every hour—both these sources of errors will be effectually eliminated. Moreover, from the fifty stars (given in Table IV.) already compared, it seems that it may be safely assumed that the differences of magnitude are obtainable with a far greater degree of precision than can be reached by eye-estimation. Also, from the repeated observation of several of Argelander's fainter stars, on different evenings, as well as of the bright stars *a Lyræ, Capella* and *Aldebaran* (given in Table V.), it seems that the deviations from mean values are less than one-tenth of a magnitude.

Table IV.

No.	Star's Designation.	Oxford Mag. κ Lyræ = 4·7.	Argelander Mag.	Heis Mag.	Houzeau Mag.
1	Lal. 33612	6·08	5·8	6	
2	Groomb. 2530	6·28	6·0	6	6
3	Groomb. 2533	5·96	5·5	5·6	6
4	Groomb. 2538	6·14	6·0	6	6·7
5	Lal. 34064	6·90	7·0 ⎫		
6	Lal. 34049	7·64	7·7 ⎭	6·7	
7	μ Lyræ	5·31	5·1	5	
8	Lal. 34132	6·08	6·1	6·5	
9	W. B. 18ʰ No. 794	5·86	5·5	6·5	6
10	W. B. 18ʰ No. 894	5·95	6·4	6	6·7
11	W. B. 18ʰ No. 934	5·68	5·8	6	6
12	a Lyræ	0·53	1	1	1
13	W. B. 18ʰ No. 972	6·30	7·0	6·7	
14	Radcliffe 3995	6·53	6·0	6·7	
15	Piazzi 18ʰ No. 153	6·24	6·5	6·7	
16	W. B. 18ʰ No. 1038	6·65	6·3	6·7	
17	Groomb. 2627	6·27	6·2	6·7	
18	W. B. 18ʰ No. 1117	6·47	7·0	6·7	6·7
19	Lalande 34853	5·79	5·5	6	6
20	Piazzi 18ʰ No. 172	6·21	6·4	6·7	
21	ε Lyræ	4·29	4·3	4·5	
22	5 Lyræ	4·61	4·6	5·4	
23	ζ² Lyræ	5·46	5·5 ⎫	4·5	
24	ζ¹ Lyræ	4·50	4·5 ⎭		

No.	Star's Designation.	Oxford Mag. κ Lyræ = 4·7.	Argelander Mag.	Heis Mag.	Houzeau Mag.
25	W. B. 18ʰ No. 1218	5·07	4·9	5	
26	Groomb. 2664	5·77	6·0	6	6·7
27	Lalande 35045	6·07	6·0	6	
28	ν Lyræ	5·45	5·5	6·5	6
29	W. B. 18ʰ No. 1402	6·61	6·5	6·7	6·7
30	W. B. 18ʰ No. 1460	5·90	6·5	6	
31	W. B. 18ʰ No. 1489	5·84	6·2	6·7	6
32	δ¹ Lyræ	6·06	6·1	6	
33	δ² Lyræ	4·54	4·5	4·5	3·4
34	Bradley 2381	6·37	7·0	6·7	
35	Lal. 35405	7·07	7·5 } 6·7		
36	Lal. 35407	7·56	8·2 }		
37	Bradley 2388	5·51	6·7	6	6·7
38	W. B. 18ʰ 1641	5·98	6·2	6·5	6
39	γ Lyræ	3·18	3·2	3·4	3
40	W. B. 18ʰ 1670	5·74	5·8	6	5·6
41	Groomb. 2727	6·01	6·9	6	
42	Groomb. 2728	5·45	6·2	6	6·7
43	W. B. 19ʰ No. 20	6·06	6·3	6	6
44	17 Lyræ	5·42	6·0	6·5	6
45	ι Lyræ	4·90	5·2	5	⸱
46	Lal. 35922	6·38	6·5 } 6		
47	Lal. 35978	6·31	6·5 }		
48	Arg. + 26° No. 3474	6·48	7·4		
49	W. B. 19ʰ No. 159	6·60	7·5 } 6		
50	W. B. 19ʰ No. 165	6·67	7·7 }		

Notes.

No. 12. The magnitude given is the mean of four separate determinations: viz. 0·36, 0·65, 0·56 and 0·56.

No. 21. The magnitude is the mean of three separate determinations, the two components being treated as one star. The individual values are 4·20, 4·27 and 4·40.

No. 39. Three separate determinations gave the magnitude as 3·10, 3·13 and 3·31. The mean is here given.

The agreement is greater with Argelander, than with Heis or Houzeau.

TABLE V.

Name of Star.		Obs. Mag.	Obs. Mag.	Obs. Mag.	Obs. Mag.	Mean Mag.	Argelander's Mag.	Average Deviation from Mean.
α Lyræ		0·36	0·65	0·56	0·56	0·53		0·09
Capella		0·47	0·42	0·60		0·50		0·07
Aldebaran		0·88	0·86	0·93		0·90		0·03
Arg. + 36 No.	3091	8·36	8·43	8·51		8·43	8·8	0·05
+ 36	3103	9·14	9·37			9·25	9·5	0·11
+ 36	3092	8·47	8·55	8·60		8·54	9·0	0·05
+ 36	3101	7·43	7·47	7·29		7·40	7·2	0·07
+ 36	3102	9·03	8·99			9·01	9·2	0·02
+ 36	3104	7·48	7·47	7·40		7·45	7·3	0·03
+ 36	3167	7·95	8·17	7·90		8·01	7·7	0·11
+ 36	3079	7·79		7·93		7·86	8·1	0·07

After the reading of the above communication to the Society, there ensued a long and interesting discussion by the Fellows present who had directed their attention to the subject.

Some doubts were expressed as to the facility, or even possibility, of the measurement of the magnitude being repeated (by the means proposed) by another observer, and with the same results, either identical or approximate. That question I have set at rest by causing certain stars to be observed by two persons. The position of extinction was not absolutely identical for both observers for the same star; but the wedge-interval between the positions of extinction for the two observers of the same star was practically the same; and this is all that is necessary, and all that can be expected in any photometric experiments, whether *nudis oculis* or with apparatus.

As an example, eight stars were observed by the two assistants, Mr. Plummer and Mr. Jenkins—the latter being comparatively new to the work—and the following are the results. It ought, however, to be premised that from the very nature of the observations there can be no previous bias on the part of the observer as to the precise position of extinction by the wedge.

No.	Star's Designation.	Observed Magnitude by P.	Observed Magnitude by J.	P.−J.
1	Lalande 33612 ...	6·05	6·08	−0·03
2	Groombridge 2530	6·10	6·02	+0·08
3	Groombridge 2533	5·92	5·77	+0·15
4	Groombridge 2538	6·03	6·16	−0·13
5	Lalande 34049 ...	7·66	7·69	−0·03
6	Lalande 34064 ...	6·84	6·90	−0·06
7	μ Lyræ	5·48	5·58	−0·10
8	W. B. xviii − 794	5·74	5·71	+0·03

The case now exhibited is the mean of five readings only; but the result, so far as it goes, shows that the two observers may possibly have a tendency to differ in their measures of light by one hundredth of a magnitude.

I have also had a further opportunity of comparing the results of three measurements of thirty out of the fifty stars referred to above. In my own opinion, they establish the exactitude of the method. As to its facility, it may be stated that during the very fine evening of the 17th inst. twenty stars were measured, each of them five times; but it would probably be better, and in the end more economical, to confine the work to about fifteen.

With regard to the homogeneity of the wedge, that element is sufficiently determined by the fact that the value of the constant κ remains sensibly persistent for all parts of the wedge; moreover, the observer's eye would necessarily detect any sensible inequality in its absorbing power, in the course of its continuous motion over the star in the focus of the eye-lens. As to what star or stars should be chosen for a standard, that is a matter to be determined hereafter; for my own purposes, I selected κ *Lyræ*, as a mere matter of convenience, for comparison with stars situated in its neighbourhood.

The practice of this Observatory will be to observe a set of stars on one fine night, taking for each the mean of five readings of the wedge. On the next fine night to repeat this same work twice, at an interval of about an hour; observing the standard star at least three times during the evening; and finally adopting the mean of the three sets as the magnitude of the star at the epoch of observation. Some months after it will be prudent to re-examine the whole with one set of five readings, by which process it may be expected that any variability in the stars will be indicated. As to an artificial standard of light, at present none seems to have been attained; it may be too much to say none is attainable. I subjoin a table of the thirty stars whose magnitudes have as yet been compared; the average deviation from the mean results is, so far, 0·08 magnitude.

TABLE VI.

Concluded Relative Magnitudes of Stars in the Constellation Lyra.

(Photometrically determined.)

Current No.	Star's Designation.	Observed Magnitude.			Mean Mag.	Average Devia-tion.	Arge-lander.	Heis.	Houzea
		1	2	3					
1	Lalande 33612	6·08	6·05	5·97	6·03	0·04	5·8	6	
2	Groombridge 2530	6·27	6·10	5·90	6·09	0·13	6·0	6	6
3	Groombridge 2533	5·96	5·92	5·90	5·93	0·02	5·5	5–6	6
4	Groombridge 2538	6·10	6·03	5·72	5·95	0 15	6·0	6	6–:
5	κ				4·7 adopted		4·7		

Current No.	Star's Designation.	Observed Magnitude.			Mean Mag.	Average Devia-tion.	Magnitude in		
		1	2	3			Arge-lander.	Heis.	Houzeau.
6	Lalande 34049 ...	7·64	7·48	7·66	7·59	0·08	7·7		
7	Lalande 34064 ...	6·90	6·65	6·84	6·80	0·09	7·0	6–7	
8	μ Lyræ	5·31	5·48	5·34	5·38	0·07	5·1	5	5
9	Lalande 34132 ...	5·98	5·76	5·70	5·81	0·11	6·1	6–5	
1C	W. B. xviii–794...	5·84	5·74	5·70	5·76	0·05	5·5	6–5	6
11	W. B. xviii–894...	5·95	6·00	5·82	5·92	0·07	6·4	6	6–7
12	W. B. xviii–934...	5·61	5·50	5·46	5·52	0·06	5·8	6	6
13	α Lyræ	0·36	0·36	0 42	0·38	0·03	1	1	1
14	W. B. xviii–972...*	6·01	6·36				7·0	6–7	
15	Radcliffe 3995 ...	6·51	6 32	6·31	6·38	0·09	6·0	6–7	6–7
16	Groombridge 2623	6·33	6·39	6·29	6·34	0 04	6·5	6–7	
17	W. B. xviii–1038...	6·64	6·85	6·70	6·73	0·07	6·3	6–7	
18	Groombridge 2627	6·21	6·17	6·30	6·23	0·05	6·2	6–7	
19	W. B. xviii–1117...	6·50	6·39	6·62	6·50	0·08	7·0	6–7	6–7
20	Lalande 34853 ...	5 79	6·00	5·98	5·92	0·09	5·5	6	6
21	Piazzi xviii–172...	6·15	6·35	6·32	6·27	0·10	6·4	6–7	
22	ε Lyræ	4·48	4·47	4·41	4·45	0·03	4·3	4–5	
23	5 Lyræ	4·72	4·75	5·07	4·85	0·15	4·6	5–4	4
24	W. B. xviii–1218	5·07	5·09	5·01	5·06	0·03	4·9	5	
25	Groombridge 2664	5·77	5·74	5·79	5·77	0·02	6·0	6	6–7
26	Lalande 35045 ...	6·07	6·01	5·92	6·00	0·05	6·0	6	
27	ν Lyræ	5·44	5·50	5·56	5·50	0·04	5·5	6–5	6
28	W. B. xviii–1402	6 61	6·43	6·36	6·47	0·10	6·5	6–7	6–7
29	W. B. xviii–1460	5·93	6·05	6·11	6·03	0·07	6·5	6	
30	W. B. xviii–1489	5·90	5·71	5·79	5·80	0·07	6·2	6–7	6

My present intention is to persevere in this research for some time to come, as being one which offers the promise of interesting results in many directions, and has been too long a desideratum in stellar astronomy. I propose, in particular, on the same principles and method, to systematically scrutinise the relative brilliancy at different parts of the Sun; employing for this purpose the De La Rue Reflector of 13 inches aperture, and provided with a Herschelian Prism in lieu of the smaller metallic plane mirror. Such observations, if continued long enough, may disclose important facts.

---

* This star was considered to be too near the horizon for accurate comparison at the time.

On the Spectra of Comets b and c, 1881, observed at the Royal Observatory, Greenwich.

(*Communicated by the Astronomer Royal.*)

Comet b, 1881.

1881, *June* 24, $8\frac{1}{2}^h$–$13\frac{1}{2}^h$.—The spectrum of Comet b was first examined on the evening of June 24, with the half-prism spectroscope mounted on the Great Equatoreal. The spectroscope was used in the "direct" position, as for the measures of displacement of lines in stellar spectra—*i.e.* with the light incident on that face of the "half-prism" which is perpendicular to the base. One half-prism was employed giving a dispersion of 18° from A to H, and a magnifying power of 14 was used on the viewing telescope. With this dispersion the spectrum from the nucleus and adjacent parts of the head appeared to be perfectly continuous, and only one band, in the green, was made out with certainty. This band was very distinct from one part of the head, but hardly traceable from any other part. Comparison with the second spectrum of carbon as given by alcohol vapour in a vacuum-tube showed that this band was situated at a distance of about 38 tenth-metres to the blue from the carbon-band at 5197 tenth-metres. Observer, W. C.

1881, *June* 25, $9\frac{1}{2}^h$–14^h.—With the spectroscope in the same position as on the previous evening, three bands were now detected in the Comet's spectrum, though the great brilliancy of the continuous spectrum nearly overpowered them, and rendered their observation a matter of difficulty. As far as could be judged, these bands were not obtained from the nucleus, but from its immediate vicinity. A comparison with the spectrum of cyanogen given by a vacuum-tube showed no instances of coincidence; but each of the three bands was seen to be coincident with the corresponding band of the spectrum of a Bunsen-flame at 5164, 4736, and 4311 respectively. No band was detected in the yellow. Observers, W. C. and M.

1881, *June* 29, 10^h–$14\frac{1}{4}^h$.—The bright bands had been isolated with such difficulty on June 24 and 25 that on this occasion the spectroscope was used as in the observations of prominences and reversed so as to give great purity of spectrum. The dispersion from A to H in this position is 5°, and the magnifying power 28. The continuous spectrum was given from a much smaller area of the Comet than on the two former nights of observations, being only 15″ to 18″ in breadth, and seen only over the nucleus an its immediate neighbourhood. Several dark lines were perceive in it, between the green and blue bands, one of these lines bein undoubtedly F. The b lines could not be made out, but the would fall just within the bright edge of the green band.

Three bright bands were seen, yellow, green, and blue, th

band in the violet not being detected. Each band was bright and fairly sharp towards the red, and became gradually fainter towards the blue end of the spectrum. They were seen bright on the bright background of the continuous spectrum from the head, but could not be traced across the nucleus. The dark lines could not be detected in the spectrum of the nucleus, but this may have been due to its being so very narrow (2″ or 3″ in breadth).

The comparative brightness of the three bands was as follows :—

Yellow band	...	3	Edge not very sharp.
Green band	...	10	Edge sharp and distinct.
Blue band	...	1	Edge sharp.

The green band was obtained from the whole of the head, and for 12′ from the nucleus down the tail.

The single-prism spectroscope was now substituted for the other for a few minutes, and the exact coincidence of the three comet-bands with the three of the Bunsen-flame noticed, the accord when the two spectra were seen side by side being as perfect in every way as the eye could judge. This spectroscope did not show the violet band, but the blue band appeared much brighter in comparison with the other two than with the half-prism spectroscope. It was estimated as brighter than the yellow band, and as quite half as bright as the green.

No dark lines were distinguished in the continuous spectrum above the green band, but several strong lines were seen below it. Daylight came on before measures of these lines could be obtained, but rough estimations of their positions showed that two of the darkest were probably the E group, and the strong double line at 5327. The D lines were not made out with certainty. Observer, M.

The half-prism spectroscope reversed, with one half-prism, as in the earlier part of the evening of June 29 was used for all the subsequent observations both of this Comet and of Comet *c* (Schäberle's).

1881, *July* 2, 12ʰ.—The continuous spectrum was faint, and no dark lines were detected in it. The green band of the Comet was seen to be perfectly coincident with the band from the Bunsen-flame. Clouds prevented further observation. Observer, M.

1881, *July* 4, 10ʰ–14ʰ.—The three bands, yellow, green and blue, were seen as usual, the first and last being very faint. No others were detected. The continuous spectrum was too narrow to show any dark lines. Observer, M.

1881, *July* 6, 10ʰ–14ʰ.—The continuous spectrum was more distinct than on July 4, but no dark lines could be detected. With the spectra of the Comet and of the Bunsen-flame arranged one above the other, and the flame adjusted until the bright sharp edge of the green band was of the same intensity in each, the resemblance between the two spectra was exceedingly striking, the three principal bands corresponding exactly in position, in brightness, and in the manner and degree in which they shaded off towards the violet. The fourth band (that in the violet) was, however, far less distinct in the Comet spectrum than in that of the Bunsen-flame—indeed, its presence in the former was only just suspected. Search was also made for a red band and with a very wide slit an extremely faint light was just suspected somewhere in the red part of the spectrum about midway between C and D. Observer, M.

The following are the determinations of the wave-lengths of the less refrangible edges of the three principal bands.

Yellow Band.

Date.	Observer.	Spectroscope.	Compared with	Width of Slit.	Wave-Length of Comet-band.
				tenth-metr.	tenth-metr.
June 29	M.	Half prism, reversed	Marsh gas	13·1	5627·0
		„ „	„	13·1	39·5
			„	13·1	19·4
July 6	M.	„	Bunsen-flame	16·8	32·4
			··	16·8	29·5
			··	16·8	29·9
			„	16·8	35·0
			MEAN	5630·4 ± 1·6	

Green Band.

Date.	Observer.	Spectroscope.	Compared with	Width of Slit.	Wave-Length of Comet-band.
June 24	W. C.	Half prism, direct	Alcohol	62·1	5151·7*
		„ „	„	62·1	59·8
25	M.		Cyanogen	9·7	67·9
			„	9·7	59·0
		„ „	„	9·7	61·5
29	M.	Half prism, reversed	Marsh gas	9·6	59·9
		„ „	„	9·6	61·8
		„ ‚	„	9 6	60·9
		Single prism	Bunsen-flame	26·1	61·5
		„	··	26·1	65·0
July 4	M.	Half prism, reversed	„	14·7	65·3
		„ „		14·7	63·7
				5·4	58·1
				5·4	61·9
		„	„	5·4	65·0

* Measure rough.

Date.	Observer.	Spectroscope.	Compared with	Width of Slit.	Wave-Length of Comet-band.
				tenth-metr.	tenth-metr.
July 6	M.	Half prism, reversed	Bunsen-flame	3·6	63·8
		„ ..	..	3·6	64·0
				3·6	64·0
				3 6	64·3
				3·6	6.|·0
				3·6	65·3
				12·4	64·0
				12·4	63·7
				12·4	63·4
				12·4	65·2
			„	12·4	64·4

Mean 5162·7 ± 0·4

Blue Band.

Date	Observer	Spectroscope	Compared	Width	Wave-length
June 29	M.	Half prism, reversed	Marsh gas	6·2	4728·2
		„ ..	„	6·2	30·2
July 6	M.	„ „	Bunsen-flame	7·9	35·5
			..	7·9	36·9
				7·9	36·0
			„	7·9	36·4

Mean 4733·9 ± 1·1

The following measures were also obtained of the dark line F seen in the continuous spectrum as compared with the $H\beta$ line from a vacuum-tube.

Date	Observer	Spectroscope.	Width of Slit.	Wave-Length of Dark Line.	Apparent Displacement.
			tenth-metr.	tenth-metr.	tenth-metr.
June 29	M.	Half prism, reversed	2·8	4860·92	+0·18
		„ „	2·8	60·42	−0·32

Thalén's wave-lengths for the less refrangible edges of the bands in the two spectra of the carbon compounds were assumed throughout, as follows :—

Band.			First Spectrum. Bunsen-flame.	Second Spectrum. Marsh Gas or Cyanogen.
			tenth-metr.	tenth-metr.
Yellow band	...	...	5633·0	5607·5
Green band	...	...	5164·0	5197·0
Blue band	...	...	4736·0	4697·0

Comet *c* 1881 (Schäberle's).

1881, *Aug.* 4, $8\frac{1}{2}^h$–$13\frac{1}{2}^h$.—The Comet showed a spectrum consisting of three bright bands approximately coincident with the three brightest bands of the spectrum of the Bunsen-flame. Eventually a very faint continuous spectrum, due to the nucleus, was detected. The green band was the only one which had a sufficiently defined edge to be measured. The yellow band appeared to be displaced towards the blue of the corresponding band of the flame spectrum by a small but unmistakeable distance—say about 8 tenth-metres. The blue band was too faint for anything to be observed concerning it further than that it showed a rough coincidence with the band from the Bunsen-flame. The whole of the head of the Comet gave the spectrum of the bands.

The relative brightness of the bands appeared to be as follows:—

Yellow band	...	...	3
Green band	...	...	10
Blue band	...	...	1

Observer, M.

1881, *Aug.* 6, $8\frac{1}{2}^h$–13^h.—The Comet's spectrum showed no change since August 4, except that it now seemed very faint. owing to the brilliant moonlight. The nucleus still gave a faint continuous spectrum, traceable from about D to F, but the spectrum of the bands was many times as bright. The whole of the head of the Comet gave the green band, but it could not be detected over the tail. The blue band was seen, but was not measurable, and no new bands could be discerned in the red, orange or violet. A narrow slit was employed on the green band, but it could not be resolved into lines. Observer, M.

1881, *Aug.* 19, 8^h–$11\frac{1}{2}^h$.—The three bands were seen as on the two previous occasions, and with a very wide slit two others were suspected, the one as far below D as the yellow band is above it—that is to say, in the neighbourhood of wave-length 6200$\pm$ tenth-metres—and the other just to the red of G or at wave-length 4310$\pm$. The continuous spectrum was now obtained not merely from the nucleus, but also from its immediate neighbourhood, and showed a total breadth of 7″, and it was visible from the red band up to the one in the violet, but no dark lines could be detected in it. The band spectrum was, however, still many times as bright as the continuous spectrum from the nucleus, at the less refrangible edges of the principal bands. The green band was given by every part of the head, even the faintest, and was clearly obtained from the tail for quite 10′ from the nucleus. A very narrow slit was used on the green band in order to detect any resolvability if possible. The band could not be broken up into lines, but a second edge was suspected about 5126. Observer, M.

The following measures were obtained of the positions of the less refrangible edges of the yellow and green bands. Of the blue band it was only ascertained that it seemed to be coincident with the band of the Bunsen-flame at 4736·0. The half-prism spectroscope reversed was used throughout, and the comparisons were all made with the spectrum of a Bunsen-flame.

Yellow Band.

Date.	Observer.	Width of Slit.	Wave Length Comet Band.
		tenth-metres.	tenth-metres.
1881, August 6	M.	33·9	5629·3
		33·9	36·3
		33·9	21·3
		33·9	33·9
		33·9	24·7
		MEAN	5629·1 ± 2·1

Green Band.

Date.	Observer.	Width of Slit.	Wave Length Comet Band.
1881, August 4	M.	15·6	5162·8
		15·6	63·0
		15·6	63·3
		15·6	64·4
		6·3	65·7
		6·3	63·9
		6·3	62·8
		6·3	61·8
		6·3	64·3
		6·3	63·1
August 6	M.	6·4	62·6
		6·4	63·1
		6·4	63·3
		6·4	64·4
		6·4	64·5
		6·4	62·8
		6·4	64·0
		6·4	65·1
		6·4	63·4
		6·4	63·6
		MEAN	5163·6 ± 0·15

The initials W. C. and M. are those of Mr. Christie and Mr. Maunder respectively.

Royal Observatory, Greenwich :
    1881, *November* 10.

On the North Polar Distances of the Cape Catalogue for 1880, and on the Greenwich and Cape Mean Systems of North Polar Distances. By A. M. W. Downing, M.A.

The chief difficulty to be encountered in discussing the North Polar Distances of the Cape Catalogue for 1880 is the uncertainty in the values of the proper motions of the great mass of southern stars. It is on this account that I have selected for comparison with this catalogue the Melbourne Catalogue for 1870, as being nearest in date of any southern catalogue. There are 911 stars common to the catalogues which are available for the comparison, and it is reasonable to suppose that when such a large number of stars is used the results will be sensibly free from error arising from the assumption of erroneous proper motions.

The assumed proper motions have been taken from the Cape Catalogue itself, or from the list of additional proper motions given at the end of the catalogue, or from the Melbourne Catalogue, or from the Cape Catalogues for 1860 or 1840: but a large number of the stars have necessarily been used with assumed proper motions. In order to show that this proceeding has not had an appreciable effect on the computed differences between the N.P.D.'s of the catalogues, when taken in groups extending over 5° of N.P.D., I have given the probable errors of a difference between the catalogue places for each group; and it appears that there is no decided increase in these probable errors in passing from stars which are observable in the northern hemisphere, and whose proper motions are known with comparative certainty, to the stars which are situated beyond the reach of northern observers.

Only one star out of the whole number has been rejected for discordance. This is Lacaille 9311, which is No. 9311 in the Cape Catalogue and No. 911 in the Melbourne Catalogue, but its place, as far as I know, is not given in any other catalogue. The difference Cape minus Melbourne amounts to −4″·61, and as its N.P.D. in each catalogue depends on a considerable number of observations, and the mean dates of observation in the Cape and Melbourne Catalogues are 1871 and 1868 respectively, it is reasonable to suppose that this star has a large proper motion in N.P.D. In fact, Lacaille's place compared with that in the Cape Catalogue gives for proper motion −0″·8. I commend this star to the attention of H.M. Astronomer at the Cape.

The places in the Cape Catalogue have been brought back to 1880, and the differences Cape minus Melbourne taken. The differences have then been collected in groups of 5°, and the means of each group and the mean N.P.D. of each group found. These mean differences have been laid down as points on cross-ruled paper, and a curve or diagram drawn through them, which may be taken to represent the systematic differences between the N.P.D.'s of the catalogues.

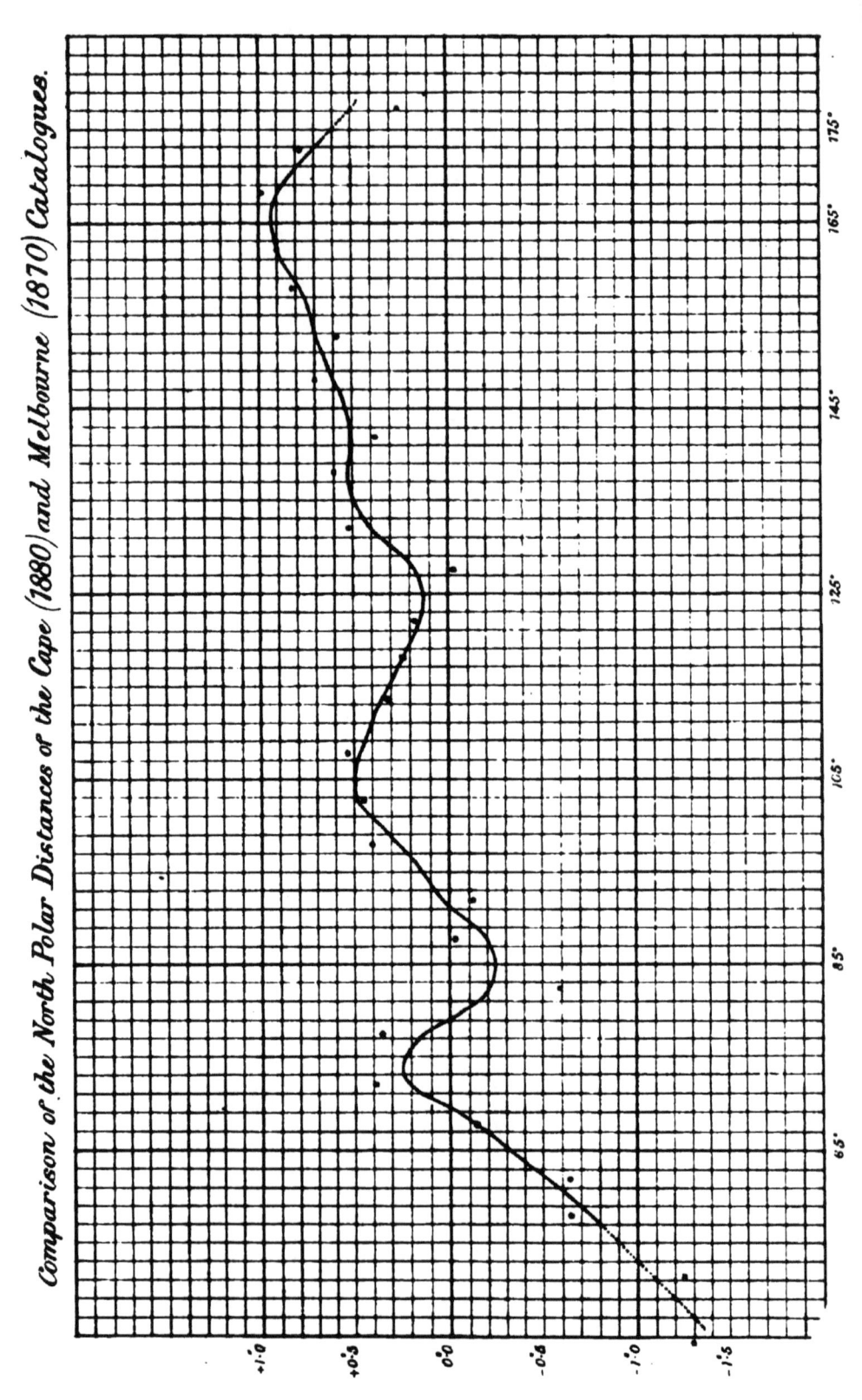

Comparison of the North Polar Distances of the Cape (1880) and Melbourne (1870) Catalogues.

The following table gives the differences, corresponding to the mean N.P.D. of each group, as computed and as read off from the curve. The column under ϵ_1 is the probable error of a single difference between the positions as given in the catalogues, and is found from the discordance of each separate difference from the mean of the group. There are not a sufficient number of observations of stars north of N.P.D. 60° in the Cape Catalogue (such observations being of course foreign to the plan of the work) to enable us to determine with any accuracy the systematic differences between the catalogues, for stars situated near the northern horizon of the Observatories :—

N.P.D.	Number of Stars.	Cape−Melbourne Means.	ϵ_1	Cape−Melbourne Curve.
° ′		"	"	"
44 38		−1·30	—	—
51 11	—	−1·25	—	—
57 41	4	−0·63	—	—
62 4	8	−0·64	±0·46	−0·51
67 57	9	−0·12	·54	−0·12
72 6	8	+0·39	·32	+0·21
77 3	12	+0·36	·56	+0·13
82 31	15	−0·60	·59	−0·22
87 34	11	−0·03	·38	−0·20
91 45	14	−0·12	·49	+0·03
97 59	7	+0·40	·32	+0·27
102 42	7	+0·45	·30	+0·47
107 27	11	+0·54	·42	+0·47
113 15	10	+0·32	·35	+0·36
117 28	37	+0·26	·50	+0·26
122 21	38	+0·18	·62	+0·15
127 32	65	−0·02	·54	+0·17
132 7	71	+0·52	·60	+0·41
137 25	61	+0·60	·49	+0·53
141 48	32	+0·38	·44	+0·52
147 59	46	+0·70	·46	+0·60
152 31	208	+0·59	·49	+0·68
157 46	130	+0·82	·52	+0·78
161 57	18	+0·91	·52	+0·91
168 8	29	+0·99	·56	+0·92
172 39	41	+0·78	·39	+0·72
177 1	15	+0·26	·23	—

The average probable error of a difference between the N.P.D.'s of the catalogues is $\pm 0'''\cdot 46$. From a comparison of the Greenwich (1864) and Melbourne (1870) Catalogues (*Vierteljahrs-schrift*, Bd. xi. S. 187) Prof. Gyldén found the probable error of a difference between the N.P.D.'s of these Catalogues, for stars situated in the zone 60°—110° N.P.D., to be $\pm o''\cdot3o$. The larger probable error found above can hardly be attributed to the use of inaccurate proper motions, for confining ourselves to stars situated between 60° and 110° N.P.D., the average probable error of a difference is $\pm o''\cdot44$. It would appear, therefore, that the probable accidental errors of the N.P.D.'s of the Cape Catalogue for 1880 are somewhat larger than those of the Greenwich Catalogue for 1864.

By reading off from the curve given above the differences corresponding to every 4° of N.P.D., and making use of the comparisons given in my papers in the *Monthly Notices* for December 1878, January 1879, and January 1881, we get the following corrections applicable to the N.P.D.'s of the several catalogues named to reduce them to the system of the Cape Catalogue for 1880; except that for the Cape Catalogue (1833) the *data* in Mr. Boss's *Declinations of Fixed Stars* have been used from N.P.D. 60° to N.P.D. 120° inclusive.

N.P.D.	Cape (1833).	Cape (1840).	Cape (1860).	Melbourne (1870).	Greenwich Mean System.
60°	$-o''88$	$+o''o8$	$-o''3o$	$-o''65$	$-o''71$
64	$-o\cdot96$	$-o\cdot1o$	$-o\cdot35$	$-o\cdot4o$	$-o\cdot74$
68	$-o\cdot78$	$-o\cdot o8$	$-o\cdot21$	$-o\cdot11$	$-o\cdot59$
72	$-o\cdot35$	$+o\cdot25$	$+o\cdot o8$	$+o\cdot2o$	$-o\cdot31$
76	$-o\cdot13$	$+o\cdot14$	$+o\cdot13$	$+o\cdot19$	$-o\cdot23$
80	$-o\cdot17$	$-o\cdot33$	$-o\cdot o5$	$-o\cdot o8$	$-o\cdot37$
84	$-o\cdot27$	$-o\cdot89$	$-o\cdot2o$	$-o\cdot24$	$-o\cdot51$
88	$-o\cdot44$	$-1\cdot32$	$-o\cdot36$	$-o\cdot19$	$-o\cdot72$
92	$-o\cdot26$	$-1\cdot19$	$-o\cdot11$	$+o\cdot o3$	$-o\cdot51$
96	$-o\cdot26$	$-1\cdot o8$	$o\cdot oo$	$+o\cdot17$	$-o\cdot45$
100	$-o\cdot27$	$-o\cdot84$	$+o\cdot15$	$+o\cdot35$	$-o\cdot3o$
104	$-o\cdot39$	$-o\cdot48$	$+o\cdot22$	$+o\cdot49$	$-o\cdot13$
108	$-o\cdot57$	$-o\cdot o9$	$+o\cdot21$	$+o\cdot47$	$+o\cdot o1$
112	$-o\cdot81$	$+o\cdot o3$	$+o\cdot1o$	$+o\cdot4o$	$+o\cdot o7$
116	$-1\cdot11$	$-o\cdot3o$	$-o\cdot14$	$+o\cdot29$	$-o\cdot1o$
120	$-1\cdot22$	$-o\cdot56$	$-o\cdot27$	$+o\cdot19$	$-o\cdot12$
122	$-1\cdot51$	$-o\cdot6o$	$-o\cdot28$	$+o\cdot16$	$+o\cdot o5$
124	$-1\cdot4o$	$-o\cdot5o$	$-o\cdot23$	$+o\cdot14$	$-o\cdot54$
128	$-1\cdot o6$	$-o\cdot19$	$-o\cdot1o$	$+o\cdot18$	
132	$-o\cdot59$	$+o\cdot o5$	$-o\cdot o6$	$+o\cdot41$	
136	$-o\cdot32$	$+o\cdot o8$	$-o\cdot21$	$+o\cdot51$	
140	$-o\cdot o3$	$+o\cdot23$	$-o\cdot18$	$+o\cdot52$	

N.P.D.	Cape (1833).	Cape (1840).	Cape (1860).	Melbourne (1870).	Greenwich Mean System.
144°	−0″.18	+0″.21	−0″.23		+0″.53
148	+0.09	+0.38	−0.09		+0.60
152	+0.61	+0.77	+0.17		+0.67
156	+0.76	+1.11	+0.30		+0.74
160	+0.75	+1.25	+0.43		+0.84
164	+0.95	+1.29	+0.59		+0.93
168	+1.11	+1.06	+0.64		+0.91
172	+0.99	+0.69	+0.56		+0.75
176	+0.75	+0.59	+0.45		+0.57

The Greenwich "Mean System" is found by taking the simple mean of the comparisons of six Greenwich Catalogues with the Melbourne Catalogue given by the Astronomer Royal on page 172 of vol. xlv. of the *Memoirs*, Besssel's refractions with Main's corrections below Z.D. 82° being used in all cases.

The above comparison with the Cape Catalogue for 1860 is very satisfactory, showing as it does that the systematic errors of the catalogues do not differ by any considerable amount, except perhaps for the circumpolar stars. It will be remembered that the south latitude of the Observatory adopted in the several catalogues was, for the 1833 Catalogue 33° 56′ 3″.20, for the 1840 Catalogue 33° 56′ 3″.25, for the 1860 Catalogue 33° 56′ 3″.56, and for the 1880 Catalogue 33° 56′ 3″.41. The refractions used were those of Bessel's *Tabulæ Regiomontanæ*, except that for the period during which the observations included in the last-mentioned catalogue were made it was found that the thermometer read 0°.55 too high, so that Bessel's mean refractions were really diminished in the proportion 0.9988 : 1. I can find no information with regard to the errors of the thermometer for the periods corresponding to the other catalogues.

The following table gives the corrections applicable to the approximation to a Mean System formed from the Cape Catalogues for 1840, 1860, and 1880, to reduce them to the Greenwich Mean System as given above:—

N.P.D.	Greenwich − Cape.	N.P.D.	Greenwich − Cape.
60°	+0″.64	96°	+0″.09
64	+0.59	100	+0.06
68	+0.49	104	+0.04
72	+0.42	108	+0.03
76	+0.32	112	−0.03
80	+0.24	116	−0.05
84	+0.15	120	−0.17
88	+0.13	122	−0.34
92	+0.08	124	+0.30

This, in common with my previous comparisons of Greenwich and Cape results, seems to indicate that there is a systematic error affecting the Cape N.P.D.'s near the north horizon, and increasing as we approach the horizon (probably the effect of uncorrected flexure), which makes the observed Z.D.'s too large. A definite decision on this point cannot, however, be arrived at with the materials at present at our disposal. But much light will be thrown on the subject by the publication of a limited fundamental catalogue, in which special attention is paid to such points as latitude, refraction, flexure, and discordance of direct and reflexion observations. I understand that the staff of the Cape Observatory are now employed on the observations necessary for the formation of such a catalogue.

The residuals given above, when laid down as points on cross-ruled paper, so nearly lie on a right line that from N.P.D. 60° to 120° they may be closely represented by the expression

$$+ 0''\!20 \sin \text{N.P.D.} + 0''\!675 \cos \text{N.P.D.} \qquad \text{(I.)}$$

If, however, it be assumed that the residuals between N.P.D. 60° and 112° arise from errors in the Cape Z.D.'s they may be very fairly represented by

$$+ 0''\!74 \sin \text{Z.D.} - 0''\!26 \cos \text{Z.D.} \qquad \text{(II.)}$$

as will be seen from the following tabular statement :—

N.P.D.	Greenwich − Cape.		Excess of	
	Formula (I.).	Formula (II.).	Formula (I.).	Formula (II.).
60°	+0″51	+0″55	−0″13	−0″09
64	·48	·51	·11	·08
68	·44	·47	·05	− ·02
72	·40	·42	− ·02	·00
76	·36	·38	+ ·04	+ ·06
80	·31	·32	·07	·08
84	·27	·28	·12	·13
88	·22	·23	·09	·10
92	·18	·17	·10	·09
96	·13	·12	·04	+ ·03
100	·08	·06	+ ·02	·00
104	+ ·03	+ ·01	− ·01	− ·03
108	− ·02	− ·05	·05	·08
112	·07	− ·10	·04	− ·07
116	·10	—	− ·05	—
120	− ·17		·00	

The last two columns give the differences of the residuals found from each of the formulæ and from the means of the catalogue comparisons given in the previous table.

With the exception of the discordance referred to above, the close agreement between the N.P.D.'s found from the Greenwich and Cape observations is very remarkable.

1881, *November* 5.

Ephemeris for finding the Positions of the Satellites of Uranus, 1882.
By A. Marth, Esq.

In October the Earth has passed through the planes of the orbits of the satellites of *Uranus* from the side on which it had been since 1840 to the other side, on which the satellites appear to move in the direction of *increasing* position-angles, and on which, after returning next spring for some months to the former side, it will remain till the year 1923. In view of the importance of securing series of observations of the satellites during the present apparition of the planet, the ephemeris is made to begin a month earlier than would otherwise be necessary.

The angle of position p_0 of the major axes, the major and minor semiaxes a and b of the apparent ellipses described by the satellites, and the latitude of the Earth above the assumed plane of their orbits, are the following :—

Greenwich Noon.		Ariel.		Umbriel.		Titania.		Oberon.		Lat. of
	p_0	a_1	b_1	a_2	b_2	a_3	b_3	a_4	b_4	Earth.
1881.	°	''	''	''	''	''	''	''	''	°
Dec. 10	15·24	14·52 + 0·67		20·22 + 0·94		33·17 + 1·53		44·35 + 2·05		+ 2·65
20	15·25	14·63	0·70	20·37	0·97	33·42	1·59	44·69	2·13	2·73
30	15·25	14·76	0·70	20·56	0·98	33·73	1·60	45·11	2·14	2·72
1882.										
Jan. 9	15·24	14·89 + 0·68		20·74 + 0·95		34·02 + 1·55		45·50 + 2·07		+ 2·61
19	15·22	15·01	0·63	20·91	0·88	34·30	1·45	45·86	1·93	2·42
29	15·20	15·11	0·57	21·05	0·79	34·53	1·29	46·18	1·73	2·15
Feb. 8	15·17	15·19	0·48	21·17	0·67	34·72	1·10	46·43	1·47	1·81
18	15·14	15·25	0·38	21·25	0·53	34·86	0·87	46·61	1·16	1·42
28	15·10	15·29	0·27	21·30	0·37	34·93	0·61	46·72	0·82	1·00
Mar. 10	15·06	15·29	0·15	21·31	0·21	34·95	0·35	46·74	0·47	0·57
20	15·01	15·27 + 0·04		21·28 + 0·06		34·90 + 0·09		46·68 + 0·12		+ 0·15
30	14·97	15·23 − 0·07		21·21 − 0·09		34·80 − 0·15		46·53 − 0·21		− 0·25
Apr. 9	14·93	15·16	0·16	21·12	0·23	34·64	0·37	46·32	0·50	0·61
19	14·90	15·07	0·24	20·99	0·34	34·43	0·55	46·04	0·74	0·92
29	14·88	14·96	0·30	20·84	0·42	34·18	0·69	45·71	0·92	1·15
May 9	14·86	14·84	0·34	20·67	0·47	33·91	0·77	45·35	1·03	1·31
19	14·85	14·71	0·35	20·49	0·49	33·61	0·80	44·95	1·07	1·38
29	14·85	14·58 − 0·35		20·31 − 0·48		33·31 − 0·79		44·55 − 1·06		− 1·36

Longitudes of the satellites in their orbits reckoned from the points where they are at their greatest northern elongations :—

Greenwich Noon.	Ariel. Long.	Diff.	Umbriel. Long.	Diff.	Titania. Long.	Diff.	Oberon. Long.	Diff.
1881.	°	°	°	°	°	°	°	°
Dec. 10	258·38	1428·47	160·09	868·76	188·35	413·55	327·70	267·42
20	246·85	·47	308·85	·76	241·90	·54	235·12	·41
30	235·32	·47	97·61	·75	295·44	·53	142·53	·40
1882.								
Jan. 9	223·79	·44	246·36	·72	348·97	·51	49·93	·38
19	212·23	·41	35·08	·71	42·48	·50	317·31	·37
29	200·64	·38	183·79	·68	95·98	·49	224·68	·36
Feb. 8	189·02	·35	332·47	·67	149·47	·47	132·04	·35
18	177·37	·33	121·14	·65	202·94	·46	39·39	·33
28	165·70	·30	269·79	·63	256·40	·46	306·72	·34
Mar. 10	154·00	·28	58·42	·62	309·86	·45	214·06	·34
20	142·28	·26	207·04	·61	3·31	·45	121·40	·33
30	130·54	·24	355·65	·60	56·76	·44	28·73	·34
Apr. 9	118·78	·23	144·25	·60	110·20	·45	296·07	·34
19	107·01	·23	292·85	·60	163·65	·46	203·41	·35
29	95·24	·22	81·45	·60	217·11	·47	110·76	·35
May 9	83·46	·22	230·05	·61	270·58	·47	18·11	·37
19	71·68	1428·21	18·66	868·61	324·05	413·48	285·48	267·38
29	59·89		167·27		17·53		192·86	

These values are to be interpolated for the times for which the positions of the satellites are required. The position-angles p and distances s are then to be found by means of the formulæ:—

$$s \sin (p - p_0) = b \sin \text{long.}$$
$$s \cos (p - p_0) = a \cos \text{long.}$$

The satellites move in the direction of *increasing* position-angles when b is positive, and in the direction of *decreasing* position-angles when b is negative, and will be at their greatest elongations ("N" in posit. p_0 and "S" in posit. $p_0 + 180°$), and at their superior and inferior conjunctions with the planet at or about the following hours, Greenwich mean time :—

Ariel.

1881.	N. d	h	S. d	h	1882.	N. d	h	S. d	h	1882.	N. d/h	h	S. d	h
Dec.	10	17·1	11	23·3	Feb.	4	3·8	5	10·0	April	3	3·1	4	9·3
	13	5·6	14	11·8		6	16·2	7	22·5		5	15·5	6	21·8
	15	18·0	17	0·3		9	4·7	10	11·0		8	4·0	9	10·3
	18	6·5	19	12·8		11	17·2	12	23·5		10	16·5	11	22·8
	20	19·0	22	1·3		14	5 7	15	12·0		13	5·0	14	11·3
	23	7·5	24	13·7		16	18·2	18	0·4		15	17·5	16	23·8
	25	20·0	27	2·2		19	6·7	20	12·9		18	6·0	19	12·3
	28	8·5	29	14·7		21	19·2	23	1·4		20	18·5	22	0·8
	30	21·0				24	7·7	25	13·9		23	7·0	24	13·3
1882.			Jan. 1	3·2		26	20·2	28	2·4		25	19·5	27	1·8
Jan.	2	9·4	3	15·7	Mar.	1	8·7	2	14·9		28	8·0	29	14·3
	4	21·9	6	4·2		3	21·1	5	3·4		30	20·5	M. 2	2·7
	7	10·4	8	16 6		6	9 7	7	15·9	May	3	9·0	4	15·2
	9	22·9	11	5·1		8	22·1	10	4·4		5	21·5	7	3·7
	12	11·4	13	17·6		11	10·6	12	16·9		8	10·0	9	16·2
	14	23·9	16	6·1		13	23·1	15	5·4		10	22·5	12	4·7
	17	12·3	18	8·6		16	11·6	17	17·8		13	11·0	14	17 2
	20	0·8	21	7·1		19	0·1	20	6·3		15	23·5	17	5·7
	22	13·3	23	19·6		21	12 6	22	18·8		18	12·0	19	18·2
	25	1·8	26	8·0		24	1·1	25	7·3		21	0·4	22	6·7
	27	14·3	28	20·5		26	13·6	27	19·8		23	12·9	24	19·2
	30	2·8	31	9·0		29	2·1	30	8·3		26	1·4	27	7·7
Feb.	1	15·3	2	21·5		31	14·6	A. 1	20·8		28	13·9	29	20·2

Umbriel.

1881.	N. d	h	S. d	h	1882.	N. d	h	S. d	h	1882.	N. d	h	S. d	h
Dec.	12	7·2	14	9·0	Feb.	4	4·2	6	5·9	Apr.	3	4·7	5	6·4
	16	10·7	18	12·4		8	7·6	10	9·3		7	8·1	9	9·9
	20	14·1	22	15·9		12	11·1	14	12·8		11	11·6	13	13·3
	24	17·6	26	19·3		16	14·5	18	16·3		15	15·1	17	16·8
	28	21·0	30	22·8		20	18·0	22	19·7		19	18·6	21	20·3
1882.						24	21·5	26	23·2		23	22·0	25	23·8
Jan.	2	0·5	4	2·2	Mar.	1	0·9	3	2·7		28	1·5	30	3·2
	6	3·9	8	5·7		5	4·4	7	6·1	May	2	5·0	4	6·7
	10	7·4	12	9·3		9	7·9	11	9·6		6	8·4	8	10·2
	14	10·9	16	12·6		13	11·3	15	13·1		10	11·9	12	13·6
	18	14·3	20	16·0		17	14·8	19	16·5		14	15·4	16	17·1
	22	17·8	24	19·5		21	18·3	23	20·0		18	18·8	20	20·6
	26	21·2	28	23·0		25	21·7	27	23·5		22	22·3	25	0·0
	31	0·7	F. 2	2·4		30	1·0	A. 1	2·9		27	1·8	29	3·5

Titania.

N. elong.		Inf. Conj.		S. elong.		Sup. Conj.	
1881.	h		h		h		h
—		—		Dec. 9	19·2	Dec. 11	23·4
Dec. 14	3·6	Dec. 16	7·8	18	12·1	20	16·3
22	20·5	25	0·8	27	5·0	29	9·2
31	13·5						
1882.		Jan. 2	17·7	Jan. 4	21·9	Jan. 7	2·2
Jan. 9	6·4	11	10·6	13	14·9	15	19·1
17	23·2	20	3·6	22	7·8	24	12·1
26	16·3	28	20·5	31	0·8	Feb. 2	5·0
Feb. 4	9·2	Feb. 6	13·5	Feb. 8	17·7	10	22·0
13	2·2	15	6·4	17	10·7	19	14·9
21	19·2	23	23·4	26	3·7	28	7·9
March 2	12·1	March 4	16·4	March 6	20·6	March 9	0·9
11	5·1	13	9·3	15	13·6	17	17·8
19	22·1	22	2·3	24	6·6	26	10·8
28	15·1	30	19·3	April 1	23·5	April 4	3·8
April 6	8·0	April 8	12·3	10	16·5	12	20·8
15	1·0	17	5·2	19	9·5	21	13·7
23	18·0	25	22·2	28	2·5	30	6·7
May 2	10·9	May 4	15·2	May 6	19·4	May 8	23·7
11	3·9	13	8·2	15	12·4	17	16·6
19	20·9	22	1·1	24	5·3	26	7·6
28	13·8						

Oberon.

N. elong.		Inf. Conj.		S. elong.		Sup. Conj.	
1881.	h		h		h		h
Dec. 11	5·0	Dec. 14	13·8	Dec. 17	22·5	Dec. 21	7·3
24	16·1	28	0·9	31	9·6		
1882.						Jan. 3	18·4
Jan. 7	3·2	Jan. 10	12·0	Jan. 13	20·8	17	5·5
20	14·3	23	23·1	27	7·9	30	16·7
Feb. 3	1·5	Feb. 6	10·3	Feb. 9	19·1	Feb. 13	3·8
16	12·6	19	21·4	23	6·2	26	15·0
March 1	23·8	March 5	8·6	March 8	17·4	Mar. 12	2·2
15	11·0	18	19·8	22	4·6	25	13·4
28	22·2	April 1	7·0	April 4	15·8	April 8	0·6
April 11	9·4	14	18·2	18	3·0	21	11·8
24	20·6	28	5·4	May 1	14·2	May 4	23·0
May 8	7·7	May 11	16·5	15	1·3	18	10·1
21	18·9	25	3·7	28	12·5		

During the period of 170 days, over which the Ephemeris extends, there will be some 230 occasions when two of the satellites pass one another at a short distance. As I have not learnt how close to the planet the satellites can be seen with some of the most powerful modern telescopes, I have not altered the limits adopted last January * for excluding the conjunctions which do not offer some fair prospect of being observable in Europe or America. I give now for the others a list similar to that printed on pp. 155 and 156, containing the computed position-angles and distances of the satellites which pass one another, for the nearest preceding and following even hour, Gr. M. T., so that the circumstances of each conjunction may be seen at a glance. The present list extends only to the end of January; the concluding portion will be communicated next month.

	G.M.T.	Ariel.		Umbriel.		Titania.		Oberon.	
	h	Pos.	Dist.	Pos.	Dist.	Pos.	Dist.	Pos.	Dist.
1881.		°	"		"		"	°	"
Dec. 15	16	14.7	14.3	8.7	7.8	20.5	15.3	—	"
	18	15.2	14.6	10.5	10.1	21.4	13.6	· —	
	20	15.8	14.3	11.6	12.2	22.5	11.7	—	
	22	16.4	13.4	12.5	14.2	24.1	9.8	—	
28	16	—		—		200.0	16.3	186.4	13.2
	18	—				200.8	15.0	187.5	14.9
	20					201.7	13.2	188.3	16.5
1882.									
Jan. 17	18	16.8	12.5	7.0	6.0			5.5	11.2
	20	17.7	10.5	9.7	9.7			6.8	12.9
	22	19.0	8.1	11.1	11.1	—		7.9	14.6
18	16	—		—		16.5	30.1	12.2	28.5
	18	—				16.7	29.0	12.4	29.9
23	20	195.3	15.0	—		199.6	16.0	—	
	22	195.8	14.6	—		200.3	14.2	—	
24	20	—		195.3	21.0	—		189.9	18.3
	22			195.6	20.7			190.5	19.9
25	0	—		195.9	20.1	—		190.9	21.5
29	10	—		197.0	16.1	190.2	13.7	—	
	12			197.5	14.3	190.9	15.5	—	
	14	—		198.2	12.3	191.5	17.4	199.0	22.9
	16			—		192.0	19.1	199.3	21.4
	18			—		192.4	20.8	199.7	19.8
31	10			16.5	17.5	—		9.4	15.4
	12	—		17.0	15.9	—		10.1	17.0

* In the Note "On the Apparent Conjunctions of the Satellites of *Uranus* with each other, 1881," published in the *Monthly Notices* of January 1881.

Note on Mr. Christie's Paper in the Monthly Notices for May 1881.
By E. J. Stone, Esq.

The remarks which Mr. Christie has inserted in the *Monthly Notices* for May 1881 are unfortunately of such a nature that they cannot close a controversy.

I certainly have not put words into Mr. Christie's mouth and charged him upon the strength of these words with fallacious reasoning. The fallacy which, in my opinion, invalidates Mr. Christie's work on these questions of refraction can be clearly stated, and it depends upon no misrepresentations of mine.

I have proved, by a re-calculation, that the residual errors given in my paper of 1867, December, are substantially correct: and that the Greenwich circumpolar observations 1857 to 1865, to the lowest altitudes observed, are well represented in mean results by refractions computed from the formula Bessel's Refractions $(1 - 0.005)$. I have also shown that the Greenwich circumpolar observations which were reduced with these refractions, 1868–1876, are well represented. I have shown that Bessel's Refractions *unaltered* do not represent these Greenwich observations. These statements are mere matters of fact; the residual errors which prove them have been given over and over again in the course of the present discussion. The agreement between the Greenwich circumpolar observations and the diminished refractions may, of course, be solely due to systematic errors in the Greenwich results; but this is improbable in itself; and, if it were true, it would not prove that the change in the refractions adopted at Greenwich in 1868 was due to an error on my part, or was made upon *slight* evidence. The evidence afforded in favour of the diminished refractions by these Greenwich circumpolar observations is, as I have elsewhere said, irresistible, unless some other alternative hypothesis can be brought forward which shall equally represent the facts, and which has *equal claims to be accepted as a possible solution of the problem of Astronomical Refraction.*

In 1868 we had no such alternative hypothesis. The broken, discontinuous, curve system with its three distinct tables of refraction, then in use at Greenwich, was considered only an unsatisfactory make-shift. The breaks of continuity, introduced into the adopted refractions, were fully recognised as impossible in nature. They had no other *raison d'être* than the facts, proved by Main, that if you would adopt Bessel's Refractions unaltered for the smaller zenith distances, and determine your colatitude under the same restrictions, then the error of the colatitude, thus determined, and the errors of Bessel's Refractions did not become sufficiently separated to be put clearly and distinctly in evidence until you reached a zenith distance of about 82°. But the error of Bessel's Refractions, in mean results, appeared to be about $0'''\cdot9$ at 82° zenith distance, whilst at 85° zenith distance the error had increased to about $2''\cdot9$. It was considered necessary, for practical applications, to represent the observations at low altitudes, and this was effected by the very simple expedient of throwing away the mean errors which Mr. Main had shown to exist in Bessel's tables. The system adopted, to effect this, was the use of three independent refraction tables—Bessel's Refractions to 82°; Bessel's Refractions × 0·9977 from 82° to 85°; and Bessel's Refractions × 0·9951 below 85° zenith distance. The arbitrary breaks thus introduced show at once that such a system cannot exist in nature. No proof of the legitimacy of the plan adopted has been attempted. But the large errors can, of course, no longer appear as "residual errors" in a discussion of observations reduced with refractions from which they have been thus arbitrarily removed; the smallness of such "residual errors" only proves that the observations under discussion are in substantial agreement with those from which Mr. Main originally determined the adopted corrections. It is the errors of refractions computed from the discontinuous, broken curve, system, abandoned in 1868, which Mr. Christie has compared in his curves and residuals with the errors which result from the use of refractions computed from the formula Bessel's Refractions × 0·995. The latter formula is continuous: it requires no greater number of disposable constants than are contained in Bessel's theory to represent in mean results all the Greenwich circumpolar observations 1857–1876 from the zenith to the lowest star observed. The refractions used by Mr. Christie are discontinuous. They are not computed from a single table: and, to secure an equal agreement between the computed refractions and the Greenwich observations, it requires three disposable constants instead of the one required by the refractions introduced in 1868. It was simply on these grounds that the diminished refractions were adopted in 1868 by Sir G. B. Airy.

It is the comparisons which Mr. Christie has instituted as *an alternative hypothesis*, to weaken the evidence in favour of the diminished refractions, between residual errors resulting from the use of refractions computed from this complex, arbitrary system with those arising from the use of Bessel's Refractions × 0·995, which constitutes the fallacy which, in my opinion, vitiates Mr. Christie's reasoning and statements on these questions of refraction. Mr. Christie may maintain that there is no fallacy in the institution of these comparisons. He is, of course, fully entitled to the benefit of his own judgment upon the point. I, however, consider it a serious fallacy—a fallacy which entirely destroys the force of his remarks on the slight evidence which these Greenwich observations afford in favour of the diminished refractions. The mere fact that it requires such a complex system with two additional constants to equally represent the observations is a proof that the evidence afforded by these Greenwich observations in favour of the diminished

refractions is indeed very strong.   At all events, the fallacy to
which I have called attention, such as it is, is not of my making.
It runs through all Mr. Christie's work on the subject of
refractions.   It is not due to words which I have put into Mr.
Christie's mouth.   I should be utterly ashamed of myself if such
a charge could fairly be brought against me.

The notice, "The Nadir Point Observation is to be taken at
$179° 40' + o^r$," to which Mr. Christie calls attention on page 344,
meant exactly what it says—viz. that the readings were to be
taken on the first revolution, or from $o^r \cdot o$ to $1^r \cdot o$.   This is shown
by the general practice to which it led.   The practice was
adopted, as I have before stated, to keep the correction for runs
small for the Nadir Point Observation.   If errors such as those
given in the Introduction to the Greenwich Catalogue exist in
the readings of the screws near $o^r \cdot o$, the supposed errors of the
screws at other readings will change from the mere effect of
these errors on the runs by such quantities as the following :—

Supposed Errors of Screws.

Runs started from	$o \cdot 5$	1	2	3	4	5
$9 \cdot 9$	$-o'' \cdot 1$	$-o'' \cdot 2$	$-o'' \cdot 5$	$-o'' \cdot 7$	$-o'' \cdot 9$	$-1'' \cdot 2$
$o \cdot 1$	$-o \cdot 1$	$-o \cdot 1$	$-o \cdot 2$	$-o \cdot 3$	$-o \cdot 4$	$-o \cdot 5$

I think that the magnitude of these differences will show that
there may be some considerable advantages in keeping the run-
correction small for an observation like the Nadir, which is
common to all the observations of a night.   If Mr. Christie cannot see the advantage of separating the effects of two distinct
sources of error, I regret it, but I cannot help it.

Mr. Christie's statement that the constant error of $-o'' \cdot 8$ in
the Nadir observations was due to error of the screws about $o^r \cdot o$,
which had suffered extra wear since 1868 from a preponderance of
some 4,000 observations taken at this part, was, I thought, sufficiently disposed of when it was pointed out that instead of 4,000
additional observations there were only 108 such observations made
with readings at and below $o^r \cdot 2$, whilst there were 72 such determinations made at readings greater than $1^r \cdot o$ during a period of
nearly two years—1868, February 1, to 1869, December 31.   This
would give less than 500 additional observations, instead of 4,000,
during the period of 1868 to 1876.   Mr. Christie, however, thinks
that these facts do not concern him.   The corrections which
have been applied to the Greenwich Observations 1874–1875 for
these supposed errors of screws are given in the following table,
which is extracted from the Introduction to the Greenwich
Catalogue, 1878 :—

9·7	−1″·38	1·4	+0″·08	3·1	+0″·38
9·8	−1·13	1·5	+0·08	3·2	+0·22
9·9	−1·17	1·6	+0·18	3·3	−0·06
0·0	−0·89	1·7	+0·12	3·4	+0·05
0·1	−0·51	1·8	+0·05	3·5	+0·31
0·2	−0·41	1·9	+0·22	3·6	+0·33
0·3	−0·42	2·0	+0·47	3·7	+0·27
0·4	−0·09	2·1	+0·55	3·8	+0·22
0·5	+0·28	2·2	+0·66	3·9	+0·21
0·6	+0·34	2·3	+0·76	4·0	+0·19
0·7	+0·28	2·4	+0·52	4·1	−0·05
0·8	+0·19	2·5	+0·39	4·2	−0·19
0·9	+0·22	2·6	+0·52	4·3	−0·03
1·0	+0·30	2·7	+0·42	4·4	−0·17
1·1	+0·23	2·8	+0·15	4·5	−0·39
1·2	+0·23	2·9	+0·15	4·6	−0·51
1·3	+0·23	3·0	+0·32	4·7	−0·74

I point out the extreme rapidity with which the error changes near 0ʳ·0, that on and after 0ʳ·5 the changes of the corrections are, comparatively speaking, small. It appears, therefore, that if these errors are due to serious wear of the screws, the serious wear has been principally confined to the threads of the screws engaged at readings below 0ʳ·5. Now, Mr. Christie has himself stated that from 1868, February 1, to 1869, December 31, there were only 283 observations of the Nadir made at and below readings of microscope A=0ʳ·4. If, therefore, we give the most liberal interpretation possible to the expression " the parts of the screw about 0ʳ·0," and understand it to mean all those parts of the screw which have been proved to be seriously defective, yet, even then, we should have less than 1,500 instead of 4,000 additional observations made near 0ʳ·0. But, in spite of these facts, Mr. Christie has thought himself justified in making such a statement as the following. " It seems a pity that Mr. Stone did not make himself better acquainted with the facts before making the confident assertion that ' whatever may have been the state of the screws near 0ʳ·0 in 1876 it certainly was not caused by the 4,000 observations made near 0ʳ·0 at my suggestion.' " It is extremely difficult to deal with such assertions in a way which can be regarded as satisfactory either to Mr. Christie or to myself.

In my opinion, neither 4,000 nor 6,000 additional Nadir observations made near 0ʳ·0 in a period of eight years would have led to any such serious wear of the screws as that indicated as existing in 1876 near 0ʳ·0. If all the observations were made near o^r·o, and the screws never worked far from this reading, such a thing might be conceivable; but in practice the screws are worked continually backwards and forwards over the whole range of the screws from o^r to 5^r·o. That the mere running down of the screws once in the course of a night's work to o^r·o should lead to any serious unequal wearing of such screws as those in use at Greenwich is to me inconceivable.

In all screws the threads near the ends of the screws are defective, and are never intended to be used. If in some of the many adjustments of the instrument the indices of the screws have been shifted relatively to the screws, and thus some of the defective threads brought accidentally into use, the apparent · wear of the screws would be accounted for; and this explanation would also show why the supposed serious wear has changed so rapidly near o^r·o, and practically became insensible at and · after o^r·5.

I believe that this affords a key to the true explanation of these errors. But, at all events, I cannot regard the way in which Mr. Christie has attempted to determine the correction required by the screws at the different readings as satisfactory. I have already shown how rapidly these supposed corrections vary with every change of starting-point for the run-correction. To attempt, therefore, an investigation of the errors of the screws or the corrections which should be applied to observations without taking the runs actually used in the reduction of the observations into account can, in my opinion, only lead to serious misconceptions with respect to the errors of the screws and no inconsiderable errors in the corrections applied to the observations.

Showers of Large Meteors.   By W. F. Denning, Esq.

I. On Oct. 29, 1881, at 9^h 58^m I saw a fine meteor, quite= ♃, descending obliquely to a point slightly below *Fomalhaut* on the S. S. W. horizon. The nucleus was distinctly pear-shaped, and it left a thick trail of luminous flakes. Path about 20°, traversed in 2 seconds. Shortly afterwards I saw two other meteors, 2nd mag., evidently from the same radiant, W. of the *Pleiades.*

The shower to which these meteors belonged is one of very special character. At the end of October and during the first few nights of November it furnishes many brilliant fireballs, and in conjunction with that closely bordering shower, the Taurids, renders the epoch a notable one for large meteors. The radiant point, as I derived it, Oct. 28-Nov. 13, 1877, from 32 paths, was at 43° + 22°, and the shower struck me as exceptional both on account of its activity and the brilliancy of its meteors. In appearance they are similar to the Taurids, the motion being somewhat slow, and the nucleus is almost invariably accompanied by a thick trail of sparks or luminous flakes. The position of the radiant is between *Taurus* and *Aries*, south of *Musca*, and fully 15° west of the true Taurids.

During the last few years many fine meteors have owed their origin to this stream, and I have collected some of the best instances together and projected the apparent paths upon a diagram as follows:—

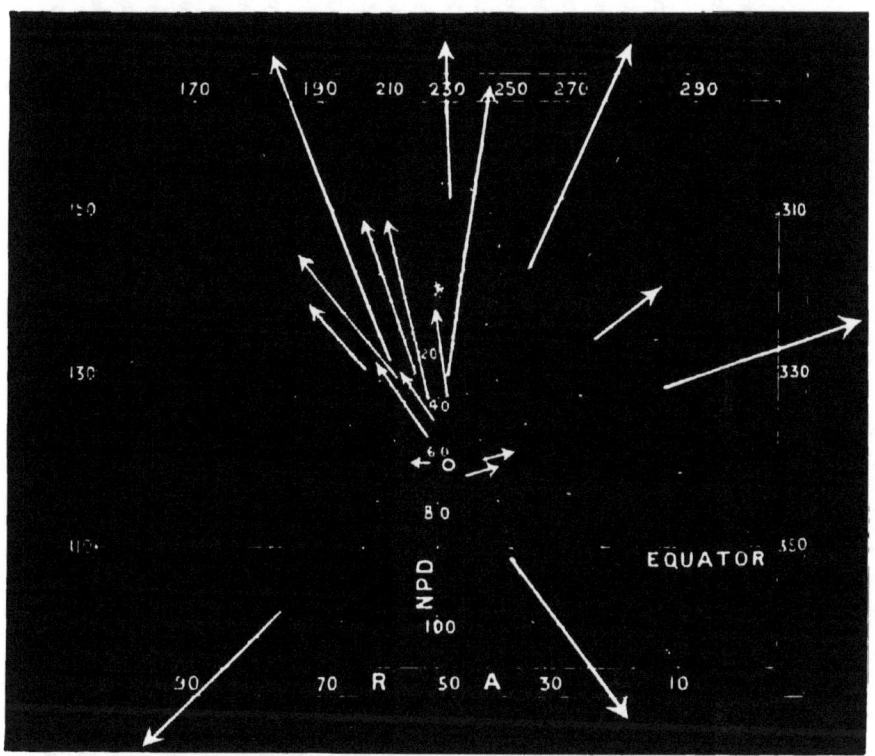

Paths of 18 Fireballs, directed from a Radiant at 48° + 22°, Oct. 25–Nov. 6.

The point of divergence is fairly precise at 48° + 22°, and considering the character of the observations the agreement is very satisfactory. Showers of small meteors very carefully recorded by one and the same observer often show a sharply defined radiant, but here, where the results of many different observers are incorporated, we can hardly expect the same close accordance. Moreover the direction of flight in the case of large fireballs, such as these, is often not very accurate, from the circumstance that the apparition is sudden and unexpected, so that the observer is often taken unawares. The glare and brilliant effect are frequently the cause of distracting the observer's

Ref. No.	Year 1800+ and Date.	Mag.	Observed Path From	To	Length of Observed Path.	Notes.	Observer and Place.
1	54, Oct. 31	♀	235+73	233+60	13°	Streak	Heis, Münster.
2	63, Nov. 6	♃	*340+23	300+10	40	Streak; curved path	Heis, Münster.
3	66, 3	2 × ♃	*63+50	185+69	54		Min., Bon.
4	66, 3	2 × ♃	*30+22	19+23	10		Crumplen, London.
5	66, 6	♃	*67+51	172+65	51		Penrose, Wimbledon.
6	69, 2	♃	*335+29	321+26	13	Swift	Zezioli, Bergamo.
7	69, 3	♃	†83−7	93−13	12	Very rd. Streak 3°	Tupman, Min.
8	69, 3	♀ ♀	*95−15	110−24	17	Very fine. Min of sparks	Tupman, Moditerranean.
9	72, 1	♀	*100+48	132+49	21	Duration 0·5 sec. ...	Tupman, Portsmouth.
10	72, 3	¼ − ☽	‡89+50	150+41	42	2·5 sec. Tail of red sparks	McClure, Glw.
11	73, Oct. 26	V. large	*56+31	105+57	42	Min of sparks ...	"F. T. S," Thruxton.
12	77, Nov. 1	♃	44+56	245+51	71	Slow. Bright train	Denning, Bristol.
13	77, 4	♃	*55+22	61+22	5	V. slow. Thick spark-train	Denning, Bristol.
14	77, 4	> ♀	‡350−1	327−17	28	2½ sec. Train 15° long.	Wood, Birmingham.
15	77, 4	♀	†10−0½	344−10	28	1 sec. streak	Backhouse, Sunderland.
16	77, 7	♂	100+59	190+51	49	2½ sec.; slow; red	aMle, York.
17	78, 3	Brilliant	*310+61	266+42	32	Streak 2 sec.	Downing, Greenwich.
18	79, 5	Fireball	*48+50	115+87	38	Train 2 sec.	Trouvelot, Cambdge., U.S.
19	80, Oct. 25	> ♀	‡55+40	76+52	18	3½ sec.; streak	Sawyer, Cambridge, U.S.
20	80, 30	> ♀	†38+18	26+22	12	Streak; 25 sec.	Denning, Bristol.
21	80, 30	> ♃	†31−1½	16−17	21	1½ sec. ...	Baxendell, Manchester.
22	80, Nov. 1	♃	*275+56	257+43	17	Train of orange flakes	Corder, Writtle.
23	81, Oct. 29	♃	*360−22	342−33	20	2 sec.; thick train	Denning, Bristol.

* Radiant well defined at 48° + 22°.   † From an apparently sub-radiant at 47° + 15°.   ‡ These conform to a radiant slightly N. in Musca.

attention from the immediate determination of the path. This is left until both the nucleus and train have died away, so that in fixing the approximate course the memory, so often an untrustworthy guide, has to be relied on. In the subjoined table the paths of 23 large meteors apparently conforming to this shower are given.

The average length of path is 28°·4 (23 meteors).

A few of these may quite possibly have been Taurids.

The catalogues of meteor observers probably contain many other fireballs belonging to this remarkable stream or its closely-allied shower, the Taurids.

The great meteor of 1859, Oct. 25, $7^h 15^m$, seen in Ireland, and estimated $= $ ☽, was from a radiant near η *Tauri* at $41° + 17°$ (or $54° + 18°) \pm 10°$, and the bright meteor of Nov. 1, 1860, $8^h 30^m$, was directed from the same shower at $55° (\pm 10°) + 20°$.* Another large meteor seen in the evening twilight on Nov. 8, 1876, agreed with this radiant near η *Tauri*, $53° + 20° (\pm 5°)$.† On Nov. 5, 1876, about 9 P.M., a large bolide giving a flash brighter than moonlight was observed at Choisi le Roy, in France, moving from a *Aurigæ* to a *Ursæ Majoris*, and leaving a streak from which the path was recorded.‡ The following night at Orsay, near Paris, A. Guillemin describes a fine meteor not much smaller than the moon's disc, which travelled very slowly southwards, in an horizontal direction, about 20° or 25° above the E.N.E. horizon. The meteor had a globular nucleus, and left a bright streak.§ From the directions of the apparent paths it is obvious these meteors were severally projected either from the radiant of the Taurids or from that of the shower adjoining it, S. of *Musca*. Both are equally remarkable for their numerous and brilliant members. They appear to attain a maximum early in November, though many meteors from nearly the same point below *Musca* are seen early in October, and the Taurids, commencing in that month, are prolonged late into November. I have observed the Muscids as an active and sharply-defined shower on Oct. 8, 15, and Nov. 13. and the Taurids with equal distinctness on Nov. 8, 13, 20 and 27. Greg's shower R_3 is evidently identical with the former. He gives the duration from Oct. 18 to Nov. 14, and the place as $43° + 26°$. Sawyer detected it on Oct. 21–22 at $47° + 27°$, and Corder has seen many of its meteors in Oct. and Nov. from the centre $43° + 15°$.

II. In the morning twilight on July 25, 1881, I saw a very fine meteor, with vivid flash and streak, in *Aquarius*. I marked its short path accurately from the streak it left. The meteor was evidently much foreshortened and close to its radiant E. of β *Aquarii*. This position agreed with one I had found as the

---

* B. A. Report on Luminous Meteors, 1880, p. 4.

† *Monthly Notices*, vol. xxxvii. p 210.

‡ *Comptes Rendus*, vol. lxxxiii. p. 862.

§ Ibid. p. 922.

result of an investigation of the fireballs of July 25–31 (*Observatory*, Sept. 1879, p. 130), and it also exibits a distinct confirmation to the fine shower seen by Sawyer at the end of July 1880, when he recorded 28 of its meteors and describes them as " very bright, generally slow." .This shower in conjunction with another fireball radiant at 303°+12°, and two rich streams of meteors at 32°+53° and 341°−14° go far to make the end of July a meteoric period of exceptional strength.

The following table includes the observed paths of 16 fireballs recorded at various stations (chiefly by foreign observers). They conform closely to the centre near β *Aquarii* :—

Ref. No.	Year 1800+ and Date.	Mag.	Observed Path From α δ	To α δ	Length of Path.	Observer or Authority.
1	64, July 30	♀	278 +38	210 +55	47	Heis, Münster.
2	64, 30	♃	267½ +30	221 +41	39	Weber, Peckeloh.
3	67, 31	♃	345 +33	12 +54	28	Zezioli, Bergamo.
4	70, 27	♃	346 − 9	355 − 9½	9	Tupman, Meditn.
5	72. 25	♃	293·6 +42·9	193·3 +52·3	63	Konkoly, O-Gyalla.
6	73, 25	♀	357·8 + 0·5	6 + 1·7	8	Nagy, O-Gyalla.
7	73, 25	♀	262·2 +17·4	212·5 +24·1	47	Nagy, O-Gyalla.
8	73, 30	♀	294 + 3	207 +28	81	Klinkerfues, Göttingen.
9	73, 30	♀	258 +10	197 +22	60	Thraen, Dingelstaedt.
10	75, 27	♀	146·3 +71	139·6 +56·3	15	Nagy, O-Gyalla.
11	75, 27	♀	273·7 +27·9	262·7 +33·4	11	Nagy, O-Gyalla.
12	78, 30	♀	330 −22	332 −27	5	Denning, Bristol.
13	79, 26	♃	16·9 +20	41·8 +26·8	24	Konkoly, O-Gyalla.
14	79, 28	♀	258·2 +14·5	233·1 +23·5	25	Konkoly, O-Gyalla.
15	79, 30	¼ = ☽	342 − 8	360 − 3	19	Payne, Suffolk.
16	81, 25	♀	336 − 5	340¼ − 4	4½	Denning, Bristol.

Radiant point at 324°−9°. Average length of course=30°·3. The shower is well defined a few degrees S. E. of β *Aquarii*; the meteors are brilliant, rather swift, and generally with streaks. It has been previously observed as follows :—

July 25–31 ...	... 324− 6	Schmidt.
1870, July 28 ...	... 326−13	Tupman.
1880, July 28–31 ...	... 330− 6	Sawyer.
28–30 ...	... 328−15	Sawyer.

Including with these the place of the fireball radiant as above 324°−9°, we derive a mean position at 326°·4−9°·8. The maximum occurs on July 28-30, and this shower undoubtedly plays an active part in rendering the last few nights of July a notable fireball epoch. It should not be confused, however, with a contemporary shower further E., 15°, which I observed in considerable strength in 1878,* and which had been thoroughly well determined by Tupman at the end of July 1870.

III. Another shower,† S. of ε *Persei,* Sept. 6-15, deserves mention as one consisting of a large proportion of brilliant meteors. It was well seen at Bristol in 1877. The meteors belong to the swift streak-leaving class, and the display apparently attains a maximum on Sept. 6-8. I have observed the following bright meteors belonging to it during the past few years :—

Ref. No.	Year 1800+ and Date.	Mag.	Observed Path From a δ	To· a δ	Length of Path.	Notes.
1	74. Sept.	6	♃	275 + 20	255 − 12	37.	Swift ; streak.
2	77,	7.	L	73 + 53	85 + 60	10·	Streak 3 sec.
3	77,	15.	♃	56 ± 0	55. − 6	6	Swift ; streak 4 sec.
4	78,	8	> ♀	246 + 21	244½ + 5	16·	Swift ; streak 4° 20'.
5	79,	7.	> 1	342 + 27	317½ + 14	28	Swift ; streak.
6	80,	6.	♃	294 + 7½	279 − 8	21·	Very swift ; streak.
7	80,	6.	L	344 + 65½	310 + 61½	15,	Very swift ; streak.

The radiant point is at about 61° +36°, as I carefully derived it in 1877. The shower is a very prominent one and is obviously the same as No. 64 (at 66° +40°, Sept. 7-15) in Tupman's catalogue and No. 147 (at 60° +32°, Sept. 8) in the list deduced by Schiaparelli from Zezioli's observations. It is best observable in the morning hours when the radiant is high. The first half of September is, similarly to the first few nights of November and the last few nights of July, a very well defined period of fireballs, and I believe that the majority of those appearing on the 6-8th are directed from this shower of ε Perseids. It will be very important in future years to look out at the several epochs referred to for further confirmation of these showers of large meteors, for there is no doubt of their annual recurrence with marked intensity.

* See *Monthly Notices,* vol. xl. p. 126 (Jan. 1880), and *The Observatory,* Sept. 1878, p. 164.

† See *Popular Science Review,* Oct. 1880, p. 335, which contains a diagram of 86 paths belonging to this shower as observed by the writer and others.

Ashleydown, Bristol :
1881, *November* 9.

Observations of Comet c, 1881. By T. W. Backhouse, Esq.

1881. Time.	Brightness of Head.	Length of Tail.	Width of Tail.	Central Line of Tail.		Mode of Obser	Hindrances.
				passes, or points to	at distance from Nucleus.		
July 27, 12 10		1¼°		at angle of 15° above β Aur.		n. e.	Twilight.
12 20		1°				(38)	,,
		1½°				(20)	,,
12 25	= 55 Aurl.: slightly fainter than 56.	2·7°	0·7°	¾ (β, π) Aur.		(2·5)	,,
29, 12 17	= 21 Lyncis.	5·8°				n. e.	
12 25		6½°	½°	½ (41, 36) Aur.	6⅓°	(2·5)	
Aug. 4, 12 50		6¾°	⅘°	½° north of 46 Aur.	6°	n. e.	Twilight.
12 55				at angle of 15° north of 57 Aur.	2½°	(3·5)	,,
5, 12 23			11' at 1° off nuc.			(20)	Haze.
12 25	Considerably brighter than 21 Lyn.					n. e.	,,
9, 13 15	Ditto, and rather fainter than κ Urs. Maj.†	shorter than on 4th. ¾°				(3·5)	Moon.
13 25	Just visible.					n. e.	..

1881.	Time.	Brightness of Head.	Length of Tail.	Width of Tail.	Central Line of Tail. passes, or points to	Central Line of Tail. at distance from Nucleus.	Mode of Obser.	Hindrances.
Aug. 10.	10 40		2°				(3·5)	Moon.
	11 0		1°				(20)	„
14,	10 50	Rather fainter than θ Urs. Maj.	4¾°				n. e.	Not very clear.
18,	10 0	.. brighter ..	8½°	1½°	about ½° f υ Urs. Maj.	8½°	„	Town smoke and light.
„					229 P. ix.	3½°	(3·5)	
	10 50			12′ at 12′ off nucleus; also at 95′ off.			(20)	
	12 15			16′ at 12′ off nucleus.			„	Moon.
19,	13 10		10¼°	1½°			n. e.	„
„					36 Urs. Maj.	5°	(3·5)	„
20,	9 40	Considerably brighter than θ or ψ Urs. Maj.					n. e.	Twilight.
	10 15		8°	1½°	β Urs. Maj.	8°	specs.	Town (?) light.
7,	8 21	(Invisible with naked eye.)	¾°	¼°	1¼ (37, 41) Com.		(3·5)	Twilight.
	8 40	Much brighter than β Comæ.					n. e.	„
28,	8 49	Not quite = η Boötis.*	2¾°	¼°	slightly np. 14 P. xiii.		(3·5)	„

* Means that the object, compared with Comet c, was more favourably situated; † less so.

My observations on this Comet were made at Sunderland.
The accompanying table gives the brightness of its head—*i.e.* of
the nucleus and the surrounding nebulosity; the length and
width of the tail, the greatest width being given unless anything
is said to the contrary; and the direction of the tail—*i.e.* of the
central or brightest line, as measured from the nucleus. When
the tail was curved, the next column gives the part to which the
direction refers. The last column but one gives the mode of
observation, n. e. indicating the naked eye; specs., spectacles to
correct my short sight (which usually made no perceptible difference); (3·5), opera-glasses, power 3·5, aperture 1½ in.;
(2·5), a smaller pair of opera-glasses; and (20) and (38), these
powers of my 4½-in. achromatic. On each night I have given
the results of the best mode of observation for that night. The
last column gives the more important hindrances to a perfect
view. Of course the low altitude of the Comet prevented it from
ever being seen so clearly as if it were high up.

The latter part of the time of the Comet's apparition the
weather was unusually cloudy.

The measurements are all taken from my drawings, or otherwise estimated.

Remarks on the Tail.

July 27.—Straight.
July 31.—Slightly curved, concave to south (2·5).
August 4.—Strongly curved (n. e.).
August 14.—I believe, slightly curved, concave to *sp.*
August 18.—Curvature not certain with naked eye, but
slight with (3·5).
August 19, 13h 10m.—Slightly curved near the nucleus, concave to *p*, as seen with opera-glasses. With the naked eye it is
straight, but as the *f* edge extends much further than the *p*
edge, it has the appearance of a slight curvature, concave to *f* 37
Urs. Maj. just within the *f* edge.
August 20, 10h 15m.—Straight. I believe I can see it nearly
to a *Urs. Maj.*, but am sure of it only to near *β*. There is
a broad extension of the tail on the *p* side in its southern
half.
August 28, 8h 49m.—Straight. The *sf* edge most definite;
14 P. xiii just within it.

Telescopic Notes.

August 18, 10h 50m.—From near the nucleus to a distance of
a degree from it, the brightest part of the Comet is a narrow ray,
almost straight, only 1′ or 2′ wide. Preceding this the tail is
very faint, except near the nucleus, and at a distance of 1° or 1½°
from the nucleus fades out altogether, the *p* edge of the tail
being, beyond that, almost a continuation of the narrow ray.

There is an indefinite projection from the nucleus towards *np*, to a distance of 12′ or 15′ from it. The *f* edge of the tail is very definite.

August 19, 13ʰ 10ᵐ.—The bright central ray broader and less definite, and more decidedly curved, concave to *p*. The *f* edge, which is very strongly curved in the same direction, is brighter than the space between it and the central ray to a distance of 25′ from the nucleus; which was not the case last night.

Notes on the Spectrum of Comet c, 1881, as seen with a Browning's Miniature Spectroscope on the 4½-in. Telescope.

July 27.—Spectrum consists chiefly of the three usual bright bands. The continuous spectrum faint, unlike the nucleus of Comet *b*.

August 19, 13ʰ 30ᵐ–14ʰ 0ᵐ.—Spectrum much the same, but brighter. The three bright bands by far the most conspicuous part. They are the only bands I can be sure of. The continuous spectrum of the nucleus begins abruptly at about half as far off the least refrangible band as that is off the middle one; and I suspect a bright line at that point.

Sunderland:
   1881, *October* 8.

Observations of Comet b, 1881, made at Stonyhurst Observatory.
By Rev. S. J. Perry.

The position of Comet *b*, 1881, has been observed here on nearly every available night since it was first visible in this latitude, but the weather has very much interfered with the completeness of the record. The observations of June 27 and July 1 were S.P. transits, and all the other positions were taken with the 8-inch Equatoreal, the comparison stars having been carefully identified.

		h m		h m s		° ′ ″
1881, June 25	G.M.T.	12 11	R.A.	5 42 51·3	N.P.D.	36 29 2·4
27		11 38		5 52 46·3		29 46.19·4
July 1		11 50		6 20 53·5		19 45 6·3
11		10 27		8 40 44·3		8 52 59·1
14		11 6		9 39 1·2		7 57 43·1
19		10 2		11 3 7·6		7 46 56·9

		h m	h m s	° ′ ″
July 20	G.M.T.	11 16	11 17 37·2	7 51 32·6
27		11 18	12 32 24·0	8 52 59·3
31		11 1	13 1 2·9	9 33 49·9
Aug. 1		11 2	13 6 36·7	9 45 34·9
Sept. 19		9 29	16 22 15·7	15 55 5·1

Stonyhurst Observatory, Whalley:
1881, November 7.

Elements of Comet b, 1881.   By H. T. Vivian, Esq.

On the appearance of the Comet in June of this year I made a series of measurements with a view to the determination, at least approximately, of its orbit. Its angular distance from two Nautical Almanac stars was measured on every available evening with a carefully-adjusted pocket sextant, the time of the observation being taken from a watch previously compared with a regulator. The stars chosen were *a Persei*, and *a Ursæ Majoris*. No correction for differences of refraction was applied, as it was not thought desirable to attempt an accuracy not within the capabilities of the instrument.

Three of the deduced positions were selected for the computation of elements, viz.:—

	G.M.T.	App. R.A.	App. N.P.D.
	h m	h m s	° ′ ″
June 24	11 30	5 37 50	40 36 31
29	10 35	6 6 37	24 38 58
July 4	10 50	6 52 17	15 5 39

and from these I have computed the following parabola:—

$$T = 1881, \text{ June } 19·0433 \text{ G.M.T.}$$

$$\varpi = 270°\ 37'\ 22'' \Big\} \text{ Apparent Equinox,}$$
$$\Omega = 271\ \ 8\ 47 \Big) \qquad \text{June 30}$$
$$i = 63\ \ 3\ 21$$
$$\log. q = 9·8803409$$

Motion Direct.

The calculated middle place compared with the observed gave:

$$C-O \quad \text{Longitude} -4'' \quad \text{Latitude} +70''.$$

In the present state of astronomical science, such an attempt as this may perhaps be considered useless, but it will serve to show that something may be done with very inadequate means, and that such observations, in the absence of better, might furnish an orbit sufficient for the recognition of the Comet on its return.

November 3, 1881.

On the Orbit of Denning's Comet. By Professor Winnecke.

From the observations, Marseilles, October 5, Dunecht, October 12, Strassburg, October 19 and 28 (made by me with the refractor of 18 Paris inches aperture), Dr. Hartwig, assistant of the Observatory, and M. Wutschichowsky, from Pulkowa, have deduced the following orbit of Denning's Comet:—

$$T = 1881, \text{Sept. } 13\cdot1697 \text{ Berlin M. T.}$$

$$\pi - \Omega = 312\degree \; 11' \; 22'' \quad \text{M. Eq. } 1881\cdot0$$
$$\Omega = 66 \quad 4 \quad 2 \qquad ,,$$
$$i = 6 \quad 52 \quad 36 \qquad ,,$$
$$\phi = 55 \quad 34 \quad 7$$
$$\log q = 9\cdot859955$$
$$\log a = 0\cdot616427$$
$$\text{Period} = 3070^d\cdot7.$$

The differences (Observed — Computed) are:—

	October 5.	October 12.	October 19.	October 28.
$\Delta\lambda$	$+1'\cdot5$	$-0''\cdot6$	$+5''\cdot0$	$+0''\cdot1$
$\Delta\beta$	$-0\cdot2$	$-2\cdot5$	$+2\cdot2$	$-0\cdot1$

The orbit of the Comet is extremely interesting by its close approximation to the orbits of Venus, the Earth, and Jupiter. M. Hartwig and M. Wutschichowsky find:—

Helio. Long.	Distance between the orbits of	
$32\degree$	♀ — ♀	0·0215
83·5	♀ — ♁	0·0359
174	♀ — ♃	0·50
222	♀ — ♃	0·164.

The Comet was by no means *bright* on October 28, when I saw it for the last time. It is therefore to be hoped that some of the very powerful telescopes in England may be directed to this the first Comet discovered in that country for many years. Its places given by the orbit are:—

12ʰ. Berl.	R.A.	Dec.	log *r*	log Δ
	h m s	° ′ ″		
1880, Nov. 14	10 35 30·7	+14 41 5	0 10699	0·01993
15	36 34·5	42 21		
16	37 35·9	43 49		
17	38 35·0	45 28		
18	39 31·7	47 18	0·12343	0·02399
19	40 26·0	49 21		
20	41 17·8	51 37		
21	42 7·3	54 5		
22	42 54·2	56 47	0 13926	0·02707
23	43 38·7	59 42		
24	44 20·6	15 2 51		
25	44 59·9	6 14		
26	45 36·7	9 50	0·15451	0·02931

Strassburg :
1881, *November* 8.

Approximate Positions of Comet b, 1881, *deduced from observations made at the Adelaide Observatory.* By C. Todd, Esq.

Date.		R. A.	Δ	Station.	✱ of Comparison.
	h m	h m s	° ′ ″		
May 28	8 0	5 1 25·	32 22 7·	Adelaide	α Columbæ.
29	7 20	5 1 51·7	31 39 39·	,,	BAC 1564.
30 }		5 2 21·8	30 51 2·	,,	,, 1615.
30 }	7 33	5 2 26·12	30 50 49·	,,	,, 1564.
31	7 8	5 2 54·6	30 0 1·	,,	,, 1615.
June 1	6 48	5 3 32·8	29 2 58·	,,	Washington 2173.
5	6 10	—	24 5 0	,,	
• 12	6 0	5 11 38·4	8 11 39·	,,	Rigel.
12	18 30	5 12 13·4	6 26 21·	,,	τ Orionis.

Physical Observations of Comet b, 1881, made at Forest Lodge, Maresfield. By Captain W. Noble.

Observing the Comet from 11ʰ 30ᵐ to 12ʰ 30ᵐ L.M.T. on the night of June 24, with a power of 42 on my 4·2-inch Ross Equatoreal, the nucleus and its surroundings presented the aspect rather indicated than drawn with absolute fidelity in the accompanying sketch numbered 1. From the brilliant nucleus there extended towards the N. and E. a line of light into the tail. This seemingly expanded into a fan-shaped aigrette to the S. and E. of the nucleus. A small aigrette of light also issued from the nucleus to the south and west of it. I noted that the very smallest stars seen through the tail lost nothing of their brilliancy. I tried my star spectroscope upon the head of the Comet with the

Comet b 1881.

(Observed by Captain Noble at Forest Lodge, Maresfield.)

Fig. 1.   Fig. 2.

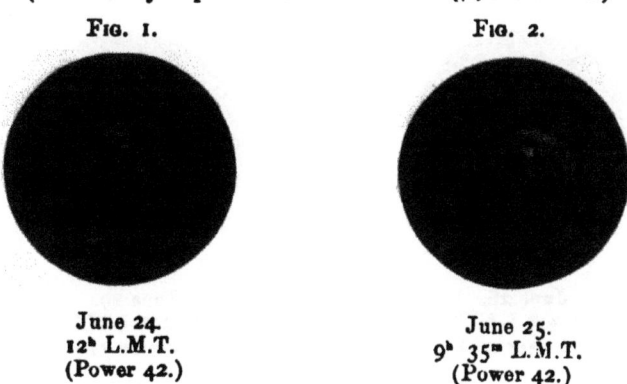

June 24.
12ʰ L.M.T.
(Power 42.)

June 25.
9ʰ 35ᵐ L.M.T.
(Power 42.)

slit as much closed as its light would permit, but failed to get anything but a continuous spectrum.

On the succeeding night, that of June 25, a most remarkable change had taken place in the appendages of the nucleus. Instead of appearing with any jets or line issuing from it, it was surrounded by two sectors of light placed unsymmetrically round it in a way which will be best understood from drawing No. 2. The nucleus itself presented the aspect of a star of the first magnitude, and seemed to be connected with the inner concave edge of the interior sector by a kind of radiating hazy luminosity. Viewing the Comet now at 10ʰ 5ᵐ L.M.T. with the star spectroscope, I at once saw a gaseous spectrum superposed on a continuous one, the most prominent line being in the bluish-green. As the night darkened the gaseous spectrum became very pronounced, and the continuous one of the nucleus lay along it as a brilliant line of light. I further noted that the strange arrangement of sectors, which I have drawn, and which earlier seemed to bound the Comet's outline, as it were, were now at some distance inside the coma.

The indifferent weather prevented any further observations until June 28, when I caught the Comet at intervals from 11ʰ 15ᵐ to 11ʰ 45ᵐ through rifts in huge drifting cumulus clouds. The nucleus had certainly diminished, and the sectoral arrangement surrounding it had changed notably in form. It gave the idea of spouting from the nucleus, but it was very fuzzy and ill-defined, and sketching was very difficult.

On the night of the 29th, which was clear, the nucleus had unmistakably diminished—or perhaps it would describe it better to say contracted, as it was decidedly less planetary and more stellar in its aspect than it was on the first night that I observed it. I noticed too (as I did on the previous night) that the nucleus was much farther within the coma than it was at the date of my earlier observation. In my sketch No. 4 it will be seen that the two sectors of light emanating from the nucleus—

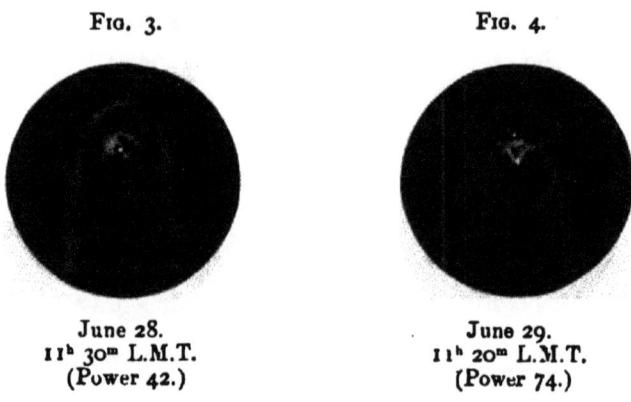

FIG. 3.   FIG. 4.

June 28.   June 29.
11ʰ 30ᵐ L.M.T.   11ʰ 20ᵐ L.M.T.
(Power 42.)   (Power 74.)

the preceding one being the brighter—give something of the effect of a sea-gull in flight; an 8-9th magnitude star, too, over which the nucleus passed almost centrally, will be seen just S. of it.

By June 30 the nucleus had shrunk very perceptibly indeed, and was more stellar than ever. With a power of 74 it looked like a 3rd magnitude star, from which pointed a hazy aigrette of light, bounded on the *s.f.* side by a dark sector. Faint traces existed of the line of light which was such a feature on the night of the 24th, but it was very short. One particularly noteworthy feature I have tried to indicate in sketch No. 5. It was this: that outside of the parabolic outline proper of the Comet was a ragged, hazy, ill-defined edge. I left home on the day after this last observation, and on my return during the second week in July the Comet had ceased to present any telescopic features of interest.

Physical Observations of Schäberle's Comet (c, 1881).

August 21, 9ʰ to 10ʰ 15ᵐ L.M.T.—The aspect of the tail was uniformly nebulous, but it was straighter and better defined on the following side. Exchanging the Power of 42 for one of 115, I saw that the nucleus proper was an exceedingly minute point, a considerable distance within the head. Nothing whatever in the shape of a sector, jet, or aigrette of light was visible. Viewed in the spectroscope, a very bright and distinct spectrum of carbon

Fig. 5.

Schäberle's Comet (c 1881).

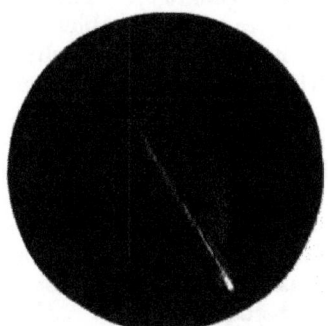

June 30.	August 24. 9ʰ 30ᵐ L.M.T.
11ʰ 30ᵐ L.M.T.	As seen in a field-glass
(Power 74.)	magnifying 4 diameters.

was seen, the three bright lines being crossed by a horizontal continuous one, presumably arising from the minute nucleus.

My sketch of this Comet represents it as viewed in a field-glass, magnifying 4 diameters, on August 24, from 9ʰ to 10ʰ L.M.T. I drew it for the strange, faint second tail, or bifurcation of the first one. Viewed with a Nicol's prism in the Equatorial, the whole coma exhibited distinct traces of polarisation.

Occultation of 13 *Capricorni by the Moon,* 1881, *October* 2.

The star disappeared instantaneously at the Moon's dark limb at 23ʰ 1ᵐ 6ˢ·3 L.S.T.=10ʰ 14ᵐ 3ˢ·5 L.M.T. Power, 154.

--- --- --- --- ---

Discovery of Comet c, 1881.   By Captain Irwin Maling, Colonial Secretary, Grenada, West Indies.

(*Communicated by the Secretaries.*)

I first observed a Comet of considerable magnitude on July 14 at 8 P.M. At that time it seemed to be about 20° from the horizon directly under the Pole star, and to be travelling a course from N.N.W. to S.W. It was passing through the constellation *Camelopardus.* The nucleus was bright, about the size of the most northerly star in *Camelopardus.* The nebulosity round the nucleus was very bright, and of a distinctly oval form; the tail brush-shaped, and pointing to the N.E. A bright star was visible south of the nucleus, and another at the lower extremity of the tail. On the 15th, at the same hour, I took advantage of a clear sky and night, and sketched the Comet, which, I am happy to say, proved to have then attained its greatest splendour. The 16th and 17th, it was too cloudy for observations. On the 18th I saw it dimly, but it had passed the star that was south of the nucleus on the 15th. I again saw the Comet on the 21st indistinctly. It had travelled westerly, and certainly with a northerly declination. On the 22nd, no observation (cloudy). The 23rd was specially favourable for observation, and I was enabled to confirm my impression of the 21st —that the Comet was travelling away from us in a northerly and westerly direction; and further that on the 18th it had attained the most southerly point of its orbit. 26th: Comet scarcely visible to the naked eye. With my glass I am able to observe that the Comet is continuing its northward and westward course. It being the rainy season here has prevented me from making more constant and accurate observations.

All these were ocular observations, only aided by a small field-glass, without any instruments or information as to any Comet being expected. Under these circumstances my conclusions may or may not be correct, and on this point I should be pleased to be informed by anyone who has had more favourable means and opportunities of observing this Comet.

Grenada:
    1881, *July* 27.

Sextant Observations of Comet b, 1881.

Two series of distances of Comet b, 1881, from fixed stars, measured with the sextant, have been received from the Rev. S. S. O. Morris, H.M.S. *Garnet*, S.E. Coast of America, and from Lieut. Bertram Gwynne, H.M.S. *Garnet*, Sandy Point Road, Straits of Magellan, respectively. The former were made on May 30, 31, June 1, 2, 3, 4, 5, 6, 7, 8, 9, 13; and the latter on May 31, June 1, 2, 3, 4, 5, 6, 7, 8, 9, 12, 13.

MONTHLY NOTICES

OF THE

ROYAL ASTRONOMICAL SOCIETY.

VOL. XLII.   DECEMBER 9, 1881.   No. 2.

EDWIN DUNKIN, Esq., F.R.S., Vice-President, in the Chair.

John McCance, Esq., Rathfern, Rayners Road, Putney Hill, was balloted for and duly elected a Fellow of the Society.

On some Systematic Errors in the Determination of the Semidiameter of the Moon from the Greenwich Observations 1750–1840. By E. J. Stone, M.A., F.R.S.

The longitudes of the Moon's centre which are contained in the Greenwich Lunar Reductions, vol. ii., appear to be affected by considerable systematic errors directly connected with the reduction of the observations from the limbs to the centre. The excess of the observed longitudes of the centre over the tabular longitudes are given separately for each limb whenever both limbs are observed near the Full. I have extracted these quantities for groups of years identical with those adopted by Sir G. B. Airy for the determination of the coefficient of the parallactic inequality; and if E_1 and E_2 denote these excesses of observed longitudes from observations of the first and second limbs respectively, the mean values of $\dfrac{E_1 - E_2}{2}$ will be found, in my table, under the heading δs. I also give the values of the coefficient of the parallactic inequality deduced directly by Sir G. B. Airy, and their values when approximately corrected for the errors of semi-diameter $\bar{c} s$.

Period.	Parallactic Inequality with adopted Semi-diam.	Correction. δs	Parallactic Inequality apparently corrected.
1750–1759	121″44	+0″15	121″60
1755–1764	121·27	+0·29	121·58
1760–1768	122·15	+0·37	122·54
1765–1773	122·63	+0·14	122·78
1769–1778	122·87	+0·50	123·40
1774–1782	123·23	+0·65	123·92
1779–1787	122·34	+2·23	124·69
1783–1791	121·15	+2·05	123·31
1788–1796	121·06	+1·03	122·15
1792–1801	121·03	+0·95	122·03
1797–1805	120·49	+1·01	121·56
1802–1810	121·17	+0·38	121·57
1806–1815	122·98	+0·01	122·99
1811–1819	124·09	−0·50	123·56
1816–1824	124·49	−0·01	124·48
1820–1829	125·82	+0·26	126·09
1825–1833	125·20	+0·54	125·77
1830–1838	124·01	+1·05	125·11
1834–1842	123·57	+0·30	125·11
1839–1847	123·24	+0·40	123·88
1843–1851	124·52	+0·70	123·67
1847–1851	125·50	+0·87	125·26
1851–1861	124·38	+0·63	125·01
1862–1876	124·76	+0·40	125·18

The result for 1851–1861 is the mean of the results for the two nine-year periods 1851–1859 and 1853–1861.

The Altazimuth Observations give

1848–1861	125″64
1862–1878	125·06.

The adopted mean semi-diameters are stated by Sir G. B. Airy to have been, $0\cdot273036 \times 3421''\cdot8$, or $15' 34''\cdot27$, for the observations 1750 December 12, to 1815 August 19; and quantities sensibly agreeing with $0\cdot27294 \times 3421''\cdot8$, or $15' 33''95$, for the observations wth the New Transit from 1816 August 7 to 839 December 31. From the year 1840 to 1861 values agreeing sensibly in mean results with Adams's value, $15' 34''\cdot68$, appear to have been adopted; whilst in and after 1862 Hansen's value, $15' 34''\cdot08$, has been adopted. An inspection of the values of δs shows at once that the forty-six-year period inequality, whose existence was first pointed out by Sir G. B. Airy in the values of the coefficient of the parallactic inequality deduced from a discussion of these observations, and which has recently been brought prominently before the Society as an inequality in the expression of the Moon's tabular longitude, is certainly directly connected with questions of semi-diameter. But it does not appear that any inequality with the same period and to the same amount is shown by our recent determinations of semi-diameter or coefficients of parallactic inequality.

The corrections δs for the groups 1779-1787 and 1783-1791 are perhaps unduly increased by the presence in both groups of an error, $7'''\cdot6$, which was found on 1874 Sep. 28. But if this observation should be rejected, the values of δs would still remain $1'''\cdot82$ and $1'''\cdot65$ respectively, and the general law would therefore remain unchanged.

It appears to me most desirable that some competent person should re-examine the calculations of the parallaxes and semi-diameters which were actually used on the work prior to 1840, when Adams's parallax was practically adopted by the application of the necessary corrections to Burckhardt's values.

There is another point of view under which the collection of these values of the coefficient of the parallactic inequality may not be without interest. The value of the mean diameter of the Moon, near the Full, found from the observations with the Old Transit 1750–1815, does not greatly differ from the value found with the present Transit-Circle. It might therefore have been expected that the coefficient of the parallactic inequality determined from these early observations would not differ very much from the value given by the recent observations. But this is not the case. The coefficient given by the observations 1750–1815 is only $122'''\cdot6$, a value which, curiously enough, differs but little from that which would be found if Encke's value of the solar parallax were adopted; but the value given by the recent observations is about $125'''\cdot2\cdot$ The difference is serious, and cannot be dismissed as due to mere personality or as certainly due, without examination and proof, to differences of instrumental means; for the diameters determined with the different instruments have been used in passing from the observed limbs to the centre, and the results should therefore be free from error of semi-diameter.

But there are indications that the numerical values of the coefficient of the parallactic inequality increase as the optical power of the instrument employed on the observations is increased. If such is the case, the effective semi-diameters near Full Moon must sensibly differ from those near quadratures. There is undoubtedly one cause for such differences in instrumental irradiation; but it is very doubtful whether this can be the principal cause of the discordance, $2'''\cdot6$, to which I have called attention. If such were really the case, we should expect the semi-diameter found from a discussion of the observations with the small instrument to be much larger than with the present Transit-Circle; and this is not the case. There is one cause which has occurred to me as a possible explanation of the observed discrepancy. The irregularities of the lunar surface subtend very sensible angles at the Earth; and these irregularities must present different aspects on the limbs when viewed under the different illuminations at Full and at quadrature; and the measured differences would most probably differ considerably with such very different instrumental means as those employed 1750–1815 and those in present use. There is an old transit instrument, with an aperture of an inch and a half, at the Radcliffe Observatory. I feel much tempted to mount the instrument and try whether there really does exist a difference between the effective semi-diameters near Full and quadrature sufficient to account for a discrepancy of $2''\cdot6$ in the values of the coefficient of the parallactic inequality as determined from observations with such an instrument and our present Transit-Circle. If such a discrepancy could be proved to exist, we should be able to infer with certainty the existence of some smaller correction in the same direction for our existing instruments; but if no such discrepancy should be found between the results, then there must be some very important error in the discussion of the early Greenwich results to have led to this constant difference.

On the Conjunctions of the Satellites of Uranus with each other, which may be observable from February to May 1882. By A. Marth, Esq.

The following list is a continuation of that printed on p. 29, and contains the computed position-angles and distances, in reference to the centre of *Uranus*, of the satellites which pass one another, for the nearest even hours, Greenwich M.T., preceding and following the times of their conjunctions.

		Ariel.		*Umbriel.*		*Titania.*		*Oberon.*	
	G.M.T.	Pos.	Dist.	Pos.	Dist.	Pos.	Dist.	Pos.	Dist.
1882.	h	°	″	°	″	°	″	°	″
Feb. 1	12			192·6	13·0	198·7	17·2	—	
	14			193·2	14·9	199·3	15·4	—	
	16		– ·	193·7	16·7	200·0	13·5	—	
15	16	195·8	13·9	—		190·1	9·9	—	
	18	196·2	12·3			191·0	11·9	—	
	20	196·8	10·2	—		191·7	13·8	—	
18	14	—		194·9	21·0	196·6	23·8		
	16			195·1	21·3	196·8	22·2		
	18	—		195·3	21·1	197·1	20·5		

	G.M.T.	Ariel.		Umbriel.		Titania.		Oberon.	
1882.	h.	Pos.	Dist.	Pos.	Dist.	Pos.	Dist.	Pos.	Dist.
Feb. 20	12	195·0	15·2	°	"	°	"	190·6	13·1
	13	195·1	15·3					190·9	13·9
	14	195·3	15·2					191·2	14·8
	15	195·4	14·9	—				191·4	15·6
21	10	13·3	8·6	15·3	11·3			—	
	12	14·0	11·2	15·4	8·9	—			
27	16	193·2	7·2	196·9	10·4	197 1	16·1		
	18	193·9	9·8	197·6	8·0	197·4	14·2		
	20	194·3	12·0	199·0	5 4	197·8	12·2		
	22	194·6	13·7	—		198·4	10·3		
March 8	18	14·8	13·9	14·3	13·7	—			
	20	14·9	14·9	14·5	15·6	—			
9	16	—		15·4	18 6	13·9	15 4		
	18	—		15·5	17·1	14·0	17·2	—	
10	14	195·9	8·3	193·5	7·0	—			
	16	196·5	5·4	194·0	9·5	—		—	
12	18	—		14 2	9·8	16·0	15·6	13·6	14 1
	20			14·4	12·1	16·2	13·7	13·8	15·8
	22	—		14·5	14·2	16·4	11·7	13·9	17·5
17	8	194·7	8·0	—		195·8	10·2	—	
	10	194·8	10·5	—		196·0	8·2	—	
19	18	—		195·0	21·2	—		194·7	19·5
	20			195·0	20·8	—		194·7	21·2
21	8			14·9	17·0	15·2	18·3	—	
	10			15·0	18·4	15·2	16·5	—	
23	18			—		195·0	32·4	195·0	34·8
	20					195 0	33·1	195·0	33·6
	22	—		—		195·0	33·7	195·0	32 3
26	8	15·0	12·8	14·9	16·9	—		15·2	16·5
	10	15·0	14·2	14·9	15·2			15·1	18·2
	12	15·0	15·0	14·9	13·2			15·1	19·8
29	8	14·8	12·4	15·4	9·9			—	
	10	14·7	10·3	15·3	12·2				
April 4	14	194·7	13·4	195·7	10·8				
	16	194·6	11·7	195·5	13·0			—	
8	12	14·3	10·3	—				17·6	10·2
	14	13·9	7·7	—		—		17·2	12·0
9	8			195·0	21·0	195 9	19·3	—	

		Ariel.		Umbriel.		Titania.		Oberon.	
G.M.T.		Pos.	Dist.	Pos.	Dist.	Pos.	Dist.	Pos.	Dist.
1882.	h.	°	″	°	″	°	″	°	″
April 9	10	—		194·9	21·1	195·7	21·0	—	
	12	—		194·8	20·9	195·6	22·7	—	
13	8	14·7	14·4	—		17·1	11·5	—	
	10	14·5	13·1			16·7	13·4	—	
	12	14·2	11·3			16·4	15·3	—	
	18	—				16·0	20·6	13·4	20·9
	20					15·8	22·2	13·2	19·3
19	16			—		194·7	33·8	194·1	34·6
	18			—		194·7	33·3	194·0	33·4
	20	—		—		194·6	32·7	193·9	32·2
20	10	16·1	9·5	13·5	11·8	—		—	
	12	15·7	11·7	13·0	9·5				
23	8	14·8	14·9	16·2	13·3				
	10	14·6	14·3	15·9	15·2	—			
25	10	16·5	8·3	—		12·1	12·3	—	
	12	15·9	10·7	—		11·5	10·4	—	
26	12	—		193·8	14·9	197·4	13 8	—	
	14	—		193·5	13·0	197·0	15 7		
	16	196·6	7·9	193·1	10·8	—			
	18	196·0	10·4	192·4	8·5			—	
May 4	8	196·0	10·9	—		—		190·8	13·1
	10	195·6	12·8	—		—		190·1	11·4

The observations of these conjunctions may serve, not only as test-observations for the theories of the motions of the satellites, but also as tests to show the power of the most powerful telescopes of the present day; and, as such conjunctions will not occur again till the year 1923, it is to be hoped that the opportunities of the coming season will not be neglected. I need scarcely add that this is also the best time for settling the question whether a fair determination of the ellipticity of *Uranus* is or is not within the reach of the best micrometrical measurements.

On the Motion of the Companion of Sirius.
By W. E. Plummer, Esq.

The history of this interesting system is too well known to need any mention here, but the recent publication in a collected form of the accurate series of observations made at the Washington Observatory makes this a fitting and easy opportunity to examine whether the observed companion is moving in an orbit resembling that in which, according to the discussion of Dr. Auwers, the perturbing body must move.

The deviation of the observed coordinates, more especially of position-angle, from those predicted by Dr. Auwers from the discussion of the proper motion of *Sirius*, has thrown some doubt on the generally received opinion of the identity of the observed companion with the perturbing body in the *Sirius* system, and recently this doubt has been somewhat strengthened by M. Flammarion, who in 1877 wrote " L'orbite apparente observée croise dès 1869 l'orbite apparente calculée et se projette en dehors suivant une toute autre courbe, qui sera plus vaste et moins excentrique."[*]

The deviations of the computed orbit from the observed positions made at Washington during the last few years are exhibited in the Table I. given below; and one of the objects of the present communication is to inquire how far M. Flammarion is justified in the conclusion he has drawn from these and similar data at his disposal.

TABLE I.

Date.	Observed Pos. Angle.	C – O	Observed Distance.	O – O
1874·23	58°·05	+ 7·08	11″·11	− 0′·19
75·28	56·38	+ 6·99	11·08	− 0·34
76·22	55·22	+ 6·47	11·19	− 0·65
77·26	53·38	+ 6·46	10·95	− 0·68
78·25	51·70	+ 6·22	10·76	− 0·79
79·20	50·13	+ 5·84	10·55	− 0·89
81·26	45·3	+ 5·9	10·0	− 1·2

I have therefore collected the whole of the observations that I could find printed, being materially assisted by the series that Messrs. Gledhill, Flammarion, and Dunér had previously formed, and endeavoured to determine the ellipse that best represented the observations; that is to say, I have treated *Sirius* as an ordinary revolving star, in which case all the sinuosities of the observed path of the larger star can be neglected, since the centre of gravity of the entire system would move in a straight line, or apparently on the arc of a great circle. By the aid of Dr. Auwers' Ephemeris, corrected by the additional formula given by Dr. Dunér in 1874, a normal position-angle for the

beginning of each year was formed. It appeared from the more
recent observations that Dr. Danér's formula, viz.

$$dP = -5\overset{\circ}{\cdot}0 - 0\overset{\circ}{\cdot}48 \, (t - 1869\cdot0) + 0\overset{\circ}{\cdot}03 \, (t - 1869\cdot0)^2,$$

might have been, with advantage, slightly corrected, by increasing the positive term depending upon the square of the interval
from 1869, but as the observations were always made near the
beginning of the year, and as the angular motion was small, it
was not feared that any source of inaccuracy would be introduced
by using the formula as given by that astronomer. These
position-angles, corrected by the formula

$$-0\overset{\circ}{\cdot}006 \, (t - 1870\cdot0)$$

to remove the effect of precession, are set down in the second
column of Table II.

To ascertain the probable error of these normal position-
angles, it was necessary to assign some definite weight to the
observations given by various observers. Not being in possession
in every case of the full particulars and circumstances under
which the observations were made, these weights have been
assigned more or less arbitrarily, and perhaps even with fickle-
ness. It was, to avoid this objection, at first intended to use
only the results obtained at the Washington Observatory, but
as this would have unhappily curtailed the period, already too
short, I have been obliged to admit all. I believe that the weight
has generally been over-estimated, and that the probable errors
given in the third column of Table II. are consequently too
small.

To determine the normal distance, I have proceeded somewhat
differently. The deviation of the observed distance from the
predicted had amounted in 1880 to quite a second of arc (as
shown in Table I.), while in 1870 the agreement was nearly
identical. The annual motion, which, therefore, in the actual
orbit differed materially from that in the theoretical, was derived
by means of an interpolating curve drawn among the observed
distances laid down to scale. The resulting values of the normal
distance with their probable errors, open to the same objection
on the score of weight as in the case of the position-angles, are
set down in the remaining columns of the table. In a few
instances in either element the number of observations was not
sufficient to determine the probable error. In the year 1881,
the place given rests upon three unpublished observations made
in this Observatory, and which Prof. Pritchard permits me to
use for the purposes of this investigation, combined with those
made at Washington, and given in the *Monthly Notices* of June
last.

TABLE II.

Date.	Normal Position Angle. 1870·0.	Probable Error.	Normal Distance.	Probable Error.
1862·0	85·13	0·09	10·18	0·042
63·0	80·97	0·41	9·64	0·117
64·0	79·76	0·59	10·35	0·157
65·0	76·52	0·25	10·18	0·227
66·0	75·61	0·38	10·49	0·347
67·0	73·73	0·52	10·36	0·398
68·0	71·01	0·36	11·17	0·078
69·0	73·80	0·88	11·22	0·012
70·0	65·48		12·06	
71·0	63·69	1·18	11·31	0·298
72·0	62·72	0·67	11·46	0·034
73·0	62·23	1·50	10·91	0·164
74·0	59·58	1·21	11·36	0·071
75·0	56·94	0·15	11·27	0·144
76·0	55·66		11·22	
77·0	53·45	0·13	11·13	0·069
78·0	52·25	0·20	10·69	0·197
79·0	49·62	0·30	10·62	0·002
80·0	48·52		10·28	
81·0	45·2		9·99	

As it is evident that no regular curve will pass through all the points determined in Table II., these were further submitted to an interpolating curve, so as derive, if possible, a series of points, which, while not differing greatly from the normal positions, might offer better prospects of successful treatment. The position-angles and distances, read from this curve, are called the "adopted position-angles" in Table III., and are the places on which all the subsequent computations have been based.

The differences from the normal positions are recorded under the headings ΔP and Δs; and it will be seen that in each coordinate, there are about eleven changes of sign out of a possible nineteen, so that the drawn curve runs sufficiently well among the observed places. The probable error has, of course, been somewhat increased, but this is immaterial, and is here given merely as a standard whereby the discrepancies that will be derived from the comparison of places computed from various elements, with the adopted coordinates, may be compared.

TABLE III.

Date.	Adopted Position Angle.	ΔP	Prob. Error of Adopted Pos. Ang.	Adopted Distance.	Δs	Probable Error of Adopted Distance.
1862·0	85·3	+0·17	0·10	9·78	−0·40	0·197
63·0	82·1	+1·13	0·59	10·02	+0·38	·204
64·0	79·6	−0·16	0·71	10·25	−0·10	·653
65·0	77·2	+0·68	0·33	10·45	+0·27	·238
66·0	74·9	−0·71	0·43	10·64	+0·15	·118
67·0	72.9	−0·83	0·55	10·81	+0·45	·507
68·0	70·8	−0·21	0 49	10·97	−0·20	·119
69·0	68·9	[−4·90]	[2·56]	11·11	−0·11	·058
70·0	67·0	+1·52		11·23	−0·83	
71·0	65·1	+1·41	1·36	11·33	+0·02	·298
72·0	63·2	+0·48	0·80	11·39	−0·07	·061
73·0	61·3	−0·93	1·65	11·40	+0·49	·286
74·0	59·3	−0·28	1·22	11·38	+0·02	·072
75·0	57·4	+0·46	0·17	11·30	+0·03	·146
76·0	55·5	−0·16		11·18	−0·04	
77·0	53·6	+0·15	0·14	11·04	−0·09	·078
78·0	51·7	−0·55	0·45	10·89	+0·20	·278
79·0	49·7	+0·08	0·32	10·71	+0·09	·086
80·0	47·7	−0·82		10·48	+0·20	
81·0	45·4	+0·2		10·06	+0·07	

It will be observed that the distances indicate a tolerably well defined maximum about the year 1873, and therefore there should be a minimum as strongly pronounced, in the annual motion about the same year. This cannot be detected in the observations, and hence it is evident that two distinct orbits can be derived, according as we assume the distances as given by observation or as computed from the formula

$$r^2 \frac{d\theta}{dt} = \kappa.$$

In presence of this difficulty, I have first assumed the period to be that required by the discussion of the proper motions of *Sirius*—namely, 50 years—and with that period have derived the following orbit :—

Elements A.

Time of Periastron Passage	$T =$	1891·51
Excentricity	$e =$	0·5046
Angle between Axis Major and Node Line ...	$\lambda =$	219 30'
Inclination	$i =$	55 0
Node	$\Omega =$	48 46
Period		50 years
Mean Distance	$a =$	8·32"

As will be seen from Table IV., the agreement between the adopted position-angles and those computed from these elements is satisfactory. That of the distances is less so. Omitting the error for 1881, for which date the distance is insufficiently determined, the sum of the squares of the discordances expressed in hundredths of seconds is 6349, while the sum of the squares of the probable errors for the same dates, given in Table III., and expressed in the same unit, is 6117. It will be interesting, therefore, to try if the distances be better represented by a different period, and also whether the limits can be fixed between which the time of revolution must lie. With this view, the period was successively reduced to 47 and to 44 years, and by the method of varying each element a new orbit was found.

	Elements B.	Elements C.
T	1891·62	1889·44
e	0·5219	0·5908
λ	217 13'	223 21'
i	56 47	58 37
Ω	48 12	45 27
a	8·50"	8·53"
Period	47 years	44 years

In the following table is given the comparison between the adopted places and those computed from the elements successively for each alternate year, more frequent comparisons not having been thought necessary :—

TABLE IV.

Date.	Adopted Position Angle.	C−O Elements A.	C−O Elements B.	C−O Elements C.	Adopted Distance	C−O Elements A.	C−O Elements B.	C−O Elements C.
1862·0	85·3	−0·2	0·0	0·0	9·78	−0·46	−0·72	−0·84
63·0	82·1	+0·3	+1·2	+0·2	10·02	−0·35	−0·54	−0·63
65·0	77·2	+0·2	+0·7	−0·2	10·45	−0·16	−0·23	−0·25
67·0	72·9	+0·1	+0·4	−0·3	10·81	−0·02	+0·03	+0·06
69·0	68·9	0·0	+0·1	−0·4	11·11	+0·04	+0·21	+0·26
71·0	65·1	0·0	0·0	−0·3	11·33	+0·01	+0·29	+0·37
73·0	61·3	−0·1	−0·1	−0·1	11·40	−0·03	+0·31	+0·42
75·0	57·4	0·0	−0·2	+0·2	11·30	−0·11	+0·30	+0·41
77·0	53·6	−0·3	−0·2	+0·2	11·04	−0·24	+0·20	+0·30
79·0	49·7	−0·9	0·0	−0·6	10·71	−0·45	+0·12	−0·06
81·0	45·4	−0·8	−0·5	−0·6	10·06	−0·62	−0·12	−0·47

It will be observed that the position-angle is sufficiently well represented in all three orbits, but that the distance disagrees by a greater amount than can be fairly attributed to the errors of the adopted places. The sums of the squares of the errors, which, on the assumption of a period of 50 years, amounted to 6349, rose, as the period was decreased to 47 and 44 years, to 12181 and 17312 respectively; from which it may be inferred that the period is not much less than 50 years, and certainly more than 47.

I have not been so fortunate in fixing a period which the time of revolution cannot exceed, for I find that if the period be greatly increased an entirely different set of elements will represent the observed places with fair accordance. For instance, if we assume the elements thus

$$T = 1857\text{·}27$$

$$e = 0\text{·}9011$$

$$\lambda = 63\overset{\circ}{} \ 50{}'$$

$$i = 75 \ 58$$

$$\Omega = 93 \ 35$$

$$a = 54\text{·}08''$$

Period   442 years.

The adopted places are represented within the following errors :---

Date.			Pos. Angle. C−O	Distance. C−O
1862·0	...	...	−0·4	−0·17
65·0	...	...	+0·5	+0·23
68·0	...	...	+0·6	+0·08
71·0	...	...	+0·2	−0·19
74·0	...	...	−0·1	−0·27
77·0	...	...	−0·5	+0·01
80·0	...	...	−0·8	+0·53

Doubtless these errors could have been reduced by continued computations, but no useful purpose would be served by prolonging the discussion in this direction. This approximate result is given here to exhibit a remarkable instance of the difficulty of determining the orbit of a double star when the arc of observation is not sufficiently extended. If, as in most instances, there had been no reason from independent sources to suspect a short period, these last elements would have been worthy of favourable consideration. There are, however, two circumstances which suggest their rejection. First, the elder Herschel or W. Struve never observed the duplicity of the bright star, and at the commencement of the century the position of the small star would have been

$$\theta = 283^{\circ}; \qquad s = 50^{'}7$$

if these latter elements are approximately correct. At such a distance from the primary it is difficult to understand how, not only Herschel, but so many astronomers failed to observe it. Secondly, with a parallax of $0^{'''}\cdot193$, which is that assigned to *Sirius* by Dr. Gyldén, the mass of the system would be more than one hundred times the mass of the Sun. This, too, is very noticeable in relation to the masses of other stars which have been hitherto approximately determined. The argument drawn from considerations of annual parallax is, however, inconclusive, because a not impossible increase in the value deduced by Dr. Gyldén would reduce the mass of the system within probable limits. And, moreover, if the parallax furnishes a reason for the rejection of the longer period, it offers as great a difficulty to the acceptance of the shorter. For, using the same value of the parallax and adopting the mean distance given in Elements A, the resulting value of the mass of the *Sirius* system is some 32 times that of the Sun. But Dr. Auwers has shown that the perturbing body has a mass of about one-half that of *Sirius*. This makes the mass of the companion about ten times that of the Sun, whence we should expect a more conspicuous object than the satellite presents.

The conclusion to which we are driven is that, notwithstanding that observations of this interesting system have been made for nearly twenty-years, or during two-fifths of the entire period of the motion of *Sirius*, they do no more than render the connection extremely probable, and will as yet not decide definitely whether this companion be the perturbing body or another member of a more complex system.   But as the Elements A differ from those given by Dr. Auwers, it is desirable to trace what effect these alterations of the elements have in explaining the irregularities of the observed Right Ascension and Declination of *Sirius*.   This is a question to which I hope to return at a future opportunity.

Oxford University Observatory:
    1881, *December* 8.

Note on Messrs. Campbell and Neison's Paper on the Parallactic Inequality, in the Supplementary Number of the Monthly Notices. By E. J. Stone, M.A., F.R.S.

I. Messrs. Campbell and Neison appear willing to leave the discussion of their treatment of the question of the irradiational enlargement of the Moon as it stands.   I gladly accept their decision on this point: I am quite content to leave the discussion as it stands.

II. But I cannot allow to pass unchallenged the statements which Messrs. Campbell and Neison have been pleased to make respecting the correction which the Meridian Observations of the Moon, 1851–1858, as given in the Greenwich volume, 1859, require for the error of adopted semi-diameter.

The adopted semi-diameter is clearly stated by Sir G. B. Airy to be Adams's value.   The following are the corrections which I have deduced from a discussion of the observed durations 1851–1858:—

Year.	No. of Observations.	Adams's value too large by
1851	7	$+0''43$
1852		-0.71
1853		-0.97
1854		$+1.17$
1855		$+0.94$
1856		$+1.77$
1857		$+0.76$
1858		$+1.02$

The deduction of these values is troublesome, and requires some care, on account of the many changes which were made in the adopted quantities of the *Nautical Almanac*, 1851–1855.   But, although the results for 1852 and 1853 are rather discordant, I can find nothing wrong in the work, and I believe they have been fairly deduced from the observations. If all the observations are given equal weight, the result is that Adams's semi-diameter is too large by $0''\cdot51$; whilst from all the observations made with the Transit-Circle 1851–1861 it would appear that Adams's semi-diameter is too large by $0''\cdot63$ to represent the observed durations.

I must leave Messrs. Campbell and Neison to explain how they justify the assertion " in all our results we have employed the observed value," whilst they assume in their work that Adams's semi-diameter is too *small* by $0''\cdot18$ to represent these observations. I need hardly point out that, unless Messrs. Campbell and Neison can prove that Adams's semi-diameter is too small by about $0''\cdot18$ to represent these observations 1851–1858, the very remarkable agreement between the values of the coefficient of the parallactic inequality, to which they appealed in the *Monthly Notices* for March 1881 as proving the existence of the forty-five-year period inequality in the expression for the Moon's longitude, no longer exists.

III. I next pass to the supposed effect of Mr. Ellis's personal error in observing the limbs of the Moon on the coefficient of the parallactic inequality 1843–1851. I give in the following table the actual differences of the longitudes of the centre as deduced from observations of each limb. This shows at once the effect of the error of the assumed diameter. It is clear that Messrs. Campbell and Neison are mistaken when they attribute the large value of the parallactic inequality given by the observations 1843–1851 to personal error in Mr. Ellis's observations of limbs.

1843–1851.

Date.		Longitude of Centre. $1\;\mathleft) - 2\;\mathleft)$.	Observer.
1843, May	13	$-0\cdot25$	E.
Aug.	9	$+1\cdot68$	M.
Dec.	6	$+0\cdot96$	H.
1844, May	2	$-2\cdot43$	R.
June	29	$+1\cdot10$	E.
Sept.	26	$+3\cdot85$	R.
1845, June	19	$+2\cdot67$	H.
Sept.	15	$-2\cdot62$	E.
1847, Mar.	31	$+1\cdot78$	R.
May	29	$+1\cdot94$	R.
July	27	$+2\cdot66$	H.
1848, Feb.	18	$0\cdot00$	R.
Mar.	19	$+5\cdot91$	H.
April	17	$+1\cdot67$	M.

Date.		Longitude of Centre.		Observer.
		$1 \; \mathbb{D} - 2 \; \mathbb{D}$.		
1848, Sept.	12	+ 1·93		H.
1849, Mar.	8	+ 0·95		H.
Aug.	3	+ 6·69		M.
1850, Mar.	27	+ 0·28		R.
Aug.	22	+ 0·42		R.
Oct.	20	+ 3·61		D.
1851, Feb.	15	+ 0·41		H.
May	14		− 0·30	M.
June	13	+ 0·98		J. W. B.
July	12	+ 0·83		E.
Aug.	11	+ 1·25		W. E.
Sept.	9		− 0·39	D.
Dec.	8	+ 2·51		R.

The observers are Messrs. Main, Henry, Richardson, Ellis, Dunkin, Breen, and William Ellis.

The error of the adopted semi-diameter is $-0''·705$ from the whole of the observations, and is $+0''·235$ from Mr. Ellis's observations alone.

The value of the coefficient of the parallactic inequality deduced directly from the observations with the adopted semi-diameter is

$$124''·52.$$

The value which would be given with the mean correction is, very approximately,

$$125''·26.$$

But, if we applied a correction deduced from Mr. Ellis's observations alone, we should still find a value

$$124''·39.$$

The value deduced by Messrs. Campbell and Neison, with a correction for the supposed forty-five-year period inequality, was

$$122''·75.$$

The excess of the observed value, even on the extreme supposition that Mr. Ellis's diameter should be alone used for the reduction of all the observations, is still $1''·6$ over the computed value. How, then, can Messrs. Campbell and Neison assert that " with this correction the discordance vanishes " ?

Note on the Variable Star D.M. +1° No. 3408. By Professor
C. Pritchard.

On the receipt, Dec. 5, of the Dun Echt Circular, No. 41,
relative to the variability of the star D.M. +1° No. 3408, after
the manner of the Algol type, I instructed Mr. Plummer to
make comparisons of its brightness by the photometric means
which I described in the *Monthly Notices* of last month.

The comparison stars taken were D.M. +2° No. 3283 (a) and
+1° No. 3411 (β). The variable star is denoted by V.

$$\text{Dec. 5, } 4^{\text{h}}\ 57^{\text{m}} \text{ G.M.T.} \quad \ldots \quad \ldots \quad V - a = \cdot 31 \text{ magnitude}$$
$$7,\ 5\ \ 9 \quad \ldots \quad \ldots \quad V - a = \cdot 30 \quad \text{,,}$$

These results were further confirmed by comparison of (a)
with (β), and each with V, thus—

$$\text{Dec. 7, } 5^{\text{h}}\ 28^{\text{m}} \text{ G.M.T.} \quad \ldots \quad \ldots \quad \beta - V = \cdot 34 \text{ magnitude}$$
$$7,\ 5\ 14 \quad \ldots \quad \ldots \quad \beta - a = \cdot 54 \quad \text{,,}$$

These last two comparisons would give $V - a = \cdot 20$ magnitude,
implying a discordance of only one-tenth of a magnitude among
these insufficient observations, made, as will be explained, under
somewhat unfortunate circumstances. In these results, regarding the brightness of the variable at the above epochs, there is
little or nothing noteworthy, inasmuch as the variation of light
is said to be of the Algol type, and neither of the times of observation is that of a minimum as recorded by Mr. Sawyer.
Assuming with Argelander that $a = 6\cdot5$ mag., $V = 6\cdot8$ magnitude. This is altogether at variance with Argelander's estimate
$5\cdot5$; an estimate certainly not by any means borne out by the
eye on Dec. 5 and Dec. 7, and which eye-observation agrees substantially with the wedge-reading. Neither do these photometric
observations agree with Mr. Sawyer's estimate of variation from
the 6th to $6\cdot7$ magnitude; but we are left in doubt as to Mr.
Sawyer's standard of magnitude.

Unfortunately, the Sun is even now far too close to the star
to admit of a very reliable determination of its brightness, or
varying brightness, and this increasing proximity will effectually
prevent further comparison for some time to come. The very
singular discordances recorded in the estimates of its magnitude, varying from $5\cdot5$ to $8\cdot0$ magnitude, by astronomers so
able as Schjellerup, Lamont, and Argelander, will render this star
an object of interest and inquiry a few months hence, when the
Sun will have quitted its vicinity. The above observations are to

be regarded simply as a record for comparison with other observers, who may also have already made similar measures.

It may interest some members of the Society who are skilled
in astrometry, to be informed that the photometric measures of
the relative brightness of all the stars in *Lyra* contained in Dr.
Heis' Catalogue, 75 in number, together with two others which
ought have been inserted on account of their brilliancy, have
now been completed at this Observatory by three sets of five
readings on two nights. The consistency of the results is encouraging, and leaves no doubt in my mind of the facility and
accuracy of the method devised. I have accordingly arranged
for the systematic prosecution of these observations for all the
stars visible *nudis oculis* from the Pole to 5° South Declination.
This will embrace all the stars recorded in Dr. Heis' Catalogue
within the limits prescribed.

Oxford University Observatory:
  1881. *December* 8.

---

*On a Method for Finding the Elements of the Orbit of a Comet by a
Graphical Process.* By F. C. Penrose, Esq.

It may be interesting to those who are not familiar with the
analytical methods of working out the orbit of a Comet to examine a process of graphical construction which will, in many
instances at any rate, suffice to give very rapidly a good approximate solution by means which demand only conversance
with the methods of practical geometry and sufficient spherical
trigonometry to reduce Declination and Right Ascension into
latitude and longitude.

I was first led to try if by graphical operations I could get a
general idea of the motion of Tebbutt's Comet (1881, *b*), and
finding that I succeeded to a greater extent than I had anticipated,
I proceeded to attack Schäberle's Comet *c*, with some diffidence,
indeed, as it seemed to be giving a good deal of trouble to the
computers, and obtained a result which I found compared favourably with the published elements. I venture, therefore, to think
that these attempts may be worth laying before the Society.

I should premise that I do not think that much reliance could
be placed on a graphical solution of the orbit of a Comet from a
few days' observation only, unless it should happen to be near
the Earth or to the perihelion; and also that four places are much
better than three, although it would be of course theoretically
possible to find the elements from three. I will begin with the
case of Tebbutt's Comet (Comet *b*, 1881).

I take four observations reduced to the ecliptic.

	G.M.T.	Longitude.	Latitude.
		° ′ ″	° ′
Cape of Good Hope	May 31·21	69 38 6	−52 9
	June 9·18	74 16 36	−38 31
Greenwich ...	June 24·48	86 19 6	+25 59
	July 13·58	102 37 33	+60 36*

The apparent places are taken, as minute accuracy is not attempted.

The plane of the paper represents the ecliptic, S being the Sun's centre. The positions of the centre of the Earth are calculated for the times given above, and laid down on the diagram at the points entered, with the dates (May 31, June 9, &c.), and from these are laid down the longitude lines (L_1, L_2, &c.). The straight line SX is drawn at 90 degrees from that of the equinoxes. In the scale used six inches represents the Sun's mean distance. For working the solution it is desirable that this scale, divided into 100 parts, should be drawn upon the edge of slip of stout paper or card. The slide rule may also be used with advantage to facilitate the necessary calculations.

The first thing to be done is to obtain a rough approximation to the place of the projection of the Comet where it passes the different longitude lines (L_1 L_2, &c.). To get this it is convenient first to form a short table of numbers proportionate to the time intervals between the dates of the observations, which are respectively about 9, 15·25 and 19·1 days.

9	15·25	19·1
10·6	18·00	22·5
11·2	19·00	23·7
11·8	20·00	25·0
12·3	21·00	26·2
12·9	22·00	27·5

---

* The observations are these, namely :—

(1)   *Cape of Good Hope.*

Cape Mean Time.	R.A.	Declination.
h m s	° ′ ″	° ′ ″
May 31, 6 10 24	75 46 25	−29 42 19
June 9, 5 31 18	77 16 24	− 15 44 53

(2)   *Greenwich Observations.*

		N.P.D.
G.M.T.	h m s	° ′ ″
June 24, Transit sub pol	5 39 39	40 36 35
	h m s	
July 13, 13 52 0	...	...

Comet observed in Transit-Circle together with 37 Camelopardi.

δ R.A. (Comet − Star) + 56 85s

δ N.P.D. (Comet − Star) − 0″4

The mere inspection of the lines L_1 L_2, &c , is sufficient to show that the Comet was passing with a direct course between the Earth and the Sun. Lay the edge of the card or paper scale across these lines and bend it round in such a curve as will cut off upon the scale numbers giving a tolerable approximation to one of the sets in the table given above. For instance, with 20 on the middle arc CB we shall find the other two places fairly represented by A on L_1 and D on L_4; but it will be seen that BA exceeds 11·8, CB being 20. But as the areas about S must be proportional to the times in the projection on the ecliptic as well as in the orbit, it follows that if BA is nearer to S, or lies more obliquely with respect to S than CB, it may exceed the number parallel to 20 on the table; whereas, for the same reason, no such course as the dotted line HK could be chosen.

In this first operation it is almost impossible that the numbers should come out exactly, and it is not worth while to occupy much time upon it. It is, however, important to get fairly proportionate numbers on each side of the middle arc, and in this case the places ABCD, which appear to give the measurements from the scale of 13·5, 20 and 26, will answer for a first start.

· The areas of the segments about S must be now examined more carefully. If the points had been correctly chosen

$$\frac{\text{area SBC}}{15 \cdot 3} = \frac{\text{area SCD}}{19 \cdot 1} = \frac{\text{area SAB}}{9} ;$$

but on comparison we shall find both SCD and SAB to be too large. As respects SCD an obvious improvement will arise from approaching nearer to S, because the lines L_3 and L_4 converge rapidly, and when the points A and C have been advanced to a and c the areas S a c and S c D will have a better relation to one another. Also if on L_1 A be shifted to a the areas S a b and S b c will be found to be more nearly adjusted, and the proposed curve now becomes a b c D.

Thus far we have only considered the longitudes and the areas resulting from them; but the latitudes should now be examined. And it will be convenient first to mark off upon each of the longitude lines, near the places where the curve crosses them, the points where the height above or below the ecliptic has such values as 40, 45, 50, 55 on L_1, 22, 24, 26, 28 on L_2, &c.—these points being given by the cotangents of the latitude.

We shall see that the amended curve cuts L_1 where this reading is about 47·5, and L_4 at about 56, whilst the horizontal distance between the points is about 52. Half the sum of these heights divided by half the horizontal distance should give a value for the tangent of the angle of inclination,

$$\frac{51 \cdot 75}{26} = \tan 63^\circ\ 20'.$$

From L_2 and L_3, by the same method, we get

$$\frac{19 \cdot 9}{9 \cdot 3} = \tan 64^\circ\ 55'.$$

Observing this difference, it would seem that b and c have been carried rather too near S, because the latitudes at B and C would have given for the value

$$\frac{24 + 13}{19},$$

giving 62° 50′; but it is undesirable to complicate and confuse the diagram, on so small a scale as is necessary to accompany this paper, with more lines than are shown upon it, and it will suffice for the present purpose to combine the two readings, viz.—

$$\frac{51\cdot75 + 19\cdot9}{26 + 9\cdot3} \text{ or } \frac{71\cdot65}{35\cdot3} = \tan 63° 50′.$$

To obtain the line of the nodes we must measure off from b or c a distance corresponding to the cotangent of the angle of inclination, and thus obtain $cn = 6\cdot82$, and we can now draw SN for the line of the nodes, which makes an angle of 1° with SX, and which, reckoned in longitude, will be 271°.

The next process is to develop the projection so as to get the true figure of the orbit by drawing through the points $a\,b\,c\,D$ perpendiculars to SN, and cutting off points upon these lines as determined by the secant of the angle of inclination—viz. such as $PF = F\,a \times \sec 63° 50′$.

It will now be convenient to prepare and transfer to tracing paper the arc of a parabola of which the focal distance at the vertex may be taken as S n.

Should it, however, not be obvious that S n is the proper distance, the way to proceed would be this. Take the points Q and R, and with twice the distance SQ or SR draw arcs of circles, which will intersect in some point beyond S. From this point draw a straight line through S back again to the figure, and it will give the approximate place for the vertex of the parabola, and also determine the parameter. With these parabolic elements construct, as before proposed, on tracing paper an arc of sufficient length to reach from P to T. Then, keeping the focus upon S, turn the tracing about until it makes the best possible coincidence with the points P, Q, R, T, and then it will point out the perihelion, the time of the passage in the developed orbit plane, and, by projection, its longitude on the ecliptic; and it will also give an improved value of the distance.

Following the indications of this small scale figure, the results would be

T	...	...	...	June 14·70
ϖ	*i.e.* Longitude of Perihelion 266·37			
q	Distance ...	...	0·73 or [9·8633]	
☊	...	..	...	... 271° 0′
i	...	...	...	... 63° 50′

I have not tried this Comet on a scale adequate to obtain the best possible graphical approximation, but there is no doubt that the elements above given could be brought nearer the results of calculation which are given below, according to Mr. Hind's parabolic elements, especially as to the date of the perihelion.

$$T \quad \ldots \quad \ldots \quad \ldots \text{ June } 16\cdot457$$

$$\pi \quad \ldots \quad \ldots \quad \ldots \quad 265\overset{\circ}{\,} 15' 44''$$

$$q \quad \ldots \quad \ldots \quad \ldots \quad [9\cdot8657]$$

$$\Omega \quad \ldots \quad \ldots \quad \ldots \quad 270\overset{\circ}{\,} 57' 46''$$

$$i \quad \ldots \quad \ldots \quad \ldots \quad 63 \ 28 \ 46$$

The other example which I give is Schäberle's Comet, *c*, 1881. Observations on five days have been considered

	G.M.T.	Longitude.	Latitude.
Oxford Radcliffe Observatory	July 31·41	95° 19′ 56″	+22° 34′ 17″
	Aug. 4·42	99 2 24	25 8 9
.	10·44	108 5 14	29 49 0
	19·53	137 33 22	36 7 0
Marseilles	Sept. 2·33	199 18 0	17 23 0*

As in the former case, SX is drawn at 90° from the line of the equinoxes, and the same scale is used as in the former diagram.

The first process is to lay down the longitude lines L_1, L_2, L_3, &c., which show that we have now the case of a retrograde orbit, and that the Comet is passing between the Sun and the Earth, and, as the latitudes are all positive, it is, within the above limits, to the north of the ecliptic.

The intervals between the first four observations are respectively 4·01, 6·02, and 9·09 days. From these we determine the following ratios, viz.:—

---

* The observations which I have followed are—

Oxford Radcliffe Observatory.

Passage of Comet *sub polo.*

	G.M.T.	Observed R.A.	Observed N.P.D. uncorrected for parallax.
	h m s	h m s	° ′ ″
July 31,	9 57 43 2	6 28 21·6	44 3 46·1
Aug. 4,	10 32 53·5	6 49 17·7	41 47 13·1
10,	10 29 50·1	7 42 55·6	38 21 32
19,	12 35 2·4	10 23 57·4	40 44 51
Marseilles.			
Sept. 2,	8 0 0	13 37 27	81 31 0

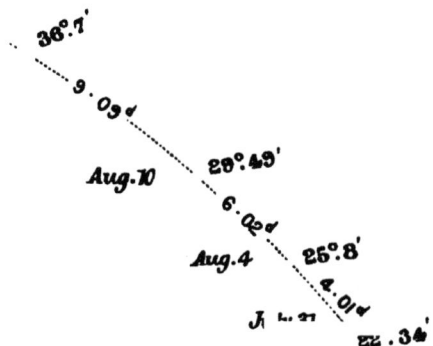

36°.7'

9.09 d

Aug.10   28°.49'

6.02 d

Aug.4   25°.8'

4.01 d

Jul.27   22°.34'

K

Lt

S   N

4·01	6·02	9·09
8	12	18·1
10	15	22·6
12	18	27·2
14	21	31·7

If we bend the scale across the longitude lines for a first approximation, we shall find that 12, 18, and 27·2 would agree with the dotted line ABCD pretty nearly; but on drawing a curve through these points and joining them with S, we should find the area ASB to be too small in comparison with BSC, whilst CSD is too large, showing that the projection of the orbit must lie more nearly in the direction $a\,b\,c\,d$. Before that line, however, can be adopted, the latitudes should be examined, and the scales of height above the ecliptic marked upon L_1, L_2, &c.

The line of the nodes can now be drawn approximately. It must pass through S and cannot be far from SN, which direction will follow from the condition that there must be an invariable proportion between the heights above the ecliptic and the length of all such base lines as $a\,n$, $b\,h$, &c., this ratio being the tangent of the angle of inclination of the orbit, and this angle will have to be determined simultaneously with the direction of line of the nodes. In getting the angle of inclination primary attention should, in this particular case, be paid to the point C, because a moderate variation in the direction of SN will affect the length of the perpendicular drawn from C in the direction CS very little, and a first trial appears to give for the tangent of the angle of inclination $\dfrac{43}{51\cdot7}$, or 39° 30′; but as this angle will not also satisfy the ratios obtained from the other points, we shall see that C must be moved onwards, as to c, where the value of the tangent becomes $\dfrac{44\cdot6}{51\cdot7}$, or 41°; and using this value for the inclination, we may proceed to obtain a more exact determination of the line of the nodes.

Looking first at the longitude line L_1, and observing that the approximate curve cuts it near a, where the value on the scale formed on L_1 is about 48·5, draw $a\,n$ proportional to the co-tangent of 41°. Then from another of the longitude lines (say L_4) draw $d\,m$, bearing the same ratio to 38, which is the height shown on the scale formed on L_4. These lines $a\,n$ and $d\,m$ may be used as radii for striking arcs of circles, as shown in the diagram, to guide the direction of SN. If the work has been correctly done, n, s and m will lie in the same straight line: if not, draw SN parallel to the line touching the two circular arcs, and take this as the corrected line of the nodes.

On each of the longitude lines, and on each side of the points

a, b, c, &c.—viz. from 48 and 50 on L_1, from 44 and 48 on L_2, and so on with regard to the other points—drop perpendiculars on SN proportional to the cotangents of the inclination, and join their extremities as NN_1, MM_1, &c. The line of the nodes will intersect these base lines at $n\,h\;m\,k$; and from the last-named points draw verticals to meet the longitude lines. These will give the corrected places for the projection of the orbit. The areas may now be re-examined, but if the work has been correctly done they will probably be found to satisfy the eye. The method above given has been somewhat more complicated than in the case of Comet b, but in that case there were observations available on each side of the ecliptic.

We may now proceed to develop the orbit from the projection, as shown by the dotted lines, by multiplying the verticals already obtained by the secant of the angle of inclination.

It is quite obvious that the perihelion distance occurs not far from the point H of the developed orbit.

If a circle be drawn through the three points G, H, and K it will show that the nearest point to S will be at or near P. With this distance for its focal length, construct a parabola as in the former case; and after twisting the figure about in the manner already described, we shall obtain the position of the perihelion P and a further correction of the distance. Correct, if necessary, the parabola to the distance so determined, and apply it again to the developed orbit, and it will be found that in this case the points, except those close to the perihelion, will lie in regular sequence outside the parabolic arc, demonstrating that this Comet's orbit must be distinctly hyperbolic. The axis of the curve passes of course through the perihelion, and if projected on to the ecliptic we obtain its longitude.

The elements obtained from this figure are

T	...	...	... August 22·3
π	...	...	... 147 30 0
q	...	...	0·627 or [9·7973]
☊	...	...	... 95 55 0
$\imath$	...	...	... 41 0 0

In the case of this Comet I have also endeavoured by means of a larger scale, and by the use of trigonometrical formulæ instead of some of the graphic work to obtain a still closer appoximation.

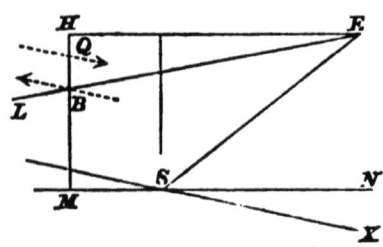

To explain this method of working, I take one of the points of the former diagram, for instance B. Let the angle

$$ESN = a$$
$$SE = r$$
$$EB = \rho$$
$$HB = x$$

Let EH be parallel to SN, and let the angle $HEB = \gamma$, and let l be the latitude of the Comet as seen from E.
It then follows that

$$\rho \tan l = MB \tan i$$
$$MB = r \sin a - x, \text{ and } \rho = x \operatorname{cosec} \gamma$$
$$\therefore (r \sin a - x) \tan i = x \operatorname{cosec} \gamma \tan l$$
$$\frac{r \sin a}{x} - 1 = \operatorname{cosec} \gamma \tan l \cot i$$
$$\text{and } x = \frac{r \sin a}{\operatorname{cosec} \gamma \cot i \tan l + 1}.$$

From this it is easy to obtain MB, MH, and SM by calculation, and it will be found more accurate to lay the points which are wanted down by scale than to obtain them graphically from the intersections of lines. If we want the point Q of the developed orbit, it is got from

$$MB \sec i.$$

When the places in the orbit have been determined in this manner and laid down on the drawing, let the approximating conic section be applied, and it will be seen where the ordinates and abscissæ require modification : this must be done by small variations of the inclination and the line of the nodes; but as these elements affect the different places differently, such modifications of each must be chosen as will best suit the general purpose. To estimate the effect of the different variations we might take the differential coefficients with respect to the angle XSN (which angle may be called ω), and also with regard to i, and apply them to each case. But it is very little more trouble to compute the places directly for slightly different values of i and ω, which has been done to form the following table. The observed places have also been corrected for parallax, &c.

			July 19·164	July 31·41	Aug. 19·53	Sept. 2·33	Sept. 21
i	41° 0′	MQ	·7416	·7419	·5836	·3269	·1130
ω	5 40	SM	·6436	·3146	·2405	·5909	·9005
i	41 0	MQ	·7514	·7468	·5827	·3253	·1123
ω	5 56	SM	·6668	·3223	·2433	·5919	·8964
i	41 15	MQ	·7544	·7482	·5801	·3248	·1123
ω	5 56	SM	·6755	·3261	·2414	·5923	·8966

From this table the effect of small changes in i and ω upon the different points can be easily ascertained and a choice made of such changes as will best suit an approximating hyperbola. It may, however, be found necessary to reconsider the conic elements, but the same table can again be used for a final adjustment. Working in this manner I have arrived at the following elements of the orbit, which give a very close agreement at all the places that I have tried, and also give sensibly areas proportional to the times.

$$T \quad \ldots \quad \ldots \quad \ldots \text{ August } 22\cdot61$$
$$\pi \quad \ldots \quad \ldots \quad \ldots \quad 146\;44\;0$$
$$q \quad \ldots \quad \ldots \quad 0\cdot629\;[9\;79865]$$
$$\Omega \quad \ldots \quad .. \quad \ldots \quad 95\;40\;0$$
$$i \quad \ldots \quad \ldots \quad \ldots \quad 40\;48\;0$$

And the hyperbola which seems to coincide best with the orbit has for its major axis twelve times the mean distance of the Sun and $e = 1\cdot052416$.

The elements which are given in the *Astronomische Nachrichten*, No. 2390,

$$T \quad \ldots \quad \ldots \quad \text{August } 22\cdot774$$
$$\pi \quad \ldots \quad \ldots \quad \ldots \quad 218\;36\;39 \text{ or } 141°\;23'\;21''$$
according to the way it is reckoned.
$$q \quad \ldots \quad \ldots \quad \ldots \quad [9\cdot80018]$$
$$\Omega \quad \ldots \quad \ldots \quad \ldots \quad 96\;25\;48$$
$$i \quad \ldots \quad \ldots \quad \ldots \quad 139\;50\;17 \text{ or } 40°\;8\;43''$$
according to the way it is reckoned.

Notes on Sketches of Comet b, 1881. By E. B. Knobel, Esq.

June 24.—The head presented a curious unsymmetrical form : an irregular fan, much developed on the following side, and with little extension on the preceding side; emanating from this in the direction of the tail, was a bright ray, as shown in the sketch made that evening. The brightest portion of the fan was that in front, or nearest the Sun. The nucleus was eccentrically situated in the fan, on the preceding side of the bright ray, which certainly did not radiate from the nucleus. An envelope, apparently of irregular form, surrounded the fan, the curve of which was not symmetrical with the head; exterior to this was the parabolic envelope.

June 29.—The ray seen June 24 had disappeared. The fan was more symmetrical in shape, and the brightest portion was on the preceding side of the nucleus. On this date, at about 11·15, the nucleus passed very close to a 7·4 mag. star, Rad. 1661.

COMET b 1881.

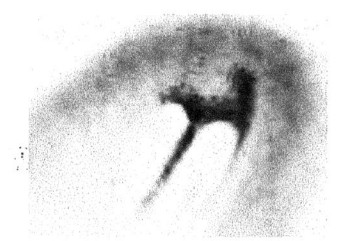

June 24 11 hrs.

June 29 12 hrs.

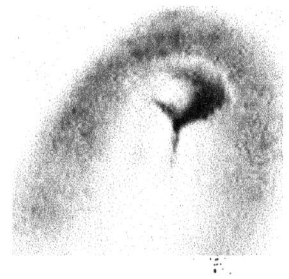

June 30 11 hrs

July 1 11 hrs.

July 3 12 hrs

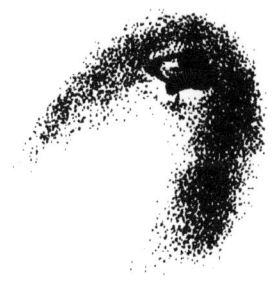

July 4 10 hrs.

July 6. 11 hrs

July 11 10½ hrs.

A series of measures of distance, taken one after the other as rapidly as possible, gave :—

$$16\overset{''}{\cdot}5$$
$$18\cdot9$$
$$20\cdot3$$
$$20\cdot6$$

The motion of the Comet was so perceptible and so rapid that accurate measures were impossible. The intervals of time between the measures were not noted.

June 30.—The appearance of the Comet had much changed ; the brightest portion of the fan was on the following side. A small ray or brush of light, slightly inclined to the direction of the tail, emanated from the nucleus. There appeared to be a bright space in front and surrounding the fan, and exterior to this a dark envelope.

July 1.—The fan appeared smaller and symmetrical; the nucleus small and stellar.

July 3.—The nucleus was very small. The fan appeared to be reduced to a curious wisp, springing from the nucleus and curved towards the preceding side.

July 4.—The fan is symmetrical and somewhat resembling the appearance on June 30, with the brightest portion on the following side.

July 6.—The fan is symmetrical, the central portion being the brightest, and resembling the curved wisp observed July 3. A slight brush of light was noticed emanating from the nucleus in the direction of the tail.

July 11.—The fan small and reduced to a very similar appearance to that noticed on July 3.

Observations of Mimas, 1881. By A. A. Common, Esq.

I give below some conjunctions of *Mimas* with end of ring, as observed with the three-foot reflector.

G. M. T.
h m

Nov. 15, at 12 43, *Mimas* was seen approaching *sf* conjunction, and level with division in ring.

12 59, not up.

13 11·5, just seen nearly up, and not seen again (though looked for during half an hour), although no difference could be seen in the sky to account for this.

Nov. 17, at 10 20, *Mimas* not yet up.

10 22, steadily seen, not up.

10 24, up to *sf* conj.

10 29, past.

Nov. 19, at 12 23, near *nf* conj.
     12 34, up ; difficult to see.
     12 39, considered past.
Nov. 20, at   5 55, *Mimas* not yet up to *sf* conjunction.
     6   8, thought to be about up.
     6 14, considered past.
Nov. 22, at   6 55, *Mimas* and *Enceladus* were nearly in a line
        drawn perpendicular to the major axis of
        ring, *Mimas* being nearest the end of
        ring.
     7 4 to 7 6, they were in a line.
     7 19, *Enceladus* much the nearest to end of ring.
     8 24, *Mimas* not yet up to *nf* conjunction.
     8 29, very near, if not up.
     8 34, now up.
     8 41, past ; hard to see.

This satellite has been seen at other times, and is generally visible when clear of ring ; but it is a most difficult object in my three-foot telescope under the best conditions, when near the end of ring, even now when it is at some little distance apparently, *Hyperion* being very much easier to observe even at conjunction ; it is therefore easily seen any fine night, when the Moon is not too near.

Ealing :
  1881, *December* 8.

Note on the Discovery of Comet c 1881 (*Schäberle*).
By W. F. Denning, Esq.

From the heading to Captain Maling's paper in the *Monthly Notices* for November, 1881, p. 49, and from the reports of the meeting of the R. A. S. in the *Astronomical Register* and *Observatory* for December, where it is mentioned that "Captain Maling discovered Comet c on the 14th of July, whereas Schäberle discovered it on the 15th of July," it would appear that there is here a claim to priority of discovery which, however, cannot be admitted on many grounds. Schäberle discovered it on the night of July 13, and determined the position as $= a$, $5^h 44^m 59^s$, $\delta, + 38° 37'$, at $14^h 47^m 30^s$, Washington mean time ; whereas Captain Maling's alleged first observation was made on the evening of July 14.

It is evident that the Comet "of considerable magnitude" seen by Captain Maling in the constellation *Camelopardus* was not Schäberle's Comet at all, but Comet *b*, discovered by Tebbutt at Windsor, N. S. W., on May 22, which occupied a position[*]

[*] On July 14 at 13^h, the nucleus of Comet *b* was near the star P. IX. 37 (mag. 4·5), in the head of *Camelopardus* (a, $9^h 18^m 38^s$, $\delta, + 81° 53' 18''$). Capt. Maling says the Comet he saw was about 20° from the horizon, directly under the polar star, which is impossible, because the altitude of *Polaris* at Grenada, West Indies, is only 12°.—W. F. D.

in the same region as that assigned to the object observed by
Captain Maling. It could not possibly have been Schäberle's
Comet, because it was far below the horizon at the time of
Captain Maling's observations; and moreover this Comet was at
the middle of July a very small telescopic object, about 1' diameter, and, being immersed in the morning twilight, must have
been utterly invisible with a small field-glass such as that used
by Captain Maling. The facts clearly show that Comets *b* and *c*
have been confused, and the former mistaken for the latter.

Ashley Down, Bristol:
  1881, December 7.

Note on Silvering large Mirrors.   By A. A. Common, Esq.

The anticipated difficulty of silvering large mirrors face
downwards in the ordinary way, caused me to adopt the plan of
silvering the mirror of my three-foot reflector face upwards, in
the cell in which it had been made. At first a fairly good film
of silver was got in this way, and a source of some anxiety was
thought to be got over, as the removal of the silver at intervals
of about one year was always contemplated; but subsequent
attempts were not so successful, and I determined to try the
other way.

As the plan I have designed and used for holding the mirror
was quite successful, and is, I think, novel, I propose to describe
it, especially as I have little doubt that the use of large reflectors
will extend. After considering the various mechanical means
that could be used to hold the mirror by the edge or back, or
both, any of which would be open to the great objection that they
might be causing strain of a most injurious kind, in a way that
would not become apparent till perhaps too late, I decided to use
the pressure of the atmosphere by the application of something
in the nature of a large sucker, from which the air could be
withdrawn to get the necessary hold on the back of the mirror,
which in this case is about thirty-seven inches in diameter, four
and a half inches thick, and weighs over four hundred pounds.

To carry out this idea, I had made a cast-iron box or cell,
round in shape and about thirty inches diameter, the rim or
side being four inches deep. The edge of this rim was turned
quite flat and true, and grooved.

For the purpose of attaching the lifting gear three lugs were
cast on the bottom, and iron eyes were fitted to them. Two
small taps were fixed in the bottom of this box, to which
could be attached flexible tubing.

One of these taps closed the connection of the box with a
small vacuum gauge, made for the purpose, of quill glass
tube in the shape of the letter U, each leg being about twelve
inches long; mercury being poured in till it stood about half way
up a rough scale on the board to which the gauge was fixed

enabled the pressure to be read off pretty closely. The other tap closed the connection of the box with the apparatus for withdrawing the air, which in the first case was my lungs; but, as there was not a back pressure valve, it was found convenient to use a small air-pump, such as is used with a seven-inch bell-glass; sufficient exhaustion could easily be got by the lungs to lift the mirror, with additional pressure for safety.

An indiarubber ring, about one and a half inch wide, and the diameter of the box, was provided to make the joint between the edge of box and the back of mirror.

The box itself might have been made much shallower, but I thought it advisable to have a precaution against the accidental breaking of the gauge or indiarubber tubes. The method of using this apparatus was as follows : the mirror was taken out of the cell and turned face down on a soft bed prepared for it, the

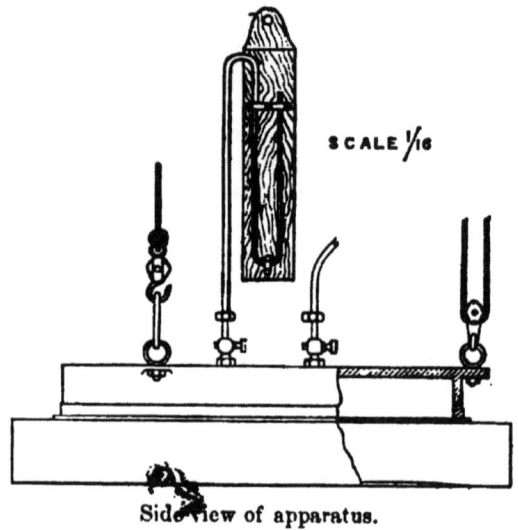

SCALE 1/16

Side View of apparatus.

indiarubber ring placed central on the back, and the box on the ring. On sucking the air out of the box the pressure made a per_fectly tight joint, and any desired degree of exhaustion could be made and kept. Two pulleys were hung from a beam in my workshop, one being attached to two of the eyes, and taking two-thirds of the weight, the other being attached to the other eye. By this arrangement the mirror could be put on edge or lowered flat or at the proper angle to enter the solution so as to drive out any air-bubbles or floating particles. All being ready, the pressure gauge, which was hung up over the box was made to register about five inches of mercury, and the mirror hoisted from its bed, turned on edge, cleaned, &c., all the operations being done in the same order as is usual in silvering smaller mirrors. The back of the mirror has a polished surface, yet the hold of the indiarubber under the pressure then used was very strong—

in fact, sufficient to hold the glass edge-up, in which position the tendency to slide off must have been very great. Of course, great care was taken to keep the water from getting on the back of the mirror and into the joint.

A sketch is given showing the arrangement, partly in section, that will complete the explanation. The application of this principle in a more perfect manner, especially for mirrors of comparative thinness, would naturally be to make the box large enough to hold the mirror, the rim acting as a support to the mirror when edgeways, and coming up to about one-half the thickness of edge, when the mirror back was close to bottom of box; then, by making a joint between the edge of mirror and rim of box by a flexible band in such a way that the mirror

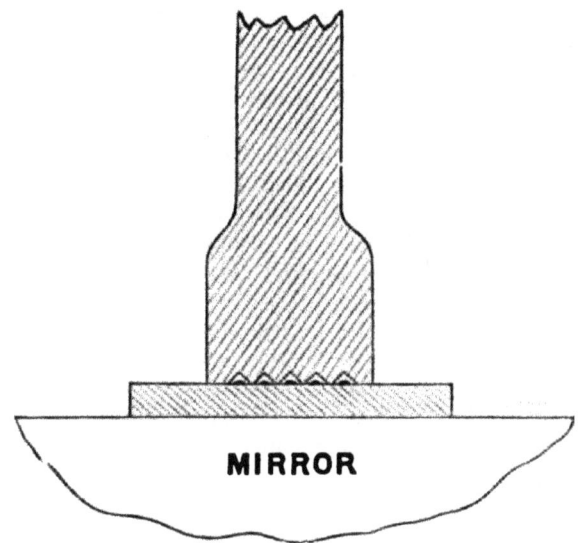

Section showing joint.—Full size.

would be free to move up or down for a short distance, in turning the whole over the pressure of air, as soon as the mirror had dropped out of the box the little necessary distance, would exactly balance or support the whole surface equally. A further extension of the idea would be to make the permanent cell that is always provided to contain the levers and plates act as the box would do, and so always be ready without shifting the mirror; and this plan I shall certainly use if I ever make a larger telescope.

Several different processes of silvering were tried, but good results could not be obtained, perhaps because the temperature at the time I made the trials was very low, being from 50° to 56° F., or from the reducing agent not being correctly made.

After some experiments, I found the following process answer

very well, and as the solutions seem to keep well, and are easily
made, it may be worth mentioning.

Separate solutions were made of nitrate of silver and caustic
potash in the proportion of 1 oz. of each to 10 oz. of water.
For the reducing agent glucose was used, and a solution of
one half oz. in 10 ozs. of water was made. These three solutions,
or any quantity in the same proportion, with liq. ammoniæ and
distilled water being ready, the proper quantity was determined on upon the basis that the above quantities would be
sufficient for 250 square inches of surface, and used in the following way :—

Ammonia was added to the solution of silver till the turbid
appearance first produced had quite cleared; the potash solution was then added, and ammonia again added till the mixture was clear. Then a weak solution of silver was added,
drop by drop, till the appearance was decidedly turbid again.
The mirror which, before beginning, had been lowered into the
dish already containing the proper amount of distilled water,
was now lifted up, and the above mixture, together with the
glucose solution, poured in and stirred well, and the mirror
carefully lowered. At a temperature of about 56° a fine film
was got in 43 minutes on the three-foot mirror. I have used
some of the same solutions at various times to silver small
surfaces, and find I can get a good film in much less time, particularly if the temperature is a little higher. No doubt for
higher temperatures some modifications would have to be made,
but the use of glucose allows a more certain determination of the
proportions proper for certain temperatures than any mixture of
sugar and acid, the active properties of which as a reducing
agent are uncertain and changeable.

Ealing :
1881, *December* 8.

On a new Form of Transit-Circle with a Prismatic Object-Glass.
By E. J. Stone, M.A., F.R.S.

The proposed form is adapted either for Transit-Circle or
Altazimuth, and it appears to me to offer many advantages. I
am quite sensible of the difficulties of securing good prisms, but
I think they are not insuperable. I have been in communication
with Mr. Grubb on the subject, and I hope to give an object-
glass of this description a fair trial as soon as Mr. Grubb is
sufficiently relieved from the pressure of work in connection with
the equipments of the Transit of *Venus* expeditions to undertake
the manufacture of the object-glass.

I take for the calculations the following indices of refraction;
but others can easily be adopted, if known, for the glass to be
used :—

	Crown Glass.	Flint Glass.
D	1·52959	1·63504
E	1·53301	1·64202
F	1·53605	1·64826

The focal length of the combination is about 100 inches (99·972 inches) The following are the radii of curvatures, crown glass in front :—

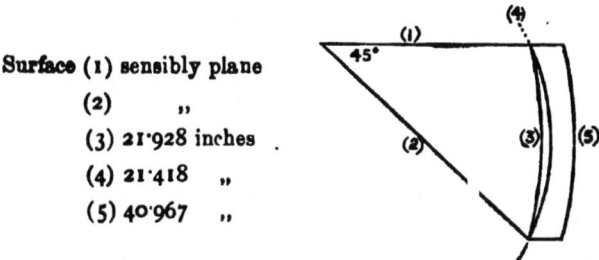

Surface (1) sensibly plane

    (2)   „

    (3) 21·928 inches

    (4) 21·418 „

    (5) 40·967 „

The spherical aberration is sensibly destroyed for light of the refrangibility of E.

The focal lengths are

D	99·973	
E	99·968	99·972
F	99·972	
C	100·028	

It is therefore an an achromatic object-glass for D, E, F. C stands out; but it occurs to me that if the wires were illuminated by red light through the opposite face of the prism, there would

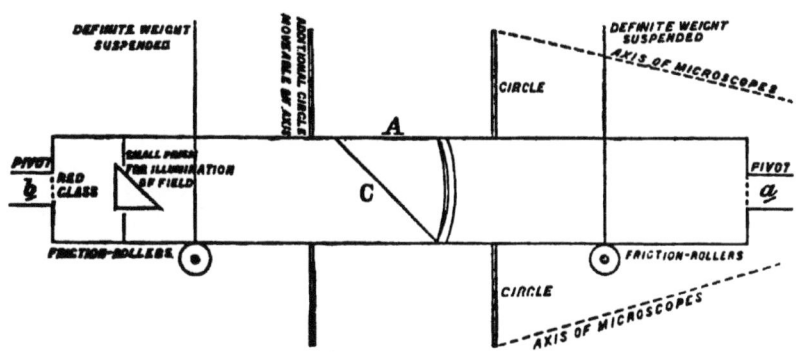

be better definition and practically no loss of light. The only light really lost by the illumination of the wires would be that outstanding about C.

H

The surfaces A and C should be sensibly plane, but, as they turn with the whole optical sytem, exact geometrical planes are not necessary for accuracy.

The eyepiece joins on at *a*, and can be partially supported by friction-rollers with a definite weight suspended to avoid flexure. The whole turns about the pivots *a* and *b*, and should be reversible. The adjustments can all be determined as for an ordinary Transit-Circle, but reflexions near the zenith would be possible.

Radiant Points of Shooting Stars observed at Bristol in the Years 1878 *and* 1879. By W. F. Denning, Esq.

In continuation of my previous catalogues of meteor showers (*Monthly Notices*, Jan. 1877 and March 1878), I send two further lists, of 20 Radiants observed between July 21 and Aug. 10, 1878, and of 47 observed between July 28 and Nov. 14, 1879. I have selected the positions of such showers as were best observed, for in observations of this character there are always a large proportion of suspected streams from which four or five meteors only have been recorded. These have been omitted in the present case—indeed, it is necessary that uncertain positions should be invariably excluded, because they only originate complications which it is most desirable to avoid.

Towards the end of July and early in August 1878 we had a succession of very clear nights, and I recorded 621 shooting stars in 34 hours of observation. Amongst the large number of radiant points resulting from their reduction, two were of very exceptional character, and have already been referred to in the *Monthly Notices* (Jan. 1880, pp. 124-127). The following list comprises the most important streams :—

Radiant Points observed July 21 to August 10, 1878.

Reference No. 1878.	Date.	Radiant Point α	δ	No. of ⨍'s	Notes and Comparisons.
1	July 31–August 1	332	+50	14	Lacertids. Meteors very swift and short.
2	July 27–30	341	−13	54	δ Aquariids. Slowish, long meteors; fine shower. = No. 59, 1877.
3	July 30–August 1	32	+53	63	χ Perseids. Swift meteors, with streaks; very fine shower.
4	July 31–August 1	12	+70	16	Meteors faint and not very swift. Heis 15° + 70°, July 28–29.
5	,,	321	+31	10	Maximum, August 1. Very slow; near ζ Cygni.
6	August 1–2	291	+70	14	δ Draconids. Swift, short meteors. Heis 292° + 70°, August 3–19.
7	July 25–31 & Aug. 10	6	+37	23	Swift, streak-leaving meteors. = No. 49, 1877.
8	July 29–August 2	333	+9	10	Meteors very slow. Corder 333° + 12°, July.
9	July 21–August 1	11 / 12	+47 / +52	26	Meteors very swift, with streaks. Radiant diffuse. Probably two showers close together.
10	July 25–26	332	+37	11	Bright, slow meteors, with long paths. S. & Z. 336° + 30°, July 28.
11	July 26–August 1	28	+36	12	Very swift meteors, with streaks. S. & Z. 32° + 35°, July 30.
12	July 26–31	333	+18	8	Swift, faint meteors. Corder 333° + 12°, July.
13	July 26–27	354	+42	7	Slowish, faint. Well defined.
14	July 31–August 1	7	+11	9	Meteors bright and very swift, with streaks. Radiant exact.
15	July 29–August 1	23	+41	7	Swift, streak-leaving meteors.
16	July 20–28	18	+59	8	Swift, streak-leaving meteors. Corder 20° + 60°, August.
17	July 27–31	28	+28	7	Swift meteors.
18	July 31–August 1	65	+60	7	Small, faint, and slow moving.
19	July 27–31	332	+27	8	Faint; very swift. S. & Z. 336° + 30°, July 28.
20	August 7–10	42½ / 44	+54 / +59	106	Perseids. Apparently a double shower, separated 5° in declination.

In addition to these I may mention showers at $96° + 72°$, $47° + 25°$, $31° + 18°$, $305° - 15°$, $50° + 75°$, $33° - 20°$, $134° + 78°$ $284° + 44°$, $22° + 13°$, $331° + 62°$, $76° + 54°$, &c., as observed with less distinctness. Indeed, though I have selected only 20 centres for the above table, I saw 41 different showers during the seven nights July 26–Aug. 2 ; and from a careful analysis of the path directions, there were indications of at least 14 additional radiants, so that the aggregate number of meteor streams in action at this particular epoch was fully 55. It is true that the end of July is rather a special period, but the number of visible streams (chiefly remarkable for their extreme feebleness) is very large on every night of the year. During my observations I recorded the number of meteors seen each night, together with the state of sky, duration of watch, &c. The figures show an exceptional abundance of shooting stars :—

1878.	Period of Observations.		Length.	$\int$'s seen.	State of Sky.
	h	h	h		
July 26	$10\frac{3}{4}$ to	13	$1\frac{1}{2}$	27	Many clouds.
27	$10\frac{1}{2}$	$14\frac{1}{2}$	4	93	Very clear.
28	$10\frac{1}{2}$	$14\frac{1}{2}$	4	44	Haze ; stars dim.
30	$10\frac{1}{4}$	$14\frac{1}{4}$	4	76	A little haze.
31	10	$14\frac{1}{2}$	$4\frac{1}{2}$	94	Very clear.
Aug. 1	$9\frac{3}{4}$	12	$2\frac{1}{4}$	42	Very clear. Cloudy after midnight.
2	10	$11\frac{1}{2}$	$1\frac{1}{2}$	20	Clear. Cloudy after $11\frac{1}{2}^h$.

Thus, during the period July 26 to August 2, 1878, I noted 403 shooting stars in watches extending in the aggregate over 22 hours.* This number was counted, notwithstanding the fact that, while registering the apparent paths of such as were accurately observed, many others must have escaped observation. On the nights of July 27 and 31, the horary rate for one observer, persistently looking eastwards, exceeded 30, and on the intervening nights I believe the numbers were equally large, but there was a good deal of haze which effectually obscured many of the smaller meteors. In 1879, I partially confirmed my results of the preceding year, for on July 28 in a two-hours watch between $11^h 45^m$ and $13^h 45^m$, I counted 45 meteors, of which several belonged to the rich showers of Aquariads and Perseids seen in 1878. In 1881, July 27, I also noted a remarkable frequency of shooting stars, but, being engaged in telescopic observations, their numbers and directions were not fully recorded on that occasion.

Observations were resumed at the end of July 1879, and continued until the middle of November, when ill health necessitated a suspension of the work. 1096 shooting stars had

---

* These figures include 7 meteors seen in a short interval on July 29.

been observed in 86 hours of watching. The results are grouped into five periods as follows :—

1879.	Hours of Observation.	Meteors Seen.	Radiants.
July 28–Aug. 12	8¼	143	5
August 21–25	16	225	12
September 14–25	19	270	13
October 14–20	24½	280	11
November 12–14	18½	178	6

Only the most active radiants are mentioned. In the following table 47 positions are given, and to these 601 meteors were found to be conformable, or nearly 13 to each centre. Many of these showers correspond with those observed in 1877, and I have pointed out these agreements in cases where they are very close.

The chief shower observed in 1879 was on Aug. 21–25 (No. 6), from a point at $291° + 60°$ (56 ♌ s), near o *Draconis*. I gave a diagram of the tracks and accompanied it with a detailed account of the shower in the *Monthly Notices*, Jan. 1880, pp. 127–128; I also noted a very active stream (No. 32) in October from a point in the south of *Aries*, $31° + 9°$, which I have described in the same number of the *Notices*. As to the other displays, I saw 38 Perseids on Aug. 9–12 from the usual radiant north of η *Persei*, and in October the Orionids (No. 33) were well seen. On November 13 I watched the sky continuously for 11½ hours ($5^h 30^m$ to 17^h), and counted 100 shooting stars; of these 18 were fine Leonids, nearly all of which made their apparition between $14\frac{3}{4}^h$ and $15\frac{1}{2}^h$. The apparent paths were noted with extreme care, and I found the radiant to be very sharply defined at $148° + 23°$ (No. 42). Two showers (Nos. 28–31 and 18–39) in *Auriga*, separated 23° in declination, were sharply defined both in September and October from the points $77° + 57°$ and $76° + 33°$, and I had seen the latter well both in October and November 1877 from the same point. The radiant immediately N. of β *Trianguli* (No. 19) in September was one of the best seen during the year; and the point of divergence is identical with one I have several times noted at the end of July and during the first half of August. I saw 25 Taurids and 19 Muscids on the nights of Nov. 12–14. They form active showers of slowish trained meteors, and are notable on account of their contemporary occurrence with the Leonids. They often supply meteors of considerable brilliancy, and will probably be frequently re-observed in future years, as two of the finest showers visible in November.

Radiant Points observed July 28 to November 14, 1879.

Reference No.	Date.	Radiant Point. a δ	No. of $\uparrow$	Notes and Comparisons.
1879. 1	1879. July 28	338 − 14	14	δ Aquariads. Slowish, long bright meteors. = Greg. 109.
2	July 28–29	32 + 53	9	χ Perseids or Perseids II. Swift; streaks.
3	July 29	30 + 37	6	Swift; streaks. A well-defined shower.
4	August 9–12	46 + 58	38	Perseids. Meteors very swift, with streaks.
5	9–12	215 + 76		Several bright meteors from a radiant here.
6	21–25	291 + 60	56	Draconids. A fine shower of bright, slowish meteors.
7	21–23	46 + 47	9	α Perseids. ⎫ = No. 63, 1877.
8	21–25	62 + 35	10	ε Perseids. ⎬ Meteors, swift with streaks.
9	21–23	61 + 50	11	μ Perseids. ⎭ = No. 83, 1877.
10	21–23	319 + 30	12	Slow. At ξ Cygni. G. & H. 315° + 31°, July 4–August 22.
11	August 23	343 + 14	6	Bright, swift meteors. Sawyer 348° + 17°, August 28–September 2.
12	August 21–25	339 − 10	16	Slow, bright meteors. Aquariads. See No. 1. = No. 59, 1877.
13	21–23	70 + 66	9	Swift meteors. Radiant close to c Camelopardi.
14	August 23	350 + 47	8	Rather slow. Well defined and exact.
15	August 21–23	266 + 47	7	Max. August 22. Slow, trained meteors.
16	21–23	5 + 17	8	Slow meteors. Tupman 7° + 13°, August 18, 1869.
17	21–25	24 + 42	9	Swift, bright meteors, with streaks. At γ Andromede.

18	September 14-25	76+32	10	Swift, with streaks. Schmidt 70°+32°, September. See No. 39.
19	15-25	30+36	16	Very sharply defined. Meteors not very swift. S.&Z. 28°+35°, Sept. 23.
20	14-25	99+43	15	Very swift, with streaks. Seen also on Oct. 15 and 20 at 105°+50°.
21	14-21	76+44	11	α Aurigids. Swift and short, with streaks.
22	14-25	31+19	14	Max. September 21, 10 /'s. Slowish. In Aries.
23	14-25	82+75	8	Not swift; streaks. Radiant exact.
24	20-21	192+79	9	Very, very slow. 8° p. β Ursæ Minoris.
25	15-25	355+18	6	Slowish, bright meteors. G. & H. 352°+17°, August 22–October 15.
26	14-25	61+48	8	Swift, no streaks. At μ Persei.
27	14-21	87+43	8	β Aurigids. ⎫ Meteors, swift with streaks.
28	14-21	76+56	7	δ Aurigids. ⎬ Beginning of No. 31.
29	14-21	50+54	7	Rapid, no streaks. Also at 31°+52° (5 slow /'s).
30	Sept. 21-Oct. 20	84—11	7	Swift, streaks. Tupman 85°−15°, August 31, 1870.
31	October 14-20	78+57	12	δ Aurigids. Not very swift; streaks.
32	14-20	31+9	37	Max. October 15, 21 /'s. An active shower of slow and generally small meteors. Corder 32°+11°, Oct. Tupman 28°+10°, Oct. 13.
33	October 15 & 20	93+17	39	Orionids. Very swift and short, invariably leaving streaks.
34	15 & 20	106+23	14	Very swift, long meteors, with streaks. Tupman 105°+24°, Oct. 12, 1869.

1879.

Reference No.	Date	Radiant Point. α	δ	No. of ☄'s	Notes and Comparisons.
35	October 4–6 & 16–20	316 + 59		17	Meteors quick, rather bright. = Heis, B 9, 315° + 59°, Oct. 3–Nov. 13.
36	October 14–20	95 + 46		11	Max. October 14. Meteors swift, with streaks. = No. 97, 1877.
37	15 & 20	7 + 51		11	Max. Oct. 15. Slow and faint. Just S. of α Cassiopeiæ. = No. 137, 1877.
38	October 20	45 + 6		8	Meteors slow. N. of α Eridani. Tupman 44° + 4°, October 14, 1869.
39	October 14–15	76 + 33		7	Not very swift. = No. 126, 1877. Tupman 77° + 30°, October 13, 1869.
40	15 & 20	114 + 62		7	Meteors bright, very swift and streak-leaving.
41	15 & 20	70 + 65		7	Meteors rather slow. At c Camelopardi. = No. 34, 1876.
42	November 13	148 + 23		18	Leonids. Meteors brilliant, with streaks. A contemporary shower seen at 125° + 55° (5 ☄'s).
43	November 12–13	133 + 70		8	Rapid meteors, without streaks.
44	November 12	62 + 21½		14	Slow. } Showers of Taurids. Meteors brilliant, with trains of ashy sparks.
45	November 13–14	58 + 21		11	Slow. }
46	12–14	46 + 21		19	Muscids. Max. November 13. Meteors slow, with spark-trails.
47	12–14	80 + 24		7	Meteors not rapid. Beginning of Taurids II.

Ashley Down, Bristol:
1881, Nov. 15.

MONTHLY NOTICES

OF THE

ROYAL ASTRONOMICAL SOCIETY.

| VOL. XLII. | JANUARY 13, 1882. | No. 3. |

J. R. HIND, Esq., F.R.S., President, in the Chair.

Harry Escombe, Esq., Durban, Natal;
Major Edward Smith Gordon, R A., Royal Arsenal, Woolwich;
Thomas Hands, Esq., B.A., 19 Castle Street, Carlisle;
Jasper Nicolls Harrison, Esq., Saling Grove, Braintree, Essex;
Alfred Morris, Esq., Sydney, New South Wales;
Samuel Hickling Parkes, Esq., King's Norton, Worcestershire;
George Elliot Ranken, Esq., B.A., 4 Philbeach Gardens, South Kensington;
Isaac Roberts, Esq., Kennessee, Maghull, near Liverpool;
Captain Moultrie Salt, 48 Clifton Hill, St. John's Wood, N.W.;
Rev. William Joseph Wilby, B.A., H.M.S. "Champion," Pacific Station;

were balloted for and duly elected Fellows of the Society.

Remarques sur la méthode proposée par M. le professeur Pritchard pour la mesure de l'éclat des astres. (Exposé d'un autre procédé pour atteindre le même but.) Par M. Loewy, membre de l'Institut.

Depuis plusieurs années, nous avons entrepris, MM. Paul et Prosper Henry et moi, une série de recherches ayant pour but la construction d'un appareil destiné à mesurer avec facilité et précision l'éclat des astres. Parmi les procédés que nous avons expérimentés se trouve celui dont s'est servi M. le professeur Pritchard; comme lui nous avons employé pour éteindre la lumière des étoiles un verre prismatique achromatisé.

Nous avons dû abandonner ce procédé après avoir constaté qu'il offrait des inconvénients sérieux, surtout quand il s'agit d'affaiblir la lumière dans une proportion considérable, en agissant par exemple sur 7 or 8 grandeurs d'étoiles. En effet, nous avons étudié par des procédés physiques la nature de la lumière après son passage à travers le verre neutre. Pour effectuer ces expériences, nous avons recueilli, en France, en Angleterre, et en Allemagne une cinquantaine d'échantillons, les meilleurs que nous ayons pu nous procurer; mais nous avons toujours constaté une coloration très notable dans les images affaiblies, même pour les lames prismatiques les plus pures et qui à la vue semblaient n'accuser aucune différence de teinte. De telle sorte qu'aucun des verres examinés ne pouvait être considéré comme possédant un pouvoir absorbant égal pour des astres différemment colorés.

Dans ces conditions il nous a donc paru dangereux de nous servir d'un procédé qui est de nature à introduire dans les recherches des erreurs systématiques.

De l'ensemble de nos expériences nous avons été amenés à conclure qu'il ne fallait pas amoindrir l'intensité lumineuse au delà d'environ 3 grandeurs d'étoiles, si l'on ne veut pas s'exposer à des erreurs sensibles.

Le procédé utilisé pour pouvoir se rendre compte de l'action du verre neutre est fort simple.

Nous avons pris deux lumières, d'une nature identique, destinées à nous fournir deux images; et nous avons ensuite, à tour de rôle, affaibli l'une des images par l'éloignement de la source lumineuse, et l'autre par l'intervention de la plaque prismatique. C'est de cette façon qu'il a été facile de distinguer des colorations très fortes et très différentes selon la nature du verre employé. Nous avons en outre, à l'aide du spectroscope, constaté l'existence de plusieurs bandes d'absorption. En résumé, aucun des verres n'a pu supporter d'une manière satisfaisante l'épreuve à laquelle nous les avons soumis; mais peut-être M. le professeur Pritchard a-t-il été plus heureux dans ses recherches pour se procurer un verre suffisamment neutre.

Pour échapper à ces difficultés, nous avons pensé à un autre procédé, à une disposition tellement simple, que nous demandons, comme M. le professeur Pritchard, si elle n'aurait pas été déjà proposée par quelque savant.

Nous avons introduit dans la lunette un diaphragme ayant l'ouverture de l'objectif et perforé d'un petit trou circulaire.

En faisant glisser ce diaphragme le long de l'axe optique, la quantité de lumière transmise sera inversement proportionnelle au carré de la distance par rapport au foyer.

Toute lunette munie d'un semblable diaphragme pourrait servir à mesurer l'éclat des étoiles: mais il est encore plus rationnel et plus avantageux de construire un instrument affecté spécialement aux mesures photométriques. Cet appareil peut être réalisé d'une façon très simple et très peu couteuse. En effet, si l'on veut déterminer la différence d'éclat des dix premières grandeurs d'étoiles, il suffit, pour atteindre ce but, d'employer une petite lunette d'environ 35 millimètres d'ouverture, et de la munir d'un diaphragme mobile percé d'un trou de $0^{mm\cdot}6$; à l'aide des lectures faites sur une échelle divisée, placeé le long de la lunette, il sera facile de calculer la différence d'éclat des astres comparés.

Toutefois, ce procédé ainsi utilisé pour la comparaison directe de 10 grandeurs d'étoiles présenterait de légers inconvénients. Selon la grandeur de l'astre, la plaque circulaire rétrécirait dans des proportions très différentes l'ouverture de l'objectif, et alors, d'une part, la quantité de lumière transmise pourrait ne pas êtro rigoureusement inversement proportionnelle au carré de la distance par rapport au foyer, et d'autre part, le disque stellaire, ne conservant pas le même diamètre, n'impressionnerait peut-être pas la rétine d'une manière identique.

Pour atténuer ces difficultés, il vaudrait mieux restreindre un peu l'amplitude de la comparaison directe, et ne se proposer que l'évaluation de 5 ou 6 grandeurs.

Pour opérer dans ces nouvelles conditions on peut employer divers procédés. Par exemple, avec la même lunette d'environ $35^{mm\cdot}$ d'ouverture, munie d'un diaphragme perforé d'un trou circulaire de $3^{mm\cdot}5$, on arrivera, en déplaçant le diaphragme, à éteindre les étoiles comprises entre la 5e et la 10e grandeur.

Mais si l'on veut, avec la même disposition, étudier des étoiles plus brillantes, il faudra placer devant l'oculaire, sous un angle de $45°$, une lame non argentée de cristal de roche ou de verre ordinaire, destinée à diminuer préalablement par la réflexion l'intensité lumineuse. Pour nous rendre compte de l'action produite par la plaque de cristal, nous avons effectué des expériences physiques directes, et nous avons trouvé que le pouvoir réfléchissant des deux surfaces planes de cristal, taillées parallèlement à l'axe, est ègal à $0^{\cdot}111$ de la lumière incidente.

On pourra, comme on le voit, étudier de cette façon les astres compris, à peu près, entre la 3e et la 7e grandeur.

En adaptant à la lunette, toujours sous un angle de $45°$, au lieu d'une lame, deux lames parallèles séparées l'une de l'autre par l'ouverture de l'oculaire, il sera possible de mesurer depuis la 5e jusqu'au delà de la 1re grandeur.

En résumé, en employant cette méthode, on peut, par les combinaisons les plus diverses, comparer l'intensité relative d'éclat de toutes les grandeurs d'étoiles. Ainsi, avec une petite lunette de 1 centimètre d'ouverture ayant un diaphragme perforé d'un trou de 1 millimètre il sera facile, en utilisant les plaques de verre, de comparer entre elles toutes les étoiles comprises entre la 7e grandeur et les plus brillantes de notre ciel.

Toutefois, pour opérer avec sécurité, il faut, pendant l'opération, soit à la main, soit à l'aide d'un mouvement d'horlogerie, maintenir l'axe au centre du champ.

Ce procédé présente un avantage marqué sur l'emploi du verre neutre : les quelques inconvénients signalés plus haut, qu'il laisse encore subsister, ne nous paraissent pas de nature à exercer une influence sensible sur les résultats que l'on peut obtenir par la méthode d'extinction : nous avons même pu constater, par l'expérience, que le disque stellaire, malgré le rétrécissement de l'ouverture, n'offre à la vue aucune différence appréciable d'aspect.

Nous étudions en ce moment une nouvelle disposition qui nous permettra d'atteindre dans des recherches photométriques la précision la plus rigoureuse, et nous espérons pouvoir en donner bientôt la description.

Note on the Employment of Photography in the Transit of Venus of 1882. By E. W. Maunder, Esq.

There is a very general feeling that the results obtained from the photographs taken during the Transit of *Venus* in 1874 are not sufficiently satisfactory to make it worth while to employ photography in the approaching Transit. There is, however, some probability that the circumstances of the Transit of this year may render it possible to attack the problem of the solar parallax in a wholly different manner to that employed in the treatment of the 1874 photographs. And as I have not seen this feature of the 1882 phenomenon adverted to in this connection, I have thought that it might be worth while to draw attention to it, though I cannot think that it has wholly escaped notice.

In measuring the 1874 photographs, the object sought to be attained was the determination of the distance between the centres of the Sun and of *Venus*. This necessitated of course the precise measurement of the position of the limbs of the Sun, and practically of the planet as well. But it was found that the limbs of the Sun were not measurable with anything approaching the required accuracy, and that *Venus* herself was also often strangely distorted. Instead of a hard, sharp, regular line, marking the frontier between light and darkness, the Sun was found to melt away gradually to nothingness, and the difficulty of determining *where* on that delicate shading was the true edge of the Sun, was often aggravated by the irregularities of a "boiling limb." The case was made worse in the English photographs by the need of arriving at the true radius of the Sun in order to find the value of the scale.

Some part of these difficulties were due, I think, to defects which might be readily obviated on a second attempt, for the uncertainty in the determination of the Sun's radius was certainly greater than that usually found in the measurement of the series now being taken at Greenwich for Sun-spot observation. But it does not seem at present possible to *ensure* the taking of Sun-pictures in which the uncertainty in the determination of the position of the limb shall not be large as compared with the uncertainty in the solar parallax. It is possible, however, that in 1882 the parallax may be inferred without the measurement of the Sun's limb at all.

The Transit of 1874 happened at a time when the solar activity was rapidly declining, and as *Venus* passed across the Sun at a great distance from the centre running almost along the fifty-ninth parallel of North latitude, to which its path was but very slightly inclined, there was no possibility of its encountering or even approaching any solar marking which might serve as a fiducial point, for at that period spots did not wander further from the equator than latitude 15°, and faculæ than latitude 20°, and the manifestations of both orders of phenomena were becoming feeble. In December, 1882, on the contrary, we shall have reached, or at all events be not far from attaining, the time of maximum activity, and *Venus* will travel along a path which crosses the forty-second parallel of South latitude at a small angle. The planet's limb, therefore, from a southern station would be seen to touch the thirty-seventh parallel. It would not be an unprecedented event if there were a spot or spots in a position to be actually occulted by *Venus*, but the probability is very considerable that spots will be found between latitudes 20° and 30°, for this is the favourite *locus* of spots at times of greatest disturbance; so that there is good reason to expect that at some time during the Transit the limb of *Venus* will be within some two or three minutes of arc of a spot. And it is even more likely still that faculæ will be found near the one limb or the other, perhaps near both, extending across the planet's path.

In a lightly-exposed and slowly-developed picture the solar details are seen with great distinctness, and not only spots, but faculæ, and even less conspicuous and defined markings still, often offer easily recognisable points, the positions of which could be undoubtedly measured with the high degree of accuracy required. If then at one or more stations a long-focus photo-heliograph were employed similar to those used by Lord Crawford and the Americans, and by this instrument the distance between several well-marked points on the Sun's surface determined, a scale would be obtained that could be used for the other photo-heliographs, and no necessity would exist for measuring the Sun's limb at all. And it would, I think, be found that *Venus* would show a far more definite and regular outline on pictures taken with a small exposure than in the more fully exposed pictures of the last Transit.

In this manner the two great difficulties attending the reduction of the photographs of 1874 would be overcome : the limb of *Venus* would be more measurable, and that of the Sun left unattempted. There would be the further advantage of the maximum distance to be measured being only three, four, or at the most five minutes of arc, instead of the whole diameter of the Sun, whilst the relative displacement of *Venus* as seen from two compared stations would generally exceed half a minute.

Then the combination of the pictures would be unrestricted, except that it would be necessary that only photographs which were taken nearly simultaneously should be compared. Thus if soon after ingress *Venus* were found to be near spots, or amongst faculæ giving one or two good points of reference, we might compare pictures taken, say, at Greenwich, Honduras, the Falklands, and Natal; and these might be combined in any pairs, the four photographs thus giving six determinations of parallax. Or near egress, photographs taken at San Francisco, at Auckland, and at Cordoba might be compared, when the three photographs, each taken with a Sun more than $30°$ high, would give three determinations of parallax, the factor in each case being fully $0\cdot70$.

This free combination of photographs would tend to in some degree eliminate the various sources of error. Thus where only three pictures were compared, three positions of *Venus* would be obtained, forming a triangle, the angles of which would be most precisely known. It is clear that no unsuspected cause of distortion, atmospheric, instrumental, or photographic, could so operate as to alter the length of the sides of the triangle—that is, the parallactic displacement of the planet—without betraying its presence by altering the angles as well. I fear the method of contacts provides no such check upon the errors to which it is liable.

If fortune so far favour the observers of the next Transit that *Venus* should occult some spots or well-marked faculæ, then the photographs will be of additional service, not only as themselves affording the means of determining afterwards the times of disappearance and reappearance, and as giving a determination of the parallax comparable in method with contact observations, while free from many of their drawbacks, but by enabling observers to note the times of the phenomena with their telescopes, with the consciousness that the particular marking observed could be afterwards identified.

But if, despite all the probabilities of the case, the Sun disappoint our legitimate expectations, and present a clear, unruffled countenance to us on the sixth of December next, then I venture to think that the following slight alterations might tend to secure an improvement in the results from the measurements of the separation of the centres. First, if collodion be used, the substitution of a wet process for a dry. The American photographs, which seem to have proved decidedly more measurable than our own, were wet plates, and the only good photograph secured by the English photographers was obtained by the same method.* Besides, the wet plates taken at Greenwich on every fine day certainly show limbs better adapted for measurement than do

---

* *Monthly Notices*, vol. xxxviii. p. 509.

a heavy percentage of the *Venus* plates. Some of the new gelatine processes might give even better results, but I have no personal experience of the behaviour of these plates under the micrometer, and I should not be inclined to expect any very good results from them. Next, a shortened exposure, and if possible a better means of making it than by the drop-slit as at present arranged. Thirdly, some means of securely clamping the instrument. The Dallmeyer photoheliographs are sadly deficient in clamping power. Lastly, a small finder added to each instrument would be a great improvement, and would have prevented the unfortunate failure of the Janssen method at Honolulu.

The most important station for photographic work would be the Falkland Islands; Santiago and Cordoba and other South American posts ranking next, as in all probability photographs. will in any case be taken, both in Europe and the United States, supplying the needed corresponding northern stations. A photoheliograph in Cape Colony and another in New Zealand would be desirable, but less essential. There is one already at work in the Mauritius, where ingress is visible at sunset, but the sun will be very low there. Cuba or Central America would also be good positions, but the post most necessary to be occupied is, as has been already stated, the Falkland Islands, there being but a poor chance of fine weather in the Straits of Magellan.

Blackheath:
1882, *January* 12.

The Relative Motions of the Great Red Spot and Brilliant Equa-toreal Spot on Jupiter. By W. F. Denning, Esq.

In the early part of the night of December 24 the sky was shrouded in thick haze, through which *Jupiter* shone very dimly; but at about $9^h 35^m$ the atmosphere became much clearer, and I was enabled to observe the markings then visible on the planet. At $9^h 43^m$, with power 150 on my 10-inch Reflector, I noted the middle of the red spot and the bright equatoreal spot crossing the central meridian together. I had anticipated from comparative observations of the two objects on the nights of December 20, 22, and 23, that they must occupy the same longitude at this time, and the fact was thus proved by actual observation.

Now, on November 19, 1880, at $9^h 23^m$, I observed the same markings in conjunction as they came to transit; and I have watched the motions of these objects, as frequently as opportunity permitted, during the interval of 400 days and 20 minutes which has elapsed between the two observed conjunctions of November 19, 1880, and December 24, 1881.

The red spot has performed 967 rotations, but the white spot has circulated around *Jupiter* 976 times, or 9 times more than the red spot. The mean revolution of the bright spot, relatively to the position of the red spot, has therefore been 44 days 10^h 42^m 13^s·3. The epochs of conjunction computed on this period, and dating from the first observation on November 19, 1880, are :—

		h	m	s
1880, November	19	9	23	0
1881, January	2	20	5	13
February	16	6	47	27
April	1	17	29	40
May	16	4	11	53
June	29	14	54	7
August	13	1	36	.20
September	26	12	18	33
November	9	23	0	47
December	24	9	43	0

The two spots have, however, both shown irregularities of motion in the interval, so that, though the mean motion is represented by the period as above, the individual conjunctions derived from observation are not precisely conformable to the above times. Apart from this, the two objects near the epochs of their conjunction could not be always followed, owing to cloudy weather or to the unfavourable position of *Jupiter*. I have, however, derived the approximate times of actual conjunction from observations close to the dates on which they occurred as follows :—

Approximately Observed Conjunction.			Derived from Observations on	Observed Period of Revolution.	Diff. from Mean 44^d 10^h 42^m.
	h	m		d h m	d h m
1881, Jan. 3	7	9	1880, Dec. 28 } 1881, Jan. 6 }	44 21 46	+0 11 4
			♃ near ☉ {	{ Four periods averaging	−0 3 31
June 29	11	53	July 10 {	44 7 11	
Aug. 13	1	1	Aug. 8	44 13 8	+0 2 26
Sept. 26	7	32	Sept. 21 } 28 }	44 6 31	−0 4 11
Nov. 10	11	·3	Nov. 14 } 15 }	45 3 31	+0 16 49
Dec. 24	9	43	Dec. 24	43 22 40	−0 12 2

Thus the observed times are very close to those computed for the conjunctions of June 29, August 13, and September 26, but the conjunction of November 10 happened 16^h 49^m later than usual, while that of December 24 showed a nearly corresponding period the other way, for the time is earlier by 12^h 2^m than that predicted. In fact the revolution of the bright spot in the former case showed an excess of 1^d 4^h 51^m as compared with the latter. Between September 28 and October 17 the bright spot exhibited a remarkable deviation from its mean motion, for its longitude became displaced to the extent of $+13°$ E. The daily rate of rotation is $878°\cdot48$ as derived by Mr. Marth from a large number of observations, and while on September 28 the longitude of the white spot was $18°\cdot4$, it increased to $19°\cdot4$ on September 30, to $23°\cdot5$ on October 3, to $27°\cdot4$ on October 11, and to $31°\cdot4$ on October 17. This shows a great retardation, while the subsequent observations in November and December prove the motion to have been slightly accelerated. These curious variations have obviously originated the large difference between the periods of the last two revolutions of the spot.

I believe the observed times of conjunction given in the above table, though depending upon eye estimates, may be relied upon to within small limits of error. I have adopted the daily motion of the brilliant spot as $=8°\cdot1$ or 13^m 24^s, as compared with the position of the red spot.

During the period July 8–December 24, 1881, I obtained 43 transits of the latter and 41 of the former. The bright spot is subject to temporary obscurations at periods which seem to be of irregular occurrence.

For the information of observers who desire further to trace the remarkable motions of these spots, I give the following computed times of conjunction for the ensuing year:—

		h	m	s
1882, February	6	20	25	13
March	23	7	7	27
May	6	17	49	40
June	20	4	31	53
August	3	15	14	7
September	17	1	56	20
October	31	12	38	33
December	14	23	20	47

It is probable that the brilliant spot has been perceptible for some years, though observers have not until recently been able to connect the observations made during several oppositions of the planet. Dr. F. Terby, of Louvain, observed a brilliant white spot N. of the *f.* end of the red spot on November 27, 1879, and January 11, 1880. One revolution of this object had evidently been completed in the interval of 45 days, and the position of Dr. Terby's spot, relatively to the red, is identical with that computed back for the brilliant spot now visible. The conjunction of November 29, 1879, was also observed by Mr. Gledhill at Halifax. He noted a "bright gap" into N. border of the great South belt and slightly preceding the middle of the red spot (November 29, 1879, 6^h 30^m). A collection of the sketches made during the last five years would undoubtedly allow some salient facts to be gleaned as to the previous history of this singular object.

Ashley Down, Bristol :
1881, *December* 26.

Observation of the Outer Satellite of Mars made at the Royal Observatory, Greenwich.

(*Communicated by the Astronomer Royal.*)

The satellites of *Mars* have been looked for on four nights during the present Opposition, viz. on Dec. 29 and 30, and January 6 and 9. On three nights the search was wholly unsuccessful, but on December 30 a single measure was secured of an object which is believed to have been *Deimos*, the outer satellite. It was glimpsed once or twice between 9^h and 11^h G.M.T., but at length it was steadily held for about three minutes, and a fairly satisfactory measure obtained.

1881, *December* 30.

Greenwich Mean Solar Time.	Greenwich Sidereal Time.	Distance in Arc.	Position-Angle.
h m s	h m s	"	° '
11 7 51	5 45 56	46·6	242 16

The position-angle may perhaps be rather small.

The satellite was seen again a little later on, but clouds came up before another measure could be secured.

It is believed that *Phobos* was glimpsed three or four times between 8^h 30^m and 9^h 0^m G.M.T., and its distance was roughly estimated as 0·8 diameter of *Mars* clear of the disk, and its position-angle as about 240°.

Deimos was far more difficult to observe than during the Opposition of 1879.

The observation was obtained with the Great Equatoreal of $12\frac{3}{4}$ inches aperture. Observer, Mr. Maunder.

Royal Observatory, Greenwich :
1882, *January* 13.

Observations of the Transit of Mercury, November 7 and 8, 1881,
made at the Melbourne Observatory.

(*Communicated by R. L. J. Ellery, Esq., Government Astronomer.*)

Notes by Mr. Ellery.

Morning fine, but sky covered with cirrus, which did not
hide the Sun, but rendered precise observations of details difficult;
the thickness of film variable. Used 8-inch telescope, with
Herschelian prism and power of 125, with yellow smoke shade.
Edge of Sun commenced to boil violently a few minutes just
prior to time of first contact, and was so disturbed as to make
the time of external contact so uncertain that it was not esti-
mated. Internal contact very good at $11^h 8^m 0^s \cdot 1$ ($19^h 58^m 44^s \cdot 06$).
As the planet passed on to the Sun's disk it appeared clearly
attached by a broad smoky band to the edge, which faded away
flickeringly to a mere filament, till at $11^h 8^m 23^s \cdot 8$ ($19^h 59^m 7^s \cdot 7$),
the planet became quite free suddenly. At egress the clouds had
almost entirely cleared away, and left the Sun quite clear for the
latter phases. At $1^h 15^m 6^s$ a thin flickering line, not at all persistent, was seen to connect planet and limb intermittently. This
quite disappeared before internal contact at $1^h 15^m 33^s \cdot 41$, which
was sharp and well defined, with no distortion of Sun's limb.
External contact at egress, $1^h 17^m 12^s \cdot 9$. The external contact was
beautifully clear, and nothing remarkable was observed; not the
slightest trace of any projection of the Sun's limb around the
planet, nor any appearance of planet outside the Sun. Whenever
Mercury was seen on the Sun's disk clear of clouds a white spot
was visible on the planet's centre, ill defined, and impressing me
with its being some optical or delusory effect.

d	h	m	s	
7	19	58	44·06	Internal contact at Ingress.
		59	7·7	Planet suddenly free of filament.
8	1	15	6	Thin flickering line connected planet and limb.
		15	33·41	Internal contact at Egress.
		17	12·9	External „

Melbourne Mean Time.

Notes by Mr. White.

Observed with the North Equatoreal of four and a half inches
aperture. At ingress a positive eyepiece with power 82 was
used, the shade being a light yellow. At egress a Herschelian
prism was employed with a negative eyepiece magnifying 84
times, the shade being a wedge of neutral-tinted glass. The
corrected times are as follows :—

d	h	m	s	
7	19	57	11·7	External contact at Ingress.
		58	47·2	Internal ,,
8	1	15	30·i	Internal contact at Egress.
		17	10·5	External ,,

Melbourne Mean Time.

The limb of the Sun became violently agitated in the neighbourhood of the planet about a minute before the time of external contact at ingress, and continued disturbed during the whole of this phase. When about two-thirds of the planet had entered on the Sun's disk, it assumed a pear shape, the narrow part being towards the Sun's edge. The time given as internal contact is when a thread of light was first seen between the two limbs. At egress the definition was exceedingly good; the contacts were formed without distortion or clinging.

Notes by Mr. Moerlin.

Instrument used, 4½-inch Equatoreal by Cooke, Herschelian prisms, negative eyepiece, magnifying power 60; the shade was of a yellow-green colour. The following are the observed phases:—

d	h	m	s	
7	19	57	23·0	External contact at Ingress
		59	43·0	Internal ,,
8	1	15	34·5	Internal contact at Egress.
		17	6·5	External ,,

Melbourne Mean Time.

The time of external contact at ingress very satisfactory, but the internal contact very likely too late, as there was a cloudiness between the edge of the Sun and the planet before a complete separation took place. The contacts at egress I consider good, no ligament or bead having been seen, but a clean and comparatively sharp contact.

Notes by Mr. Turner.

Instrument used, 12-inch Newtonian Reflector, full aperture, Herschelian positive diagonal eyepiece, power 71, with neutral shade.

Phenomenon of ingress not observed, sky being very hazy. During the entire transit the sky was more or less hazy, *Mercury* appearing as a small round spot, for the most part somewhat ill-defined.

At egress it cleared up considerably, and definition was all that could be desired—sharp, clear, and steady.

At $1^h 15^m 21^s$, Melbourne mean time, a ragged filament was thrown out between the planet and the Sun's limb; this lasted about two seconds, the filament then assuming a solid form.

At $1^h 16^m 4^s$ the planet was slightly past internal contact; *exact* internal contact could not be determined.

External contact took place at $1^h 17^m 7^s$; this was a most reliable observation, the Sun's limb being sharp and steady.

Transit of Mercury, 1881, November 7-8.
By John Tebbutt, Esq.

Everything turned out favourable for the observation of this important phenomenon. The instrument employed in the observations was the 4½-inch Equatoreal Refractor with its full aperture. The eyepiece was a negative one, combined with Cooke's diagonal prism, the magnifying power being 120 diameters. The lens next to the prism was smoked, and a shade of a brownish-green colour was fitted on between the eye and the eye-lens. By this combination the Sun's limb was straw-coloured against a dark sky-background. The instrumental conditions were, indeed, almost precisely those under which the Transit of *Venus* was observed here in 1874. Owing to the new three-inch transit instrument being engaged in prime vertical work the observations for time were made with the old two-inch transit. Good sets of observations were obtained on the evenings both preceding and following the transit, and these were supplemented by a transit of *Regulus* just before the ingress. Filmy clouds began to form about the time of ingress, and gradually increased in density, but not so as to interfere with the observations. The mean time of the first external contact was noted as $7^d 20^h 20^m 43^s\cdot3$, but this observation was unsatisfactory owing to the fact that the indentation produced by the planet in the Sun's limb was very marked when first caught in the telescope. At $20^h 22^m 9^s\cdot8$, the limbs of the Sun and planet near the point of contact being pretty well defined, two or three rapid undulations of light occurred between the closing cusps, and the thread of light was at once established. Six and a half seconds later the band of light had become very distinct. With the exception of the momentary disturbance just mentioned there was nothing to interfere with the observation of the contact. There was no distortion of the planet; there was no black drop, no triangular black ligament like that experienced by me in May, 1878, nor was there even the shadowy connection observed at the contacts of *Venus* in 1874. The planet was frequently watched during its transit; it maintained its circular form throughout, but was not always equally black. At $20^h 45^m$ I for the first time remarked a faint whitish spot in the centre of the planet's disk; it was observed on several occasions subsequently, more particularly towards the egress. At $1^h 28^m$ during a slight tremor of the images, I noticed that this spot lengthened out, and formed for a few moments an ill-defined greyish streak across the disk. This phenomenon I believe to be an optical one, and not connected in any way with the planet itself. I watched carefully for the halo or ring sometimes seen round the planet during transit, but failed to detect any trace of such a phenomenon. I was equally unsuccessful in my search for a satellite. Shortly after ingress the planet passed near two groups of Sunspots, and near the centre of the disk there was a large double spot, surrounded with an extensive but light penumbra. The details of this object were very marked and beautiful. In a smaller telescope the nucleus of this spot appeared round and unusually black, and by an inexperienced observer might at first have been mistaken for the planet itself. At the time of egress the Sun's limb, owing to the light clouds having become more dense, was somewhat fainter, but remarkably steady and well defined. The clearly defined limbs of the Sun and planet were seen quite up to the moment of contact, there being not the slightest distortion of either of them. There was no ligament, indeed, of any kind, either black or shadowy. At 8^d 1^h 38^m $54^s\cdot5$ the thread of light between the limbs was very fine; three seconds later it was broken and the internal contact was certainly complete. I estimated the planet to be centrally bisected at 1^h 39^m $43^s\cdot5$, and the notch in the Sun's limb vanished at 1^h 40^m $36^s\cdot5$. At no time did I succeed in observing the planet outside of the Sun's limb. Considering that shortly after the egress the clouds thickened so as to obscure the Sun, I may regard myself as greatly favoured in the work of observation. Seeing, too, that our Australian Observatories are the only ones from which a view of the whole transit could be obtained, the results arrived at will possess a greater interest. In conclusion I here present a *résumé* of the principal phases of the phenomenon in local mean time:—

	d	h	m	s
First external contact	Nov. 7	20	20	43·3
First internal contact	7	20	22	9·8
Last internal contact	8	1	38	57·5
Last external contact	8	1	40	36·5

Observatory, Windsor, N. S. Wales:
   1881, *November* 12.

Note on the Transit of Mercury, 1881, *Nov.* 8.
By Dr. L. S. Little.

At sunrise *Mercury* was already some distance within the Sun's limb. The sky was cloudless, and continued so until after the Transit was over, with the exception of rare passing cumuli.

The air was unusually warm and damp; the thermometer was at 74° at the time the Transit was over, which is higher than it had been for nearly three weeks. On this account, perhaps, the definition of the Sun's surface and spots was never very good. It was certainly best soon after sunrise.

The instrument used was a five-inch Refractor, by Casella, reduced to four inches, with diagonal eyepiece, negative power of 150, and dark neutral-tint shades.

Mercury appeared blacker than the darkest part of the large Sunspot, but not so startlingly black as *Venus* appeared during Transit. He was always surrounded by a darkish halo, which seemed well defined, extending to a distance about equal to the planet's semi-diameter. With no power could any spots on the planet be detected. He was seen throughout as a uniform, slightly purple, black disk.

The time of internal contact I found it impossible to determine at all accurately, but give it as noted at the time. The time of external contact was definite, and is simply the time when no portion of *Mercury* was any longer visible on the Sun's disk. I could see no trace of him after external contact.

Several photographs were taken with a lens of thirty-feet focus, but they are bad.

		h m s	
Internal contact at Egress	...	15 36 12	G. M. T.
External „	...	15 37 44	„

The time was taken by a sidereal chronometer, the error of which was determined on the morning and in the evening of the Transit.

Shanghai, 1881.

On an Object seen near Comet b, 1881, on June 10, 1881.
By W. Bone, Esq., M.D.

(From two letters to the Secretaries.)

(1.)

On June 10, 1881, whilst measuring the position of the Comet, then visible here at $5^h 52^m$ mean time of place, I noticed a peculiar discordance in each succeeding measure, and at length found that the star (?) from which I was measuring was a rapidly-moving body. At first I was inclined to believe it the result of refraction, but this should have affected both Comet and star nearly equally. On more careful inspection I found it was somewhat discoid, but its light, although bright, was diffused and hazy. It moved through 6' of arc in $34^m 34^s$ of time, in a northerly direction. I immediately telegraphed down full particulars to the Melbourne Observatory, and asked for instructions. Bad weather prevented me from searching for it next morning, and in the evening I could not succeed in again picking it up, neither could I find it where seen on the preceding evening. I never received any answer from the Melbourne Observatory; but when in Melbourne a few days since I called there, and on reminding Mr. E. J. White of the circumstance, he said that Dr. Gould had stated he saw the nucleus split into two *about* that time; but I have since ascertained that it was so observed at Cincinnati on July 6.

This struck me as so remarkable that, acting on the principle of your society, " *Quicquid nitet notandum,*" I determined to send you my record of the observation.

The approximate position of the body at that time was

$$\text{R. A.} \quad \begin{array}{ccc} \text{h} & \text{m} & \text{s} \\ 5 & 18 & 8 \end{array}$$

$$\text{Dec. S.} \quad \overset{\circ}{14} \overset{'}{18} \overset{''}{0}$$

If this Comet threw off an appendage on July 6 it might possibly have also done so shortly before I saw this body; and this may be a common phenomenon with comets, but noticed now in consequence of the larger aggregate amount of attention bestowed upon them by the increased number of observers. I am making all possible preparations for the approaching Transit of *Mercury*, and should be glad of any hints with regard to the Transit of *Venus* that might possibly render the efforts of an amateur of some degree of scientific value. I trust by that time to have my 8-in. telescope by Grubb erected.

Castlemaine, Victoria :
    1881, *October* 22.

(2.)

Enclosed is the original telegram referred to in my communication to you a fortnight since :

Telegram to E. J. White, Esq. or R. L. J. Ellery, Esq, Observatory, Melbourne. Transmitted 1881, *June* 10, 8ʰ 10ᵐ P.M.

" Whilst measuring Comet to-night found what I thought was star, discoid, and travelling south 6′ arc in 34ᵐ time. Declination South, 14° 24′; Right Ascension, 5ʰ 18ᵐ 30ˢ at 6.45 mean time. No asteroid in that place. Could not be refraction. Travelled 24ˢ in Right Ascension in same interval. Appeared like circular Comet. Reply. Will you search to-morrow morning ?—W. BONE."

I have since seen in *Nature* of August 11, 1881, p. 342, a notice of Dr. Gould's having observed a similar phenomenon about two days after I had seen it, and the position given in my telegram to Mr. White at the Melbourne Observatory (which was never answered) makes it appear that it was actually another Comet we observed on those dates.

I should place the magnitude at about 2·5, for it was visible to the naked eye in first twilight. Its travel was perceptible between each set of comparisons, and amounted to 6′ in 34ᵐ of time in Declination, moving 24ˢ in R.A. in the above time.

Castlemaine, Victoria :
    1881, *November* 6.

Observations of the Transit of Mercury on November 7 and 8, 1881.
By A. V. Nursinga Row, Esq., of Vizagapatam, India.

(*Communicated by request by Piazzi Smyth, Edinburgh.*)

The above well-known and liberal-minded amateur astronomer, computer, and meteorologist, commenced preparations for the above Transit by computing the times of the principal phases for his place of observation by Mr. Woolhouse's formula in the Appendix to the *Nautical Almanac* for 1836. In these computations he assumed the longitude of his Observatory at 5^h 33^m 30^s East; the geographical latitude $= 17°$ $42'$ $9''$ North; and the geocentric latitude $= 17°$ $35'$ $37''$. The computed times of the phases visible at Vizagapatam are:—

	h	m	s	
Middle of Transit ...	18	30	50	
Internal contact at Egress	21	9	58	Vizagapatam Mean Time.
External contact at Egress	21	11	41	

When the Sun rose on November 8, Civil reckoning, Mr. Nursinga Row was at the eye-end of his telescope, an excellent 6-inch objective one by Messrs. Cooke & Sons, equatoreally mounted under a revolving dome, and was prepared to observe the remainder of the Transit, the planet having entered on the disk of the Sun, and having almost completed half of its course at sunrise. He believes accordingly that he saw that the middle of the Transit occurred at 6^h 30^m $50^s·4$ A.M. with a least distance for the centres of $3'$ $52''$. But at this stage of the proceedings there came pouring into his Observatory such a stream of his English lady and gentlemen friends, that Mr. Nursinga Row confesses he rather lost his presence of mind as an observer. The egress of the planet was afterwards beheld; and found to agree closely with the time computed beforehand.

The Merope Nebula.   By Lew's Swift, Esq.

This nebula, so easily seen by some, and which is not at all discernible by others, even with the largest and best telescopes, is *par excellence* the greatest enigma in observational astronomy. That an object of such easy and unmistakable visibility through a 3-inch telescope should be wholly invisible through the $18\frac{1}{2}$-in. Refractor of the Chicago Observatory, seems at first thought to be at variance with the generally received opinion that the larger the telescope, the brighter an object appears.

My experience with this nebula has been as follows:—In 1874, while searching for comets, I ran upon it, and having never heard of a nebula in the *Pleiades*, strongly suspected that

K

it was a new comet, which illusion the following night quickly dispelled, as the object was stationary. I wrote the fact to Mr. Burnham, who replied that it was a variable nebula, discovered by Tempel in 1859. Feeling a great interest in the object, I, from that time to the present, have not failed, on every favourable occasion, to carefully observe it for the purpose of satisfying myself as to the fact of its variability, with the result that for seven years it has shown to my eye and instrument no sign whatever of change. My object in inviting the attention of the readers of the *Notices* to this subject is, however, not to discuss this question of variability, but its existence, which to me is as palpable as that of the great nebula in the *Triangles* (M. 33), which, except in brightness, it greatly resembles. It is a far easier object to observe than the great nebula closely following ζ *Orionis* (H.V. 28=G.C. 1227). Though this latter nebula is catalogued as single, I make it quadruple.

I am told that no larger telescope will show it, even as a single nebula, from its close proximity to the overpowering light of the star. The question here arises—Why will not a large telescope be as effectual as a small one in revealing faint nebulæ close to very bright stars ? The reason is obvious, though at a mere glance, apparently contrary to the well-known principle of optics, that two objects, in close juxtaposition, maintain the same relative brightness, whether the telescope be large or small. To me this law holds good only when star is compared with star, or nebula with nebula, and not when one body is a star which cannot be magnified, and the other a nebula which can. Then, too, with large telescopes, eyepieces of higher power are generally used and the field correspondingly contracted, and so, of course, the opportunity for contrasting the faint light of the nebula with the surrounding dark sky is diminished if not entirely lost.

Professor Hough and Mr. Burnham are unable to see, or even glimpse, the *Merope* nebula with the great Refractor of the Chicago Observatory, and they therefore doubt its existence. It is quite natural for those trained observers, with such an instrument, to call into question the reality of a nebula which they are unable to see—in fact, find darkness only where observers with small telescopes see a nebula. They ascribe the nebulous appearance to a halo surrounding *Merope*; but why do hundreds of people observe it around and following *Merope*, and not around and following *Alcyone* and the other bright stars of the group? It is certainly very strange that of all the stars in the heavens, *Merope* alone should show nebulosity immediately following it!

The object is quite large, and appears triangular with the corners rounded off. Though quite faint, I have seen it with a 2-in. aperture on my 4½-in. telescope, using my comet eyepiece, power 25.

Mar. 22. 6 h.   Mar. 26, 7 h.   Mar. 28, 6½ h.

Mar. 30. 6¾ h.   Mar. 31, 6⅘ h.   Apr. 5, 6¼ h.

del.   Hollewe

If Messrs. Hough and Burnham will contract the aperture of their telescope to from 4 to 6 inches, and use a power of about 30, they will see as a reality what they now believe to be a myth.

Rochester, N.Y.:
1881, December 2.

Observations of Venus in the Spring of 1881.
By W. F. Denning, Esq.

In the months of March and April 1881 *Venus* became a splendid object in the evening sky, and I undertook a series of observations chiefly in those months, with a view to recover the delicate markings recorded by some earlier observers. The first observation was made on December 10, 1880, but the planet was near the horizon, and her diameter only 13″·8, so that nothing definite was seen.

1880, *December* 20, 3^h 40^m–4^h 5^m.—*Venus* well defined with power 200. There were spots of the most certain character, though extremely faint.

1881, *January* 6, 3^h 50^m–4^h 5^m: power 200.—Some minute markings or centres of shading, apparently giving the surface a mottled aspect, just barely discerned. A few light specks were apparent towards the circular contour of the W. limb, where the brightness of the disk was very conspicuous.

January 31, 5^h. No dark markings distinguished.

February 16, 5^h–5$\frac{1}{2}^h$.—*Venus* splendidly defined. The N. horn evidently the sharpest. There were cloudy condensations distributed over the planet's surface except around the W. border, where the brightness was very intense. I suspected crater-like objects of very minute type on the terminator, and a shading running from the N. horn about one-third round, and parallel to the bright interior edge of the planet.

March 1, 4$\frac{1}{2}^h$–5$\frac{1}{2}^h$.—Faint dusky patches again seen, but they were extremely delicate, and only caught during moments of superb definition.

March 22, 5^h–7^h.—No distinct spots seen, though at times I suspected minute shadings elongated in latitude between terminator and W. limb. No spots or crater-like objects on the terminator, which is evidently not serrated as some observers have described it. The cusps were markedly bright; so was the surface round the W. limb, but the terminator was much shaded. The rippling appearance of the planet, especially when the air is unsteady, naturally gives the impression of a jagged terminator and mottled aspect of the entire disk as noted by some observers. But though to-night the planet's disk was very closely examined for detail, these appearances could not be certainly descried. The terminator showed some gentle undulations, but there was an entire want of that remarkably rugged outline depicted in some sketches.

March 26, 6⅓ʰ–7¼ʰ.—Definition not nearly so good as last night, when the view was nearly perfect. Disk apparently mottled with grey patches and intervening luminous veins or striations, but these may have been due to tremor. I noticed that the cusps were very brilliant and sharp, extending considerably beyond the half-circle, and obviously different from the ordinary Moon-crescent, but in the case of two bodies presumably so distinct in physical characteristics, we cannot expect close agreement. Atmospheric refraction in the case of a planet enveloped in deep and dense air-strata must necessarily diffuse sunlight over a greater area of the surface than in the case of a planet with a non-appreciable atmosphere like the Moon, which can only be illuminated over those parts directly exposed to the Sun. *Venus*, however, would always be reflective more than half hemisphere, and this must cause the anomalous prolongation of the cusps so often noticed and possibly account for the visibility of the whole circle of the planet when near her inferior conjunction.

March 28, 6ʰ–7ʰ.—A bright, small spot detected just within N. cusp, and a very faint cloudy area extending from the terminator towards W. limb in S. hemisphere. Also a grey shading in N. hemisphere, running from the terminator. Definition was splendid, power 290, and Barlow lens increasing it to 400, gave excellent views. The mottled aspect of the planet was far less obvious to-night; indeed, I usually find it difficult when definition is good.

March 30, 0ʰ 30ᵐ.—Definition execrable; indeed, it was invariably found that observations when attempted with the Sun high above the horizon were utterly useless. The atmosphere is too brilliantly illuminated by the Sun's rays and its tremors rendered too strikingly apparent to allow even a lustrous object like *Venus* to be seen with good effect, and it is surprising that such a time is often recommended as eminently suited for such work. We must admit that the glare of *Venus* is considerably moderated in the presence of sunshine, but this effect, though essential in some degree, is brought about by the highly luminous condition of our atmosphere, which unfortunately then reveals its disturbing elements to the utmost degree. During the observation of to-day *two* crescents were seen in the field of view while employing a Kellner comet-eyepiece, power 38, and Barlow lens. There was a large and faint crescent almost central in the field, and a small bright crescent (the real image of the planet) situated slightly to the W. of it. The reputed observations of a satellite of *Venus*, described in astronomical text-books, immediately recurred to me, but it was palpable that the large crescent now seen was a mere spectral appearance. Curiously enough, the two crescents were turned the same way; in fact, one seemed an exact counterpart of the other as regards phase. The smaller one was estimated one-sixth the apparent diameter of the larger. The eyepiece was rotated without any displacement in the relative positions of the objects and then removed from the instrument. On looking into the tube at the small diagonal plane, the explanation became obvious. The sunshine streaming into the main aperture of the telescope fell partially upon the small sliding tube carrying the eyepiece, and formed a bright crescent upon the west side, this being feebly reflected in an inverted form through the eyepiece, and thus the "ghost" was originated. The explanation was extremely simple, and I have no doubt that the alleged observations of a satellite of *Venus* made in the last century were capable of a similar solution—indeed, it is hard to think that the origin of such illusions can escape discovery if carefully sought after.

March 30, $6\frac{1}{2}^h$–7^h.—Bright spot still at N. horn, and the cloudy diffused spot over S. hemisphere. The terminator evidently undulatory in its real figure, and I carefully noted it with power 400. There was a dark notch in it near N. horn, and near the bright spot before alluded to. The latter is extremely small, and looks like a crater, though I could not be certain of this.

March 31, $6\frac{1}{4}^h$–$6\frac{3}{4}^h$.—Appearance of *Venus* somewhat similar to that depicted last night, except that the markings appeared to have gone slightly westward. The bright spot and dark notch near the N. horn were again seen, though the former was not so distinct as on the previous nights. Definition was good with 200 and 290, and very fair with 400.

April 1, 0^h 30^m.—Observed *Venus*, but definition was very unsatisfactory, and no details could be made out. The spectral crescent was again seen as on March 30.

April 2.—High wind, but the sky very clear. *Venus* could not be observed with high powers owing to the constant vibrations of the telescope.

April 5, 6^h–$6\frac{1}{2}^h$.—The crescent evidently much narrower. There is a faint shading over the apparent N. hemisphere and an indentation at N. cusp, which is very plain, though to-night it appears further from the extremity of the cusp than on the 30th and 31st, but it may not be identical. I strongly suspected the disk to be variegated with alternating light and grey markings, and the terminator to show minute bright specks. I repeatedly had the impression of one situated between the N. cusp and centre of terminator. It looked like a longish curling speck just merging out of the dark contour of the unilluminated hemisphere. The two cusps were notably bright: indeed, their intensity is striking compared with the region near the terminator, which is invariably much shaded. The difficulty in speaking positively as to the mottled or granulated aspect of the planet's disk and the occurrence of craterlike objects on the borders of the terminator arises from two causes—viz. their minute character and the unsteadiness of the image brought about by atmospheric undulations. The surface of the planet can, it is true, never be viewed absolutely free from that rippling or tremulous motion originated by the continual transit of air-waves across the disk; and it is not difficult to understand the indistinctness and uncertainty with which minute objects such as these are viewed under the influence of these moisture-laden currents. *Venus* by her excessive lustre is much affected by such phenomena, but some excellent views of this planet may be obtained at about the time of sunset. My own observations indicate that the most suitable period of observation is from half an hour before to half an hour after sunset, when I have usually had no difficulty in obtaining a sharply-defined image of the planet.

Briefly summarising the foregoing results, it would appear that there are dark shadings upon the planet, and light areas with occasional bright spots near the cusps. The latter are very bright: so are the limbs; while the interior region is less luminous, and there is a gradual shading off towards the terminator, which does not show a distinctly serrated or jagged aspect, though its contour is evidently undulatory, with sometimes an indentation sufficiently obvious to command attention. As to the craterlike objects suspected near the terminator, I believe they are illusory, and caused by the rippling of the telescopic image. It is difficult to conceive that such objects could be perceived upon *Venus*, unless we assume that she has little or no atmosphere, which is impossible, when we consider that the prolongation of the cusps and other phenomena distinctly affirm its existence. There is no doubt that this planet requires very delicate observations, and that her configuration is by no means as devoid of interest as is frequently asserted.

It will be seen by the sketches, and it is a fact I several times noted during my observations, that the positions of the spots compared at similar times on consecutive nights showed a slight movement westward. This approximately confirms the rotation of $23^h 21^m$ given by Cassini and others. The axis appears to be greatly inclined, for the direction of the spots was from about SSE. to NNW. referred to the line of the cusps.

Bianchini, who observed many dark spots on the planet in 1726–7 with a telescope of 66 feet focus and $2\frac{1}{2}$ inches aperture, gave an erroneous rotation period of 24 days 8 hours. But his observations are reliable, though he put a wrong construction upon them. The markings which he noted as occupying the same positions at an interval of 24 days 8 hours had completed 25 circuits around the planet in the meantime. An observer who notes the progressive westerly movement of the spots as compared at nearly the same times on successive nights would certainly infer a period nearly agreeing with that formerly deduced by Bianchini, but when the markings are watched for two or three hours on the same night their swifter motion becomes evident. There are many difficulties, however, in the way of tracing the spots during long and regular periods.

Ashley Down, Bristol:
1881, *November* 10.

Observations of Jupiter's Satellites made at the Observatory, Madras.

(*Communicated by N. R. Pogson, Esq., Government Astronomer.*)

Date. 1862.	Madras Mean Time. By Obser. h m s	By N. A. m s	Phenomenon and Phase.		Telescope. Maker.	Power.	Initials of Obseiver.
Feb. 15	10 56 35·0	56 23	I. Ec. D.	e	Tulley	180	N. R. P.
18	12 4 53·3	2 10	III. Ec. D.	c	,,	134	..
Mar. 3	9 12 7·8	11 44	I. Ec. D.	e	,,	180	
	9 25 48·8	26 0	II. Oc. R.	b	,,	180	
	11 39 9·8	40 0	I. Oc. R.	b		180	
10	11 5 7·9	5 25	I. Ec. D.	e	,,	134	
May 4	10 1 44·6	1 48	I. Ec. R.	b	Lerebours	66	,,
1866. Sept.20	8 11 46·7	12 9	I. Ec. R.	b	,,	66	,,
1868. Jan. 8	7 37 30·7	39 21	III. Ec. D.	c	,,	66	,,
Oct. 28	8 17 48·3	17 16	II. Ec. R.	b	,,	106	,,
	8 42 0·3	46 0	I. Sh. E.	c	Simms	166	,,
1869. Jan. 13	8 43 27·5	43 14	I. Ec. R.	b	Tulley	100	,,
Feb. 21	7 19 13·2	19 6	I. Ec. R.	b	Lerebours	106	,,
Mar. 5	7 14 21·9	14 20	II. Ec. R.	b	Simms	102	E. I. P.
	7 14 21·2	14 20	II. Ec. R.	b	Lerebours	106	N. R. P
Dec. 10	7 57 11·8	57 23	I. Ec. R.	b	Simms	165	,,
1870. Feb. 1	10 26 46·2	26 41	I. Ec. R.	b	,,	105	E. I. P.
	10 26 32·3	26 41	I. Ec. R.	b	Lerebours	106	J. P. B.
2	7 40 12·5	38 36	II. Ec. D.	e	Simms	105	E. I. P.
	7 39 27·7	38 36	II. Ec. D.	c	Lerebours	106	N. E. P.
	7 39 49·7	38 36	II. Ec. D.	e	Tulley	100	N. R. P.
	9 54 25·6	55 55	II. Ec. R.	b	Simms	105	E. I. P.
	9 54 29·6	55 55	II. Ec. R.	b	Tulley	100	N. R. P.
9	10 15 57·2	14 24	II. Ec. D.	e	Simms	105	J. P. B.
	10 14 32·5	14 24	II. Ec. D.	e	Lerebours	106	E. I. P.
	10 15 13·0	14 24	II. Ec. D.	e	Tulley	100	N. R. P.
10	6 51 16·6	51 38	I. Ec. R.	b	Simms	105	E. I. P.
	6 51 22·0	51 38	I. Ec. R.	b	Lerebours	106	J. P. B.
	6 51 32·6	51 38	I. Ec. R.	b	Tulley	100	N. R. P.
17	8 46 50·0	47 35	I. Ec. R.	b	Simms	105	E. I. P.
	8 47 33·7	47 35	I. Ec. R.	b	Lerebours	106	N. E. P.
	8 47 38·2	47 35	I. Ec. R.	b	Tulley	100	N. R. P.

Date. 1870.	Madras Mean Time. By Obser.	By N. A.	Phenomenon and Phase.		Telescope. Maker.	Power.	Initials of Observer.
	h m s	m s					
Feb. 20	10 41 38·7	39 26	III. Ec. D.	*e*	Lerebours	63	G.W.B.
	10 41 16·3	39 26	III. Ec. D.	*e*	Tulley	52	E. I. P.
27	7 0 22·6	2 25	II. Ec. R.	*b*	Simms	163	R. F.C.
	7 0 52·7	2 25	II. Ec. R.	*b*	Lerebours	106	N. E. P.
	7 0 55·7	2 25	II. Ec. R.	*b*	Tulley	134	N. R. P.
Mar. 6	9 35 38·1	38 43	II. Ec. R.	*b*	Lerebours	63	E. I. P.
	9 35 40·1	38 43	II. Ec. R.	*b*	Tulley	100	R. F.C.
1871. Jan. 5	11 32 27·4	32 46	I. Ec. R.	*b*	Tulley	100	N. R. P.
21	9 51 53·3	52 18	I. Ec. R.	*b*	Simms	105	N. E. P.
	9 51 53·4	52 18	I. Ec. R.	*b*	Tulley	100	N. R. P.
23	9 8 57·7	10 3	III. Ec. R.	*b*	Simms	105	N. E. P.
	9 7 21·3	10 3	III. Ec. R.	*b*	Tulley	100	N. R. P.
27	9 49 6·6	50 26	II. Ec. R.	*b*	,,	100	,,
	11 28 56·5	24 0	I. Tr. I.	*b*	Simms	102	N. E. P.
	11 26 9·3	24 0	I. Tr. I.	*b*	Tulley	187	N. R. P.
	12 26 57·0	25 0	I. Sh. I.	*b e*	Simms	102	N. E. P.
	12 25 59·9	25 0	I. Sh. I.	*b*	Tulley	187	N. R. P.
	13 38 38·8	39 0	I. Tr. E.	*b e*	Simms	102	N. E. P.
	13 39 48·0	39 0	I. Tr. E.	*e*	Tulley	187	N. R. P.
28	8 34 5·7	33 0	I. Oc. D.	*b e*	Simms	163	N. E. P.
	8 33 34·2	33 0	I. Oc. D.	*b e*	Tulley	250	N. R. P.
	11 47 29·1	47 51	I. Ec. R.	*b*	Simms	163	N. E. P.
	11 47 34·4	47 51	I. Ec. R.	*b*	Tulley	134	N. R. P.
30	10 42 8·4	39 15	III. Ec. D.	*e*	,,	134	..
Feb. 3	7 39 57·4	38 0	II. Oc. D.	*e*		134	
	12 24 58·5	26 11	II. Ec. R.	*b*		134	
4	10 22 53·8	23 0	I. Oc. D.	*e*		100	
	13 43 20·1	43 29	I. Ec. R.	*b*		100	
6	8 12 2·4	12 28	I. Ec. R.	*b*		100	
	10 2 13·0	59 0	III. Oc. D.	*c*		100	
13	10 7 50·3	8 11	I. Ec. R.	*b*		100	
21	6 53 33·7	55 25	II. Ec. R.	*b*	..	138	
28	9 29 33·9	31 1	II. Ec. R.	*b*	,,	100	
Mar. 7	6 45 44·3	43 2	III. Ec. D.	*e*	Simms	120	,,
	6 55 18·7	52 0	II. Oc. D.	*e*	..	120	
8	10 24 7·0	24 32	I. Ec. R.	*b*	..	120	..
14	10 46 55·8	43 47	III. Ec. D.	*e*	,,	120	,,
	10 46 37·6	43 47	III. Ec. D.	*e*	Lerebours	165	N. E. P.

Date. 1871.	Madras Mean Time. By Obser. h m s	By N. A. m s	Phenomenon and Phase.	Telescope. Maker.	Power.	Initials of Observer.
Apr. 1	9 9 20·8	10 50	II. Ec. R. *b*	Simms	105	N. R. P.
19	6 49 49·5	46 32	III. Ec. D. *e*	„	105	„
	6 49 34·5	46 32	III. Ec. D. *e*	Lerebours	164	N. E. P.
·	9 31 2·0	31 27	III. Ec. R. *b*	„	164	„
1874. Jan. 29	11 18 31·9	17 5	II. Ec. D. *e*	Simms	195	N. R. P.
	11 19 19·0	17 5	II. Ec. D. *e*	Lerebours	160	E. I. P.
1875. Apr. 7	12 4 22·4	6 0	I. Sh. E. *b m e*	Lerebours	160	N. R. P.
	12 14 53·5	19 0	I. Tr. E. *b m e*	Simms	163	E. I. P.
	12 15 9·6	19 0	I. Tr. E. *b m e*	Lerebours	160	N. R. P.
8	9 34 27·2	37 0	I. Oc. R. *b*	Simms	163	E. I. P.
	9 35 15·7	37 0	I. Oc. R. *b m e*	Lerebo: rs	160	N. R. P.
13	9 18 20·6	12 0	II. Sh. I. *b e*	„	160	„
	9 36 25·4	26 0	II. Tr. I. *b*	Simms	163	E. I. P.
	9 25 55·6	26 0	II. Tr. I. *b m e*	Lerebours	160	N. R. P.
	11 39 57·2	43 0	II. Sh. E. *b*	Simms	163	E. I. P.
	11 38 52·2	43 0	II. Sh. E. *b m e*	Lerebours	160	N. R. P.
	11 46 45·6	50 0	II. Tr. E. *b e*	Simms	163	E. I. P.
	11 44 14·6	50 0	II. Tr. E. *b m e*	Lerebours	160	N. R. P.
15	9 8 57·6	8 34	I. Ec. D. *e*	Simms	163	E. I. P.
	9 8 47·0	8 34	I. Ec. D. *e*	Lerebours	106	N. R. P.
	11 18 27·9	21 0	I. Oc. R. *b*	„	106	„
27	8 23 41·0	25 1	III. Ec. R. *b*	Simms	163	E. I. P.
	8 23 58·5	25 1	III. Ec. R. *b*	Lerebours	106	N. R. P.
29	11 10 1·0	11 55	II. Ec. R. *b*	Simms	163	E. I. P.
	11 9 25·4	11 55	II. Ec. R. *b*	Lerebours	108	N. R. P.
May 1	9 32 25·0	32 41	I. Ec. R. *b*	Simms	163	E. I. P.
	9 32 35·7	32 41	I. Ec. R. *b*	Lerebours	106	N. R. P.
4	8 51 31·0	42 0	III. Oc. D. *e*	Simms	163	E. I. P.
	8 48 38·2	42 0	III. Oc. D. *e*	Lerebours	160	N. R. P.
	12 20 24·8	22 26	III. Ec. R. *b*	Simms	163	E. I. P.
	12 20 33·8	22 26	III. Ec. R. *b*	Lerebours	160	N. R. P.
6	10 23 30·7	26 0	II. Oc. D. *b m e*	Simms	163	E. I. P.
	10 25 46·6	26 0	II. Oc. D. *b m e*	Lerebours	160	N. R. P.
	13 45 49·0	48 29	II. Ec. R. *b*	Simms	163	E. I. P.
	13 45 48·8	48 29	II. Ec. R. *b*	Lerebours	160	N. R. P.
8	11 26 29·8	26 42	I. Ec. R. *b*	Simms	105	E. I. P.
	11 26 26·9	26 42	I. Ec. R. *b*	Lerebours	106	N. R. P.

Date. 1875.	Madras Mean Time. By Obser. h m s	By N. A. m s	Phenomenon and Phase.	Telescope. Maker.	Power.	Initials of Observer.
May 11	12 3 24·3	0 0	III. Oc. D. *b m e*	Simms	163	E. I. P.
	12 3 57·2	0 0	III. Oc. D. *b m e*	Lerebours	160	N. R. P.
	14 3 26·0	4 0	III. Oc. R. *b*	Simms	163	E. I. P.
	14 3 24·7	4 0	III. Oc. R. *b*	Lerebours	160	N. R. P.
	14 21 38·0	19 25	III. Ec. D. *e*	Simms	105	E. I. P.
	14 22 53·5	19 25	III. Ec. D. *e*	Lerebours	106	N. R. P.
15	9 54 4·0	55 0	II. Tr. E. *e*	Simms	163	E. I. P.
	9 54 39·8	55 0	II. Tr. E. *b m e*	Lerebours	160	N. R. P.
	10 32 22·0	32 0	I. Oc. D. *b m e*	Simms	163	E. I. P.
	10 32 41·1	32 0	I. Oc. D. *b m e*	Lerebours	160	N. R. P.
	11 8 44·3	16 0	II. Sh. E. *b e*	Simms	163	E. I. P.
	11 9 17·4	16 0	II. Sh. E. *b m e*	Lerebours	160	N. R. P.
	13 20 33·2	20 49	I. Ec. R. *b*	Simms	105	E. I. P.
	13 20 26·5	20 49	I. Ec. R. *b*	Lerebours	106	N. R. P.
17	7 49 10·7	49 21	I. Ec. R. *b*	Simms	105	E. I. P.
	7 50 5·7	49 21	I. Ec. R. *b*	Lerebours	106	N. R. P.
22	9 49 26·2	44 0	II. Tr. I. *b m e*	Simms	163	E. I. P.
	9 51 1·4	44 0	II. Tr. I. *b m e*	Tulley	262	N. R. P.
	10 0 4·7	14 0	III. Sh. E. *b m e*	,,	262	,,
	11 23 8·9	19 ·0	II. Sh. I. *b*	Simms	163	E. I. P.
	11 23 5·0	19 0	II. Sh. I. *b*	Lerebours	160	N. R. P.
	12 11 52·9	13 0	II. Tr. E. *b m c*	Simms	163	E. I. P.
	12 11 28·2	13 0	II. Tr. E. *b m c*	Lerebours	160	N. R. P.
	12 18 55·0	18 0	I. Oc. D. *b m e*	Simms	163	E. I. P.
	12 19 46·2	18 0	I. Oc. D. *b m c*	Lerebours	160	N. R. P.
23	10 18 53·6	18 0	I. Sh. I. *b m e*	Simms	163	E. I. P.
	10 18 39·4	18 0	I. Sh. I. *b c*	Lerebours	160	N. R. P.
	11 40 11·9	42 0	I. Tr. E. *b m e*	Simms	163	E. I. P.
	11 39 24·8	42 0	I. Tr. E. *b m e*	Lerebours	160	N. R. P.
	12 27 56·4	31 0	I. Sh. E. *h m e*	Simms	163	E. I. P.
	12 27 51·8	31 0	I. Sh. E. *b*	Lerebours	160	N. R. P.
24	6 44 56·0	45 0	I. Oc. D. *h m e*	,,	160	,,
	9 43 8·6	43 35	I. Ec. R. *h*	Simms	105	E. I. P.
	9 43 13·1	43 35	I. Ec. R. *b*	Lerebours	106	N. R. P.
June 9	8 10 17·2	43 0	III. Ec. R. *b*	,,	106	,,
	8 10 18·2	43 0	III. Ec. R. *h*	Simms	106	E. I. P.
1876. May 19	10 41 58·2	41 36	I. Ec. R. *h*	,,	143	E. I. P.
	10 42 8·1	41 36	I. Ec. R. *b*	Lerebours	106	N. R. P.

Date. 1876.	Madras Mean Time. By Obser. h m s	By N. A. m s	Phenomenon and Phase.	Telescope. Maker.	Power.	Initials of Observer.
May 24	9 18 36·6	20 44	II. Ec. R. *b*	Simms	145	E. I. P
	9 18 49·1	20 44	II. Ec. R. *b*	Lerebours	106	N. R. P.
25	8 13 29·8	12 0	III. Oc. D. *b e*	Simms	105	E. I. P.
	8 13 16·4	12 0	III. Oc. D. *bm e*	Lerebours	165	N. R. P.
	10 34 7·4	38 47	III. Ec. R. *b*	Simms	145	E. I. P.
	10 35 36·9	38 47	III. Ec. R. *b*	Lerebours	105	N. R. P.
26	10 15 0·6	13 0	I. Oc. D. *bm e*	Simms	145	E. I. P.
	10 13 31·7	13 0	I. Oc. D. *b m e*	Lerebours	160	N. R. P.
	12 36 18·7	35 43	I. Ec. R. *b*	Simms	145	E. I. P.
	12 35 46·5	35 43	I. Ec. R. *b*	Lerebours	160	N R. P.
27	7 17 58·4	20 0	I. Tr. I. *b m e*	Simms	145	E. I. P.
	7 36 32·8	20 0	I. Sh. I. *b e*	„	163	E. I. P.
May 27	9 30 29·6	32 0	I. Tr. E. *b m e*	Simms	225	E. I. P.
	9 30 39·8	32 0	I. Tr. E. *b m e*	Lerebours	160	N. R. P.
	9 42 43·3	46 0	I. Sh. E. *b m e*	Simms	163	E. I. P.
	9 42 40·8	46 0	I. Sh. E. *b e*	Lerebours	160	N. R. P.
June 11	10 52 39·4	52 48	I. Ec. R. *b*	Simms	145	E. I. P.
	10 52 39·3	52 48	I. Ec. R. *b*	Lerebours	105	N. R. P.
Aug. 3	8 55 51·7	56 36	II. Ec. D. *e*	Simms	163	E. I. P.
	8 57 20·2	56 36	II. Ec. D. *e*	Lerebours	106	N. R. P.
	9 46 28·2	46 0	I. Oc. D. *b m e*	Simms	163	E. I. P.
	9 46 53·0	46 0	I. Oc. D. *b*	Lerebours	106	N. R. P.
1878. Aug. 20	10 15 9·9	13 50	I. Ec. R. *b*	Simms	105	B. B. H.
Sep. 19	8 55 13·3	57 0	I. Oc. D. *b e*	„	163	E. I. P.
	8 55 36·1	57 0	I. Oc. D. *b e*	Lerebours	160	N. R. P.
	12 23 50·2	23 42	I. Ec. R. *b*	Simms	105	E. I. P.
	12 23 17·9	23 42	I. Ec. R. *b*	Lerebours	106	N. R. P.
21	6 52 36·8	52 31	I. Ec. R. *b*	Simms	105	E. I. P.
	6 52 4·3	52 31	I. Ec. R. *b*	Lerebours	106	N. R. P.
28	8 47 47·7	47 53	I. Ec. R. *b*	Simms	105	E. I. P.
	8 47 59·3	47 53	I. Ec. R. *b*	Lerebours	106	N. R. P.
Oct. 1	11 27 56·9	28 38	II. Ec. R. *b*	Simms	102	E. I. P.
	11 28 3·6	28 38	II. Ec. R. *h*	Lerebours	106	N. R. P.
2	6 41 40·4	41 49	III. Ec. D. *e*	Simms	73	„
5	7 7 50·2	8 0	I. Oc. D. *b m e*	„	160	E. I. P.
	7 7 34·9	8 0	I. Oc. D. *b m e*	Tulley	100	N. R. P.
14	7 7 38·8	7 33	I. Ec. R. *b*	Simms	102	C. A. P.
	7 7 47·8	7 33	I. Ec. R. *b*	Lerebours	106	N. R. P.

Date. 1870.	Madras Mean Time. By Obser.	By N. A.	Phenomenon and Phase.	Telescope. Maker.	Power.	Initials of Observer.
	h m s	m s				
Oct. 19	10 55 2·7	10 56	I. Oc. D. *b e*	Lerebours	160	E. I. P.
21	9 2 50·8	2 55	I. Ec. R. *b*	„	106	C. A. P.
	9 2 54·3	2 55	I. Ec. R. *b*	Tulley	100	N. R. P.
	10 8 1·2	13 0	IV. Oc. D. *b e*	Simms	163	E. I. P.
	10 8 8·0	13 0	IV. Oc. D. *b e*	Lerebours	160	N. R. P.
1880. Dec. 13	7 16 32·3	15 0	I. Oc. D. *e*	Simms	165	S. B. H.
	10 44 8·9	44 34	I. Ec. R. *b*	„	102	„
	10 44 13·1	44 34	I. Ec. R. *b*	Lerebours	105	E. I. P
1881. Feb. 5	7 0 46·0	0 0	I. Tr. I. *b m e*	„	160	E. I. P.
	8 14 24·9	10 0	I. Sh. I. *b e*	„	160	„

Date. 1862.	Observer.	Notes.
Feb. 15	N. R. P.	Faded away gradually for about 10ˢ.
18	„	Fading for about 4ᵐ, and uncertain for about 4ˢ.
March 3	„	First satellite fading for 9ˢ at eclipse.
	„	Second satellite about 1ᵐ in getting clear of the disk.
	„	The night windy and cloudy during the third observation.
10	„	Fading for about 23ˢ.
May 4	„	Increasing about 1ᵐ 15ˢ.
1866. Sept. 20	N. R. P.	Increasing for about 15ˢ.
1868. Oct. 28	N. R. P.	Second satellite increasing for about 1ᵐ 15ˢ.
	„	Last trace of shadow emersion of first satellite.
1869. Feb. 21	N. R. P.	Pretty sudden, but nearly 2ᵐ regaining ordinary brilliancy.
March 5	„	About 1ᵐ increasing to full brightness.
1870. Feb. 1	—	Neither observer satisfied.
9	N. R. P.	Faded away slowly for nearly 2ᵐ, and uncertain to about 10ˢ.
	J. P. B.	Uncertain to 5ˢ.
20	G. W. B.	Uncertain to about 17ˢ.
1871. Jan. 5	N. R. P.	Faint and slowly increasing for a minute or more.
21	„	Increasing through full 6ᵐ; uncertain to 10ˢ.
27	„	Full brilliancy of second satellite acquired in less than 3ᵐ after reappearance.
	N E. P.	First satellite about 3ᵐ entering on disk.
	N. R. P.	Uncertain to about 30ˢ.

Date. 1871.	Observer.	Notes.
Jan. 27	N. E. P.	Shadow immersion of first satellite : first and last contacts differed by 2ᵐ 38ˢ.
	N. R. P.	Uncertain to about 30ˢ.
	N. E. P.	Transit emersion of first satellite : first and last contacts differed by 2ᵐ 45ˢ.
28	N. R. P.	First contact and last disappearance differed by 52ˢ.
30	,,	Fading away for full 6ᵐ before.
Feb. 6	,,	Uncertain to about 30ˢ.
1874 Jan. 29	N. R. P.	Fading visibly for about 2ᵐ 30ˢ, and uncertain to about 3ˢ.
	E. I. P.	Uncertain to 2 or 3 seconds.
1875. April 7	N. R. P.	Transit emersion of first satellite : satellite sharp and white on large north belt.
13	E. I. P.	Transit immersion of second satellite : completely on disk 1ᵐ 57ˢ after.
	,,	Shadow emersion of second satellite : completely gone 1ᵐ 35ˢ later.
15	N. R. P.	First satellite fading away for about 1ᵐ 30ˢ before disappearance.
	,,	First satellite clear of the disk 58ˢ after first reappearance.
27	,,	At least 6ᵐ in acquiring its full light.
29	,,	About 3ᵐ increasing to full brightness.
May 4	,,	Third satellite increasing for 9ᵐ after reappearance.
6	,,	About 4ᵐ increasing to full brightness.
8	E. I. P.	About 3ᵐ ,, ,,
11	,,	Third satellite quite out a little over 4ᵐ later.
	N. R. P.	,, ,, ,,
15	E. I. P.	Transit egress of second satellite : first seen 3ᵐ 31ˢ before.
	N. R. P.	Increasing for 3ᵐ or 4ᵐ.
22	E. I. P.	Completely on disk 1ᵐ 45ˢ later.
23	N. R. P.	Completely gone 1ᵐ 30ˢ later.
1876. May 24	,,	Well out when first seen : 1ᵐ increasing.
25	E. I. P.	First contact and complete disappearance of third satellite differed by 14ᵐ 5ˢ.
	N. R. P.	The interval between the first contact and the complete disappearance of the third satellite was 16ᵐ 19ˢ.
	,,	Reappearance of third satellite : equal to fourth at 10ʰ 42ᵐ 14ˢ, and of full brightness at 10ʰ 46ᵐ 43ˢ.
26	E. I. P.	Increasing for about 4ᵐ.
27	,,	First and last contacts differed by 2ᵐ 12ˢ.

Date.	Observer.	Notes.
1876.		
June 11	N. R. P.	Increasing for about 3^m.
Aug. 3	E. I. P.	Second satellite fading for about 5^m.
	N. R. P.	Second satellite reappeared from occultation 2^m or 3^m earlier.
1878.		
Sept. 19	E. I. P.	First and last contacts at disappearance of first satellite differed by $2^m\,40^s$.
	N. R. P.	First and last contacts at disappearance of first satellite differed by $6^m\,58^s$. Sky hazy and definition bad.
	„	First satellite increasing for about $2^m\,30^s$ after reappearance.
21	E. I. P.	Increasing for about 3^m.
28	„	Increasing for about $2^m\,30^s$.
Oct. 1	„	„ $1^m\,30^s$.
14	„	Sky very hazy.
21	E. I. P.	First and last contacts differed by $7^m\,55^s$.
	N. R. P.	„ „ 6^m.
1880.		
Dec. 13	S. B. H.	First satellite last seen. Passing clouds about.
	„	Increasing for about $2^m\,30^s$.
	E. I. P.	„ „
1881.		
Feb. 5	E. I. P.	*Jupiter* too low for further observation.

Most of the observations were made by the Astronomer and various members of his family, and a few by friendly amateurs, who, being present at the time, kindly lent their assistance.

The observations, mostly recorded with sidereal clocks, have been reduced to Madras mean time, and the *Nautical Almanac* times of the various phenomena added in a separate column to show the difference between observation and theory. The small letters, *b* for beginning or first contact, *m* for middle or bisection, and *e* for end or last contact, have been used to signify the phase noted.

The instruments used were the eight-and-a-half inch Equatoreal by Messrs. Troughton & Simms; the six-and-a-quarter inch Equatoreal by Messrs. Lerebours & Secretan; and the excellent five-foot telescope by Tulley, originally constructed for Captain W. H. Smyth, R.N., in 1829, afterwards the property of J. Lee, Esq., LL.D. of Hartwell, and known as the Smythian Telescope. The magnifying powers employed are also given.

The names of the observers, indicated by their initials, were:—

Norman Robert Pogson, Esq., C.I.E., Government Astronomer.

Norman Everard Pogson, late Assistant Government Astronomer.

Elizabeth Isis Pogson, present Assistant Government Astronomer.

Charles Albion Pogson, an intelligent little boy of ten, who could record the clock very correctly.

Captain James Palladio Basevi, R.E., of the Great Trigonometrical Survey of India.

George W. Barclay, Esq., Assistant-Superintendent, Government Telegraph Department, in charge of Madras Office.

Robert Fellowes Chisholm, Esq., F.R.I.B.A., Consulting Architect to Government, Madras.

Major Bertie B. Hobart, R.A., Military Secretary to the Governor of Madras.

Surgeon-Major Samuel Bradshaw Hunt, Indian Medical Department.

The reductions of occultations of planets and stars, and of other casual phenomena observed at Madras since 1861, are in hand, and will follow shortly.

Madras Observatory:
    1881, *June* 9.

Observations of Jupiter's Satellites made at the Stonyhurst Observatory. By the Rev. S. J. Perry.

Satellite.	Phenomenon.	G.M.T. h m s	In excess of N.A.	Observer.	Remarks.
1881, Jan. 6 II.	Ec. R. First seen	9 53 14·2	−0 31·88	S. J. P.	Definition very bad, wind very high. Full brilliancy.
	Full light	58 37·7		"	¾ of III.
19 I.	Ec. R. First seen	9 32 2·4	+0 15·4	"	Good passing clouds.
	Full light	33 59·7		"	
31 II.	Ec. R. First seen	7 3 29·0	−·0 42·0	"	Very good. Auroral clouds.
	Full light	5 11·3		"	
Feb. 8 III.	Ec. R. First seen	6 11 13·0	−5 17·0	W. C.	Very good.
	Full light	16 32·8		"	
11 I.	Oc. D. Ext. cont.	6 25 49·5		"	Good. Thin clouds passing.
	Bisection	28 42·0		"	
	Last seen	30 53·5		"	
July 20 I.	Ec. D. Light fading	14 31 22·5		"	Very good. Dark screen in eyepiece.
	Last seen	37 45·3	+1 41·3	"	
Sept. 14 I.	Tr. I. First cont.	9 45 23·2		"	Poor. Very unsteady
	Bisection	46 24·8		"	
	Int. cont.	49 41·4		"	
19 II.	Oc. R. First seen	10 35 54·5		S. J. P.	Definition fair.
	Bisection	37 53·3		"	
	Last cont.	39 47·9		"	

Satellite.	Phenomenon.	G.M.T. h m s	In excess of N.A.	Observer.	Remarks.
1881, Oct. 8	I.	Oc. R. Bisection	8 59 58·3		W. C.	Good. Dancing slightly.
		Last cont.	9 2 6·8		"	
14	I.	Tr. I. Ext. cont.	11 19 28·7		"	Good, but unsteady.
		Bisection	24 6·4		"	
		Int. cont.	27 18·8		"	
15	I.	Ec. D. Fading	7 50 32·0		"	Very good. Definition rather poor.
		Half brilliancy	51 29·0		"	
		Last seen	52 5·5	+0 32·5	"	
	I.	Oc. R. Bisection	10 42 52·2		"	Very good. Definition excellent.
		Ext. cont.	45 40·0		"	
16	I.	Sh. E. Int. cont.	7 18 16·2		"	Observation only fair. Shadow very small. Definition poor and unsteady.
		Bisection	19 58·7		"	
		Last cont.	20 53·9		"	
	I.	Tr. E. Int. cont.	7 55 21·4		S. J. P.	Limb unsteady. Definition good.
		Bisection	58 37·7		"	
		Ext. cont.	8 1 37·9		"	
17	II.	Ec. D. Fading	16 3 57·2		W. C.	Very good. Dark screen used.
		Half brilliancy	5 25·7		"	
		Last seen	7 29·0	+0 21·0	"	

	Satellite.	Phenomenon.	G.M.T. h m s	In excess of N.A.	Observer.	Remarks.
1881, Oct. 29	I.	Ec. D. Fading	11 38 44·4		W. C.	Very good. Dark screen used.
		Half brilliancy	39 5·2		"	
		Last seen	40 49·2		"	
	I.	Oc. R. First seen	14 12 23·1		"	Very good. Dark screen used.
		Bisection	13 12·9		"	
		Ext. cont.	14 14·7		"	
30	I.	Sh. I. Bisection	9 1 35·3		"	Good. Sh. exceedingly small, but black.
		Int. cont.	3 33·8		"	
	I.	Tr. I. Ext. cont.	9 16 44·0		"	Very good. Definition excellent.
		Bisection	18 37·2		"	
		Int. cont.	21 3·3		"	
Nov. 20	II.	Tr. I. Ext. cont.	9 33 27·6		J. R.	Good.
		Bisection	37 11·2		"	
		Int. cont.	42 3·8		"	
	II.	Sh. I. Int. cont.	10 9 20·6		W. C.	
	II.	Tr. E. Int. cont.	12 9 36·9		J. R.	Thin clouds over planet at times. Light much diminished, but definition good.
		Bisection	12 37·4		'	
		Ext. cont.	16 42·9		"	
22	II.	Ec. R. First seen	7 33 47	−0 36·3	S. J. P.	Good. *About ⅔ of Sat. I.
		Half brilliancy	35 25·0		"	
		Full brilliancy*	36 44·5		"	

Satellite.		Phenomenon.	G.M.T. h m s	In excess of N.A.	Observer.	Remarks.
1881, Nov. 22	I.	Tr. I. Ext. cont.	8 51 26·5		S. J. P.	Definition rather poor. Sat. well seen on disk, but not on band.
		Bisection	55 51·3		"	
		Int. cont.	58 6·7		"	
	I.	Sh. I. Bisection	9 15 50·8		"	Shadow not very black. Definition rather poor.
		Full on	17 11·7		"	
25	III.	Tr. I. First cont.	9 51 42·1		W. C.	Sky hazy, but definition good.
		Bisection	58 7·7		"	
		Int. cont.	10 6 42·5		"	
29	II.	Oc. D. Ext. cont.	6 40 56·0		S. J. P.	Hazy. Definition poor.
		Bisection	44 0·0		"	
		Int. cont.	46 6·2		"	
	II.	Ec. R. First seen	10 8 18·2	−0 42·8	"	Very good observation. Definition very good.
		Half brilliancy	10 4·5		"	
		Full brilliancy	11 45·8		"	
	I.	Tr. I. Ext. cont.	10 38 46·0		"	Excellent. Very bright on disk.
		Bisection	41 1·3		"	
		Int. cont.	43 20·6		"	
	I.	Sh. I. First seen	11 6 12·4		"	Sh. very large and black. Very good observation.
		Bisection	7 5·7		"	
		Full on	2 31·8		"	

	Satellite.	Phenomenon.	G.M.T. h m s	In excess of N.A.	Observer.	Remarks.
1881, Nov. 29	I.	Tr. E. Bisection	12 51 55·9		S. J. P.	} Sky hazy.
		Ext. cont.	53 49·3		,,	
Dec.	I.	Tr. I. First cont.	⁚ 2 2·9		J. R.	} Definition good.
		Bisection	5 17·4		,,	
		Int. cont.	10 24·1		,,	
	I.	Sh. I. Bisection	5 35 14		S. J. P.	} Definition good. Observation fair. Few thin clouds.
		Full on	37 39·9		,,	
	I.	Tr. E. Int. cont.	7 10 40·6		,,	} Thin clouds passing.
		Bisection	14 17·8		,,	
		Ext. cont.	17 6·3		,	
13	III.	Oc. R. First seen	7 30 13·6		,,	} Definition good. Sky rather misty.
		Bisection	33 32·9		,,	
		Ext. cont.	38 1·2		,,	
	III.	Ec. D. Fading	9 3 52·5		,,	} Definition good.
		Half brilliancy	6 59·9		,,	
		Last seen	13 5·1	+4 1·0	,,	
	III.	Ec. R. First seen	10 36 8·7	−7 29·3	,,	} Definition good.
		Half brilliancy	40 27·3		,,	
		Full light	42 24·5		,,	

	Satellite.	Phenomenon.	G.M.T. h m s	In excess of N.A.	Observer.	Remarks.
1881, Dec. 13	II.	Oc. D. First cont.	11 11 16·5		W. C.	Definition good.
		Bisection	13 47·5		"	
		Last seen	18 12·5		"	
14	I.	Oc. D. Ext. cont.	11 17 52·9		J. R.	Definition good.
		Bisection	21 39·6		"	
		Last seen	23 45·4		"	
	I.	Ec. R. First seen	14 17 16·6	−0 3·4	"	Definition good. Sky rather misty.
		Full brilliancy	20 45·1		"	
22	II.	Tr. I. First cont.	8 1 0·5		"	Very good.
		Bisection	2 51·2		"	
		Int. cont.	4 32·0		"	
	I.	Tr. I. First cont.	10 24 6·2		J. R.	Very good.
		Bisection	25 53·9		"	
		Int. cont.	27 32·7		"	

Occultations of Stars by the Moon, seen at the Stonyhurst Obs. By the Rev. S. J. Perry.

	Phenomenon.	Moon's limb.	G.M.T.	Observer.	Remarks.
1881, Feb. 11	Disapp. 3 Cancri	Dark	7 52 47·5	W.C.	Very good. Dark screen in eyepiece.
	Reapp. "	Bright	9 8 4·2	"	Excellent. Dark screen in eyepiece.
May 4	Disapp. 5 Cancri	Dark	9 12 59·84	"	Very faint, but very good. Dark screen in eyepiece.
	Reapp. "	Bright	10 16 45·18	"	" "
July 19	Dip. 8 Arietis	"	12 30 4·40	"	Thin clouds. Good. Dark screen in eyepiece.
	Reapp. "	Drk	13 11 5 90	"	Excellent.
Sept. 14	Dip. η Tauri	Bright	11 35 9·70	"	Very good.
	Reapp. "	Dark	12 10 21·80	"	Haze round the Moon
Oct. 14	Dipp. f Geminorum	Bright	13 9 58·90	"	Excellent.
	Reapp. "	Dark	13 54 57·70	"	"
15	Disapp. 29 Cancri	Bright	13 43 21·60	"	"
	Rep. "	Dark	14 53 38·20	"	"
17	Disapp. 14 Sextantis	Bright	17 59 15·20	"	Very good.
Nov. 29	Disapp. 19 Piscium	Dark	11 24 7·70	S.J.P.	"

Computed and Observed Times of Contacts of the Transit of Mercury at Castlemaine, Victoria, Lat. 37° 4' 11" *S.; Long.* 9^h 36^m 55^s *E.* By W. Bone, Esq. M.D.

(*Communicated by the Secretaries.*)

First External Contact.   Local mean time.

            h m s   h m s
Computed, 19 53 18·6 ; Observed, 19 53 54·32.
First Internal Contact.
Computed, 19 55 1·4 ; Observed, 19 55 36·92.
Second Internal Contact.
Computed, 1 12 36·7 ; Observed, 1 12 36·5.
Second External Contact.
Computed, 1 14 19·4 ; Observed, 1 14 8·8.

Taken with 4·7-inch (Wray) Equatoreal. Power used, 200. Barometer, 29·25 (1,000 feet above sea); Thermometer, shade, 71° F.

Beading very noticeable at ingress. Observation taken for mean of beading, which was about 4^s: less noticeable at egress. The time of first external contact was most carefully noted, but shows a large difference from computed time. Sky clear at each observation.

The Satellites of Mars. By John Watson, Esq.

The evening of December 22, 1881, was very clear, and having seen from Mr. Marth's Ephemeris that both the satellites of *Mars* were favourably situated, I began at 7 P.M. to map the small stars in the neighbourhood of the planet as seen with my 12-inch Cooke Achromatic mounted equatoreally in the open air. I found a power of 235 to be as good as any, and by simply hiding the planet behind the sharply-defined edge of the diaphragm of a fine Huyghenian eyepiece, the more remote *Deimos* was seen following the planet at 8 P.M., and distant about two and a half diameters of *Mars* from the edge of the primary.

Various powers were tried up to 700 (with which this instrument works well at times), but the state of the atmosphere was such that no advantage was gained by increase of power above 235. At 8^h 40^m I saw *Phobos* for the first time preceding the planet, and distant, by careful estimate, one diameter of *Mars* from the edge of the planet's disk. Later in the evening the atmosphere turned hazy.

I send this communication because very few observations of these objects are recorded in this county, and it was a matter of interest to have watched them alternately on opposite sides of the planet, and without any special contrivance.

Vane House, Seaham Harbour :
  1882, *January* 4.

Errata.

Vol. xlii. p. 40, *for* July 29 *read* July 31.
Vol. xlii. p. 52, heading of fourth column, *for* Parallactic Inequality apparently corrected, *read* Parallactic Inequality approximately corrected.
Also, in the fourth column of the table upon the same page,

opposite group	1834–1842 *for*	125·11 *read*	123·88
„	1839–1847 „	123·88 „	123·67
„	1843–1851 „	123 67 „	125·26
	1847–1851 „	125·26 „	126·42

MONTHLY NOTICES

ROYAL ASTRONOMICAL SOCIETY.

VOL. XLII. FEBRUARY 10, 1882. NO. 4.

J. R. HIND, Esq., F.R.S., President, in the Chair.

Herbert J. Bell, Esq., Royal Alfred Observatory, Mauritius; Adam Hilger, Esq., 192 Tottenham Court Road, W.; and The Rev. Henry George Bonavia Hunt, Trinity College, Mandeville Place, W.;

were balloted for, and duly elected Fellows of the Society.

REPORT OF THE COUNCIL TO THE SIXTY-SECOND ANNUAL GENERAL MEETING OF THE SOCIETY.

The following table shows the progress and present state of the Society :—

	Compounders	Annual Subscribers	Non-resident	Mathematical Society	Total Fellows	Associates	Patron	Grand Total
December 31, 1880 ...	220	363	3	5	591	41	1	633
Since elected	+3	+19	...	...	...	+5	...	...
Deceased	−4	−6	...	...	...	−3	...	...
Removals	+3	−3	...	...	...	...	...	...
Resigned	...	−8	...	...	...	...	...	...
Expelled	...	−1	...	...	...	...	...	...
December 31, 1881 ...	222	364	3	5	594	43	1	638

M

Mr. Barrow's Account as Treasurer of the Royal

RECEIPTS.

	£	s.	d.	£	s.	d.
Balance at Bankers, Jan. 1, 1881	317	14	10			
„ in hand of Assistant Secretary on account of Turnor Fund	12	5	6			
				330	0	4
Dividend on £7,500 Consols	109	4	5			
„ £5,700 New 3 per cent. Stock ...	83	7	3			
„ £7,500 Consols	110	3	2			
„ £5,700 New 3 per cent. Stock ...	83	14	5			
				386	9	3
Received on account of Subscriptions :						
Arrears	154	7	0			
257 Contributions for 1881	539	14	0			
5 Contributions for 1882	10	10	0			
24 Admission Fees	50	8	0			
21 First Contributions	30	9	0			
				785	8	0
6 Composition Fees				126	0	0
Sales of Publications :						
At Society's Rooms, 1881	56	15	11			
At Williams & Norgate's, 1880	76	8	5			
				133	4	4
				£1,761	1	11

Astronomical Society, from Dec. 31, 1880, *to Dec.* 31, 1881.

EXPENDITURE.

	£	s.	d.	£	s.	d.
Salaries:						
Editor of *Monthly Notices*	60	0	0			
Assistant Secretary	225	0	0			
				285	0	0
Income Tax and House Duty				10	10	0
Fire Insurance				7	16	6
Printing: Spottiswoode & Co.	504	13	6			
„ H. Richardson...	1	6	8			
				506	0	2
Lithography and Engraving				25	0	3
Turnor Fund: Purchase of Books for Library ...				33	6	1
Binding Books in Library				43	0	3
Lee Fund: Mrs. Harris				10	0	0
House Expenses	37	11	0			
Wages	17	14	0			
Stamps and postage	54	14	0			
Carriage of books and parcels	3	6	6			
Stationery and office expenses	4	0	4			
Expenses of meetings	22	10	0			
Coals and gas	40	18	8			
Fittings, repairs, &c.	12	16	10			
Sundries	5	10	4			
				199	1	6
Mrs. Jackson-Gwilt's annuity				8	19	0
Cheque book, and deductions on cheques ...					11	0
Due to Assistant Secretary on Petty Cash account Jan. 1, 1881						
Balance at Bankers', credited in pass book, Dec. 31, 1881	601	5	1			
Country Cheque not credited	2	2	0			
Balance in hand of Assistant Secretary on account of Turnor Fund	13	2	9			
on Petty Cash account	10	3	2			
				626	13	0
				£1,761	1	11

Examined and found correct, Jan. 10, 1882.

ROBT. J. LECKY,

J. RAND CAPRON.

Assets and present property of the Society, January 1, 1882 :—

	£	s.	d.	£	s.	d.
Balance at Bankers', Dec. 31, 1880, as credited in pass book	601	5	1			
Country Cheque not credited	2	2	0			
Balance in hand of Assistant Secretary on account of Turnor Fund	13	2	9			
„ on Petty Cash account	10	3	2			
				626	13	0
Due on account of subscriptions :						
5 Contributions of 4 years' standing ...	42	0	0			
10 „ 3 „ ...	63	0	0			
28 „ 2 „ ...	117	12	0			
64 „ · ...	134	8	0			
Various amounts	12	12	0			
	369	12	0			
Less 5 Contributions paid in advance ...	10	10	0			
				359	2	0
Due for Publications from Messrs. Williams & Norgate (for sales during 1881)				48	10	4

£7,500 Consols, including the Lee Fund (£300), the Turnor Fund (£450), and the Horrox Memorial Fund (£100).

£5,700 New 3 per cent. Stock, including Mrs. Jackson-Gwilt's gift (£300).

Astronomical and other MSS., Books, Prints, Instruments, &c.

Unsold Publications of the Society.

Four Gold Medals.

Report of the Auditors.

We, being two of the duly appointed Auditors, beg to lay before this General Meeting of the Royal Astronomical Society the following Report :—

1. We have examined the Treasurer's account, and an account of the assets and property of the Society, and have found and certified the same to be correct.

2. The receipts and expenditure for the past year are as stated in the Treasurer's account.

3. The cash in hand on December 31, 1881, including the balance at the bankers', amounted to 626*l.* 13*s.*

4. The funded property of the Society is the same as at the end of last year, and is in a satisfactory state, and the books,

instruments, and other effects have been examined and found in a satisfactory condition, so far as their safe keeping is concerned.

5. We have laid on the table a list of the names of those Fellows who are now in arrear for sums due at the last Annual General Meeting, with the amount due against each Fellow's name.

ROBT. J. LECKY,
J. RAND CAPRON.

Stock in hand of volumes of the *Monthly Notices* :—

Vol.	At Society's Rooms	At Williams & Norgate's	Vol.	At Society's Rooms	At Williams & Norgate's
I.	77	1	XXIII.	31	...
II.	77	2	XXIV.	24	...
III.	...	...	XXV.	7	...
IV.	...	...	XXVI.	10	...
V.	...	...	XXVII.	3	...
VI.	44	1	XXVIII.	75	1
VII.	2	...	XXIX.	55	2
VIII.	141	2	XXX.	68	4
IX.	24	3	XXXI.	99	2
X.	177	1	XXXII.	122	...
XI.	186	1	XXXIII.	106	...
XII.	12	2	XXXIV.	83	2
XIII.	152	3	XXXV.	66	3
XIV.	110	3	XXXVI.	39	...
XV.	127	2	XXXVII.	41	4
XVI.	110	3	XXXVIII.	105	3
XVII.	137	1	XXXIX.	108	3
XVIII.	167	...	XL.	122	3
XIX.	60	...	XLI.	125	5
XX.	31	...	Index to Monthly Notices	594	...
XXI.	19	...			
XXII.	34	...			

In addition to the above volumes of the *Monthly Notices*, the Society has a considerable stock of separate numbers of nearly all the volumes. With the exception, however, of Vols. XXXVI. to XLI. no complete volumes can be formed from the separate numbers in stock.

Stock in hand of volumes of the *Memoirs* :—

Vol.	At Society's Rooms	At Williams & Norgate's	Vol.	At Society's Rooms	At Williams & Norgate's
L. Part 1	6	...	XXIV.	163	...
L. Part 2	42	...	XXV.	175	...
II. Part 1	56	...	XXVI.	179	1
II. Part 2	22	...	XXVII.	432	...
III. Part 1	71	...	XXVIII.	392	...
III. Part 2	90	...	XXIX.	418	...
IV. Part 1	83	3	XXX.	169	...
IV. Part 2	91	3	XXXI.	149	2
V.	109	4	XXXII.	166	2
VI.	128	3	XXXIII.	172	1
VII.	153	3	XXXIV.	171	8
VIII.	132	3	XXXV.	112	7
IX.	139	3	XXXVI. (with M.N.)	206	12
X.	151	...	XXXVI. (without)	4	...
XI.	159	...	XXXVII. Part 1	353	9
XII.	166	...	XXXVII. Part 2	301	8
XIII.	173	...	XXXVIII.	292	2
XIV.	376	3	XXXIX. Part 1	264	4
XV.	145	1	XXXIX. Part 2	270	5
XVI.	172	...	XL.	297	2
XVII.	153	2	XLI.	468	3
XVIII.	155	1	XLII.	263	3
XIX.	158	1	XLIII.	277	2
XX.	160	2	XLIV.	275	4
XXI. Part 1	314	...	XLV.	365	2
XXI. Part 2	99	...	XLVI.	738	...
XXI. 1 & 2 (together)	66	1	Index to Memoirs	659	2
XXII.	160	1			
XXIII.	155	1			

Instruments belonging to the Society.

No. 1. The *Harrison* clock.
 „ 2. The *Owen* portable circles, by Jones.
 „ 3. The *Beaufoy* circle.
 „ 4. The *Beaufoy* transit instrument.
 „ 5. The *Herschel* 7-foot telescope.

No. 6. The *Greig* universal instrument, by Reichenbach and
    Ertel. The transit telescope, by Ultzschneider and
    Fraunhofer, of Munich.

  „   7. The *Smeaton* equatoreal.

  „   8. The *Cavendish* apparatus.

  „   9. The 7-foot Gregorian telescope (late Mr. Shearman's).

  „ 10. The variation transit instrument (late Mr. Shear-
       man's).

  „ 11. The universal quadrat, by Abraham Sharp.

  „ 12. The *Fuller* theodolite.

  „ 13. The standard scale, by Troughton and Simms.

  „ 14. The *Beaufoy* clock, No. 1.

  „ 15. The *Beaufoy* clock, No. 2.

  „ 16. The *Wollaston* telescope.

  „ 17. The *Lee* circle.

  „ 18. The *Sharpe* reflecting circle.

  „ 19. The *Brisbane* circle.

  „ 20. The *Baker* universal equatoreal.

  „ 21. The *Reade* transit.

  „ 22. The *Matthew* equatoreal, by Cooke.

  „ 23. The *Matthew* transit instrument.

  „ 24. The *South* transit instrument.

  „ 25. A sextant, by Bird (formerly belonging to Captain
       Cook).

  „ 26. A globe showing the precession of the equinoxes.
       The *Sheepshanks* collection :—

  „ 27. (1) 30-inch transit instrument, by Simms, with level
       and two iron stands.

  „ 28. (2) 6-inch transit theodolite, with circles divided
       on silver ; reading microscopes, both for altitude
       and azimuth ; cross and siding levels ; magnetic
       needle ; plumbline ; portable clamping foot and
       tripod stand.

  „ 29. (3) $4\frac{4}{10}$-inch achromatic telescope, about 5 feet 6
       inches focal length ; finder ; rack motion ; double-
       image micrometer ; two other micrometers ; object-
       glass micrometer ; one terrestrial and ten astro-
       nomical eyepieces, applied by means of two
       adapters ; equatoreal stand, and clock movement.

  „ 30. (4) $3\frac{1}{4}$-inch achromatic telescope, with equatoreal
       stand ; double-image micrometer ; one terrestrial
       and three astronomical eyepieces.

  „ 31. (5) $2\frac{3}{4}$-inch achromatic telescope, with stand ; one
       terrestrial and three astronomical eyepieces.

  „ 33. (7) 2-foot navy telescope.

  „ 34. (8) Transit instrument of 45 inches focal length ;
       with iron stand, and also Ys for fixing to stone
       piers ; two axis levels.

  „ 35. (9) Repeating theodolite, by Ertel, with folding
       tripod stand.

No. 36. (10) 8-inch pillar sextant, by Troughton, divided on platinum, with counterpoise stand and artificial horizon.

„ 37. (11) Portable zenith telescope and stand, 2¾-inch aperture and 26 inches focal length; 10-inch horizontal circle and 8-inch verticle circle, read to 10″ by two verniers to each circle.

„ 38. (12) 18-inch Borda repeating circle, by Troughton, 2½-inch aperture and 24 inches focal length; the circles divided on silver, the horizontal circle being read by four verniers, and the vertical circle by three verniers, each to 10″.

„ 39. (13) 8-inch vertical repeating circle, with diagonal telescope, by Troughton and Simms; circle divided on silver, reading to 10″; a 5-inch circle at eye-end reading to single minutes; horizontal circle 9 inches diameter in brass, reading to single minutes.

„ 40. (14) A set of surveying instruments, consisting of a 12-inch theodolite for horizontal angles only, reading to 10″; two sets of adjusting plates; tripod stand with enclosed telescope; heavy stand for theodolite; Y piece of level; two large and three small ground-glass bubbles divided; level collimator, object-glass 1⅝-inch diameter and 16 inches focal length; micrometer eyepiece, comb, and wires; mercury bottle and trough.

„ 41. (15) Level collimator with object-glass 1⅞-inch diameter and 16 inches focal length; stand, rider-level, and fittings.

„ 42. (16) 10-inch reflecting circle, by Troughton, reading by three verniers to 20″; counterpoise stand; artificial horizon with mercury; two tripod stands.

„ 43. (17) Hassler's reflecting circle, by Troughton, with counterpoise stand.

„ 44. (18) 6-inch reflecting and repeating circle, by Troughton and Simms, contained in three boxes, two of which form stands. Circle divided on silver, reading to single minutes; two inside arcs divided to single degrees, 150 degrees on each side; artificial horizon and mercury.

„ 45. (19) 5-inch reflecting and repeating circle, by Lenoir, of Paris.

„ 46. (20) Reflecting circle by Jecker, of Paris, 11 inches in diameter, with one vernier reading to 15″.

„ 47. (21) Box sextant; reflecting plane and level.

„ 48. (22) Prismatic compass, by Troughton and Simms.

„ 49. (23) Mountain barometer.

„ 50. (24) Prismatic compass, by Thomas Jones, mounted with a cylindrical lens.

„ 51. (25) Ordinary 4½-inch compass with needle.

No. 52. (26) Dipping needle, by Robinson.

„ 53. (27) Compass needle, mounted for variation.

„ 54. (28) Magnetic intensity needle, by Meyerstein, of Göttingen; a strongly fitted brass box with heavy magnet; filar suspension.

„ 55. (29) Box of magnetic apparatus.

„ 56. (30) Hassler's reflecting circle, by Troughton; a 10½-inch reflecting and repeating circle, with stand and counterpoise, divided on platinum with two movable and two fixed indices; four verniers reading to 10″.

„ 57. (31) Box sextant and glass plane artificial horizon, by Troughton and Simms.

„ 58. (32) Plane 2⅜-inch speculum, artificial horizon, and stand.

„ 59. (33) 2½-inch circular level horizon, by Dollond.

„ 60. (34) Artificial horizon, roof, and trough; the trough 8¼ by 4½ inches : tripod stand.

„ 61. (35) Set of drawing instruments, consisting of 6-inch circular protractor and common protractor, T-square : one beam compass.

„ 62. (36) A pentagraph.

„ 63. (37) A noddy.

„ 64. (38) A small Galilean telescope with object-glass of rock crystal.

„ 65. (39) Five levels.

„ 66. (40) 18-inch celestial globe.

„ 67. (41) Varley stand for telescope.

„ 69. (43) Telescope, with the object-glass of rock crystal.

„ 70. Portable equatoreal stand.

„ 71. Portable altazimuth tripod.

„ 72. Four polarimeters.

„ 74. Registering spectroscope, with one large prism.

„ 76. Two five-prism direct-vision spectroscopes.

„ 78. 9¼-inch silvered-glass reflector and stand, by Browning.

„ 79. Spectroscope.

„ 80. A small box, containing three square-headed Nicol's prisms; two Babinet's compensators; two double-image prisms; three Savarts; one positive eyepiece, with Nicol's prism; one dark wedge.

„ 81. A back-staff, or Davis' quadrant.

„ 82. A nocturnal or star dial.

„ 83. An early non-achromatic telescope, of about 3 feet focal length, in oak tube, by Samuel Scatliffe, London.

„ 84. A Hollis observing chair.

„ 85. Double image micrometer, by Troughton and Simms.

„ 86. 4½-inch Gregorian reflecting telescope, by Short,

with altazimuth stand and 6-inch altitude and azimuth circles and two eyepieces.

No. 87. 3¼-inch Gregorian reflecting telescope with wooden tripod stand.

„ 88. Pendulum with 5-foot brass suspension rod, working on knife edges, by Thomas Jones.

„ 89. A Rhabdological Abacus. A contrivance invented by Mr. H. Goodwyn, consisting of a box filled with compartments, in which are square rods covered with numbers, which can be arranged so as to facilitate the labour of multiplying high numbers.

„ 90. An Arabic celestial globe of bronze, not quite 6 inches in diameter.

„ 91. Astronomical time watchcase, by Professor Chevallier.

„ 92. 2-foot protractor, with two moveable arms, and vernier.

„ 93. Beam compass, in box.

„ 94. 2-foot navigation scale.

„ 95. Stand for testing measures of length.

„ 96. Artificial planet and star, for testing the measurement of a fixed distance at different position-angles.

„ 97. 12-cell Leclanché battery.

„ 98. 2 feet 6 inch navy telescope with object-glass 2½ inches, by Cooke, with portable wooden tripod stand.

„ 99. 12-inch transit instrument, by Fayrer & Son, with level and portable stand.

„ 100. 9-inch transit instrument, with level and iron stand.

„ 101. Small equatoreal sight instrument, by G. Adams, London.

„ 102. Sun-dial, by Troughton.

„ 103. Sun-dial, by Casella.

„ 104. Sun-dial.

„ 105. Box sextant, by Troughton and Simms.

„ 106. Prismatic compass, by Schmalcalder, London.

„ 107. Compass, by C. Earle, Melbourne.

„ 108. Prismatic compass, by Negretti and Zambra.

„ 109. Dipleidoscope, by E. Dent.

„ 110. Abney level, by Elliott.

„ 111. Pocket spectroscope, by Browning.

„ 112. Small brass astrolabe.

„ 113. Double sextant, by Jones.

„ 114. Two models, illustrating the effects of circular motions.

„ 115. A cometarium.

„ 116. A pair of 18-inch globes.

The following instruments are lent, during the pleasure of the Council, to the undermentioned persons :—

No. 4. The *Beaufoy* transit instrument, to the Observatory, Kingston, Canada.

 „ 12. The *Fuller* theodolite, to the Director of the Sydney Observatory.

 „ 22. The *Matthew* equatoreal, to Mr. Brett.

 „ 23. The *Matthew* transit, to Captain Noble.

 „ 74. Registering spectroscope, with prism, to Mr. Lecky.

 „ 78. The 9¼-inch reflector, to Mr. Neison.

From the *Sheepshanks* collection :—

No. 30. (4) 3¼-inch equatoreal and stand, to Mr. Sadler.

 „ 34. (8) Transit instrument, to the Rev. Professor Pritchard.

 „ 35. (9) Repeating theodolite, to the Sydney Observatory.

 „ 69. (43) Telescope, with rock-crystal object-glass, to Dr. Huggins.

During the past year the instruments which are at present in the Society's rooms have been examined by a Committee appointed by the Council. The Committee have compared the instruments with the printed list, and have reported them to be in a satisfactory condition.

The Library.

The cataloguing of the books in the library has been steadily continued, and it is expected that the work will be completed in the course of the present year.

The Gold Medal.

The Council have awarded the Society's Gold Medal to Mr. David Gill, for his Heliometer Observations of *Mars* at Ascension, and for his discussion of the results. The President will lay before the Society the grounds upon which this award has been founded.

Publications of the Society.

Vol. XLVI. of the *Memoirs* has been published during the past year. It contains the following papers :—

David Gill. Account of a determination of the solar parallax from observations of *Mars*, made at Ascension in 1877.

A. A. Common. Particulars of the mounting of a 3-foot reflector.

George M. Seabroke. Third catalogue of micrometrical

measures of double stars made at the Temple Observatory, Rugby.

A. C. Ranyard. Observations of the total solar eclipse of 1878, July 29, made at Cherry Creek Camp, near Denver, Colorado.

OBITUARY.

The Council regret that they have to record the loss by death of the following Fellows and Associates during the past year :—

Fellows :—G. S. Almond.
　　　　　Célestin Baume.
　　　　　W. R. Birt.
　　　　　J. A. Cockburn.
　　　　　Samuel Courtauld.
　　　　　Rev. J. M. Heath.
　　　　　Rev. W. H. Hennah.
　　　　　Thomas Hopkirk.
　　　　　H. W. Jeans.
　　　　　C. H. Pinches.
　　　　　Richard Webster.

Associates :—Carl Bruhns.
　　　　　　Alfred Gautier.

PIERRE JOSEPH CÉLESTIN BAUME was born in 1819 and was one of the founders of the well-known house of Baume Brothers, at Les Bois, Switzerland. The business of the house was the manufacture of a superior class of Geneva watches, and, having made continued improvements, Mr. Baume conceived the idea of submitting his watches to the judgment of the English firms, and in 1846 came to England for this purpose. He was one of the first to introduce Geneva watches into this country. His success exceeded his utmost expectations, and a business was immediately established in London, which is still carried on.

He was connected with various philanthropic and other societies, and in 1861 was elected a member of the Society of Arts. He was one who helped to found the French Hospital for foreigners in London, which was opened in 1867, and with which he was intimately connected until the day of his death.

He died suddenly at his residence in London on September 27, 1880. He was elected a Fellow of the Society on May 8, 1863.

WILLIAM RADCLIFF BIRT was born on July 15, 1804. His first writings were astronomical, some observations upon the period of the variable star β *Lyræ* having been communicated by him to this Society so long ago as 1830, and published in the first volume of the *Monthly Notices.*   He also communicated subsequently to the Society some observations on the variability of a *Cassiopeiæ.*

In the years 1839–43 Mr. Birt was employed by Sir John Herschel in the reduction, arrangement, and projection, of the numerous series of barometric observations that had been collected by him, and the results of which formed a series of Reports to the British Association on the reduction of meteorological observations.   In the Report for 1843 Sir John Herschel traced two well-defined atmospheric waves which passed over the British Isles and the West of Europe—one in September 1836, and the other in December 1837.   At the conclusion of the Report he also noticed the large fluctuations which Mr. Birt had observed in 1842, especially the symmetrical wave which occupied thirteen days in November for its complete rise and fall; and the British Association entrusted Mr. Birt—under the direction of the magnetical committee and the immediate superintendence of Sir John Herschel—with the investigation of these waves, and especially that of November.   This led to the publication by Mr. Birt of five Reports on atmospheric waves, in the volumes of the British Association for 1844–48.   In 1853 he published his *Handbook on the Law of Storms : being a Digest of the principal Facts of Revolving Storms,* which was intended for the use of captains of ships.   A second edition was issued in 1879.

At the meeting of the British Association at Swansea in 1848 Mr. Birt was requested to undertake the reduction and discussion of the electrical observations made at Kew, and the results formed a Report of nearly ninety pages in length, which was published in the volume of the Association for 1849.

About 1859 he published several notes and papers upon the illumination of *Geminus* and other lunar craters ; and his name will always be connected chiefly with his observations upon the surface of the Moon.   He was the author of five Reports upon mapping the surface of the Moon, which were published in the volumes of the British Association for 1865–69.   In these he employed a notation of his own for the symbolising and cataloguing of objects, his desire being to afford unmistakable identification of the smaller features, such as minute craters, mountains, rills, &c.   In the last Report (1869) he thus referred to the work accomplished: " Four areas of the Moon's surface, each of 5° in extent, both of longitude and latitude, have been carefully and critically surveyed, not so much for the determinations of positions as in an examination of the physical aspects of 100 square degrees of the Moon's surface by means of the comparison and measurement of photograms, combined with observation at the telescope, by several observers in concert. Outlines of the objects thus surveyed have been laid down on the orthographic projection, on a scale of 200 inches to the Moon's diameter.   The area thus surveyed includes 443 objects."

At the conclusion of the Report it is stated that the number of objects on the Moon's surface registered in accordance with the plan proposed in the first Report (1865) is 2099, of which "769 only have been published—viz. 492 in the Reports of the Committee, and 277 in Mr. Birt's monograph on the 'Mare Serenitatis.'"

He devoted much attention to the question of lunar activity and the detection of change on the Moon's surface, and the discussions relative to the crater *Linné* and the spots on *Plato* will still be well remembered. From about 1861 till Dr. Lee's death he made frequent use, at the invitation of Dr. Lee, of the Hartwell Equatoreal for the purpose of lunar observation. He was a frequent writer, up to the time of his death, on lunar matters, and was a constant contributor to the *Astronomical Register*. He collected together in 1874 under the title *Contributions to Selenography* several notices and monographs on lunar formations, which he had published by subscription during the preceding four years. When the Selenographical Society was founded, four years ago, he was elected its first president, and was re-elected every succeeding year. He did not observe, himself, after 1877, owing to age and weakness. He died on December 14, 1881, and, although his weakness had been steadily increasing, his death was unexpected and almost sudden.

He was elected a Fellow of the Society on January 14, 1859. About two years ago he presented to the Society twelve manuscript volumes containing the portion already completed of the lunar catalogue of the British Association Committee.

SAMUEL COURTAULD, of Gosfield Hall, Essex, was born at Albany, in the State of New York, on June 1, 1793, but his parents, who had gone to the United States to establish some manufacture, brought him to England in his infancy. Following in the steps of his father, who had been one of the first to introduce the industry of "silk-throwing" into Essex, Mr. Courtauld was able, by his genius and perseverance, to develop his business of a silk-throwster into the more difficult one of the manufacture of crape; and the present extensive factories founded by him now form a not unimportant part of British silk-manfacture.

Mr. Courtauld was possessed of considerable mechanical ingenuity, and with a power of research which enabled him to thoroughly exhaust his subject. Though a self-educated man, he had a profoundly scientific mind, capable of acute and sound reasoning; and without being versed in details, he delighted in the discussion of the theories of astronomy and physical science. It was this intellectual power, coupled with a vigorous understanding and an indomitable will, that made him so successful a man in all he undertook.

Mr. Courtauld's name is historically connected with the

well-known agitation for the abolition of church rates, referring to which, the *Times* remarked that "had his death happened some thirty or forty years ago a popular hero would have passed away, but he had lived to be almost forgotten, reposing as he had done for nearly thirty years on his laurels."

He died March 21, 1881, after two months' illness, in the eighty-eighth year of his age.

He was elected a Fellow of the Society on November 8, 1867.

WILLIAM HENRY HENNAH was born at Dalston in 1848, and was educated at private schools. He entered King's College, London, in 1871. His profession was that of a schoolmaster. Much of his spare time was devoted to astronomy, and, as he possessed a good telescope he was enabled to interest his friends in the subject, upon which also he lectured. After an acute attack of rheumatic fever, which lasted only ten days, he died on September 25, 1881.

He was elected a Fellow of the Society on May 9, 1873.

THOMAS HOPKIRK was born in London on August 16, 1819. His father, who had seen service in both the Royal and Merchant Navies, wished his son to follow a naval career, and he was accordingly placed under Mr. Riddle, as tutor, from whom he acquired the knowledge of mathematics that he afterwards found so useful to him. He spent two or three years at sea, but soon acquired a dislike for a nautical life and settled down at home as a mathematical tutor. His principal occupation for many years was the preparation of youths for all branches of the military service, but chiefly for the Artillery and Engineers. As a tutor he was very successful, and his reputation was well established. Almost worn out by hard work, he retired from his professional labours in 1861, at the early age of 42, and spent several years in foreign travel. He died, after a prolonged illness, at Norwood, on March 26, 1881. His death was hastened by an accidental fall, which induced paralysis.

He was elected a Fellow of the Society on December 14, 1849.

HENRY WILLIAM JEANS was born at Portsea in 1804. He left school at the early age of thirteen, and was articled to a solicitor in that town. In 1824 the late Dr. Inman, of the Royal Naval College at Portsmouth, appointed him to take charge of the chronometers in the Observatory at the Dockyard; and a few years afterwards he was made assistant-master in the College. When the College was abolished in 1837 he proceeded to Cambridge, and entered as a pensioner at St. John's College. In the College examination in 1838 he was placed in the first class. Shortly afterwards the College at Portsmouth was re-established, and Mr. Jeans' services were required, so that he left Cambridge without taking a degree. During part of the time he was at Cambridge, he had to examine the officers in mathematics, and this necessitated a journey to Portsmouth every month. He was afterwards appointed mathematical master in the College at Portsmouth, and held this post till 1866, when he retired to Langstone House, near Havant, where he resided till his death. For some time he was mathematical master in the Royal Military Academy at Woolwich, and he was an examiner of merchant officers in nautical astronomy under the Trinity Board. At Langstone he built and endowed a small chapel, which is now connected with the Rectory at Havant. He died on March 23, 1881.

He was the author of the following works, which were chiefly intended for the use of naval students :—(1) *Plane and Spherical Trigonometry.* Two Parts, Portsea, 1842. A second edition appeared in 1847-8, and a sixth edition of Part I. in 1873. (2) *Problems in Astronomy, Surveying, and Navigation* (1849). (3) *The Theory of Nautical Astronomy and Navigation* (1853). New editions appeared of Part I. in 1870, and of Part II. in 1868. (4) *Handbook of the Stars*, of which the third edition appeared in 1868.

He was elected a Fellow of the Society on March 13, 1840.

CONRAD HUME PINCHES was born in January 1820, and was the son of Mr. William Pinches, who, for more than forty years, conducted a school in Ball Alley, Lombard Street. The enthusiastic devotion to his work which the father displayed communicated itself to his sons, three of whom—Dr. C. H. Pinches, Mr. William Pinches, and Mr. Edward E. Pinches—entered the scholastic profession, and were at one time simultaneously at the head of three schools in different parts of London, containing in the aggregate more than 400 pupils.

Dr. C. H. Pinches was educated at his father's school, which he left in 1836; and in 1838 he was junior master in a school at Pentonville. In 1840 he became one of the assistant masters in the Clapham Grammar School, at the head of which was the Rev. Charles Pritchard, now Savilian Professor of Astronomy at Oxford. To his connection with this school Dr. Pinches always attributed much of his success, and a warm and intimate friendship sprang up between him and Mr. Pritchard which was broken only by death.

At the end of 1843 Dr. Pinches, acting on the advice of Mr. Pritchard, began work as a schoolmaster on his own account, taking possession of Clarendon House, Kennington Road, at that time a small school containing about twenty boys. The school rapidly increased, and new class- and lecture-rooms, and a laboratory, &c., were built. Dr. Pinches continued to conduct this school with great success till his retirement in 1871. The school was generally full, and contained sometimes over 150 pupils. When the Oxford Local Examinations were established, in 1857, he took a leading part in establishing a committee for making the necessary arrangements for conducting these examinations in London, and undertook for a time the laborious duties of honorary secretary. Although not one of the original founders of the College of Preceptors, he joined it very shortly after its establishment, and was closely connected with it until his death. He was elected treasurer in 1873.

Between 1863 and 1866 he kept terms at the Middle Temple, and was called to the bar. He was the author of a manual on Elocution.

In 1850 he married Sarah Ann East, who died a few months before him. Five sons and two daughters survive him.

He was elected Fellow of the Society on May 13, 1859, and was a regular attendant at the evening meetings.

CARL CHRISTIAN BRUHNS was born at Ploen, in Holstein, on November 22, 1830. He was the son of a locksmith, and was sent to the school of his native place, it being intended that he should be a mechanician. In the spring of 1851 he went to Berlin. His desire for knowledge, which in his boyhood had been checked by bad health, here found more scope. Through some pieces of work which as a mechanician he had to perform at the Berlin Observatory he became known to Encke the Director, who soon found out his extraordinary capacity as a computer. Encke supplied him with the means of further improving himself; and by strenuous efforts he in a very short time supplied the chief deficiencies in his knowledge, and entered the university as a student.

In 1852 Encke appointed him Second Assistant in the Observatory, and in 1854, on Galle's leaving, he was made First Assistant. He graduated at the university in 1858, the title of his dissertation being *De planetis minoribus.* He won also an academic prize with an essay on astronomical refraction; and in 1858 became a Privatdocent in the university. In 1860, at Encke's suggestion, he was appointed Extraordinary Professor of Astronomy at Leipzig in succession to D'Arrest; and when Möbius died he was made Ordinary Professor and Director of the Observatory. The existing observatory being inadequate, Bruhns was commissioned by the Government to build a new one, and selected the site and prepared the plans for the present Observatory, which was built under his supervision. The numerous observations made at this Observatory, both by Bruhns and his assistants, bear witness to the energy which he devoted to it. He discovered six comets, and applied his remarkable talent for calculation to the determination of the orbits of many comets and minor planets. During his twenty years' professorship he had as his pupils many astronomers whose names are already famous.

When in 1862 General von Baeyer founded the geodetical survey of Central Europe, Bruhns, with Nagel and Weisbach as colleagues, was appointed Commissioner for Saxony; and the latitudes and longitudes of the trigonometrically important points in Saxony were determined under his direction. Up to the time of his death he took the greatest interest in geodetical work, to which he devoted much of his time.

The establishment of a network of meteorological observations, extending over the whole of Saxony, was entirely his work. Eleven volumes of results testify to his great activity in this direction. Recognising that a science like meteorology could only be really advanced by the united cooperation of civilised countries, he urged, and successfully carried out, his project for an International Meteorological Committee. His last creative effort in meteorology was the establishment in Leipzig of the Bureau for Weather Prognostics. Bruhns took the most active interest in geographical science. He was also desirous that science should be diffused as much as possible, and made many popular communications upon astronomy and meteorology to different societies.

Besides his numerous astronomical papers, which consist chiefly of observations or calculations, his editorship of geodetical publications, and his eleven volumes of meteorological results, he published a history and account of the Leipzig Observatory, a Life of Encke, and other works. Conjointly with several others, he edited the great biographical work, *Alexander von Humboldt*. The calendar of the Statistical Bureau, the astronomical portion of which was edited by Bruhns, always contained a popular essay on some astronomical subject. Bruhns also published a very convenient table of seven-figure logarithms of numbers, and trigonometrical functions to every ten seconds, which is well known in this country.

He was possessed of great talent for organisation, as was displayed in his arrangements for the German expeditions to observe the Transit of *Venus* in 1874.

Personally, Bruhns was extremely popular, and his loss is keenly felt by his more intimate friends. He died rather suddenly on July 25, 1881. He had been unwell for some time, but it was only shortly before his death that his illness was regarded as serious.

He was elected an Associate of the Society on November 8, 1878.

BARON HERCULES DEMBOWSKI, of Milan, to whom the gold medal of the Society was awarded, in 1878, for his researches upon double stars, died at Albizzate, in Upper Lombardy, on January 19, 1881.

About the year 1852 he commenced, in his own private Observatory at Naples, a series of observations on double and multiple stars, and, being sufficiently favoured by fortune, was able to cultivate science on his own means, dedicating to it the last thirty years of his laborious life. In his Observatory at Naples the telescope he used was a dialyte of Plössl of only five inches aperture, mounted equatoreally, but unprovided with clock-motion or a position-circle. So great was his skill, and such was the accuracy of his eye, however, that the observations made by him are not inferior to any results made at the same time, even with the most perfect instruments.

In 1870 he returned to Milan, and constructed at Cassano Magnago, near Gallarate, a new Observatory, more adapted to his wants, and equipped with an excellent Refractor, of seven inches aperture, by Merz, and a Meridian Circle, by Starke. He made a complete revision of Struve's Dorpat Catalogue ; and his observations are not less remarkable for their number than for their excellence. A full account of his work, which was published chiefly in the *Astronomische Nachrichten*, was given by Dr. Huggins, in his address on presenting to him the gold medal of the Society (*Monthly Notices*, xxxviii. pp. 249–253), and it is unnecessary, therefore, to refer to it further here. Of its amount Dr. Huggins said : " If all his observations, which are now scattered through some seventy numbers of the *Astronomische Nachrichten*, were to be collected in one volume, the catalogue would not be unworthy to stand beside the most valued and extensive catalogues of double stars we possess."

Only a small portion, however, of Dembowski's observations has been published, but the complete series has been left to his heirs, with full power to make use of them for the benefit of science. In the *Atti della R. Accademia dei Lincei* for December 4, 1881, Schiaparelli strongly urges the Academy to undertake the publication of these observations, which, he states, are written out in order, and almost ready for the press. They would occupy, he estimates, four quarto volumes, or 1,500 pages in all. In the same number of the *Atti*, immediately following Schiaparelli's remarks, are two letters, from Otto Struve and Mr. S. W. Burnham, both urging the importance of the publication of these observations, and bearing the highest testimony to the accuracy and value of Dembowski's work. Struve writes : " Dans le volume publié par moi en 1878 sur mes propres observations des étoiles doubles et multiples, chaque page témoigne combien les mesures de M. de Dembowski, m'ont été utiles dans les recherches sur les mouvements dans les systèmes stellaires. Et pourtant c'est à peine la quatrième partie de ses mesures qui m'a été accessible, avec beaucoup de difficultés, par les publications occasionnelles, dispersées sur de nombreux volumes des *Astronomische Nachrichten* et en d'autres recueils périodiques. Une publication complète, et dans un ensemble soigneusement rédigé, des observations du Baron Dembowski ne pourrait donc manquer de porter des fruits encore beaucoup plus riches à l'étude de l'astronomie sidérale. Dans ces vues il suffira de signaler le fait déjà autrement connu que, dans ce trésor, il se

trouve entre autres une répétition complète, après un intervalle
d'environ 40 ans, des mesures de toutes les étoiles doubles
formant l'objet principal de l'ouvrage de feu mon père, connu
sous le nom des *Mensuræ micrometricæ.*" Struve speaks of
Dembowski's observations as being about 20,000 in number, and
says that eight years ago he had the pleasure of seeing his
manuscript journals, and admiring their order and excellent state.

With regard to the accuracy of Dembowski's work, Mr.
Burnham, who, in his letter, offers to prepare the manuscript for
the press, writes :—" That he was the best observer who ever lived,
in his special department, and in micrometer work generally,
will not be questioned by any astronomer who has had occasion
to investigate this field." Struve's words are : " S'étant pro-
curé, de ses propres économies restreintes, des instruments de
force très modique, il a, en simple particulier, labouré sans
relache le même champ de travail pendant 30 ans, en s'appli-
quant continuellement à porter ses mesures au plus haut degré
de perfection ;" and in his presidential address, Dr. Huggins
laid great stress on " the earnestness with which he sought to
attain the greatest precision possible to him."

He was elected an Associate of the Society on November 8,
1878.

JEAN ALFRED GAUTIER was descended from one of the old
families of Geneva. One of his ancestors, Jean Antoine Gautier,
had occupied himself with astronomy, and on the occasion of the
total eclipse of the Sun on May 12, 1706, made the first observations from which the longitude of Geneva could be deduced.

Alfred Gautier was born on July 19, 1793, and was educated
at the College and Academy of Geneva, among his professors
being M. A. Pictet, Lhuillier, and Deluc. He applied himself
especially to mathematics, and, while still quite young, went to
Paris, attracted by the reputation of Laplace, Lagrange, Legendre,
&c. He profited greatly by the lectures he attended there ; but,
although he applied himself chiefly to the exact sciences, he did
not neglect literary studies. In 1812, at the age of 19, he took
the degree of Licentiate in Science at the University of Paris,
and in 1813 that of Licentiate in Letters. He devoted the next
few years to the production of a work entitled *Essai Historique
sur le Problème des Trois Corps,* which obtained for him the title
of Doctor, and attracted some notice at the time. In this
volume of nearly 300 quarto pages the author gives, in a critical
form, the complete history of all that had been written on the
reciprocal action of bodies in space, adding original investigations of his own.

After completing his studies at Paris, he visited England,
where his reputation had preceded him. He formed numerous
friendships, and became especially intimate with John Herschel,
who was to become famous like his father. The year that
Gautier passed in England exercised a marked influence on all

his life, by the correspondence which he always kept up with the men of science whom he then met.

On his return to Geneva, in 1819, he was appointed honorary professor at the Academy, and charged with a course of lectures on astronomy, to which he added, from 1821, a course on higher mathematics. In the seventeen years during which he held the position of professor he contributed greatly to develop a taste for science among his auditors, and four of the present professors in the Academy were pupils of his. As Professor of Astronomy Gautier had under his direction the little Observatory which was founded in 1773, and which occupied a site not far from where the present Observatory stands. But this establishment was almost useless, and Gautier could not begin observations there until 1824, when it had been put in repair and equipped with a Repeating Circle of 20 inches diameter by Gambey.

In this rudimentary observatory he performed several pieces of work, the most important being the determination of the longitude and especially the latitude of Geneva, which previously was not known to within four seconds. The latitude had been determined by Mallet and Pictet to be 46° 12′ and 46° 11′ 32″; and Gautier, as the result of a long series of observations, made in the years 1825–28, found it to be 46° 11′ 59″·4.

Gautier was the chief mover in the foundation of the new Observatory. In 1829 he set to work to obtain a new establishment, and supported his application in various ways, especially by a petition from the watchmakers, who were naturally interested in having at Geneva the means of obtaining the exact time. The petition was signed by sixty firms.

Gautier was then a member of the Representative Council, and on June 24, 1829, the sum of 30,000 francs was voted for the erection of a new building, and a further sum of 25,000 francs for the purchase of a Transit instrument and Equatoreal. Gautier himself drew the plans of the observatory, and he had the satisfaction of seeing it completed in 1830. A description of it, by him, appeared in the seventh volume of Quetelet's *Correspondance Mathématique et Physique.*

Just at this moment, when the activity of Gautier seemed at length to have obtained free scope, a sad infirmity suddenly checked his career. He was attacked by an affection in his sight, which rendered direct observation impossible. He then accepted the help of an assistant, L. F. Wartmann, towards whom he always retained a sincere attachment, and who had already afforded most valuable assistance to him in the construction of the Observatory.

Notwithstanding this assistance, Gautier resolved to yield to circumstances, and, with a conscientious modesty which was characteristic of him, decided to retire from an office which he could no longer fill to the best advantage. He waited, therefore, with impatience the return of his pupil, M. E. Plantamour, from Germany in 1837 in order to give up to him the direction of the new Observatory, as well as his Chair of Astronomy.

The infirmity which led to Gautier's resignation did not, however, check his scientific activity, and he occupied himself with investigations connected with physical astronomy and meteorology. He followed with admiration the great works inaugurated by Humboldt and Gauss for determining the laws of terrestrial magnetism.

He always kept himself well informed with respect to the advance of science, on which he regularly wrote *comptes rendus*; and for half a century he was one of the most zealous *collaborateurs* of the *Bibliothèque Universelle de Genève*.

This activity never ceased, and during 1880 several articles were published that were signed by him. He was elected a member of the *Société de Physique et d'Histoire Naturelle* of Geneva in 1818, and regularly attended its meetings, making frequent communications to it. In the year of his death, at the age of eighty-eight years, he gave an account to the Society on June 9 of the contents of the Annual Report of the Greenwich Observatory.

He retained all his faculties to the last, and his memory was quite unimpaired. He was closely connected with other Geneva Societies besides the *Société de Physique*, and was greatly interested in the trade and industries of the town.

Gautier was married twice, but had no children. He lived in retirement in the country at Chougny, and devoted much of his time to good works. He was remarkable for his extreme modesty, which led him to keep himself in the background on all occasions, and for the amiability of his character. He died, after a short illness, on November 30, 1881. His two nephews, Emile and Raoul Gautier, have inherited from him his scientific tastes, and their pursuits are the same as those of their uncle.

In the *Bibliothèque Universelle de Genève* for 1824 and 1825 he published, under the title *Coup d'œil sur l'état actuel de l'Astronomie pratique en France et en Angleterre*, several articles relating to English astronomy. The subject of the first is the Greenwich Observatory; the second relates to the Dublin Observatory and to the most recent researches upon the annual parallax of stars; the third contains a description of the Observatories of Oxford and Cambridge, and also an account of the mathematical teaching at Cambridge. The latter possesses now a good deal of interest, as Gautier visited Cambridge at the time when the mathematical reforms, which have had so great an effect on the studies of the University, were being inaugurated; and he was brought into close contact with Whewell, Peacock, &c. The last article relates to the Observatories and Institutions of Scotland. These articles with others were published as a separate volume in 1825.

Gautier was elected an Associate of this Society on January 11, 1822, within two years of its foundation, and was the oldest Associate. He was also the oldest foreign member of the Cambridge Philosophical Society, having been elected soon after its foundation.

PROCEEDINGS OF OBSERVATORIES.

The following Reports of the proceedings of Observatories during the past year have been received by the Council from the Directors of the several Observatories.

Royal Observatory, Greenwich.

The work at the Royal Observatory during the past year has been of the same character as in previous years, the greatest attention having been given as usual to observations of the Moon on every practicable opportunity, both with the transit circle and altazimuth.   The principal planets have also been regularly observed on the meridian whenever they passed before 15^h; and the brightest of the minor planets during the interval included in the daily ephemerides given in the *Berliner Jahrbuch.* Fifteen observations of Comet *b* 1881, and four of Comet *c* 1881, were obtained on the meridian *sub polo.* Azimuths and zenith distances of Comet *b* were also observed with the altazimuth on ten days. All the *Nautical Almanac* stars visible in this latitude, additional stars used in the determination of clock-errors, and others whose places are required for special purposes have been observed on the meridian, the number of separate stars included in the Catalogue for 1881 being about 1,040.

Three determinations of the flexure of the transit circle telescope, two in May 1881 and one on 1882, January 2, have been made.   The values found are respectively $+0''\cdot13$, $+0''\cdot18$, and $+0''\cdot03$, the mean of which is nearly the same as that found from four determinations in 1880.   No flexure-correction has been applied to the observations during the year.

The observations of the altazimuth-mark fixed on the parapet of the Royal Naval College, referred to in the last Annual Report, have been found to be very accordant; and they are used in combination with the observations of a star, in forming the adopted zero of azimuth.

The calculations in every department of the Observatory are in a very forward state.   The mean R.A. and N.P.D. of all objects observed with the transit circle are completely reduced to the end of 1881, and are ready for entry into the ledgers, preparatory to the formation of the Annual Catalogue of stars, and for the more advanced reductions of the observations of the planets.   The complete copy of the observational sections of the volume for 1881 is nearly ready for the printer, waiting only for the verification of some minor points.

The spectroscopic observations have been made as usual

with the half-prism spectroscope mounted on the S.E. equatoreal. In solar work most attention has been paid to the spectra
of Sunspots, and the amount by which the lines between *b* and
F were thickened over spots has been recorded on 24 days. The
examination of the chromosphere for prominences has therefore
not been made as often as usual. Prominences were observed on
21 days, and were usually found to be very numerous. On no
occasion was the Sun observed to be free from them.

The displacement of the F or *b* line has been measured in the
spectra of 64 stars, nine of these not having been previously
observed. This work was somewhat interrupted during the
summer by the appearance of the two bright comets of the year,
b and *c*, the spectrum of the former having been examined on
six nights, and that of the latter on three. All the spectroscopic
observations have been completely reduced, and the copy for
press prepared.

Photographs of the Sun have been taken on 173 days during
1881; on only two of these days was the Sun's disk free from
spots. The photographs have been measured in duplicate to the
end of 1881, the measures reduced, and the copy for press prepared to the middle of September.

Observations of occultations of stars by the Moon, and the
phenomena of *Jupiter's* satellites have been observed frequently.
The satellites of *Mars* were looked for on several occasions, and
a measure was secured on December 30 of a faint object near
the planet, which was believed to be the outer satellite, *Deimos*.
Phobos was suspected to have been seen on the same evening,
but no measure was obtained.

The principal extra work which has fallen upon the Observatory during the past year has been the reduction of the
extensive series of observations of the solar eclipse of 1880,
December 31. These calculations were, however, completed
without sensibly interfering with the ordinary work of the computers, and the result appears to be satisfactory. More than
usual care was required in the calculation of the refraction,
owing to the very low altitude of the Sun during the latter portion of the observations. From the care taken it is believed
that no appreciable uncertainty exists in the adopted refraction
correction, which in no case exceeds 9″ in its differential effect
on the two cusps or limbs observed.

The volume of the *Greenwich Observations* for 1879 was distributed at a much earlier date than usual. The printing of the
volume for 1880 is nearly completed, and it is hoped that this
volume will be ready for distribution in the ensuing spring. The
separate copies of the Results of the Spectroscopic and Photographic Observations for 1880 have been distributed in advance
of the volume.

To obviate the inconvenience which has been felt for some
years for want of sufficient shelf-room to accommodate the
numerous additions to the library, a new building has been

erected on some vacant ground near the Magnetic Observatory, under the superintendence of the Director of Works of the Admiralty. By this additional accommodation many of the books which have been necessarily deposited in small supplementary libraries, will be brought together and arranged in a more convenient position for reference.

The retirement of Sir G. B. Airy, K.C.B., from the post of Astronomer Royal, which he had held for the long period of forty-six years, with so much credit to himself and so much advantage to the Observatory and science generally, has occasioned several changes in the staff. Mr. Dunkin, for many years Senior Assistant, has been appointed Chief Assistant; Mr. Downing has been promoted to be a First Class Assistant; and the vacancy thus made in the Second Class Assistants has been filled up by the appointment of Mr. Hollis, who entered on his duties at the end of last November. Sir G. B. Airy retired on 1881 Aug. 15.

Cambridge Observatory.

The work at this Observatory continues to be mainly directed to the observation with the Transit Circle of the stars in the zone lying between 25° and 30° of North Declination. A good deal has been done during the past year in the way of gleaning up zone stars of which the normal number of 3 observations had not yet been taken. To this end 2177 observations have been made, many of which are of stars so minute as to baffle an observer unless under very favourable conditions.

The work of reduction goes on steadily and as rapidly as circumstances will allow.

The labour bestowed on the zone observations already begins to bear fruit, in enabling us to supply other astronomers with star-places required for their researches. We were able, on June 11, to furnish Lieut.-Col. Tupman with the mean places for 1879·0 of four stars which had been compared with Brorsen's Comet. One of these had been observed twice in the course of the zone observations, and each of the others three times, and the individual results agreed satisfactorily.

Again, on November 14, we supplied M. Bossert and M. Schulhof, of the Paris Observatory, with the mean places for 1875·0 of 16 stars which were required for their researches on the orbit of the Comet of 1812. One other star, the place of which M. Bossert asked for, is just beyond the limits of our zone, and had not been observed here.

All the necessary observations for obtaining the clock and instrumental errors were of course taken throughout the year.

The observations of *Polaris* made with the Transit Circle in 1879 have been carefully discussed, with the view of obtaining corrections of the assumed N.P.D. and of the colatitude of the

Observatory. In this discussion corrections have been applied for flexure and for errors of division of the Circle. The adopted correction for flexure is

$$-0''936 \sin z,$$

where z is the zenith distance, considered positive when S. and negative when N. The corrections for errors of division as obtained from an investigation made in March and April 1880 are the following :—

$$-0''165 \text{ for the Upper Transit,}$$
$$\text{and} \quad +0'461 \text{ for the Lower.}$$

The final result obtained for the N.P.D. of *Polaris* agrees almost identically with that of the corrected Berlin Standard Catalogue. The correction to the assumed colatitude is found to be

$$+0''58,$$

and the corrected colatitude is $37° 47' 8''·98$.

A similar discussion of the observations of standard clock stars taken in 1879 has been made. The above-mentioned correction for flexure has been applied, and the following corrections for errors of division have been employed, viz.: Correction to the observed arc from the zenith

$$\text{to } 55° \text{ N.P.D.} \quad +0''29$$
$$\text{to } 70 \quad ,, \quad +0·36$$
$$\text{to } 90 \quad ,, \quad +0·16$$

the corrections for intermediate arcs being found by interpolation.

The result shows that in order to agree with the places in the Berlin Catalogue, the N.P.D. of standard clock stars, as found by employing the assumed colatitude, requires to be increased by the mean value $0''·54$.

Hence, if the Berlin places are correct, the value of the colatitude should be

$$37° 47' 8''94.$$

The mean of this result and that found from the observations of *Polaris* already mentioned is

$$37° 47' 8''96,$$

which may be regarded as the value of the colatitude given by the observations of 1879.

It may be mentioned that the colatitude similarly deduced from the observations of 1878 is

$$37° \; 47' \; 8''{\cdot}94.$$

In the interest of other observers, it may not be amiss to mention that on November 22 a discrepancy was found between the line of collimation as determined in the usual manner—viz. with the positive eyepieces in both collimators and that obtained when the plane glass cap was substituted for the eyepiece in the North collimator. The two results were brought into agreement by interposing screens of tissue paper between the lamps and the eyepieces. The light is decidedly better and the image sharper when both eyepieces are in place than when the plane glass caps are substituted.

Satisfactory observations of the two bright comets of 1881 were made, both with the Transit Circle and with the Northumberland Equatoreal, in the months of June, July, and August.

A parabolic orbit of Comet *b* was calculated by Mr. Graham, which represents very closely the observations made at Oxford and Cambridge.

The usual meteorological observations have been regularly taken, and communicated every morning by telegraph to the Meteorological Office.

The Observatory, Dunsink, County Dublin.

During the past year the South Equatoreal has been employed in Parallax research, in continuation of the work described in former years. A series of observations for the parallax of μ *Cephei* is almost completed, and considerable progress in the investigation of the parallax of Σ 2486 has been made. The reconnoitring observations described in former Reports have also been continued.

The results of the meridian observations of red stars made in 1875–76 and 1878–80 were ready for the printer in the beginning of July last, but the printing is progressing very slowly. Part IV. will, therefore, probably not be ready before March.

Both Comets *b* and *c*, 1881, were observed on the meridian as often as the bad weather would allow. In September was commenced a series of observations of upwards of 1,000 stars situated between −2° and −23° Declination, and kindly communicated to Mr. Dreyer by Professor Schönfeld as needing re-observation, either as being suspected of having proper motion or because earlier determinations disagree. These observations are progressing well.

Circle B (not before examined) was investigated last spring. Both the excentricity and the division errors are remarkably small.

Royal Observatory, Edinburgh.

Prof. Piazzi Smyth states that in consequence of the new arrangements at the Home Office for transacting Scottish business, inquiries have been in progress with regard to the Observatory, and that until these are brought to a conclusion he thinks it undesirable to make any report.

Glasgow Observatory.

Apart from the ordinary operations connected with time-signalling and meteorology, the printing of the Star Catalogue has been the chief object of attention during the past year. It is expected that the work will be ready for distribution next summer. The number of stars is 6415. A comparison of the Glasgow results with those of Bessel (1825) and Lalande (1800) has indicated some decided cases of proper motion.

Kew Observatory.

The magnetical and meteorological observations, which principally form the work of the Observatory, have been continued as in former years, and experimental work and the verification of instruments have also largely occupied the attention of the staff.

Eye observations of the Sun by means of a small portable $2\frac{3}{4}$-inch refracting telescope with a magnifying power of 42 diameters have been made on 187 days. The Sun's surface was observed to be free from spots on three of those days.

In August last an application was received from Major Herschel, by authority of the India Office, for permission to make certain experiments with the pendulums formerly used by Captains Basevi and Heaviside in their Indian pendulum operations, and for the loan of the instruments with their accompanying appliances, with facilities for prosecuting the experiments at the Observatory.

These requests were granted, and during September, October, and part of November, operations were continuously carried on, both in the Pendulum Room and the Experimental House. The necessary transits for the determination of the clock errors were made by the staff of the Observatory.

Liverpool Observatory, Bidston, Birkenhead.

The general routine work of this Observatory has not been much altered for several years past, but increasing attention has been directed year after year to the causes of change in the rates of chronometers on their passage from one seaport to another.

During the past year a large number of observations taken at sea have been received and discussed, with the view of showing the degree of accuracy with which Greenwich time by chronometers can be checked by the known position of a landmark when time signals from Observatories are not available.   On the voyage from Liverpool to Valparaiso through the Straits of Magellan, the only Observatory time signals for the use of the mariner are those at Lisbon and Rio de Janeiro.   These signals are given once a day, Sundays excepted, but ships are often not detained long enough in these ports for the navigating officers to avail themselves of them.   Cross-bearings and Admiralty charts are therefore now had recourse to most frequently for checking the errors of chronometers, and through the encouragement of the directors and the courtesy of the officers of the Pacific Steam Navigation Company, copies of the observations taken on board their ships have been deposited at this Observatory.   On reducing and arranging them so as to show the error of each chronometer, as found from time to time on the voyage, the unexpected efficiency of the results obtained has led to the introduction of printed forms for recording the observations in a systematic way, and such forms are now being filled up by the officers of the above-named company.

It has been stated in former Reports that each ship of this company carries three chronometers, the rates of which are found for every five degrees of Fahrenheit from 45° to 95° inclusive.   The rates applicable to the temperatures to which the timekeepers are exposed are added to the errors of the instruments daily.   At the end of the voyage the differences between the accumulated errors so obtained and the errors on Greenwich mean time, found by Observatory time signals, are divided by the number of days that the ship was at sea.   These corrections applied to the old rates produce new rates, which on the succeeding voyage are found to be far more nearly correct than any rates that can be obtained at the Observatory during the two or three weeks that these ships are usually detained in port.   It has also been found that these rates cannot be altered with advantage during the voyage from observations obtained at sea ; but by tabulating the apparent errors of longitude for each chronometer, as found at sea by every available means, a record is obtained which is of great practical use to the navigating officer.   From the records of such observations deposited at this Observatory in 1878–9, it was found that on the voyage to and from Valparaiso the chronometers appeared to go faster on the outward passage between Liverpool, Lisbon, and Rio, and slower on the passage home between those places, than they did on the average for the whole voyage.   Subsequently it was found, by the American determination of the longitudes of the Observatories of Lisbon and Rio, that the former was $8^s\cdot6$ and the latter $5^s\cdot4$ more westerly than those given in the *Nautical Almanac.*   The run from Liverpool to Lisbon only takes about 8 or 9 days, consequently the chronometers would appear to gain about one second a day more on the voyage out, and lose about one second a day more on the voyage home, between Liverpool and Lisbon. The passage to Rio takes from 20 to 25 days, and the gaining rates were found to be about half a second a day larger on the voyage out than they were on the voyage home between Liverpool and Rio de Janeiro.

By correcting the rates of chronometers for change of temperature, and keeping a systematic record of their performance at sea, the mariner has the necessary data for predicting their most probable rates for a future voyage. Without this information, which can only be obtained at sea, it is impossible from observations on shore to predict the rates so accurately. The navigating officer therefore has the best possible means of rating chronometers entirely in his own hands. He cannot, however, find the corrections to the rates due to change of temperature from any observations that it is practicable for him to make at sea; but when his chronometers have been tested in three definite temperatures at this Observatory he is furnished with the means of obtaining these corrections, and once obtained they are found to remain the same for very long periods. Examples may be found in the records of this Observatory in which no sensible change has taken place in the thermal error for periods of over ten years.

Radcliffe Observatory, Oxford.

The principal astronomical instruments of this Observatory are a Transit Circle of 5 inches aperture by Troughton and Simms, and a Heliometer of 7½ inches aperture by Merz, with a mounting by Repsold.

The Transit Circle has been in regular and systematic use during the past year.

The number of observations made with it is as follows:—

Transits	1745
Circle Observations	2182
Observations of Sun (Solstices and Equinoxes)	37
Observations of Moon	73
„ Comet *b*	16
„ Comet *c*	6

Microscopes have been mounted, and an examination of the division errors of the circle has recently been made. The corrections thus found have not been inconsiderable, and their application appears to have reduced the systematic errors within very small limits indeed. The Nadir Points have been determined exclusively, as at the Cape, with the Nadir reflecting eye-piece; but a very considerable number of observations of stars by reflexion, north and south of the Zenith, have been made to check the systematic errors of the instrument. From 37 Northern stars, with 109 reflexion observations and 113 direct observations, the value of $R - D$ is only $-0'''\cdot193$; whilst from 44 Southern stars, with 113 reflexion observations and 196 direct observations, the value of $R - D$ is $+0'''\cdot166$. The errors of mean Nadir Points $-0'''\cdot10$ and $+0'''\cdot08$ are very small indeed; and, if these results are confirmed by the observations of other years, the Radcliffe Transit Circle compares, in this respect, most favourably with existing meridian instruments.

The colatitude found from the observations made during the year 1880 is

$$38^{\circ}\ 14'\ 24''\cdot84,$$

a result nearly identical with that found by Johnson from a discussion of the observations made with the Mural Circle.

The Astronomical and Meteorological Results for 1880 are now passing through the press.

The spectra of the Comets b and c; 16 observations of phenomena of *Jupiter's* satellites; and 5 occultations of stars by the Moon, have been made with the 7-inch Equatoreal which is mounted in the grounds.

The Heliometer has been dismounted during a considerable part of the year 1881. A new driving-circle and a new reversing circle have been supplied by Mr. Simms. The instrument is now in good order. The dome is, however, unfortunately very heavy to move, and the extreme difficulty of keeping the object-glass quite free during the observations is a serious hindrance to the work with this instrument. It is hoped, however, that the instrument will be brought into systematic use before long.

The photographic meteorological work has been carried on successfully during the past year; but the photographic processes, the tabulation of the results, and the discussion of the results when tabulated, press very severely upon the limited staff available at this Observatory. It would be a matter of great regret to many scientific men if these meteorological observations, which have been continued for many years at this Observatory, should be abandoned; but unless some assistance, such as that afforded by the Meteorological Commission, out of the funds placed at their disposal, to several other Observatories, is afforded to this Observatory, the question of abandoning these continuous meteorological observations will have to be taken seriously into consideration.

Oxford University Observatory.

The work of this Observatory has proceeded on the same lines as heretofore. A somewhat elaborate inquiry has been made as to the amount of dependence which can properly be claimed for the photographic measurements made on the Moon;

the results have been communicated to the Society, and are now in course of printing in their *Memoirs*. Some subsequent and further investigations have been made in the same direction, and the substance has been printed in the *Observatory* for February 1882. These latter appear to offer a further confirmation, if that were needed, of the great accuracy which is attainable from the photographic method. It is highly satisfactory to observe that the American astronomers, in their photographic method applied to the Transit of *Mercury* over the Sun in 1878, have arrived at a similar conviction of the accuracy of this mode of research.*

The results of the investigation into the Physical Libration of the Moon have been given in the Memoir already referred to. In the main they do not differ from those obtained by Drs. Wichmann and Hartwig. But, inasmuch as the probable error of the longitude determination of the Libration is considerably greater than that of the latitude, in all the researches hitherto conducted, and as another system of measuring the photographs has been devised, by which these differences of probable error can be removed, a new and more elaborate system of measures has been instituted, which, it is hoped, will lead to a more unimpeachable conclusion. Herein evidently is seen the great value and convenience of photography, which admits, at any time, of a variety or repetition of measurement.

A method new in its application, and simple in execution, for the photometric measurement of the relative brightnesses or magnitudes of the stars has been communicated to the Society. It has been found to be apparently very accurate and consistent, and it is now in course of application to all stars visible to the naked eye from the Pole to −6° Declination.

Stonyhurst College Observatory.

In the course of the past year the Sun was observed at Stonyhurst 163 times, the chromosphere and sunspots receiving equal attention. The scale for the drawings of the entire disk is generally 10½ inches to the diameter, but a considerable number of partial drawings have been made on an enlarged scale of 30 inches to the diameter. The entire chromosphere has been measured 46 times.

Owing to the badness of the weather, not more than eight occultations of stars by the Moon have been secured; but better fortune has attended the observation of *Jupiter's* satellites, 46 separate phenomena having been recorded. Comets and meteors have also occupied considerable attention during the last few months.

A new star spectroscope is now in the hands of Mr. Hilger,

* *Amer. Jour. of Science*, vol. xxii. Nov. 1881.

with which it is proposed to make a systematic examination of the spectra of faint stars and nebulæ.

Temple Observatory, Rugby.

The work of this Observatory during the year 1881 has chiefly consisted of the measurement of position and distance of known or suspected binaries, 210 of which have been completed. The measures made during the last three years have been printed in vol. xlvi. of the *Memoirs* of the Society. Some few spectroscopic measures for the motion of stars in the line of sight have been made, and also an examination of the spectrum of Comet *b*.

The educational work, which takes precedence of all other, has considerably increased, and the average of instrumental knowledge attained by boys has much exceeded that of former years.

Mr. Barclay's Observatory, Leyton, Essex.

The Observatory and instruments remain as in former years.

The ordinary routine work, astronomical and meteorological, has been unremittingly carried on.

The double-star observations have been continued.

The various comets have been observed, and several observations of the satellites of *Saturn* have been made.

Mr. Campbell's Observatory, Arkley, Barnet.

The principal work of the Observatory during the year has been the taking of fifty-three additional Transits of the Moon, with a view to the determination of the parallactic inequality.

Fifty of these are double transits, the crater and limb being taken as they follow each other over the threads of the instrument.

The limb transits are taken in hopes of ascertaining, by comparing them with the crater transits, the effects of irradiation, of irregularities on the limb and of errors of semi-diameter. This will make in all one hundred and thirty-three crater and sixty-seven limb transits. All these have been completely reduced and compared to the tables.

Considerable progress has been made in the discussion of these results in order to determine the value of the parallactic inequality independently of the semi-diameter.

It is hoped that the weather will permit of a greater number of transits being taken during the coming year.

A new method of measuring the errors of Collimation and

Azimuth has been adopted with satisfactory results. This method is an adaptation of the one followed at Greenwich, kindly explained to us by Mr. Dunkin. It will form the subject of a paper which, it is hoped, will be communicated to the Society on some future occasion.

Mr. Common's Observatory, Ealing.

During the past year the 3-foot telescope has been used in observations of Comet *b*, 1881. Photographs were obtained on the night of June 24 and on subsequent nights. Encke's Comet was found on August 27. The month of November was exceptionally fine, and measures and observations of *Hyperion* and *Mimas* were obtained. December was very bad, the only fine night being the 22nd, when *Phobos* and *Deimos* were both seen and measured.

A Transit-room has been built and provided with a $3\frac{1}{4}$ inch Transit instrument, and an Equatoreal Reflector of $6\frac{1}{2}$ inches aperture with unsilvered glass mirrors for viewing the Sun, has been erected in a revolving house.

Colonel Cooper's Observatory, Markree.

With the Meridian Circle and the chronograph were observed double stars south of the equator. It is intended that these observations shall be made in all the four positions of the telescope, but as the instrument cannot be reversed without several alterations in the Transit-room, observations have hitherto been made in only two positions—*i.e.* the object-glass and eyepiece have been changed.

Measures of double stars were taken in the first part of the year 1881 by aid of the large Refractor. At the end of the year drawings of planets and observations of *Jupiter's* satellites and other phenomena were made.

With the Comet-Seeker red stars and variable stars have been observed. The observations of the latter were, however, to a great extent made useless by the overcast sky.

Some progress has been made with the recalculation of the orbits of the binary stars, and much time will be devoted to this subject in the future. It is hoped that these observations will be brought to a conclusion in a few years more, and will be published in a volume with the double-star observations.

During the past year the repairs of the Observatory have been finished, and much time has been taken up in adjusting the instruments, arranging the new library, and directing the workmen.

A sketch of the History of Astronomy in Ireland is in progress. It dwells in particular upon the activity of this Obser-

·vatory, and upon the unpublished researches that have been carried on here.

The meteorological registers have been kept as heretofore by an assistant. The rainfall of the last half century is being discussed, and it is believed that traces of periodicity coinciding with the periodicity of sunspots have been detected. · The connection between the amount of rain and the force of the wind is also being examined.

The Earl of Crawford's Observatory, Dun Echt.

Beyond furthering the reduction and printing of observations connected with the last Transit of *Venus*, the work of the Observatory was almost entirely restricted to the observation of comets, of which nine were visible in the course of the year. The places determined are distributed as follows :—

		No. of Obs. in 1881.
Comet *e* 1880 (Swift)	2
„ *f* „ (Pechüle)	3
„ *a* 1881 (Swift)	5
„ *b* „ (Gould, &c.) ...	19
„ *c* „ (Schäberle)... ...	·5
„ *d* „ (Encke)	2
„ *e* „ (Barnard)	⨿
„ *f* „ (Denning)	4
„ *g* „ (Swift)	8

It would be conferring a favour if the Council would decide as to the notation of comets, and as to the desirability of entirely giving up the use of the small letters ; if so, these Comets would be 1880 V. and VI. and 1881 I. to VII.

With a slight exception, all the observations have been reduced. Drawings of the jets issuing from the nucleus of Comet *b* were made on a number of nights ; these and some observations of the spectra of Comets *b* and *c* still await publication. Some coloured stars and nebulæ were examined with the spectroscope, and a few unsuccessful sweeps were made on the plan devised by Prof. Pickering, and in a slightly different form by the late Prof. d'Arrest. In this connection it may be mentioned that the double nebula found at Dun Echt on Nov. 18, 1881, had been discovered by Mr. Burnham in 1873, but owing to the former observations having been given among double-star measures, they were overlooked until attention was drawn to them, in the first place by Mr. Hind.

During the year 29 Circulars, Nos. 16 to 44, were printed at the Observatory and widely distributed. With the exception of No. 41, which gives particulars of Mr. Sawyer's Variable,

D.M. + 1°, 3408, they are all chiefly concerned with the Comets of the year. In all, twenty sets of elements and many Ephemerides are given. Valuable computations have been communicated by Messrs. Chandler, Oppenheim, and Schulhof. Most of the information received from America was sent by means of the "Science Observer" code, invented by Messrs. Ritchie and Chandler. It is believed that an examination of the working of this code as exemplified in interchanging early news of comets between the "Science Observer," Boston, U.S., and Dun Echt Observatory will show that all the chief difficulties which formerly beset the exact and economical transmission of astronomical data by telegraph have been surmounted by this self-checking code. As an example may be mentioned Comet *a*, 1881 (Swift), which was discovered at Rochester, N.Y., on the morning of May 1. Information received through the kindness of the late Astronomer Royal and the Smithsonian Institution enabled it to be observed at Dun Echt on May 2, and eventually elements and an Ephemeris computed in America were forwarded by the code and published at Dun Echt on May 9, while like results from Dun Echt observations were circulated in the United States a day later.

Particulars of two specially interesting objects pointed out by Professor Pickering are given in Circular 43: the star D.M. + 36°, 3987 (place for 1880, 20^h 12^m 32^s·7, + 37° 3′·2), which has a bright blue band in its spectrum, first seen on Nov. 24, and a very small planetary nebula (place for 1880, 20^h 6^m 26^s·38, + 37° 3′ 25″·2) discovered on Nov. 25.

The arrangements respecting the time-gun, time and meteorological observations remain as before.

At the end of October the engagement of Mr. H. J. Carpenter terminated. In him the Observatory loses a most reliable computer.

Mr. Edward Crossley's Observatory, Bermerside, Halifax.

There is very little to report with regard to the work of this Observatory. During the year very bad observing weather has been experienced. The usual observations of double stars and the phenomena of *Jupiter's* satellites were attempted, but the results are neither numerous nor very good. Much time, however, has been devoted to the 3½-inch Transit Circle, with a view to test its powers and to obtain a correct value for our latitude. Four sets of 24 stars have already been observed and reduced, and it is hoped with the return of fine weather to obtain a satisfactory result.

Mr. Huggins' Observatory, Upper Tulse Hill.

During the past year photographs have continued to be taken of the spectra of stars.

On the night of June 24 a good photograph was secured, after an hour's exposure, of the more refrangible portion of the spectrum of Comet *b*. The head of the comet was brought upon the slit; and afterwards through the other half of the slit, for the purpose of comparison, a spectrum of *Arcturus* was taken. The spectrum of the comet consisted of a continuous spectrum and two groups of bright lines.

The continuous spectrum could be traced from about F to a little distance beyond H. In this continuous spectrum were distinctly present G, *h*, H, K, and several others of the Fraunhofer lines. The spectrum was, therefore, certainly due to reflected solar light. The more refrangible bright group consisted chiefly of two bright lines, having the wave-lengths 3883 and 3870. These lines are accompanied by a faint luminosity, shading off towards the more refrangible limit of the spectrum. The second bright group occurs between G and *h*, and commences with a line, wave-length between 4220 and 4230.

A small increase of brightness was suspected between *h* and H, and may indicate a third fainter bright group.

The two strong groups correspond with similar groups which appear under certain circumstances in the spectra of the compounds of carbon. Professors Liveing and Dewar have recently shown that these groups indicate certainly the presence of nitrogen in combination with carbon.

This photograph, therefore, gives us information of the presence of nitrogen, in addition to carbon and hydrogen, which were shown to exist in comets by eye-observations made at this Observatory in 1868.

It is worthy of notice that the more refrangible bright group is very intense, showing that the cometary light of this refrangibility (though invisible to the eye) is very bright relatively to that of other parts of the spectrum. In a photograph which was taken on June 25, with an exposure of an hour and a half, when the circumstances were less favourable, this bright group is very strong in the negative, while the continuous spectrum and other groups are so extremely faint as to be only doubtfully suspected.

Some attempts have been made during the past year to apply photography to the spectra of some nebulæ, but hitherto without success.

The Earl of Rosse's Observatory, Birr Castle.

During the year 1881 astronomical observations were made at Birr Castle Observatory on 79 nights, in the course of which

33 drawings of the planet *Jupiter*,
3   „   „   „   „   *Saturn*,
19 „   „   „   „   *Mars*,
8   „   „   „   Comets *b* and *c*, 1881,
4   „   „   Lunar Craters,
6   „   „   the central part of the *Orion* Nebula,
were made with the aid of the 3-foot Reflector.

During the year 29 observations of nebulæ were also made with the 6-foot Reflector, and sketches were obtained.

Lunar Radiant Heat.—Determinations were made on a good many nights, and physical experiments for ascertaining the constants and special qualities of the galvanometer and thermo-piles used were carried on during the summer.

Of the 33 drawings of the planet *Jupiter* mentioned above, some were made during the Opposition of 1880–81, and are included with those made in November and December 1880 in a set of 25, just about to be published in the *Transactions* of the Royal Dublin Society. The rest, with a few additional ones which we hope to get before the close of the present season, are likely to follow before long. The drawings of the Comets *b* and *c*, 1881, are being published by the same institution.

Meteorological observations have been continued as in the previous year.

Colonel Tomline's Observatory, Orwell Park, Ipswich.

The work of this Observatory has been confined almost entirely to the comets of the year. Owing to the unfavourable weather experienced during the later months and to other causes, but few opportunities occurred for the observation of the smaller comets, but good series of the larger ones were obtained. The reductions of these and of the arrears mentioned in the last Report have been completed and all are now ready for publication. It is much to be regretted that many of the positions depend either on old Catalogue places of stars or, still more doubtfully, on micrometrical comparisons with such, the Observatory not being furnished with a meridian instrument suitable for the observation of the comparison stars.

In view of the value to science of the observation of the forthcoming Transit of *Venus*, Colonel Tomline has allowed the observer to volunteer for that service, and during some portion of the ensuing year the work of this Observatory will be intermitted in consequence.

Royal Observatory, Cape of Good Hope.

The following Report includes the work of the past two years.

The Cape Catalogue for 1850, containing 4,715 stars (chiefly B.A.C. stars of South Declination), is now completed in MS., but has been withheld from publication till a comparison with Mr. Stone's great Catalogue for 1880 (only recently received) has been made.

The whole of the occultations of stars by the Moon observed here since 1834 have been collected and reduced, and the equations which result from comparison with the *Nautical Almanac* places have been computed.

Professor Simon Newcomb has undertaken the computation of the places of the Moon from Hansen's *Tables* for the epochs of occultations observed previous to 1860, and this part of the work is reported by him to be nearly finished.

Advantage has been taken of the receipt of Newcomb's *Catalogue of Standard Clock and Zodiacal Stars,* to re-compute the apparent places of the occulted stars from the data of that catalogue. Thus the completion of the longitude operations will permit the publication of between 600 and 700 observed occultations rigidly compared on a uniform system with Hansen's *Tables of the Moon,* and extending over a period of nearly 50 years. The same work will contain a series of observations of *Jupiter's* satellites extending over the like period, besides a long list of unpublished observations of double stars. After many delays, caused first by the breakdown of the submarine cable, and afterwards by the destruction of the land-lines by the rebel Kaffirs in the Transkei, the longitude operations mentioned in the last Report to the Society (*Monthly Notices,* vol. xl. p. 232) were begun in May 1881, and have been steadily continued since. Passing over the many difficulties and hindrances that occurred both before and during the operation, it is sufficient here to report the following work accomplished.

1881, *March, April, May.*—Very complete determinations of personal equation in time determinations and sending and receiving signals.

June.—Six nights' exchanges—Mr. Finlay at Durban, Mr. Maclear at the Cape.

July.—Mr. Finlay, on his way to Aden, exchanged signals with the Observatory from Delagoa Bay and Zanzibar, connecting the intermediate stations Inhambane and Quillimaine by chronometers.

August and September.—Mr. Finlay was seriously ill at Aden during the latter half of August and the beginning of September.

In the beginning of August Mr. Gill, accompanied by Mr. Maclear, went to Durban, and organised the necessary arrangements for exchanging signals through the Aden-Durban cable, 4,000 miles in length. At the end of August Mr. Gill returned to the Observatory, the necessary preliminaries having been satisfactorily arranged. Signals were exchanged between Aden and Durban on sixteen nights, on ten of which a complete set of time determinations were obtained at both ends, either immediately before or after the exchange of signals; but only on three nights, from interruptions by cloud, were complete sets of time determinations secured at both ends both before and after the exchange of signals.

[A complete set of time determinations is defined as one polar star above and one below pole, and four time-stars, "Lamp West," and a similar series, "Lamp East." Time determinations without reversal are not included.]

October and part of November.—Mr. Finlay returned to Durban, determining again the longitudes of Zanzibar, Mozambique, and Delagoa Bay by exchanges with Durban, and connecting Quillimaine with Mozambique and Delagoa Bay by chronometers.

November and part of December.—Four nights' observations were secured by Messrs. Finlay and Maclear at Durban for relative personal equation in time determination, and sixteen sets (eighty coincidences in all) for personal equation in observing mirror signals. [For these latter determinations the galvanometers of both observers were put in circuit side by side at Durban. The sidereal clock was retained at Aden, and signals sent from it by Mr. Prosser, of the Eastern Telegraph Company —who was specially trained by Mr. Finlay for the work—were recorded by the mean time clock at Durban by both observers, each employing his own galvanometer.]

On December 7 Mr. Finlay left Durban and returned to the Royal Observatory on December 11.

Since the beginning of December the weather at Durban has been persistently cloudy, except on December 23, when complete sets of time determinations both before and after the exchanges were secured. Three more such sets of observations and exchanges will enable Mr. Maclear to return here, and another series of personal equation determinations will complete this protracted but important operation.

The warmest thanks are due to Sir James Anderson, Chairman of the Eastern Telegraph Company, who has granted the free use of the cable for the work, and to the officials of the Company at Aden, Delagoa Bay, Durban, and Mozambique for their courtesy and hospitality.

In conjunction with Major Lemesurier, R.E., the longitudes of several points in the Transvaal have been established. The accepted longitude of Pretoria was found to be $0°\ 47'$ (41 miles on the earth's surface) too far east.

The Transit Circle has been employed as follows :—

1. In the observation of a list of 389 time-stars and 88 southern circumpolar stars used on the expedition.

[It is proposed at a very early date to publish the results of the observations of the circumpolar stars for use of observers in the Transit of *Venus* expeditions of 1882.]

2. In the observation of 331 stars which are to be employed as standard stars in the observation of the zones founded on Professor Schönfeld's *Durchmusterung*, from N.P.D. $90°$ to $120°$, together with forty-seven northern stars and twenty-eight more southern stars for purposes connected with refraction. Corresponding observations of the same stars are being made at Leiden. All these observations, already far advanced, will be concluded during the present year, each star being observed at least twelve times in both coordinates.

3. A long-sustained series of observations of a and β *Centauri.*

4. A list of about thirty faint stars of which occultations by the Moon were observed in Australia, whose places were determined at the request of Professor Auwers.

5. A number of stars observed with various comets during the past years.

6. Stars whose occultations by the Moon have been observed here.

7. Stars employed as comparison stars in researches with the Heliometer for stellar parallax. The total number of observations made has been as follows:—

	1880	1881
Transits	3824	4749
Observations of Z.D.	3165	2016

The completion of the above programme requires 1,567 observations in R.A. and 2,616 observations in N.P.D.

The division errors of the Transit Circle have been rigorously investigated for each degree. The new screws of the Circle microscopes and of the declination micrometer have been rigidly investigated. The resulting errors are very small.

Proposals have been made by H.M. Astronomer for increasing the range through which observations can be made by reflection; at present stars cannot be observed by reflection beyond 42° Z.D. These proposals of necessity involve considerable alterations in the arrangement of the collimators, but the desired end is so essential that it is hoped these proposals will be duly sanctioned.

The Equatoreal has been employed in observing the following comets:—

1880. Comet *a*.—With ring micrometer and 6·9-inch Equatoreal, by Finlay. Gill on six nights.

1881. Comet *b*.—Observed by Dr. Elkin (Mr. Gill absent in England) on six nights with the Heliometer.

1881. Comet *c*.—Observed on fifteen nights (August 31 to September 27) by Gill and Elkin with the Heliometer, and further continued on fourteen nights (September 30 to October 18) with the 6·9-inch Refractor and parallel-wire micrometer.

The observations of Comet *a*, 1880, are published in the *Monthly Notices*, vol. xl. p. 626; those of Comets *b* and *c*, 1881, will be published so soon as satisfactory observations of the comparison stars have been made with the Transit Circle.

Extensive alterations have been made in the Equatoreal. The object-glass, previously of inferior quality, has been repolished by Merz, of Munich, and greatly improved in its performance. A new micrometer has been adapted by Messrs. Repsold of Hamburg, in which bright wires in a dark field or dark wires in a bright field are given at pleasure from a single lamp, the intensity and colour of illumination being under complete control by the observer. The same lamp also illuminates the readings of the micrometer head, the position circle, and the declination circle. The whole is complete, convenient, and satisfactory in every respect. Apart from deficiencies of the driving clock, the Equatoreal is now a first-rate instrument of its size, but the need of an adequate Equatoreal is a pressing want, which it is hoped will soon be supplied. The observation of occultations of stars by the Moon and of the eclipses of *Jupiter's* satellites was necessarily discontinued during the alterations in the Equatoreal, but a very large number of such observations has nevertheless been secured. The Heliometer belonging to. Mr. Gill has been actively employed by him in conjunction with Dr. Elkin since March 1881, in investigations on the parallax of α *Centauri* and eight other interesting southern stars. By the present programme these observations will be brought to a conclusion in the beginning of 1883. Up till now the full number of observations planned for (over 300) has been secured, and it is found practically that the amount of work undertaken is the maximum that can be sustained by two observers with one Heliometer.

The reductions are in a forward state. The ledgers are formed for 1879 and 1880, and the reductions are completed up till July 1881, and reduced to time of true transit to the end of 1881. The examination of the reductions in 1881 has necessarily been much delayed by the absence of the Chief-Assistant and Second-Assistant on the longitude operations.

The oppositions of the minor planets *Victoria* and *Sappho* in July, August, September, and October 1882, afford a remarkably favourable opportunity for determining the solar parallax by Galle's method. In a circular on the subject Mr. Gill shews that if observers secure even two-thirds of the accuracy attained by Brünnow in the measurements of difference of declination, the solar parallax will be found with a probable error of $\pm$ $0''\cdot009$ from 100 corresponding observations of *Victoria* or *Sappho* in both hemispheres.

A programme of the proposed observations of these planets has been addressed to those Observatories likely to take part in the work, and it is hoped that very extended cooperation will be secured.

The system of Time Signals has been extended and improved.

Arrangements have been made for a tidal survey of the coast of South Africa.

Self-recording tide and wind gauges are now being erected in Table Bay and East London. Similar instruments are under construction for Durban and Port Elizabeth.

The meteorological observations made at the Royal Observatory in the years 1879 and 1880, together with those made in different parts of the Colony, have been printed in the Reports of the Cape Meteorological Commission for these years.

Extended experiments have been made at the Observatory on the effects of different exposures of thermometers.

NOTES ON SOME POINTS CONNECTED WITH THE PROGRESS OF ASTRONOMY DURING THE PAST YEAR.

The Minor Planets.

Only one minor planet (220) was discovered during the year 1881. Herr J. Palisa, however, detected an object on February 23, which he announced as a new planet; but the elements of its orbit were afterwards found to resemble very closely those of *Juewa*, leaving no doubt that the latter planet had been accidentally reobserved.

(220) was discovered by Herr Palisa at the Observatory of Vienna on May 19, 1881. Its mean distance (2·402) shows that its relative position lies among the inner group of minor planets whose orbits are nearest to that of *Mars*.

The following planets discovered by Herr Palisa in 1879 and 1880 have been recently named :—

(205)	Martha	(212)	Medea
(207)	Hedda	(214)	Aschera
(208)	Lacrimosa	(216)	Cleopatra
(210)	Isabella	(218)	Bianca
(211)	Isolda	(219)	Thusnelda.

From an interesting memoir by Herr Hornstein, communicated to the Vienna Academy, it appears probable that all the largest of the minor planets have now been discovered, and of those having a diameter exceeding 25 geographical miles the number is extremely small, most of them having been discovered before the year 1860. From this research it seems that the number of asteroids with a diameter less than five miles is also very small, especially in that portion of the minor planet zone nearest to *Mars*. In the outer zone nearest to *Jupiter* there may be a more considerable number of those minute bodies, in reality too faint to be observed in the telescopes usually employed in the search for minor planets. Omitting the few of comparatively large magnitude, the general average diameter of the remaining members of the group appears to be between five and fifteen miles, a much smaller value than had been hitherto assumed from previous investigations. E. D.

The Comets of 1881.

Seven comets have appeared during the year 1881, one of which was the periodical comet of Encke. A few very brief details of each are given in the following notes :—

1. *Swift, a* 1881.—This comet, the first of 1881, was discovered by Mr. Lewis Swift, at Rochester, New York, in the night of April 30. At the time of its discovery the comet was situated about 10° north of *a Andromedæ*, and was a difficult object to observe, owing to the morning twilight. On the receipt of the telegram announcing its discovery, it was soon picked up at Dun Echt, Vienna, and other European Observatories.

According to the elements calculated by Herr Zelbr, the comet passed its perihelion on May 20. These elements bear some resemblance to those of a comet observed in China B.C. 69.

2. *Tebbutt, b* 1881.—The discovery of this interesting comet was first announced in Europe by telegrams from South America—the first from Dr. Gould ; the second sent by order of the Emperor of Brazil. This comet was, however, first observed at Windsor, N.S.W., by Mr. Tebbutt on May 22, when it was easily visible to the naked eye, its nucleus being of equal brightness to a star of the fifth magnitude. On May 23 it was observed at Melbourne by Mr. Ellery ; on May 25 by Dr. Gould at Cordoba ; and on May 29 by M. Cruls at Rio de Janeiro. At the Cape of Good Hope the comet was seen at Graham's Town by Mr. Eddie on May 27 ; and at the Cape Observatory by Dr. Elkin, who has remarked that between May 31 and June 4, the nucleus equalled a star of the second magnitude, and was accompanied with a tail of 5° or 6° in length.

This comet was very extensively observed at the principal Observatories in the northern hemisphere, and many details of the observations are given in the *Astronomische Nachrichten*, the *Monthly Notices*, and other astronomical journals. According to the elements calculated by M. Dunér and M. Engström, from a series of observations extending from May 23 to August 13, the perihelion passage of Comet *b* occurred on June 16, about a week before its appearance above the north horizon of Europe. Some interesting sketches of the envelope and nucleus by Captain Noble and Mr. Knobel are given in the November and December numbers of the *Monthly Notices*.

The spectrum of the comet was favourably observed at the Observatories of Paris, Greenwich, and Brussels, by Professor C. A. Young, Dr. Huggins, and by other observers. It appeared to be continuous, with three bright bands in the yellow, green, and blue. Dr. Huggins succeeded in photographing the spectrum of the comet, which showed two groups of bright lines in the violet and a continuous spectrum extending from about F to some distance beyond H. Dr. Huggins remarks that he has measured the photographs of the comet's spectrum and has found for the two strong bright lines in the more refrangible group the wave-lengths 3883 and 3870, and he has no doubt that these lines represent the brightest end of the ultra-violet group which, under certain circumstances, is noticed in the spectra of the compounds of carbon. The wave-lengths of the second and fainter group lie between 4220 and 4230. An excellent illustration of this photograph, exhibiting the two sets of bright lines and the solar spectrum, is given in the *Report* of the British Association for 1881, and in the *Proceedings* of the Royal Society, vol. xxiii. page 2. M. Janssen and Professor H. Draper have successfully taken some excellent photographs which exhibit many interesting features in the structure of the envelope and tail.

3. *Schäberle, c* 1881.—The third comet of the year was discovered by Mr. Schäberle on the night of July 13, at the Observatory of Ann Arbor, U.S. Though not relatively so bright as Comet *b*, it was a conspicuous object to the naked eye in the month of August, when observers had the rare opportunity of witnessing two comets with tails several degrees in length at the same time. Schäberle's comet was observed in Europe a few days after its discovery, at Strassburg on July 17, at Dun Echt on July 19, and at Arcetri on July 21, when it presented a nucleus equal in brightness to a star of the sixth magnitude, and a tail about one degree in length. Its maximum brightness occurred on August 21, about the time of perihelion, when the tail was about ten degrees long. This comet was very favourably observed at most of the principal Observatories. Its spectrum consisted of three bright bands approximately coincident with the three brightest bands of the spectrum of the Bunsen-flame, and a faint continuous spectrum.

4. *Encke, d* 1881.—From the orbit and ephemeris of Encke's comet, calculated by Dr. O. Backlund, of Pulkowa, it appeared that this interesting periodical comet would pass its perihelion on November 18, 1881. By the aid of this ephemeris the comet was detected at Strassburg by MM. Winnecke and Hartwig on August 20; by M. Tempel, at Arcetri, on August 21; at Pulkowa, on August 24, when it was scarcely visible in the refractor of 14 inches aperture; by Mr. Common, at Ealing, on August 27, and by Mr. Lohse, at Dun Echt, on August 28. Mr. Common, observing with his three-foot reflector, remarks that the comet was about 2' in diameter, very faint, with slight indications of an increased brightness in the centre.

5. *Barnard, e* 1881.—This comet was discovered by Mr. E. E. Barnard, at Nashville, Tennessee, on September 19. It was in perihelion on September 13. Dr. Hartwig observed it on October 3, with the great refractor of the Strassburg Observatory, when it appeared as a bright round nebulous mass of about the eighth magnitude. The elements bear no resemblance to those of any former comet.

6. *Denning, f* 1881.—This comet was discovered by Mr. W. F. Denning at Bristol on the morning of October 4, when it appeared like a small bright nebula, round, with a condensation of light in the centre. From observations made at Marseilles on October 5, at Dun Echt on October 12, and at Strassburg on October 19 and 28, Dr. Hartwig and M. Wutschichowsky have found that this comet moves in an elliptic orbit with a period

of 3070·7 days. Professor Winnecke considers that there is a great probability that it is identical with the comet discovered by Goldschmidt at Paris in 1855, and supposed by him to be De Vico's. The orbit of the comet is very interesting, from its close approximation to the orbits of *Venus*, the *Earth*, and *Jupiter*; it may therefore be affected by considerable perturbations from each of these planets. Its perihelion passage took place on September 13.

7. *Swift, g* 1881.—The seventh comet of 1881 was discovered by Mr. Lewis Swift on November 16, four days before it arrived at perihelion. It was observed at Harvard College by Mr. Wendell on November 17, 19, and 20; but through some confusion in the transmission of the telegram announcing the discovery, no information relating to this comet reached Europe till the publication in the Dun Echt Circular, No. 40, of a set of elements determined by Mr. S. C. Chandler, from Mr. Wendell's observations, which had been cabled from America on November 22, by means of the *Science Observer* code of signals. The comet was picked up at Dun Echt on November 22, at Königsberg and Strassburg on November 25, and on the two following days at Rome, Paris, and O'Gyalla.

In addition to these seven comets, a faint object, supposed to be a comet, was noticed on May 12, by Mr. Barnard, at Nashville, Tennessee, very near *a Pegasi*. It was again seen on the following night; but, though carefully looked for at several American Observatories, no trace of it could be found afterwards.

E. D.

The Cape Catalogue for 1880.

This, the most extensive southern catalogue hitherto published, contains the places of 12,441 stars, derived from observations made with the Cape transit circle from Jan. 1, 1871, to April 30, 1879. During the whole of this period the observing was steadily directed to the formation of a catalogue of well-distributed stellar zero-points for those portions of the sky which are beyond the reach of the observers in the principal northern observatories. But a considerable number of stars, which can be well observed at the northern observatories, were also observed at the Cape as a check upon the existence of any systematic errors in the work.

Mr. Stone adopted, as the basis of his working catalogue, Lacaille's Zone-Catalogue of 9.766 southern stars. But as some stars of the sixth magnitude, and many stars of the seventh magnitude, are contained in Brisbane's Catalogue of 7,385 stars which had not been observed by Lacaille, all these sixth magnitude stars, and all the seventh magnitude stars which did not fall in parts of the sky already well covered with Lacaille-stars, were subsequently included in the working catalogue. Finally, a large stereographic projection of the southern hemisphere was prepared, and upon it were projected the places of all the stars which had been already observed, and wherever lacunæ appeared within the limits of 115° N.P.D. to 180° N.P.D. efforts were made to fill them up by observing stars of rather a lower magnitude than the seventh of Lacaille's scale.   The Cape Catalogue for 1880 ought, therefore, to contain a pretty exhaustive *Durch-musterung* of stars down to the seventh magnitude inclusive within the above-mentioned limits of N.P.D.

In order to avoid, as much as possible, the danger of leaving the work a mere fragment, it was split up into zones which were, to a considerable extent, complete in themselves; and when the stars of any zone were not sufficiently well observed in one year, they were carried into the working catalogue for the following year, and thus a sufficient number of stars were observed in different years to check any systematic instrumental changes.

As a rule, each star was observed three times in each element.

The Right Ascensions of clock-stars used were those of the Standard Lists of the Greenwich Observatory for the different years.   The Right Ascensions of polar stars for the determination of azimuthal errors were derived from Mr. Stone's paper on the "Places of Eight Close Southern Polar Stars 1860 to 1900," or from the results of the observation of stars near the South Pole made in 1871.

The latitude of the Observatory adopted in the formation of the Catalogue is the mean of those used for the Cape Catalogues for 1840 and 1860, and the refractions used in the reductions were computed from the *Tabulæ Regiomontanæ*. It appeared, however, that during the years 1871–1879 the thermometer whose readings were taken for the calculation of the temperature-corrections read too high by 0°·55, so that the refractions used were really those of the *Tabulæ* diminished in the proportion of 0·9988 : 1. The few observations made below 85° Z.D. were reduced with the mean refractions of the *Fundamenta* multiplied by 1·003282. It appears, however, that the use of Bessel's refractions unaltered would not change the results of the Catalogue by as much as two-tenths of a second for any of the southern stars.

A comparison between the Catalogue places and those of the *Nautical Almanac* for 1880 shows a close agreement between the Cape and *N. A.* Right Ascensions, which might have fairly been expected from the use of the same fundamental Right Ascensions for the determination of clock-errors.   But the perfectly independent North Polar Distances also, as a rule, agree very closely, and satisfactorily show the accuracy of the refractions used.

The great majority of stars in the Catalogue are reduced to mean place without the application of any proper motion, but the mean date of observation is given in every case, so that any proper motion that hereafter is found to exist can be applied

without difficulty.   As yet our means of determining the proper
motions of southern stars with any accuracy are very limited.
Lacaille's zone observations, as at present reduced, are useless
for this purpose, and the more recent southern catalogues are
not sufficiently remote in epoch for a satisfactory determination.
Mr. Stone's great Catalogue will be invaluable for this purpose
as time goes on, and if in another hundred years the stars contained in the Catalogue shall be reobserved and compared with
their places as given in it, "the astronomers of that period will
know more about the values of the constants of precession, the
proper motions of individual stars, including the motion of the
solar system in space, and the connection of systems of stars
with one another, than we, of this generation, are ever likely to
know." Happily, however, it is not necessary to wait till the
year 1980 in order to make use of the Catalogue.   To the now
large number of active southern observers it will be of immediate
use and of incalculable value, and astronomers of every clime
will find that it adds greatly to the materials available for prosecuting their researches in various directions.   A. M. W. D.

Newcomb's Standard Stars.

This work contains a catalogue of 1,098 standard clock and
zodiacal stars, which may be used as reference stars for investigations connected with the lunar and planetary theories, especially
in the reduction of the older occultations.   It originally included
only time stars, and stars occultations of which by the Moon had
been well observed; but afterwards it was found advisable to
greatly enlarge the work, so that it may be considered as
including two classes of stars:
(1) All the standard stars of the *American Ephemeris*, omitting for the most part those added for field work.
(2) All stars to the sixth magnitude, inclusive, which can be
occulted by the Moon, together with stars below the sixth
magnitude which had been observed by Bradley.
The equinox to which the Right Ascensions of all the stars are
reduced is that of Professor Newcomb's paper on the "Right
Ascensions of the Equatorial Fundamental Stars" which appeared as an appendix to the *Washington Observations* for 1870.
In the formation of the star places the author has been able to
use Auwers' reduction of Bradley's observations, so that the
definitive positions of this Catalogue are entitled to considerably
greater weight than either Gould's Right Ascensions or Boss's
Declinations, as these astronomers had to depend on the places for
the epoch 1755 of the Bradley stars as given in the *Fundamenta*.
The Declinations of the stars in the Catalogue are substantially
founded on those given by Boss, the most important modification arising from the substitution of Auwers' reduction of
Bradley (referred to above) for that of Bessel.   And the author

has derived his definitive positions by making use of most of the great catalogues, depending chiefly, of course, on observations made at Greenwich, Pulkowa, and Washington. In the case of Washington he has been able to use the results of observations made during the years 1866 to 1873 with the transit circle. May we hope soon to see a Star Catalogue compiled from the observations made with this powerful instrument?

In the case of the stars observed by Bradley, the positions and other data are given for the two fundamental epochs, 1755·0 and 1850·0. The positions of the fundamental time stars are also given for 1900. In the case of stars between 10° and 30° from the Pole, the data are given for 1755, 1800, 1850, and 1900. For stars yet nearer the Pole, the epochs 1775, 1825, and 1875 are added. At the end of the volume there is given, in a separate table, the R.A.'s of time stars for 1800, and for quinquennial epochs, 1830–1900.

Professor Newcomb also adds formulæ for reducing the star places to any epoch between 1750 and 1900, from the data of the Catalogue, by Taylor's theorem, Hill's formulæ for the secular variation of the annual motion and proper motion of the stars, and Struve's values of m and n for the epochs for which the positions of the stars are given in the Catalogue.

This work is therefore admirably adapted to be of the greatest assistance to those who have to reduce lunar or planetary observations made during the last or the present century, and it will, no doubt, be extensively employed by all such persons.

A. M. W. D.

Professor Hall's Observations of Double Stars.

The volume of *Washington Observations* for 1877 contains, in an Appendix, an important series of Observations of Double Stars made by Professor Asaph Hall mostly with the 26-inch refractor in the years 1875 to 1880. In the introduction Professor Hall gives a full description of his method of observing, with a discussion of the errors of the filar micrometer employed.

The first portion of the work consists of observations of a series of thirty double stars selected by M. Otto Struve as standards for the comparison of micrometrical measurements. The whole number of these observations is 296; the mean distance of the thirty stars is 17″·6; the probable error of a single distance $= \pm 0″·059$, and of a single angle $= \pm 0″·075$, both very small.

In order to apply a geometrical test to the observations, careful measurements were made of the Trapezium in *Orion* and of the multiple stars Σ 2703 and Σ 311—that is to say, of four stars in a trapezium, three stars in a slightly obtuse triangle, and three stars nearly in a straight line. The results indicate no important systematic errors; but Professor Hall concludes his

P

discussion of the Trapezium in *Orion* by saying, " Probably some compensation of these errors will occur when all the parts of the figure are measured at the same hour angle, and in future observations of this kind, it would be interesting to measure some of the parts at quite different hour angles."

The observations with the 26-inch refractor consist of micrometric measurements of 368 objects, mostly double stars observed by the Struves, and a few others, chiefly those discovered by Mr. Burnham. Nearly all the observations depend on four settings of the position-circle, and on two measurements of the double distance. A few of the early observations of the distances depend on four measurements; but Professor Hall soon found that two measurements give all the accuracy necessary in a single night.

The distances of the following difficult objects were measured by Prof. Hall at Washington and Mr. Burnham at Chicago, about the same epoch, and may therefore be compared.

	Hall.	Epoch.	Burnham.	Epoch.
γ^2 Andromedæ	0″38	1877·11	0″43	1878·65
Aldebaran	31·272	1878·02	30·45	1877·89
β Leporis	3·16	1877·12	2·68	1878·00
ϵ Hydræ	3·317	1878·33	3·46	1878·22
ω Leonis	0·45	1878·34	0·63	1878·11
μ^2 Boötis	0·732	1876·44	0·68	1878·41
	0·730	1879·54		
μ^1 Herculis	0·88	1878·5	1·12	1878·5
α^2 Capricorni	1·11	1878·71	1·06	1878·53
O Σ 413	0·56	1878·71	0·74	1878·53
τ Cygni	1·09	1878·76	1·06	1878·41
μ Cygni	3·73	1878·84	3·88	1878·42

The comparison shows for the majority of cases a satisfactory agreement. It is, however, noticeable that Prof. Hall finds no difference in the distance of μ^2 *Boötis* in 1876 and 1879, and that his value is rather larger than that found by other double star observers.

A valuable series of observations of the companion of *Sirius* is given, made in the years 1866 and 1872 to 1879, with the 9·6-inch and 26-inch refractors; there is also appended a micrometric discussion of the faint stars near the Ring Nebula in *Lyra*. The work terminates with a series of measurements of 35 double stars, made in 1863 with the 9·6-inch equatoreal at the instigation of Captain Gilliss, to compare the accuracy of angles and distances determined with a filar micrometer, with those computed from differences of Right Ascension and Declination observed with the transit circle. E. B. K.

The Transit of Venus 1882.

The Report of the International Conference on the Transit of *Venus*, held in Paris last October, gives the following list of stations to be occupied for observation of the coming Transit of *Venus*, together with the names of the Chiefs (in brackets) :—

Argentine Republic.—
   2 stations (Bœuf)

Brasil.—
   Itapeva
   Pernambuco
   Rio Janeiro
   Antilles (?)
   Straits of Magellan (?)

Chili.—
   Santiago

Denmark.—
   St. Thomas or Santa Cruz

France.—
   Cuba (D'Abbadie)
   Martinique (Tisserand)
   Florida (Perrier)
   Mexico (Bouquet de la Grye)
   Chili (Le Clerc)
   Santa Cruz (Fleuriais) ⎫
   Chubut (Hatt)   ⎬ Argentine Republic
   Rio-Negro (Perrotin)   ⎭

Germany.—
   Argentine Republic
   Straits of Magellan or Falkland Islands
   United States (2)

Great Britain.—
   Bermuda
   Jamaica
   Barbadoes
   Cape of Good Hope (3)
   Madagascar
   New Zealand
   Falkland Islands
   Sydney
   Melbourne

Holland.—
   Curaçoa or St. Martin

Mexico.—
   Chapultepec

Portugal.—
   Lorenzo Marques

Spain :—
   Cuba (2)
   Porto Rico

No information has been received as to the stations to be occupied by Austria, Italy, and the United States.

The British stations will each be provided with two 6-inch refractors, the French with one 8½-inch and one 6¼-inch, the Dutch with one 6½-inch, the Spanish with a 6-inch and a 4-inch, and the Brazilian with refractors of 11¾ inches, 9¾ inches, or 6½ inches.

The opinion of the members of the Conference was not favourable on the whole to the employment of photography, after the experience of the late Transit, the satisfactory results obtained from the American photographs having been made public too late to allow of proper time for consideration and preparation. Photographs will, however, be taken at two of the French stations. The Conference has drawn up instructions for observation of the contacts, and it recommends that, in addition to such observations, micrometer measures should be made on the double-image principle, either with heliometers or Airy double-image micrometers.

As regards the steps to be taken after the Transit, it was, after some discussion, arranged that an International Conference should meet in 1883. to come to an understanding on the means to be taken for making use in the best and quickest manner of the Transits of 1874 and 1882, and to consider whether an International Bureau should be formed for obtaining these results.   W. H. M. C.

Observations of the Transit of Venus 1874.

The volume of " British Observations of the Transit of *Venus* 1874," edited by Sir G. B. Airy, has been recently published. It contains the full reports of the observations of the Transit itself made by the five British expeditions to the Hawaiian Islands, Egypt, Rodrigues, Kerguelen Island, and New Zealand respectively, together with the reductions and results of transit and altazimuth observations for local time, longitude and latitude. The observations and reductions are printed in detail for the district of the Hawaiian Islands, and these may serve as a specimen of the work at the other stations, for which abstracts only of the transit and altazimuth results are given. In every case the actual observations of the Transit are given in detail, and the equation of distance of centres is formed for the observed time of internal contact as interpreted from each observer's report, and also for each measure of cusps or limbs with the double-image micrometer. The determination of the longitudes of the several stations, whether by telegraph, chronometer-runs, or observations of the Moon with the transit or altazimuth, formed a large part of the work, and constitutes not the least interesting part of the volume. The whole of the reductions were carried out, under the superintendence of Sir G. B. Airy, by Col. Tupman, assisted by the various observers and by computers. No results of the measurement of photographs of the Transit are printed, but in an appendix a general statement of the photographic operations is given.

It had originally been intended to include in this volume the observations made in Australia, India, and other colonies, which had been communicated to the Society and were handed over to

Sir G. B. Airy for the purpose. But in the end difficulty was experienced in obtaining authority to publish them at the expense of the Government, and the observations in question have quite recently been returned to the Society. The Council have them under consideration with a view to their publication as soon as possible in the *Memoirs*.

The American observations of the Transit with their reduction are to be issued in four parts, with the following arrangement of subjects:—

Part I. General account of the operations, and reduction and discussion of the observations of the Transit of *Venus*.

Part II. Observations in detail made at each station, with their reduction.

Part III. Discussion of the longitudes of the stations.

Part IV. Measures of the photographs with their reduction and discussion.

The numerical reductions have been made, under the direction of the Commission, by Messrs. D. P. Todd, Townsend, and Ritter, and the whole is edited by Professor Simon Newcomb.

Part I. now issued contains: (i.) History of the operations, selection of stations, eight in number (Wladiwostok, Nagasaki, Peking, Crozet Islands, Kerguelen Island, Tasmania, New Zealand, and Chatham Island), and instrumental equipment, the chief feature in the latter being the long-focus photoheliographs with heliostat mirrors. (ii.) Positions of stations, constants of instruments, and *personnel* of the parties. (iii.) Discussion of the photographic operations, including the accurate determination of the distance from the photographic plate to the optical centre of the objective and mirror combined, and equations of distances of centres of Sun and *Venus* from the measures of each photograph. (iv.) Optical observations of the Transit, with a discussion of personal equations from observations of the model Transit. In Appendix I. a proposal is brought forward for the correction of atmospheric dispersion in photography by giving an eccentricity to one of the lenses of the object-glass; and in Appendix II. are given the observers' records of their contact observations.   W. H. M. C.

The Solar Parallax as derived from American Photographs of the Transit of Venus 1874.

The total number of photographs of the Transit of *Venus* 1874 taken by the American parties is 213, distributed as follows:—

Northern Stations.		*Southern Stations.*		
Wladiwostok	... 13	Kerguelen	... ...	8
Nagasaki	... 45	Hobart Town, Tasmania ...		37
Peking ...	... 26	Campbell Town, Tasmania		32
		Queenstown, N.Z....	...	45
		Chatham Island ...	...	7

The distance between centres of *Venus* and the Sun (*s*) and the position-angle of the line joining the centres (*p*) were measured on each photograph, and from the equations of condition in *s* and *p* thence formed Mr. D. P. Todd has deduced,[*] by the method of least squares, the value of the solar parallax and corrections to the tabular difference of R.A. and Dec. of *Venus* and the Sun (δA and δD) as follows:—

	From Measures of Distance.	From Measures of Pos. Angle.	Mean.
π	$8''888 \pm 0''042$	$8''873 \pm 0''060$	$8''883 \pm 0''034$
δA	$+ 1·181 \pm 0·202$	$+ 1·109 \pm 0·109$	$+ 0·075 \pm 0·006$
δD	$+ 2·225 \pm 0·070$	$+ 0·637 \pm 0·224$	$+ 2''083 \pm 0''067$

The value of π thus found gives for the Sun's mean distance 148,103,000 kilometres=92,028,000 miles, if we adopt the dimensions of the Earth given by Col. A. R. Clarke in his "Geodesy."

The tabular R.A. and Dec. of *Venus* relative to the Sun have been computed from the data of Mr. G. W. Hill as formed by him from Hansen's Tables of the Sun and his own Tables of *Venus*. The following are the corrections to reduce these to the data used by Sir G. B. Airy in the reduction of the British observations as interpolated from the *Nautical Almanac*, in which Leverrier's Tables of the Sun and *Venus* and parallax 8″·95 are adopted :—

	Ingress.	Egress.
δA	$- 3''62$	$- 3''54$
δD	$- 0·95$	$- 0·98$

Thus, as compared with Airy's tabular quantities, the corrections to relative R.A. and Dec. of *Venus* given by the American photographs are :—

$$\delta A \quad + 4''71 \qquad \delta D \quad + 3''05.$$

Mr. Todd finds for the probable error of a single photograph in distance $0''·88$, and in position-angle $3'·45$.   W. H. M. C.

The Solar Parallax from Meridian Observations of Mars.

In September 1876 a circular was sent out from the Washington Observatory inviting the co-operation of Northern and Southern Observatories in a scheme of meridian observations of *Mars* and comparison stars at the Opposition of 1877, on Win-

---

[*] *American Journal of Science*, 1881 June; *Observatory*, 1881 July; *Copernicus*, 1881 September.

necke's plan, for determination of the parallax. In response to this Prof. Eastman has received observations from Melbourne, Sydney, the Cape, Leiden, and Cambridge, U.S., which he has discussed, together with the observations at Washington, in. Appendix III. to the "Washington Observations" for 1877. The following are the results for the Mean Solar Parallax which he obtains from the several combinations:—

Washington—Melbourne	$8\overset{''}{\cdot}971 \pm 0\overset{''}{\cdot}032$	(19 nights)
Washington—Sydney	$8\cdot885 \pm 0\cdot055$	(11 nights)
Washington—Cape	$8\cdot896' \pm 0\cdot073$	(9 nights)
Leiden —Melbourne	$8\cdot969 \pm 0\cdot026$	(27 nights)
Cambridge —Melbourne	$9\cdot138 \pm 0\cdot050$	(10 nights)

Four of the eleven observations at Sydney, and two of the nine at the Cape have been rejected as being abnormal. The mean of the remaining seventy results from all the stations, with regard to the weights is

$$8\overset{''}{\cdot}980 \pm 0\overset{''}{\cdot}017.$$

But Prof. Eastman considers that the value found from Cambridge and Melbourne is too large, and, as this may be due to the method of observing transits for N.P.D. over wires inclined about 6° to the horizontal, which was adopted at Cambridge, he rejects all the Cambridge observations and obtains

$$8\overset{''}{\cdot}953 \pm 0\overset{''}{\cdot}019.$$

In a discussion of the Leiden and Melbourne observations published in 1879 (*Astronomische Nachrichten*, No. 2288), Mr. Downing found for the Solar Parallax $8'''\cdot960 \pm 0'''\cdot051$, a result which agrees substantially with that found by Prof. Eastman.

With regard to the large value of the Solar Parallax found from meridian observations of *Mars*, Prof. Eastman remarks that the prescribed method of observing (with two parallel wires separated by a space slightly less than the diameter of *Mars*) was fully carried out at only two stations and partially at one, and he considers that the method has never had a fair trial. As regards the possibility of a systematic error due to the chromatic dispersion of our atmosphere, which has been suggested by Mr. Gill, Prof. Eastman has tried experiments with *Jupiter* which, as far as they go, show that there would be no error from this cause depending on the illumination of the field.   W. H. M. C.

Observations of the Transit of Mercury 1878.

The Washington Observations for 1876 contain a series of Reports on telescopic observations of the Transit of *Mercury* 1878 made by the staff of the U.S. Naval Observatory. The observations are fully discussed in two categories—

1. Telescopic observations of contacts, by Prof. J. R. Eastman.
2. Phenomena attending observations of contacts, by Mr. H. M. Paul.

Prof. Eastman has discussed, not only the results obtained by his colleagues of the Naval Observatory, but also all the complete observations that had been received from other observers throughout the country. These number 109, and are arranged in tabular form, giving full geographical particulars of the observing stations, the instruments employed, the Washington mean time of observed contacts, and geocentric times of contact taken from the American Ephemeris. Weights 1 to 3 are assigned to the observations according to the experience of the observer, the size of the telescope, the magnifying power employed, the atmospheric condition, the phenomena observed as given in the notes, the accuracy of the local time when used, and the authority for the longitude of the station.

The comparatively few observations which have the highest weight 3, are assumed as representing the true phenomena. Comparing these results for the four contacts with the computed times taken from the American Ephemeris which depend upon the earlier theory of Le Verrier, they give the following differences :—

Contacts.				C−O
				s
I.	...	...	...	+ 77
II.	...	...	...	+ 84
III.	...	...	...	+ 110
IV.	...	...	...	+ 119

But taking the data of the *Nautical Almanac*, which depend upon Le Verrier's Tables in vol. v. of the *Annales de l'Observatoire*, and adopting the semi-diameter of the Sun as used in the American Ephemeris, which is 2″ smaller than that of the *Nautical Almanac*, these results become :—

Contacts.				C−O
				s
I.	...	...	...	+ 2
II.	...	...	...	+ 9
III.	...	...	...	+ 25
IV.	...	...	...	+ 34

Professor Eastman remarks that in the present state of our knowledge with regard to the Theory and Tables of *Mercury* it is very difficult to interpret satisfactorily the results of the comparison between the observed and the computed times of contact; and after considering the correction to his former Tables published by Le Verrier in 1859 in the volume referred to above, he concludes that it is "very desirable that the Theory of Mercury should again be thoroughly examined, employing the latest and best values of the necessary data, and that the corrections should only be obtained from comparison with all the trustworthy available observations to date."

In his discussion Professor Eastman finds that the methods of observing in nearly all the cases may be included in three classes, and his remarks have an important bearing upon the forthcoming Transit of *Venus.*

1. Many observers endeavoured to determine the times of contact by noting the occurrence of similar phases of the Transit before and after real contact, and deducing the true time from the similarity of the phases observed to those figured in the instructions which had been issued by the Naval Observatory. This method generally failed, because of the apparently slow motion of the planet and of the manifest inability to determine the likeness of phases.

2. A large number of observers endeavoured to fix upon the time of geometrical contact as the true time of second and third contacts—that is, when the outlines of the Sun and planet exactly coincide at one point. The method failed because of the difficulty in observing such a coincidence of position with regard to the black limb of the planet and the bright limb of the Sun, while in some instances it failed on account of the appearance of the "black drop."

3. The third class of observers fixed upon the time of *second* contact at the moment when the first flash of light appeared to dart between the cusps of sunlight as they closed round the planet. When the "black drop" was seen, the instant of its rupture was taken as the true time of contact. The time of third contact was taken at the moment when the thin line of light between the limbs of the Sun and planet was first broken. When the "black drop" was seen, the moment of its formation was taken as the true time of third contact. The first flash of light between the cusps at second contact, and the first break in the thin line of light at third contact are sharply-marked phenomena, and the times of contact obtained by different observers by this method agree better than by any other method. In nearly all cases the rupture of the "black drop" at second contact, and its formation at third contact are well-marked phenomena and easily observed. Professor Eastman further adds that in any future observations it is very desirable that this method should be strictly followed in all observations of interior contacts.

Mr. H. M. Paul has given a lengthy and exhaustive discussion of all the phenomena attending observations of contact as recorded by the observers. These he tabulates according to the physical appearance noted, giving the probable error of the time observation of such appearance:—

Table A. The first or last glimmer of actual sunlight between the planet and the cusps. The probable error appears to be small.

Table B. The closing or breaking of the line of light, which appears to have been noted by some observers in addition to the phenomena of Table A. The probable error is about 4^s.

Table C. Breaking or forming of the "black drop." Here the probable error is about 3^s for second and third contacts.

Tables D, E, F, and G give discussions of observations of the particular phases I. II. and III. depicted in the diagram to Professor Newcomb's instructions; the probable errors in some of which appear to be large.

Table H discusses apparent geometrical contact, and reveals a great discordance in the observations, the probable error for second contact being 15^s·26 and for third contact 13^s·30.

Tables I and K are devoted to geometrical contact, where black drop was neither seen nor mentioned.

Table L discusses geometrical contact with accompanying black drop. Here the probable error is very large, being 28^s·05 for second contact.

The remaining tables are devoted to "Interior Contacts with no Description of Phase," a tabular account of "The Black Drop" from quotations from the Reports, and "Special Observations and Peculiar Appearances observed."

Two series of micrometer measures of the diameter of *Mercury* on the Sun were made by Professor Young and Professor Pritchett, the mean result of the former being 11$''$·74, and of the latter 11$''$·39.

The remainder of the discussion is devoted by Mr. Paul to "External Contacts and some Phenomena connected with them," and to the "black drop," both subjects being treated in an able and exhaustive manner. The conclusion to which he comes is similar to that of Professor Eastman—viz. "if the black drop shows itself, then the experience of this Transit plainly shows that the phase to be watched for and timed is the breaking or forming of this ligament;" in no case will it be of any advantage to try and form any estimate of the time of apparent geometrical contact.

E. B. K.

Observations of the Total Solar Eclipse of 1878, *July* 29.

A quarto volume of more than 400 pages issued by the United States Naval Observatory during the past year contains Reports of about sixty observers of the total eclipse of 1878 July 29, as well as Reports by two observers of the eclipse of 1880 Jan. 11, which was visible in California, but as the Sun only had an altitude of about 11° above the horizon at the time of totality during the later eclipse, and the total phase only lasted about thirty seconds, the corona was seen under very unfavourable circumstances, and no physical observations of importance appear to have been made.

The volume is illustrated by thirty lithographic plates, mostly drawn to the same scale as that adopted in the Eclipse Volume of the *Memoirs*—viz. 1¼ inches to the diameter of the dark Moon. The drawings and the photographs of the corona taken during the eclipse of 1878 all concur in showing that the corona extended to a great distance from the Sun's limb in the eastern and western equatoreal regions; while in the neighbourhood of the Sun's northern and southern poles there were groups of separate curving rays, similar to those seen in the polar regions of the corona, visible during the Eclipse of December 1871. The drawings have not been oriented with the Sun's axis vertical upon the page, but with a few exceptions, which are all noted, the highest part of the drawing corresponds with the highest part of the corona as seen above the horizon from the place of observation. From a comparison of the drawings it is easy to see that the great equatoreal bundles of rays were not quite symmetrically situated with respect to the Sun's axis of rotation.

The editing of the volume and the production of the plates have been carefully superintended by Professor Harkness, who has given on plates Nos. 19 and 20 diagrams showing the extension of the corona on several photographs taken with different exposures at Creston and La Junta. The photographs were viewed by transmitted light and contour lines traced corresponding to the boundary of the photographic action. From these drawings Professor Harkness has constructed two curves, corresponding to the intensity of the light of the corona at various heights above the Sun's limb in the equatoreal and polar regions, which show that the intensity of the light of the corona decreased much more rapidly in passing away from the Sun's limb in the polar than in the equatoreal regions. The volume also contains valuable evidence with respect to the polarisation of the corona, and other matters.   A. C. R.

The Spectrum of Comet b 1881.

The observations on the spectra of the comets of last year would appear to uphold the views of the nature of the cometary material which have prevailed since, in 1868, the bright-band spectrum was shown to agree with that of some compounds of carbon. The observations of the spectra of some twenty comets since that date have shown a general agreement of three bright bands, but with some divergences as to their exact positions and maxima of brightness. Considering that in different comets, and, indeed, in the same comet from night to night, the conditions of temperature, density, and supply, and probably chemical nature of the light-giving stuff, must be in constant change, it is to be expected that corresponding modifications of the band spectrum will show themselves. These minor divergences, except so far as they arise from difficulties of observation, probably indicate real differences existing in the cometary material. The spectroscopic results of the brighter comets of the past year should therefore be regarded as representing the *particular comet only at the time of observation.* It will be sufficient, therefore, in this place to refer to the main facts. The observations of Young, Vogel, Konkoly, and others, as well as those made at the Greenwich Observatory, would show that the positions of the bands agree with those seen in the so-called "flame spectrum" of carbon, though there is reason for regarding this spectrum as due to a hydrocarbon, and the same form of spectrum may be obtained from an induction spark under suitable conditions for low temperature. This coincidence has been made more certain by Young's observations of three fine lines in the middle band, which are also seen in the spectrum of the Bunsen flame. Bright lines were also seen by Konkoly. Suspicions of the violet carbon band were obtained at Greenwich and by Konkoly, and also of a less refrangible band between C and D.

On the evening of June 24 Mr. Huggins obtained a photographic spectrum of the head of Comet *b.* This contained two bright groups, the more refrangible one consisting mainly of two bright lines λ 3883 and λ 3870 ; the other group began at a wave-length between 4220 and 4230. These correspond to two groups frequently present in the spectra of the compounds of carbon, which Professors Liveing and Dewar have shown to indicate the presence of nitrogen ; we must, therefore, now add this element to carbon and hydrogen, and probably oxygen, as the principal constituents of the matter of this comet. There was also a suspicion of a third group between h and H.

Besides these groups there was in the photograph a continuous spectrum extending from about F to a little beyond K. In this spectrum the Fraunhofer lines G, h, H, K, and many others, are certainly present, showing without doubt that this spectrum was due to reflected solar light. At Greenwich and elsewhere some Fraunhofer lines were detected in the visible part of the spectrum, and Prof. Wright and other observers detected polarisation in a plane, pointing to the Sun as the source of light. It may be, however, that besides reflected sunlight some part of the bright continuous spectrum of the head of the comet was due to emitted light.

Dr. Draper, a few days after June 24, obtained a photograph of the comet's spectrum in which are seen three bright groups and a continuous spectrum, but no Fraunhofer lines.

The tail of the comet appeared, taking the whole of the recorded observations, to shine by reflected solar light; but also some of the light-emitting stuff of the comet was at times carried into the tail, and could be traced to a greater or less distance from the head.

Prof. Vogel has since examined the spectra of the gases set free by heat from certain meteorites in vacuo. His results agree with those of Prof. Graham in 1867 and of Prof. Wright in 1875. He finds the spectra of the hydrogen and oxygen compounds of carbon simultaneously to be present, and under certain conditions of temperature the hydrocarbon spectrum to predominate to such an extent as to give a spectrum identical with that observed by him in the light of Comet *b*.   **W. H.**

The Publications of the Potsdam Observatory.

The Astrophysical Observatory of Potsdam was established in 1874, when Professors Spörer and Vogel were appointed Observers, with Dr. O. Lohse as First Assistant, but it was not till the spring of 1877 that observations could be made at the new institution. In the interval Prof. Spörer continued the observations of Sun-spots which he had carried on at Anclam from 1861 to 1874—at first in a temporary Observatory in Potsdam, and afterwards, from the summer of 1876, in the grounds of the new Observatory. Prof. Vogel and Dr. Lohse meanwhile carried on their work at the Berlin Observatory.

The instruments at the Potsdam Observatory comprise a refractor by Schröder, of $11\frac{3}{4}$ inches aperture, with equatoreal mounting by Repsold and spectroscopes by Schröder, Hilger, &c.; an 8-inch refractor by Grubb, equatoreally mounted; a $5\frac{1}{4}$-inch Steinheil equatoreal, formerly used at Anclam by Prof. Spörer for his Sun-spot observations; and a Photoheliograph with $6\frac{1}{4}$-inch objective and 10-inch Heliostat mirror by Schröder, with mounting by Repsold.

Two volumes of the Potsdam publications have now been issued, in which Prof. Spörer's observations of Sun-spots at Anclam since 1871, and the work done by Prof. Vogel and Dr. Lohse at Bothkamp and at Berlin, are incorporated with the more recent results at Potsdam.

The results of Prof. Spörer's Sun-spot observations are given for the years 1871–1879, in continuation of the Anclam results 1861–1871, given in Publication XIII. of the Astronomische Gesellschaft (1874), and in a continuation published by the Prussian Government in 1876 (Leipzig, W. Engelmann). The observations were made at Anclam till 1874 June, and afterwards at Potsdam with the $5\frac{1}{4}$-inch Steinheil, the positions of the spots being determined by means of a glass diaphragm with a network of cross-lines. They are supplemented by observations by Konkoly at 'O-Gyalla and Secchi at Rome, and also by photographs taken by Lohse in 1874 and 1875, so that the gaps are filled up as far as possible. From the autumn of 1879 the 8-inch Grubb refractor was used in place of the $5\frac{1}{4}$-inch Steinheil. The results are arranged so as to exhibit the position of each spot on the Sun's disk (pos.-angle and dist. from centre) for each day of observation, together with the respective heliographic longitudes and latitudes. A series of diagrams is given showing the distribution of spots over the Sun's surface for each rotation, and drawings on a larger scale are added, exhibiting changes from day to day in the more remarkable spots. In the discussion of results a comparison is made of the rotation-angles for 78 spots with the formula:—

$$\text{Daily rotation} = 8°548 + 5°798 \cos. \text{lat.}$$

and from this Prof. Spörer concludes that the large deviations from the formula are always towards the west, indicating that a descending current has brought down with it the larger velocity of the higher regions of the Sun's atmosphere. The formula for the rotation-angle would be explained by supposing that ascending currents coming from a certain depth bring to the surface the smaller linear velocity of the lower level. The frequency and the mean heliographic latitude of spots are also discussed, each spot receiving a number in Prof. Spörer's scale according to its size and duration. Unfortunately, these relative numbers are not printed, and we do not find any indication of the areas of spots.

In Vol. I. part ii. Dr. Lohse gives his observations of *Jupiter* and *Mars*, with two plates of drawings made at Bothkamp and Berlin. From observations in 1872 and 1873 of a spot on *Jupiter* in latitude 30° S., extending over a space of 151 days, he deduces a rotation-period of $9^h 55^m 19^s·6$, which is $5^s·9$ less than the mean of those found by Airy, Mädler and Schmidt. As regards the physical condition of *Mars*, he inclines to Mr. Brett's view that the planet is still very hot, and that thus clouds cannot be formed in the atmosphere except at the poles. The white patch seen at the pole would thus be a cloud-cap, and not snow, as is commonly supposed. Dr. Lohse thinks that there would be great difficulty in accounting for the remarkable brightness of the white patch if it be supposed to be due to reflection from snow, seeing that the light would be greatly enfeebled by its passage twice through the atmosphere of *Mars*.

Prof. Vogel's researches on the solar spectrum in Vol. I. comprise maps of the spectrum extending from wave-length 5400 to 3900, together with measures of the positions of the lines. These maps contain a much larger number of lines than Ångström's charts, the spectroscope with which they were made being a powerful instrument with a train of four compound prisms and a half-prism at each end. The wave-lengths of the lines are inferred from those of Ångström's tables by formulæ of interpolation, the position of each line having been measured four or five times with the micrometer. For the blue and violet portions of the spectrum from F to H, the measures were made on photographs, two and a half compound prisms being employed between F and G, and three simple prisms of dense flint between

G and H. An important feature of the work is the care taken in the estimate of the breadth and intensity of each line. As regards the existence of bright lines in the solar spectrum corresponding to lines of oxygen, Prof. Vogel considers that these are merely spaces between dark lines, which seem bright by contrast; and he attaches very little importance to the apparent coincidence found by Prof. H. Draper, as the dispersion used by him was small, and the oxygen lines are diffused. He thinks, however, there may be real bright lines in the spectrum, in particular those pointed out by Cornu at w. l. 3881·5 and 3885·5.

In connection with Prof. Vogel's work, Dr. Gustav Müller, Assistant at Potsdam Observatory, has made two maps of the solar spectrum, representing the lines visible with moderate and small dispersions respectively. The first gives the spectrum from B to H, as seen with a single large flint prism of 60°; and the second shows what is seen with a small spectroscope, the field of view of which includes nearly the whole spectrum at one time. These maps have been prepared for convenience in identifying lines, the spectrum of small dispersion being specially suitable for comparison with star-spectra.

The observations of Comet *b* 1881, form the concluding section of Vol. II. These consist of drawings showing the changes in the appearance of the head on eleven nights, from June 23 to July 15, and of measures of the bands in the comet's spectrum, agreeing generally with the results found by other observers.   W. H. M. C.

Professor Pickering's Researches on Variable Stars.

In an important paper on the "Dimensions of the Fixed Stars," published in the Proceedings of the American Academy, Professor Pickering has discussed the variability in the light of stars of the *Algol* type—that is to say, of stars which, during the greater part of the time, remain unchanged in brightness, but at regular intervals lose, in the course of a few hours, a large part of their light, and regain it with equal rapidity. The various hypotheses which have been advanced to explain the phenomena, whether by volcanic eruption, collision, Sun-spots, or inequality in the photosphere of the revolving star, are rejected in favour of the hypothesis of a uniformly bright star, accompanied by a revolving dark satellite, which causes in every revolution a partial eclipse to us of its primary. From the minimum light of the star the minimum diameter of the satellite is computed, and from a full discussion of this point a radius of 0·764 of the radius of *Algol* is found for the satellite; and assuming that the brightness of *Algol* equals that of our Sun, Professor Pickering finds that its diameter$= 0''·006$, and the diameter of the orbit of the satellite will be about $0''·028$· These small quantities preclude any micrometric measurements being made. Upon these assumptions a circular orbit is found, the elements of which satisfy the observations fairly well. Professor Pickering has discussed all the available observations of minimum of *Algol* from 1783 to 1879, with this curious result, that a change appears to have taken place between 1830 and 1850. There seems during this interval to have been a change of four or five seconds in the period, and besides this there has been a small but gradually increasing diminution in the period throughout the century.

In a second communication to the same Academy, and under the title of "Variable Stars of Short Period," Professor Pickering has discussed a class of Variable Stars whose "light is continually varying, but the changes are repeated with great regularity in a period not exceeding a few days."

The most natural explanation of the variation of a star of short period is that it is due to its rotation round its axis, and this explanation the author accepts and discusses. The variation in the light of *Iapetus*, the outer satellite of *Saturn*, is instanced as an example, this view, of course, possessing no novelty.*

Assuming that the axis of revolution is perpendicular to the line of sight the author shows that the variation of light L may be approximately represented by the formula

$$L = a + b \sin v + c \cos v + d \sin 2v + e \cos 2v,$$

where a = the mean light, v = the angle of rotation, b and c are constants depending upon the comparative brilliancy of the two hemispheres, each of which is supposed to be of uniform intensity, but one brighter than the other; d and e depend on a supposed deviation of the body from the form of a solid of revolution. The paper contains a thorough discussion of the application of this equation to represent the variation in light of all the stars in this class. The four variable stars of short period—ζ *Geminorum*, β *Lyræ*, η *Aquilæ*, and δ *Cephei*—have been observed with such care that their variations are known with precision. As the variation is periodic, the time is denoted by the angle v, such that 360° shall correspond to one period or revolution of the star.

The conclusions arrived at with regard to the above stars are, that ζ *Geminorum* appears to be a surface of revolution, of which one side is about four-fifths of the brightness of the other; in the other three stars the darker side is more than half as bright as the other; and that the difference in the case of β *Lyræ* amounts only to 10 per cent. In other words, if the effect is due to spots we must conclude that they cover only $\frac{1}{10}$ of the hemisphere in the case of β *Lyræ*, and about $\frac{2}{5}$ in the cases of η *Aquilæ* and δ *Cephei*. It is also shown that β *Lyræ* is much elongated,

---

* Dr. Derham wrote *à propos* of "new stars:"—"Some persons think they may be such stars as have one side darker than the other (as one of *Saturn's* satellites is supposed to have), and so appear only when the bright side is turned towards us, and disappear as the darker takes place." *Astro-Theology,* 5th ed. 1726.

the ratio of its axes being as 5 : 3 ; while the other two stars, η *Aquilæ* and δ *Cephei*, have this ratio as 6 : 5. The dark portion of β *Lyræ* is at one of the ends, and it appears to be symmetrically situated as regards the larger axis. The dark portions of η *Aquilæ* and δ *Cephei* are placed somewhat preceding an end—that is, they are turned towards the observer before the end has been directed to him. For this reason the time from minimum to maximum is greater than from maximum to minimum.

These results are of very great interest, and the able discussion of the subject contained in the paper renders it an important and valuable contribution to our knowledge. The paper concludes with a consideration of the hypothesis, and its consequences, that if there be a common origin of the stars of the Milky Way, a general coincidence in their axes of rotation seems not improbable.

Prof. Pickering has communicated to the American Academy a further paper on "Photometric Measurements of the Variable Stars β *Persei* and D.M. 81° No. 25." A good series of observations of *Algol* was made, including seven minima, with a probable error for a single minimum of 3·8 minutes. Ceraski's Variable D.M. 81° No. 25 was observed with the photometer on the 15-inch Refractor. The entire number of measures is 273 sets, or 3276 settings. Five minima were observed with an average probable error for a minimum of 1·3 minute. The greatest change in light amounted to 0·02 magnitude a minute, or at the rate of 1·2 magnitude an hour.

As this is a star of the *Algol* class, its variation in light is considered to be due to an eclipsing satellite, and the conclusions arrived at from a full discussion of the observations with regard to this remarkable variable are that " the eclipse is total (that is, that the star is entirely covered by the satellite), and that the light during the minima is due to one of the two following causes: first, that the satellite is self-luminous, and that the light at the time of the minimum is that received from the satellite, the star itself being completely obscured ; or the satellite consists of a cloud of meteors, so scattered that about 0·110 of the light could pass through the central portions."

Though astronomers may hesitate to accept these conclusions as proved, there can be little doubt that the valuable researches of Prof. Pickering deserve their best and most careful consideration.

E. B. K.

Standards of Stellar Magnitude.

In the summer of 1880 the American Association appointed a committee, consisting of the leading astronomers of the United States, for the purpose of selecting a series of Standards of Stellar Magnitudes. This committee, of which Prof. E. C. Pickering was the chairman, at once brought the subject before the Councils of the Astronomische Gesellschaft and the Royal Astronomical

Society to secure their co-operation in obtaining a series of standards which might be generally accepted by astronomers. The Council of the Astronomische Gesellschaft unanimously agreed to co-operate, and appointed Dr. Schönfeld as their representative, and a committee was also appointed by the Royal Astronomical Society.

The objects of the committee may be briefly expressed: first, to select a series of stars suitable as standards; secondly, to secure measurements of their relative light by various independent methods; and thirdly, from the results to deduce their magnitudes according to any scale that may be deemed best.

In the Report recently issued by the American Committee the plan of operations is thus very clearly put :—

"Stars may be conveniently divided, according to their brightness, into three classes—

"1. Lucid stars, or those brighter than the sixth magnitude;

"2. Bright telescopic stars, from the sixth to the tenth magnitude;

"3. Faint telescopic stars, fainter than the tenth magnitude.

"It is proposed that the first of these classes be assigned to the Royal Astronomical Society, the second to the Astronomische Gesellschaft, and the third to the American Association. In accordance with this scheme the following plan is recommended for the fainter stars.

"The standard stars to be so selected that they will form twenty-four groups near the equator, and at approximately equal intervals in Right Ascension. Each group to consist of a series of stars decreasing in brightness by differences of about half a magnitude from the tenth magnitude to the faintest object visible in the largest telescopes. The groups to be located by bringing a star visible to the naked eye into the field of the telescope, waiting for two minutes, and then forming a chart of the zone ten minutes wide passing through the centre of the field of the telescope during the next four minutes. This zone will therefore be defined as the region from 5' N. to 5' S. of the bright star, and from 2ᵐ to 6ᵐ following it. The stars to be selected from this zone."

The advantages of the system proposed are that an observer in any part of the earth and at any season will find comparison stars conveniently situated for observation.

The plan proposed is at present only under consideration, and final action has not yet been taken. Before the scheme is definitely settled, it would perhaps be desirable that the important question of the adoption of a uniform light ratio for naked-eye as well as telescopic stars should be carefully considered.

Our knowledge of stellar photometry has been materially advanced by the researches which have been conducted with so much energy and ability at the Harvard College Observatory, and which have been mainly directed to naked-eye stars. The

American Association may be congratulated on having brought before astronomers a more comprehensive scheme for the extension of our knowledge of stellar magnitudes in a definite and tangible form, and endorsed by such distinguished names as those appended to the Report.

E. B. K.

Physical Observations of Jupiter.

During the last three oppositions this planet has received much attention from observers. The red spot has maintained its conspicuous appearance, and this, with numerous other markings, has developed some remarkable phenomena. Mr. Marth finds (*Monthly Notices*, May 1881, p. 367) that the average rotation period of the red spot between the last months of 1878 and the first months of 1881 was $9^h 55^m 34^s\cdot47$, equal to a daily rate of $870°\cdot42$; but the motion of the spot perceptibly slackened in the interval, and the rate of retardation is increasing. The white equatoreal spots have also been carefully watched, and the observations of the most prominent of these indicate a mean period of $9^h 50^m 6^s\cdot6 = 870°\cdot48$ daily rate, between October 1880, and the present time. Mr. Denning has been following this object closely, and has obtained observations of 70 transits. He remarks that between Nov. 19, 1880, and Dec. 24, 1881, it completed 976 rotations, while the red spot had performed 967. The swifter motion of the bright equatoreal spot enables it to complete a revolution of *Jupiter* relatively to the red spot in $44^d 10^h 42^m 13^s\cdot3$. It gains $13^m 24^s = 8°\cdot1$ daily on the red spot. Other singular phenomena have been noticed. In October 1880 there was an eruption of dark spots from a belt some short distance N. of the equator, and these rapidly increased and developed longitudinally until, gradually losing their distinctive features, they formed a new belt on the planet. The spots of this group moved with great celerity while visible, the rotation period being only $9^h 48^m$, or $7\frac{1}{2}^m$ less than the red spot. The latter object is now (Jan. 1882) of a much lighter tint than formerly, and very dark markings have appeared at its E. and W. extremities, while the N. margin of the spot is much darker than the centre. The general features of the planet appear to have remained permanent during the last few years. The history of the red spot previous to the last months of 1878 is at present uncertain on account of insufficient evidence. In case observations or sketches are available, which are not yet in the possession of the Society, it would be desirable that they should be communicated to the Society, so that the development of the spot may be traced as far back as possible.

Backlund's Researches on Encke's Comet.

In a Paper "Zur Theorie des Encke'schen Cometen," published in the Memoirs of the Petersburg Academy, Dr. O. Backlund has made an interesting contribution towards a determination of the absolute perturbations of Encke's Comet by *Jupiter*. When the late Dr. von Asten resolved to construct the theory of the motion of Encke's Comet upon a foundation of absolute perturbations, the possibility of the successful accomplishment of such a formidable undertaking might well be considered questionable. Hansen had, indeed, shown in his "Mémoire sur le calcul des perturbations qu'éprouvent les comètes," published in 1850, how the convergency of the series which represent the perturbations may be in some respects greatly increased by adopting a principle of partition according to which the coordinates of the comet are expressed in different portions of the orbit by different variables. But the practical difficulties in applying his developments in that part of the orbit of Encke's Comet where it approaches the orbit of *Jupiter* appeared so very great that Hansen himself recommended for this part of the orbit the computation of special perturbations by one of the usual methods. The attempt to determine the absolute perturbations promised to become feasible, since Gyldén took advantage of the resources of the theory of elliptic functions for the purpose of increasing further the convergency of the series in the development of the perturbations.

In order to show that the obstacles in the way could be surmounted, Asten computed and published in 1872 the absolute perturbations of the first order for the portion of the orbit between the true anomalies 170° and 180°. However, not only practical considerations, but certain theoretical doubts, occasioned by the acceleration of the comet's mean motion, induced Asten not to go on with his determination of the absolute perturbations, but first to compute the special perturbations since 1848, and to combine them with the preceding ones in one investigation so as to represent the whole series of observations which had been made since 1818. The results of his laborious researches, which are highly interesting in several respects, were published in 1878.

After Asten's early and lamented death in the same year, Dr. Backlund has undertaken the continuation of these researches, and, besides computing the special perturbations of the comet during its last revolution, has resumed the attempt to determine the absolute perturbations produced by *Jupiter*. His present Memoir refers to the part of the orbit between the true anomalies 180° and 190°, or from aphelion to the point where the distance of the comet from the Sun is about 3·78. The treatment of the problem differs, in some respects, considerably from that adopted by Asten; for in such an untilled field the best way of working is only found out by practical

experience. By means of the two investigations the most difficult portion of the perturbations of the first order may be
considered to have been determined, and the computations for
the rest of the orbit are partly finished, so that the absolute
perturbations (of the first order) of the mean anomaly, of the
logarithm of the radius vector and of the sine of the latitude,
which are produced by *Jupiter*, will be known before long.

*Mr. G. W. Hill's Paper on Gauss's Method of Computing Secular
Perturbations.*

In the astronomical papers prepared for the use of the
American Ephemeris and Nautical Almanac Mr. G. W. Hill
has published a paper with this title, the object of which is to
reproduce, in a form practically useful for astronomers, the
results given by Gauss, in his celebrated memoir, *Determinatio
Attractionis* &c.   Mr. Hill writes :—" Gauss investigates the
expressions for the components of the attraction of a certain
species of elliptic ring on a point, which can be advantageously
employed in computing the secular perturbations of a planet,
at least the parts of them which are of the first order, with
respect to the disturbing forces. This method merits attention,
because with it we can secure almost absolute accuracy, at the
cost of a comparatively small outlay of labour. Moreover, it is
capable of being applied with success to all asteroids, and even
to such refractory cases as the periodic comets. Yet I can find
but two published investigations where it has been employed.
. . . . . This, perhaps, is due to the circumstance that the memoir
of Gauss does not contain all the formulæ needed in the application. A double integration being necessary, Gauss has con
sidered only that in respect to the eccentric anomaly of the
disturbing body, and, having regard to elegance only, has not
reduced his equations to the forms giving the utmost brevity of
calculation. Hence I propose to give an exposition of the
method, with the additional formulæ required."

Gauss's Memoir is a mathematical one, of great beauty, and
it is perfect as regards mode of investigation and style ; but the
method is not presented in such a manner that it admits of being
readily applied by an astronomer to an actual case. Mr. Hill,
by this paper, places Gauss's process within the reach of any
computer possessed of ordinary mathematical knowledge, and he
has thereby conferred a benefit upon astronomy. It may be
mentioned that in Gauss's investigation a certain cubic equation
has to be solved, the roots of which are denoted in Gauss's
notation by G, G', G'', and that the solution depends upon certain complete elliptic integrals, the square of the modulus of
which is $G' + G'' \div (G + G'')$.

Mr. Hill, after giving Gauss's investigation in an extended
form, and with the addition of all the details necessary for its

application to an actual numerical case, reproduces, at the end of his paper, the formulæ by themselves, in the form in which they are required by the practical calculator, and he works out as an example the computation of the secular perturbations of *Mercury* produced by the action of *Venus*, the final results agreeing very closely with those obtained by Leverrier.

Apart from its astronomical importance, Gauss's memoir is memorable on account of its containing the method known by his name for the calculation of complete elliptic integrals by the process of the arithmetico-geometric mean. Mr. Hill replaces this by a process which is equivalent to the use of Landen's transformation, and he also gives three tables, the values in which were deduced from Legendre's tables of elliptic integrals. In an appendix, however, in which he gives a second method of solving the cubic equation above referred to, he also gives Gauss's own method of approximating to the values of the integrals. The latter would appear to be the more convenient of the two in practice, as no tables except ordinary logarithms are required; and this was found to be the case by Professor Adams, who employed both sets of formulæ in his calculations on the secular variation of the node of the orbit of the November meteors, the results of which were published in the *Monthly Notices* (vol. xxvii. p. 247). It may be remarked that by an oversight Mr. Hill does not mention Professor Adams among those who have practically applied Gauss's method.

The translation of Gauss's memoir into a form accessible to the practical astronomer seems to have been very ably performed by Mr. Hill, and the formulæ are put in a convenient shape for actual computation. The comparison with Leverrier's results for *Mercury* disturbed by *Venus* is valuable, and it would be interesting to work out the secular variations of the orbits of the other principal planets by the same method.

The Stresses caused in the Interior of the Earth by the Weight of Continents and Mountains.

In the course of the year Mr. G. H. Darwin has communicated to the Royal Society a memoir having the above title, that will be published in the *Philosophical Transactions*, but of which only an abstract has as yet appeared.

The author considers the subject of the solidity and strength of the materials of which the Earth is formed from a new point of view.

The existence of dry land proves that the Earth's surface is not a figure of equilibrium appropriate for the diurnal rotation. Hence the interior of the Earth must be in a state of stress, and as the land does not sink in nor the sea-bed rise up, the materials of which the Earth is made must be strong enough to bear this stress; and we are thus led to consider how the stresses are distributed and what are their magnitudes. In this memoir the author solves a problem of this kind for the case of a homogeneous incompressible elastic sphere, and applies the results to the case of the Earth.

If the Earth be formed of a crust with a semi-fluid interior, the stresses in that crust must be greater than if the whole mass be solid, very far greater if the crust be thin.

The strength of the elastic solid is estimated by the difference between the greatest and least principal stresses when it is on the point of breaking—or, in other words, by the breaking stress-difference. The most familiar examples of breaking stress-difference are when a wire or rod is stretched or crushed until it breaks, then the breaking load divided by the area of the section of the wire or rod is the measure of the strength of the material. Stress-difference is measured by tons per square inch.

On evaluating the stress-difference arising from given ellipticity in a rotating spheroid of the size and density of the Earth, it appears that if the excess or defect of ellipticity above or below the equilibrium value were $\frac{1}{1000}$, then the stress-difference at the centre would be 8 tons per square inch; and that, if the sphere were made of material as strong as brass, it would be just on the point of rupture. Again, if the homogeneous Earth, with ellipticity $\frac{1}{232}$, were to stop rotating, the central stress-difference would be 33 tons per square inch, and it would rupture if made of any material excepting the finest steel.

In the second part of the paper it is shown that the great terrestrial inequalities, such as Africa, the Atlantic Ocean, and America, are represented by an harmonic of the 4th order; and that, having regard to the mean density of the Earth being about twice that of superficial rocks, the height of the elevation is to be taken as about 1,500 metres.

Four tons per square inch is the crushing stress-difference of average granite, and accordingly it is concluded that at 1,000 miles from the Earth's surface the materials of the Earth must be at least as strong as granite. A very closely analogous result is also found from the discussion of the case in which the continent has not the regular wavy character of the zonal harmonics, but consists of an equatoreal elevation with the rest of the spheroid approximately spherical.

From this we may draw the conclusion, either that the materials of the Earth have about the strength of granite at 1,000 miles from the surface, or that they have a much greater strength nearer to the surface.

This investigation must be regarded as confirmatory of Sir William Thomson's view, that the Earth is solid nearly throughout its whole mass. According to this view, the lava which issues from volcanoes arises from the melting of solid rock, existing at a very high temperature, at points where there is a diminution of pressure, or else from comparatively small vesicles of rock in a molten condition.

On Small Displacements of the Plumb-line.

In a Report on the measurement of the lunar disturbance of gravity presented to the British Association at York, Mr. G. H. Darwin and Mr. Horace Darwin have described an instrument for detecting and measuring small changes in the direction of the force of gravity, and the results obtained by means of it.

The instrument consisted of a heavy pendulum, with a contrivance for rendering evident exceedingly minute displacements of the pendulum with reference to the surface of the Earth. The following is a description of the principle involved :—

To the bottom of the pendulum bob is fastened a short silk fibre, and the other end is attached to the edge of a galvanometer mirror. A second short fibre, of equal length with the former, is also attached to the edge of the mirror, and its other end is fastened to a support which is fixed on the basement of the structure which supports the pendulum.

The mirror is thus hung by a bifilar suspension, one half of the suspension being attached to the pendulum, and the other half to the basement.

When the pendulum moves relatively to the basement, the rotation of the mirror about a vertical axis exhibits that component of the motion of the pendulum which is perpendicular to the plane through the two silk fibres. The rotation of the mirror may be observed as in a galvanometer. In order to render the instrument sensitive to small deflections of the plumb-line, it is necessary that the two silk fibres should be brought into close proximity. The sensitisation was effected by means of fine screw adjustments.

Instruments involving this principle were erected in the Cavendish Laboratory at Cambridge in two different forms: in the first, the support for the pendulum consisted of a massive stone gallows; in the second, and better instrument, it was a stout copper tube standing upon a large block of stone. The authors succeeded in eliminating the disturbing influence of purely local tremors by hanging the pendulum in a mixture of alcohol and water, and this did not interfere with the sensitiveness of the instrument to perturbations of longer period.

An apparatus was introduced for determining the deflection of the pendulum relatively to the basement in terms of the angular displacement of the galvanometer mirror.

It was found that a deflection of the pendulum through $\frac{1}{300}$th of a second of arc was rendered distinctly visible, and this deflection corresponded with a movement of the bottom of the pendulum by $\frac{1}{40000}$th of a millimetre, perpendicular to the silk fibres. Many precautions were taken for the purpose of equalising the temperature of the instrument, and to avoid the disturbing effects of the weight of the observer's body. The sensitiveness to this latter kind of disturbance is illustrated by the fact that the instrument gave different readings when a person stood at distances of 16 feet and of 17 feet from the pendulum ; this experiment rendered evident the change in the state of elastic distortion of the soil due to exceedingly small changes in the distribution of the disturbing weight. It is, however, to be noted that the jumping and stamping of a person near the instrument produced no sensible effect on the reading, provided his mean position remained unaffected.

The authors conclude from their first set of experiments that stone columns are highly sensitive to the warping influence of changes of temperature, and of small stresses; and that a massive stone structure may be sensibly tilted over by the percolation of small quantities of water round its foundations. It was on account of disturbance due to temperature that they adopted copper, surrounded by a large mass of water, as the support for the pendulum in the second set of experiments.

By means of a series of observations it was found that the pendulum has a diurnal period of· oscillation in the meridian, with a range of a fraction of a second of arc. Besides this, the mean diurnal position of the pendulum was found to travel on the whole in one direction for weeks together, although the path of pendulum was often interrupted by apparently capricious reversals. No observations have hitherto been made with regard to disturbances in the prime vertical.

The pendulum is never at rest, but the image of the scale in the observing telescope is found to be continually waving to and fro in an uncertain manner. Nevertheless, there are periods, lasting sometimes for several days, of abnormal agitation in the image, and again of abnormal quiescence.

The authors propose to modify the form of the instrument, and to carry their observations further. They are of opinion that the sensitiveness is amply great enough to measure the lunar attraction, but that, at least on the Earth's surface, it is impossible to obtain any basement sufficiently immovable for that purpose.

In the second part of the paper some account is given of the work, in the same field, of Zöllner, d'Abbadie, Plantamour, Siemens, and of others, and the authors' experience is in agreement with the observations of their predecessors.

In the last part they comment on the various forms of instrument which have been adopted for the detection of small changes of level. They suggest precautions to be adopted by astronomical observers for the protection of their instruments, and they speculate on the physical significance of the facts which they have observed.

Longitude of the Morrison Observatory, Glasgow, Missouri.

The volume of *Washington Observations* for 1877 contains in an Appendix a telegraphic determination of the longitude of the Morrison Observatory, Glasgow, by Prof. Eastman and Mr. Pritchett.

The arrangement for the exchange of signals was to allow the Washington clock to break the closed circuit from Washington to Glasgow, each second for three minutes, recording each break on the chronograph at Washington and on the chronograph at Glasgow. Then the Glasgow clock was allowed to break the circuit in the same way for three minutes. The hour and the first and last minutes of the recorded signals of the Washington clock were telegraphed to Glasgow, and similar data in relation to the Glasgow clock were received from Glasgow, and this completed the exchange of signals.

The clock-corrections at the Naval Observatory were obtained by Prof. Eastman with the Transit Circle from observations of stars of the American Ephemeris, the adopted clock corrections and rates being obtained by the method of least squares for all the observed clock-corrections. The corrections for personal equation were derived from observations with Prof. Eastman's personal equation apparatus. These corrections depend upon five nights' observations during which the telegraphic signals were exchanged.

At the Morrison Observatory the Transit Circle by Troughton and Simms has a 6-inch object-glass and 24-inch circles read by four microscopes. The clock used in the time observations in exchange of signals is by Frodsham, and had been running for two years, on a very small and constant rate. The time signals were exchanged on June 16, 17, 19, 21, and July 2, 1880. Unfortunately, the clock stopped on June 23. After cleaning, it resumed almost exactly the same rate on which it had been running before. Mr. Pritchett's personal equation was determined with one of Professor Eastman's machines. The clock-corrections were obtained from observations of 12 or 14 stars on the above dates, the resulting values, determined by least squares, having small probable errors.

In the final result there is a satisfactory accordance between the observations of each night at each station;* but it is noticeable that whereas the mean difference of longitude from signals Glasgow to Washington is 1^h 3^m $6^s \cdot 081$, the mean difference from signals Washington to Glasgow is 1^h 3^m $5^s \cdot 638$, evincing a small systematic difference. The mean of these results, with a further small correction, gives 1^h 3^m $5^s \cdot 926$ as the longitude of the Morrison Transit Circle, forming an important addition

---

* In the Washington record of the Washington clock, June 16, there is an evident misprint of 44^m for 48^m.

to the valuable series of telegraphic determinations of longitude so successfully carried out by the United States Naval Observatory.   E. B. K.

Universal Time and the Selection of a Prime Meridian.

In September 1880 M. Otto Struve presented to the St. Petersburg Academy of Sciences a Report on a paper by Mr. Sandford Fleming, presented to the Toronto Institute, on "Time, Reckoning and the Selection of a Prime Meridian," and also on Professor Cleveland Abbé's "Report on Standard Time to the American Meteorological Society."

A strong desire had been felt by the United States and British North America to have some uniform system of time, particularly for railway and telegraphic purposes. Professor Abbé advised that, for the use of the United States alone, time should be reckoned from a meridian six hours west of Greenwich: that is to say, about the meridian of New Orleans. The American Meteorological Society further considered it desirable that in the future a universal time reckoned from the meridian 180° from that of Greenwich should be generally introduced.

Mr. Fleming also proposes that this meridian 180° from Greenwich should be adopted throughout the world as the prime meridian, and that time reckoned from it should be called "cosmopolitan," to distinguish it from "local" time, and that this time should be used for scientific purposes, and also for certain requirements of civil life. Mr. Fleming propounds two questions:—

1. Does the prime meridian proposed appear to be the most convenient one to be adopted by all civilised nations?

2. If this meridian be objected to, can another be suggested which has a better prospect of being accepted throughout the world?

In 1870 M. Otto Struve delivered a discourse to the Russian Geographical Society on the question of the adoption of a prime meridian. At first sight the simplest solution seemed to be to adopt that of Greenwich as the zero of time. There were many reasons in favour of it, the principal perhaps being that the large majority of nautical charts were constructed for that meridian, and that in long voyages nearly all seamen were in the habit of reckoning their longitude from Greenwich. Still, considering that this would occasion longitudes being reckoned in different parts of Europe and Africa with different signs, and that there might be some hesitation as to its adoption by French and other seamen, M. Struve decided to recommend the meridian 180° from Greenwich as the best, and likely to be the most generally acceptable, for the following reasons:—

1. The only continent it crosses is the most eastern extremity of Asia, inhabited by a scanty and half-civilised population.

2. This meridian is that where navigators are now in the habit of changing their date one day ; thus the commencement of a new date would be coincident with the zero of cosmopolitan time.

3. It would not materially affect the present custom of seamen and geographers, except by the simple addition of twelve hours or 180° to all the longitudes.

4. It would not involve any material alteration in the Ephemerides commonly used by seamen, except the alteration of noon for midnight and *vice versa* in the British *Nautical Almanac*.

One advantage to Canada and the United States would be that cosmopolitan time would differ so widely from local time, that there would be little chance of the two being confounded.

M. Otto Struve therefore advises the Imperial Academy of St. Petersburg to decide in favour of this prime meridian.

The proposition of Mr. Fleming that cosmopolitan time should be used for certain purposes as civil time, suggests difficulty and probable confusion. A further suggestion of his is to a certain extent necessitated by what is proposed—namely, the abolition of the division of the day into two series of twelve hours, and replacing it by one consecutive series of twenty-four hours. This proposition is strongly advocated by M. Struve.

A great advance was made when the majority of astronomers accepted the commencement of the fictitious year proposed by Bessel, when the Sun's mean longitude has a certain value, as the Epoch to which to refer all observations, and similarly a most important advantage would be gained by the unanimous adoption of a prime meridian as the zero of universal daily time. The subject is one of great importance and worthy of the serious consideration of astronomers.*   E. B. K.

The Smithsonian Report on Astronomical Observatories.

Among the contributions to the Smithsonian Report for 1880 will be found a valuable and interesting account of all the known astronomical Observatories, prepared under the auspices of the Smithsonian Institution. The object of the compilers appears to be to present a yearly summary of the state and progress of Astronomy, thus giving a record of current astronomical work as illustrated by the independent annual labours of each of the 220 Observatories, the names of which, at least, will be found in the list. It can hardly be expected that a work of this kind can be prepared without many omissions or errors, of which we have noted several important ones ; but most of these will probably disappear in the second annual issue. The information obtained

* " Uniformity in nomenclature and modes of reckoning in all matters relating to time, space, weight, measure, &c., is of such vast and paramount importance in every relation of life as to outweigh every consideration of technical convenience or custom."—Sir John Herschel, *Outlines*, Art. 147.

by a reference to this work is not confined to a statement of the observations made during the preceding year, or of the proposed subjects of observation for the ensuing year, but, in many instances, brief historical notes are given, with the names of the past as well as the present directors of the older Observatories. It also gives the *personnel* of each establishment, the instruments employed in the observations, and the character of the observations made or contemplated. In the revised issue intended to be included in the Smithsonian Report for 1881, it is to be hoped that many of the frequent blanks found in the first edition will be filled up, and if astronomers generally will respond to the invitation of the Secretary of the Smithsonian Institution, by taking the trouble to point out any errors or omissions observed, and by forwarding them to him, together with such information respecting either the past history or present condition of their respective Observatories, the value of this excellent work of reference will be much increased.

E. D.

Papers read before the Society from February 1881 to
February 1882.

1881.

Mar. 11.   Observations of Comet *e*, 1880 (Swift), made at Dun
Echt Observatory with the filar micrometer of the
15·06-in. refractor.   The Earl of Crawford.

Effect on the Moon's movement in latitude produced
by the slow change of position of the plane of the
ecliptic.   Sir G. B. Airy.

·Observations of the phenomena of the satellites of
Jupiter, with a few transits of the red spot; also a
few observations of the brighter satellites of *Saturn*,
made at the Observatory of Mr. Edward Crossley.
J. Gledhill.

Note on the comparative brightness of the nebula of
Orion.   E. B. Knobel.

Comparative observations of disappearance and reappearance of *Jupiter's* satellites.   Rev. S. J. Johnson.

Note on a paper by Mr. Denning in the December
number of the *Monthly Notices*.   J. L. McCance.

A reply to Mr. Christie's paper on Mr. Stone's alterations of Bessel's refractions.   E. J. Stone.

The partial eclipse of the Sun, 1880, Dec. 31.   J.
Rand Capron.

Second catalogue of radiant points of meteors.   E.
F. Sawyer.

On the determination of the value for the parallactic
inequality in the motion of the Moon.   J. Campbell and E. Neison.

Apr. 8.   Observations of the red spot on *Jupiter*.   J. Tebbutt.

Note on the inequality of the Moon's latitude which
is due to the secular change of the plane of the
ecliptic.   Prof. J. C. Adams.

Note on the durations observed by the French
observers at the Transit of *Venus*, 1874.   E. J.
Stone.

The planetary nebula in *Cygnus*.   S. W. Burnham.

On the solar parallax derived from observations of
Mars at Ascension in 1877 (abstract).   D. Gill.

May 13.   On a simple approximate method of calculating the
effect of refraction upon the distance and position-
angle of two adjacent stars.   Prof. R. S. Ball.

On the inclination of the Zodiacal light.   T. W.
Backhouse.

Fall of a meteorite on March 14, 1881. Prof. A. S. Herschel.

The nebula near '*Merope*. Prof. G. W. Hough and S. W. Burnham.

Transit times of the spots on *Jupiter*. W. F. Denning.

Addition to a paper on the effect of the Moon's movement in latitude produced by the slow change of position of the plane of the ecliptic. Sir G. B. Airy.

Note on the physical libration of the Moon. E. Hartwig.

Note on the flexure of the Greenwich Transit Circle. W. H. M. Christie.

Further remarks on Mr. Stone's alterations of Bessel's refractions. W. H. M. Christie.

On the supposed difference in the refractions north and south of the zenith of Melbourne. A. M. W. Downing.

Observations of the companion of *Sirius* made at the United States Naval Observatory, Washington. Prof. A. Hall.

Note on the effect of atmospheric dispersion on the determination of the solar parallax by meridian observations of *Mars*. E. J. Stone.

Note on some points connected with the determination of the coefficient of the parallactic inequality. E. J. Stone.

Ephemeris for physical observations of *Jupiter*, 1881–82. A. Marth.

Note on photographs of the nebula in *Orion*. Dr. H. Draper.

June 10. The geographical position of the Observatory of the John C. Green School of Science at Princeton, New Jersey. Prof. C. A. Young.

Ephemeris for finding the positions of the satellites of *Neptune*, 1881–82. A. Marth.

Observations of the solar eclipse of 1880, Dec. 31, made at the Royal Observatory, Greenwich. The Astronomer Royal.

On the comparison of the computed and observed motions of the components of the binary system ξ *Ursæ Majoris*. Prof. C. Pritchard.

On the Moon's photographic diameter, and on the applicability of celestial photography to accurate measurement. Prof. C. Pritchard.

On the determination of the Moon's libration. A. Marth.

Proper motions, chiefly in the southern hemisphere, obtained from a comparison between the positions of the Cape Catalogue, 1880, and other published catalogues. E. J. Stone.

| Nov. 11. Observations of Comet *b*, 1881, made at Monte Video. Lieut. B. Gwynne.

Note on Mr. Gill's Ascension observations. G. F. Hardy.

Apparent places of Comet *b*, 1881, from observations made with the North equatoreal of the Melbourne Observatory. R. L. J. Ellery.

Observations and orbit-elements of Comet II., 1881. J. Tebbutt.

Comet *c*, 1881 (Schäberle). T. W. Backhouse.

Elements of Comet *b*, 1881. H. T. Vivian.

Approximate positions of Comet *b*, 1881, deduced from observations made at the Adelaide Observatory. C. Todd.

Observations of Comet *b*, 1881, made at Stonyhurst Observatory. Rev. S. J. Perry.

Physical observations of Comets *b* and *c*, 1881, made at Forest Lodge, Maresfield. Capt. W. Noble.

Ephemeris for finding the positions of the satellites of *Uranus*, 1882. A. Marth.

Note on sketches of Comet *b*, 1881. E. B Knobel.

On the North Polar Distances of the Cape Catalogue for 1880, and on the Greenwich and Cape Mean Systems of North Polar Distances. A. M. W. Downing.

On the spectra of Comets *b* and *c*, 1881, observed at the Royal Observatory, Greenwich. The Astronomer Royal.

On a simple and practicable method of measuring the relative apparent brightnesses or magnitudes of the stars, with considerable accuracy. Prof. C. Pritchard.

Showers of large Meteors. W. F. Denning.

Observations of Comet *b*, 1881, made at Sandy Point, Straits of Magellan. Lieut. B. Gwynne.

Observations of Comet *b*, 1881. Rev. S. S. O. Morris.

Observations of Comet *c*, 1881, made at Grenada, West Indies. Capt. J. C. Maling.

Note on Mr. Christie's paper in the *Monthly Notices* for May, 1881. E. J. Stone.

On the orbit of Denning's Comet. Prof. A. Winnecke.

On the value of the parallactic inequality. J. Campbell and E. Neison.

Dec. 9. Observations of *Venus* in the spring of 1881. W. F. Denning.

Radiant points of shooting stars observed at Bristol in the years 1878 and 1879. W. F. Denning.

On the motion of the companion of *Sirius*. W. E. Plummer.

Note on the variable star D.M. + 1° No. 3408. Prof. C. Pritchard.

On the method for finding the elements of the orbit of a comet by a graphical process. F. C. Penrose.

On the conjunctions of the satellites of *Uranus* with each other which may be observable from February to May 1882. A. Marth.

On a new form of transit circle, with a prismatic object-glass. E. J. Stone.

Note on Messrs. Campbell and Neison's paper on the parallactic inequality, in the supplementary number of the *Monthly Notices*. E. J. Stone.

Note on silvering large mirrors. A. A. Common.

Observations of *Minas*, 1881. A. A. Common.

Note on the discovery of Comet *c*, 1881 (Schäberle). W. F. Denning.

On some systematic errors in the determinations of the semi-diameter of the Moon from the Greenwich Observations 1750–1840. E. J. Stone.

1882.
Jan. 13.

The relative motions of the great red spot and brilliant equatorial spot on *Jupiter*. W. F. Denning.

Observations of the Transit of *Mercury*, 1881, November 7-8, made at Windsor, New South Wales. J. Tebbutt.

Observations of the Transit of *Mercury*, 1881, November 7-8, made at the Melbourne Observatory. R. L. J. Ellery.

Observations of the satellites of *Mars*. J. Watson.

Phenomena of *Jupiter's* satellites and occultations of stars by the Moon observed at Stonyhurst Observatory. Rev. S. J. Perry.

Observations of *Deimos* made at the Royal Observatory, Greenwich. Astronomer Royal.

The *Merope* nebula. Lewis Swift.

Note on the employment of photography in the Transit of *Venus*, 1882. E. W. Maunder.

Remarques sur la méthode proposée par M. le Professeur Pritchard pour la mesure de l'éclat des astres. M. Loewy.

The Transit of *Mercury* observed at Castlemaine, Victoria. Dr. W. Bone.

Note on the Transit of *Mercury*, 1881, November 8. Dr. L. S. Little.

Madras observations of *Jupiter's* satellites. N. R. Pogson.

Observations of the Transit of *Mercury* on November 7-8, 1881, at Vizagapatam. A. V. Nursinga Row.

*List of Public Institutions and of Persons who have contributed to
the Library &c. since the last Anniversary.*

Her Majesty's Government in Australia.
Her Majesty's Government in India.
The French Government.
The Norwegian Government.
The Lords Commissioners of the Admiralty.
British Association for the Advancement of Science.
British Horological Institute.
Geological Society of London.
Institute of Actuaries.
Meteorological Office.
Meteorological Society.
Photographic Society of Great Britain.
Physical Society of London.
Royal Asiatic Society.
Royal Geographical Society.
Royal Institution.
Royal Observatory, Greenwich.
Royal Society of London.
Royal United Service Institution.
Selenographical Society.
Society of Arts.
University College, London.
Zoological Society of London.
Bristol Museum and Library.
Cambridge Philosophical Society.
Dublin, Royal Irish Academy.
Royal Dublin Society.
Edinburgh, Royal Society.
Glasgow Philosophical Society.
Kew Observatory.
Leeds Philosophical and Literary Society.
Liverpool Free Public Library.
Liverpool Literary and Philosophical Society.
Liverpool Observatory.
Oxford, Radcliffe Library.
Oxford, Radcliffe Observatory.
Rugby School Natural History Society.
Adelaide, Royal Society of South Australia.
Amsterdam, Royal Academy of Sciences.
Ann Arbor Observatory.
Batavia, Magnetical and Meteorological Observatory.
Batavia, Royal Society of Natural History.

Berlin, Physical Society.
Berlin, Royal Academy of Sciences.
Berlin, Royal Observatory.
Berlin, Royal Prussian Geodetic Institute.
Berne, Central Meteorological Institute.
Bologna, Academy of Sciences.
Bombay, Royal Asiatic Society.
Bordeaux, Society of Physical and Natural Sciences.
Boston, American Academy of Arts and Sciences.
Boston Scientific Society.
Brussels, Royal Academy of Sciences.
Brussels, Royal Observatory.
Buda-Pesth, Hungarian Academy of Sciences.
Calcutta, Asiatic Society of Bengal.
Cape of Good Hope, Royal Observatory.
Cherbourg, National Society of Sciences.
Chicago, Dearborn Observatory.
Chicago, Public Library.
Cincinnati Observatory.
Coimbra, Observatory of the University.
Copenhagen, Royal Academy of Sciences.
Dijon, Academy of Sciences.
Geneva, Society of Physics and Natural History.
Genoa, Society of Literature and Science.
Göttingen Observatory.
Göttingen, Royal Society of Sciences.
Haarlem, Musée Telyer.
Helsingfors, Finnish Society of Sciences.
Kasan, Imperial University.
Leghorn, Technical and Nautical Institute.
Leipzig, Astronomische Gesellschaft.
Leipzig, Royal Saxon Society of Sciences.
Lisbon Geographical Society.
Lisbon, Royal Academy of Sciences.
Madison, Washburn Observatory.
Melbourne Observatory.
Melbourne, Royal Society of Victoria.
Milan, Royal Observatory.
Moncalieri Observatory.
Montpellier, Academy of Sciences.
Mozambique, Society of Geography.
Munich, Royal Bavarian Academy of Sciences.
Naples, Royal Academy of Sciences.
Neuchatel, Society of Natural Sciences.
New York, Astor Library.
Palermo Observatory.
Paris, Academy of Sciences.
Paris, Bureau des Longitudes.
Paris, Dépôt Général de la Marine.
Paris, Ecole Polytechnique.

Paris, International Committee of Weights and Measures.
Paris, Mathematical Society of France.
Paris Observatory.
Paris, Philomathic Society of France.
Philadelphia, American Philosophical Society.
Philadelphia, Franklin Institute.
Potsdam, Astrophysical Observatory.
Prague Observatory.
Pulkowa Observatory.
Rome, Central Meteorological Office.
Rome, Italian Society of Sciences.
Rome, Italian Spectroscopic Society.
Rome, Pontifical Academy dei Lincei.
Rome, Royal Academy dei Lincei.
St. Petersburg, Imperial Academy of Sciences.
San Fernando Observatory.
Stockholm Observatory.
Stockholm, Royal Swedish Academy of Sciences.
Sydney, Government Observatory.
Sydney, Royal Society of New South Wales.
Tasmania, Royal Society.
Tiflis, Physical Observatory.
Toronto, Canadian Institute.
Toronto, Meteorological Office.
Toulouse, Academy of Sciences.
Toulouse Observatory.
Turin, Observatory of the Royal University.
Turin, Royal Society of Sciences.
Vienna, Imperial Academy of Sciences.
Vienna, Imperial Observatory.
Washington, American Ephemeris Office.
Washington, Smithsonian Institution.
Washington, United States Coast Survey.
Washington, United States Naval Observatory.
Yale College.
Zurich, Geodetic Commission of Switzerland.
Zurich, Society of Natural History.
Editors of the American Journal of Science.
Editors of the American Journal of Mathematics.
Editor of the Analyst.
Editor of the Astronomical Register.
Editor of the Astronomische Nachrichten.
Editor of the Athenæum.
Editors of the Bulletin des Sciences Mathématiques, &c.
Editor of the English Mechanic.
Editor of the Journal of Science.
Editor of Die Naturforscher.
Editors of the Observatory.
Editors of La Revue Scientifique.
Editor of Sirius.

Mons. A. d'Abbadie.
Dr. A. Abetti.
Sir G. B. Airy.
Mons. O. Backlund.
Exors. of Mons. C. Bernaerts.
S. W. Burnham, Esq.
C. E. Burton, Esq.
J. Rand Capron, Esq.
Rev. G. T. Carruthers.
Signor G. Celoria.
G. F. Chambers, Esq.
T. Cooke, Esq.
The Earl of Crawford.
Dr. W. Doberck.
Mons. H. Faye.
Sig. G. Ferrari.
S. Fleming, Esq.
Mons. W. Fonvielle.
W. Godward, Esq.
Major E. S. Gordon.
Signor J. Guccia.
G. T. Gwilliam, Esq.
Dr. H. Gyldén.
G. Hamilton, Esq.
Prof. W. Harkness.
Prof. E. S. Holden.
Prof. J. C. Houzeau.
H. A. Howe, Esq.
Dr. N. de Konkoly.
S. P. Langley, Esq.
Dr. A. Lindhagen.
Dr. O. Lohse.
Prof. Loomis.
Dr. G. Lorenzoni.
J. L. McCance, Esq.

Dr. W. Meyer.
Prof. S. Newcomb.
A. V. Nursingrow, Esq.
O. T. Olsen, Esq.
Prof. Th. von Oppolzer.
Prof. J. A. C. Oudemans.
H. M. Parkhurst, Esq.
Rev. J. Pearson.
Prof. G. Petrosemolo.
Prof. E. C. Pickering.
Prof. E. Plantamour.
J. Prosser, Esq.
Dr. W. T. Radford.
A. Ramsay, Esq.
Rev. W. J. B. Richards.
E. Roberts, Esq.
The Earl of Rosse.
H. C. Russell, Esq.
E. Sang, Esq.
Signor G. V. Schiaparelli.
Mons. Th. Schwedoff.
John N. Stockwell, Esq.
Prof. O. von Struve.
John Tebbutt, Esq.
D. P. Todd, Esq.
L. Trouvelot, Esq.
Dr. H. C. Vogel.
Prof. E. Weiss.
Prof. F. Winnecke.
D. Winstanley, Esq.
Dr. R. Wolf.
W. S. B. Woolhouse, Esq.
W. Wray, Esq.
Prof. A. W. Wright.
Prof. C. A. Young.

ADDRESS

Delivered by the President on Presenting the Gold Medal of the Society to Mr. David Gill.

GENTLEMEN,—

I have now the pleasant duty to state to you the general grounds upon which your Council have felt justified in awarding the Society's Medal this year to Mr. David Gill, H.M. Astronomer at the Cape of Good Hope, for his Heliometric observations of *Mars* at Ascension, and discussion of his results.

The planet *Mars* has occupied no inconsiderable place in the successive attempts at the solution of the grand problem, the determination of the Earth's mean distance from the Sun, upon which depend all measures of absolute distances and dimensions beyond our Moon. The earliest real approximations to the value of the solar parallax were obtained, as is well known, in 1672, through the intervention of this planet, which, in the summer of that year, was at one of its close oppositions, and therefore most favourably situated for the object in view. The first of the Cassini's had perceived the advantage that might be taken of the near approach of *Mars* to the Earth to ascertain the amount of his parallax, and thence of the Sun's, by comparing observations made simultaneously, or nearly so, at distant points upon the Earth's surface ; and, accordingly, when Richer was sent on an astronomical expedition to the island of Cayenne, in French Guiana, in 1672, it was arranged that observations of this planet should be specially made for comparison with others to be made at Paris. In Cassini's Memoir on this subject he refers to several methods of utilising observations of *Mars* for the determination of his parallax : one of which is now commonly known as the meridional method, while another corresponds in principle with that which we are accustomed to term the diurnal method. By the meridional method, or by comparisons of the meridional altitudes of *Mars* at Cayenne and Paris, he inferred that the solar parallax was 9½ seconds, or, having regard to all the observations, that it might be taken indifferently from 9¼ seconds to 9⅔ seconds.

It is gratifying to us as Englishmen that this important advance in the knowledge of the solar parallax from the exceptionally favourable Opposition of *Mars* in 1672 is not due alone to the French astronomer, but that our countryman Flamsteed arrived quite independently at nearly the same result. In the history of his own life he tells that whilst he was inquiring for the planet's appulses to the fixed stars, by the help of Hecker's Ephemerides he found that in September 1672 the planet *Mars*, then newly past his perihelion and opposition to the Sun, would pass amongst three contiguous fixed stars in the water of *Aquarius*, and that, by reason he was then very near the Earth, this would be the most convenient opportunity that would be afforded for many years for determining his parallax, and consequently that of the Sun. A notice which he sent to Oldenburg was printed in No. 86 of the early "Transactions of the Royal Society," on August 19, 1672; and he further states that Oldenburg having before sent his admonition into France, the gentlemen of their Academy took care to have it observed in several places. But we have seen that there was independent action in France. Flamsteed nearly missed the opportunity of contributing to the knowledge of the Sun's distance, being called away from home on, as he says, the very day when he had designed to begin his observations; but he succeeded in observing *Mars* with the instruments of his friend Townley, and again on his return to Derby. These observations are found at pp. 15 and 16 of the first volume of the *Historia Cœlestis*: his conclusion therefrom was that the Sun's horizontal parallax could not be more than 10″.

On many occasions, as we all know, during the two centuries following, the planet *Mars* has been observed for the purpose of investigating the solar parallax, usually on the meridional method, without any result which could be properly called definite. Professor Harkness, of Washington, in a recent summary of results from various methods, gives as limiting values by observation of *Mars* on the meridional method 8″·84 and 8″·96, and on the diurnal method, 8″·60 and 8″·79. I have no intention of occupying your time by noticing in detail the successive attempts to measure the solar parallax through the intervention of *Mars*, but the first efforts of Flamsteed and Cassini will always possess exceptional interest.

Mr. Gill's investigation for which the Medal has been awarded forms part of the 46th volume of our *Memoirs*. He remarks that about a year after the last Transit of *Venus*, when observers began to compare notes, and attempts were made to select corresponding phases at the contacts, a doubt began to arise in the minds of many astronomers whether we should not again repeat the experience of the Transit in 1769, and find the observations capable of so many interpretations as almost to preclude the possibility of an unprejudiced and final discussion; and that in consequence his attention had been directed to the opportunity afforded by the close Opposition of *Mars* in September 1877, as a means of arriving at an independent determination of the solar parallax. In 1874, in conjunction with the present Earl of Crawford and Balcarres, he had attempted a new method of finding the solar parallax by combining the suggestions of Sir George Airy as to employing the diurnal method with those of Professor

Galle, with respect to utilising the minor planets, the Mauritius expedition allowing of the Heliometer being brought to bear upon the necessary observations. He was convinced that a very good determination of the parallax might be made by the diurnal method of extra-meridional transits and the high degree of precision which had been found to attend the Heliometric measures of the minor planet *Juno* at Mauritius in 1874 induced the anticipation that the employment of the same instrument in the case of *Mars* might lead to a more accurate result than any which would be likely to follow by the method of transits. Without supposing that the angular distance between a star and a disk like that of *Mars*, could be measured with the same precision as that between a star and such a stellar-looking object as a minor planet, he yet expected that a greater degree of precision would be attained than on any other known method of observation.

Under these circumstances Mr. Gill applied to the Earl of Crawford (then Lord Lindsay) for the loan of the Heliometer, with which he had had such satisfactory experience during the expedition organised by that nobleman for the observation of the last Transit of *Venus*, and his request, he tells us, met with a most ready compliance and the most effective assistance in carrying out the preparations for an expedition which Mr. Gill contemplated either to St. Helena or Ascension. In the autumn of 1876 application was made to the Government Grant Committee of the Royal Society, but some hesitation being felt as to voting the sum required to one object, however important, the Committee appear to have advised a reference to the Government, with the view to the expense being independently provided for. Here, however, the Council of the Royal Astronomical Society intervened, after receiving an application from Mr. Gill, and the requisite sum of 500*l*. was voted by the Council, three of the Fellows becoming security for the repayment of 250*l*. to the Society if that amount should not be obtained from some other source; and Mr. Gill has acknowledged his indebtedness to the scientific spirit and generous help in the last difficulties attending the organising of his expedition rendered by Lord Crawford, Dr. Warren de la Rue, and the President of the Royal Society.

Having decided to observe at Ascension (it is understood on the advice of Lieut. Neate, R.N.), Mr. Gill left England on June 15, 1877, and reached Ascension on July 13; no time was lost in erecting the Observatory, which was ready for work four days later. The instruments had all been landed without breakage or accident, mainly through the instructions issued by the chief proprietor of the line of mail steamships to his officers to afford every aid in the matter of transport; indeed, I shall, I believe, be divulging no secret when I add that the generous assistance rendered by Sir Donald Currie, M.P., in a scientific matter in which the Council of the Royal Astronomical Society had taken such active interest, called forth from them a cordial vote of thanks to Sir Donald Currie on the occasion. The site originally fixed upon for the Observatory was changed for reasons which are fully detailed by Mr. Gill in his Memoir. A site at the south-west extremity of the island was ultimately selected, and on August 4 the Heliometer was in position in the new locality now called "Mars Bay." A single set of evening and morning observations of *Mars* during the night of July 31 had been obtained at the first site, and from August 4 to October 4 observations were continued at Mars Bay at every opportunity.

Mr. Gill left Ascension on January 9, and arrived in England on January 24.

In his investigation as it appears in the *Memoirs* many details were omitted, on the suggestion of the Council, from the original manuscript of the work, as possessing comparatively minor interest, but the Society is in possession of the original observations of *Mars* as they were forwarded from time to time from Ascension, and is also made the depository of the five volumes containing these original observations, where they will be available for future reference.

The Heliometer was fully described in vol. ii. of the Dun Echt Observatory publications. For determination of time Mr. Gill provided himself with a 30-inch transit instrument, which was mounted in a tent lent by the Admiralty from those used in the Transit of *Venus* expedition at Mokattam; with a Reflecting Circle by Troughton, and a Prismatic Circle by a Berlin optician. The methods of determining time are described, and the details of determination of latitude and longitude of the two Ascension stations are contained in Mr. Gill's original manuscript. In his reductions he uses the following, which he considers amply exact enough for the purposes in view:—

	h m s	Lat.
Garrison, Ascension	Long. 0 57 42 W.	−7 55 50
Mars Bay, Ascension	0 57 39 W.	−7 59 15

Before leaving England Mr. Gill had applied to the Directors of the principal Observatories in various parts of the world requesting meridian observations of the comparison stars intended to be employed in his observations of *Mars*. His application met with a most liberal response, fifteen series of observations in all having been received. The Observatories thus contributing to the success of the enterprise were those of Albany, U.S., Berlin, Cambridge, U.S., Cordoba, Greenwich, Königsberg, Leyden, Leipsic, Melbourne, Oxford, Paris, Pulkowa, and Washington.

The methods of observation with the Heliometer were very similar to those adopted in the case of the *Juno* observations at Mauritius, though some changes were made in one or two important points; the chief difference consisting in the use of a reversing prism placed in front of the eyepiece which could be rotated about its axis, by which "all the observations can be

s

made precisely as if the line joining all stars, or the planet and stars, had always a constant relation to the vertical axis of the observer's eye. This was always kept in mind, and all the measures were made with the prism so turned as to make the line joining the two objects under observation apparently horizontal. Other special precautions were adopted to ensure the utmost possible precision in the measures." Mr. Gill's Memoir enters into full particulars of instrumental adjustments, the formation of instrumental distances, corrections for refraction, &c.

An important and laborious part of the work consisted in the Heliometric triangulation of the comparison stars, with the view to adding additional precision to their meridian observation, according to the plan which Mr. Gill had notified in a communication on his proposed expedition presented to the Society in April 1877. Definitive mean and apparent places of the stars were deduced from a comparison of the results, proper motions where sensible being inferred from a discussion of all the old observations. Corrections to Right Ascensions depending upon magnitude were also investigated. Preliminary corrections to Leverrier's Tables of *Mars* were ascertained. Chapter XIII. of the Memoir contains the heliometric measures between July 31 and Oct. 3. The formation of the equations of condition is then explained, with reference to possible errors that may affect the comparison between an observed and tabular distance, including one—an effect of refraction—suggested to Mr. Gill by Sir George Airy. A table is given showing all the complete combinations that can be framed subject to conditions of strict symmetry in the morning and evening observations as well as regards identity of method in which the corresponding observations were made as in the arrangement of the equations. It is right to say that all important parts of the heavy calculations involved have been executed in duplicate. In Chapter XV. we have particulars as to the deduction of Mr. Gill's final results. The equations are treated upon two systems, by the method of least squares, the first and most laborious assigning for the Sun's parallax

$$8\overset{''}{\cdot}78 \pm 0\overset{''}{\cdot}012.$$

The second, involving a comparatively simple method of reduction, gave

$$8\overset{''}{\cdot}784 \pm 0\overset{''}{\cdot}013.$$

Mr. Gill concludes that whatever method of reduction be employed the result must be practically the same—viz., $8''\cdot78 \pm 0''\cdot012$, subject to a possible small correction due to sources of error, which he mentions.

This value for the solar parallax he presents as the definitive result of his investigations, and combining it with Listing's value of the equatorial radius, he infers for the mean distance of the

Earth from the Sun 93,080,000 miles. He concludes his Memoir with a comparison of results of the various methods of deducing the solar parallax, and reverts to an opinion he had long held, that perhaps the most promising method to ensure precision is by heliometric observation of the minor planets, or such of them as approach nearest to the Earth. He thinks that if a Heliometer of 6 or 7 inches aperture, in a portable form, were available, it might be conveyed from time to time, when exceptionally favourable oppositions of these bodies take place, to such stations as would be best situated for a particular opposition; and that "in this way, without undue loss of time, advantage might be taken of the most favourable opportunities, and in course of a few years such a series of determinations would be obtained as would set at rest this most important and difficult question."

I should remark that in what has preceded I have purposely confined myself to a brief outline of Mr. Gill's investigation, without touching upon any feature of it which can well be questioned by other astronomers; but in such discussions there will in most cases be some differences of opinion, especially in the formation and treatment of the equations of condition and deduction of probable error.

As a piece of admirable work, carried to its conclusion with the unremitting energy and great zeal which Mr. Gill has always evinced, I believe you will concur in the opinion of your Council that he has well deserved the award of the Medal.

My Lord Crawford,—In transmitting this Medal to Mr. Gill, I ask you to assure him of the high estimation with which the Royal Astronomical Society regards his energetic efforts in the cause of the science to which we are devoted, and to express our earnest wishes that continued health may enable him to apply them in the wide field which is open to him as H.M. Astronomer at the Cape, with equal discrimination and as great success as heretofore.

The Meeting then proceeded to the election of the Officers and Council for the ensuing year, when the following Fellows were elected :—

President.

E. J. STONE, Esq., M.A., F.R.S., Radcliffe Observer.

Vice-Presidents.

J. C. ADAMS, Esq., M.A., LL.D., F.R.S., Lowndean Professor of Astronomy, Cambridge.
W. H. M. CHRISTIE, Esq., M.A., F.R.S., Astronomer Royal.
J. R. HIND, Esq., F.R.S., Superintendent of the *Nautical Almanac.*
H. J. S. SMITH, Esq., M.A., LL.D., F.R.S., Savilian Professor of Geometry, Oxford.

Treasurer.

FRANCIS BARROW, Esq., M.A.

Secretaries.

J. W. L. GLAISHER, Esq., M.A., F.R.S.
E. B. KNOBEL, Esq.

Foreign Secretary.

The EARL OF CRAWFORD AND BALCARRES, F.R.S.

Council.

Sir G. B. AIRY, K.C.B., M.A., LL.D., D.C.L., F.R.S.
JAMES CAMPBELL, Esq.
ARTHUR CAYLEY, Esq., M.A., LL.D., D.C.L., F.R.S., Sadlerian Professor of Pure Mathematics, Cambridge.
A. A. COMMON, Esq.
G. H. DARWIN, Esq., M.A., F.R.S.
A. M. W. DOWNING, Esq., M.A.
EDWIN DUNKIN, Esq., F.R.S.
WILLIAM HUGGINS, Esq., LL.D., D.C.L., F.R.S.
GEORGE KNOTT, Esq., LL.B.
ALBERT MARTH, Esq.
EDMUND NEISON, Esq.
A. COWPER RANYARD, Esq., M.A.

MONTHLY NOTICES

OF THE

ROYAL ASTRONOMICAL SOCIETY.

Vol. XLII.   March 10, 1882.   No. 5.

E. J. Stone, Esq., M.A., F.R.S., President, in the Chair.

Samuel Jefferson, Esq., 17 Virginia Road, Leeds ; and
James Leigh, Esq., Watford House, King's Norton, near
Birmingham ;
were balloted for, and duly elected Fellows of the Society.

*Notes on M. Loewy's Remarks relative to the Wedge-Extinction
Method of Stellar Photometry.* By Prof. C. Pritchard, F.R.S.

In the *Monthly Notices* for January last, M. Loewy has done
me the honour of criticising a method devised by me for the
photometric measurement of the relative brightness of the stars,
which I somewhat fully described in the *Monthly Notices* of
November 1881. He states that he himself and his two associates MM. Paul and Prosper Henry have for several years
been engaged in considering various photometric methods
devised for the same purpose ; and, among other methods, that
which I have proposed, and, indeed, to a very considerable extent carried out with what seems to be success. M. Loewy
states that the French astronomers abandoned this method. I
would here remark that, in a question of this sort, it seems
desirable to give numerical instances of a sufficient number of
observations, in order that other astronomers may form an
independent opinion, and clearly see for themselves the merits
or demerits of the question at issue. This I have endeavoured
to do in the paper of November last, giving many results as
they were obtained direct from the heavens and the photometer.
The results there given, though quite satisfactory, were first

attempts; and could be, and have been, improved by larger experience and additional precautions.

Besides this, the method adopted at Oxford is, I thought and still think, peculiar; and is most probably by no means that which, in its completeness, was tried and is said to have failed at the Paris Observatory. I shall soon explain more at large what I mean as I proceed.

M. Loewy objects to the plan of extinction by a wedge, because (unless I mistranslate or misunderstand his remarks) he found that the reducing of the light by means of the wedge to the extent of seven or eight magnitudes, introduced serious inconveniences.

Now, in my plan of using the wedge, I have uniformly avoided any such process as comparing the light of two stars differing by seven or eight magnitudes, *by means of the wedge alone.* Further, M. Loewy adds that they found it necessary to confine themselves to differences of magnitude not exceeding three magnitudes in the use of the wedge alone. To this I agree *in practice*; but, with the wedge in my possession, I could extend the use of it much further ; certainly to six magnitudes without the "introduction of sensible errors."

But what is done, in effect, is this. A star of a magnitude intermediate between those which it is proposed to examine, is selected. In *Lyra*, for instance, κ *Lyræ* was selected about magnitude 4½, and other stars in its neighbourhood, from the first to the eighth magnitude, were examined and compared with it, so that there could be no great or violent differences in the lights compared.

Again, I do not trust to the wedge-extinction process alone ; but in such a case as comparing κ *Lyræ* with α *Lyræ*, the brighter star (α) would be observed in a telescope with an aperture of (say) two inches, whereas the fainter star (κ) would be observed with an aperture of 4 inches ; thus removing at once an amount of light from the brighter star, equivalent to a magnitude and a half, leaving the wedge to act upon a difference of light amounting to about 2½ magnitudes only, which, in my present wedge, implies the use of less than an inch and a half : and even this amount of wedge interval might be still further reduced, by diminishing the aperture of the telescope to one inch, as is often done in practice at this Observatory. In fact, the telescope used has an aperture of four inches, reducible to three inches, two inches, and one inch, at pleasure. But my experience is not in favour of too violent a reduction of aperture.

Hence it will be observed that my method of photometry is a mixed method of aperture and wedge, each continually testing and verifying the other—a great gain, as I conceive. Thus, I submit, I have successfully met M. Loewy's natural objection to the use of a wedge beyond reasonable limits, however perfect it may be. Probably, however, I did not sufficiently explain this point of practice in my November paper.

The next objection to the process of wedge-extinction urged by M. Loewy is apparently more formidable. It applies to the material itself of the glass wedge. He states that the Paris astronomers, after a trial of some fifty specimens of neutral-tinted glass, invariably found that they were differently absorbent for different colours. If that were universally the case, and if this difference of absorbing power were decidedly marked, then, I presume, the method of wedge-extinction must be abandoned. But my experience is contrary to that of the French astronomers.

I have submitted the wedge—or, rather, the two wedges—to a scrupulous and exhaustive trial. Their effect on a Sun spectrum I have found to be uniformly absorptive for all the colours, extending from the violet to beyond B in the red. A little beyond B, noticeable absorption commences. But this is just the part of the spectrum where the light of ordinary stars becomes very feeble, and, consequently, the wedge is trustworthy; excepting, it may be, in the case of a star of a decidedly red colour. But this is, unfortunately, one of the cases where neither the eye nor any photometer that I know can be fully trusted in the comparison of different intensities of light. In a delicate question of this sort, I thought it best to consult one of the highest authorities in this country, Dr. Huggins. He at once referred me to a passage, written by himself, in the *Philosophical Transactions* for 1866, in which he says: "An examination of the neutral-tint glass, with a prism, showed that the absorptive power of the glass, for all refrangibilities in the brighter portion of the spectrum, was very nearly uniform." That eminent physicist was good enough to examine, also, a few other specimens of neutral-tinted glass in his possession, with a substantially similar result; and he adds, in a private communication to me, "In the case of the stars visible to the naked eye, among which there are no very red stars, the wedge would give results quite comparable probably with those of any other method, because any small errors arising from the want of absorptive uniformity would doubtless be much smaller than the unavoidable errors of estimation coming in from other causes."

This cautious expression of opinion, after due and skilful examination on Dr. Huggins' part, entirely coincides with my own experience, after the use of the wedge some thousand times. Thus, I think, M. Loewy's second objection is satisfactorily removed.

M. Loewy then proceeds to explain another method which he had devised, but which he, not unnaturally, doubts if it be new. It is to virtually reduce the aperture of the telescope by means of a moveable pierced diaphragm, placed inside the telescope, and near to the eyepiece. He is quite right in the suspicion that the method is not new. Our well-remembered colleague Mr. Dawes constructed and used a similar apparatus, described in the *Monthly Notices*, vol. xxv. p. 229. But I am not aware that Mr. Dawes obtained any considerable results, either by that means, or by another means in which, like myself, he employed variable apertures and wedge combined. The cause, or causes, of failure—for a practical failure it seems to have been—are not difficult now to see. Most probably M. Loewy will avoid them.

I have tried the method of small apertures, but I abandoned it in despair—possibly I was impatient. With a wedge, used unskilfully enough, some five years ago, I was equally impatient and equally unsuccessful. But a patient re-consideration of the whole case has now landed me in what I believe to be a valuable and trustworthy photometric method.

I now proceed to what I regard as an actual test of the reliability and value of the method employed in this Observatory from which may be formed a safe and practical conclusion in regard to the points raised by M. Loewy.

Professor Pickering has devised a photometer of a peculiar construction, depending on double refraction. It resolves itself at last into the comparison or equalisation of lights of two stars placed in a very skilful way in juxtaposition. I need not say that this is the very antithesis of the method I employ. Results obtained by means of the two instruments and the two methods, if they be employed on the same star, must, on comparison, necessarily reveal facts. Now, it so happens, that Professor Pickering has recorded his results of the measurement of the difference of brightness of twenty-six stars of the *Pleiades*. Out of these twenty-six, fifteen have already been observed in Oxford,* as well as some twenty others of this group, not observed at Harvard College. The Oxford observations are the results of three nights' measurement of each star—five readings of the instrument being taken on each night. Finally, the means are taken, and the average deviations of the fifteen separate measures from their general mean, are recorded in the subjoined tables. An inspection of these results will enable astronomers to judge of what has been so far achieved, and what sort of promise the method or methods hold out, for the photometric estimation of relative star-magnitudes.

---

* Since the writing of this paper, the photometric measurement of forty-three stars in the *Pleiades* has been completed. Of these, forty have been also carefully measured as to their relative coordinates with the Duplex Micrometer. in order to compare them with Bessel's Heliometer measures. The results, it is expected, will be communicated to the Society in May next.

TABLE I.

Comparison of the Magnitudes of Fifteen Stars in the Pleiades that have been Photometrically obtained by Observations made at Harvard College and at Oxford.

Bessel's Designation.	Observed Differences of Magnitude from Merope.			Mean difference of Mag.	Average Deviation.	Resulting Oxford Mag. Merope= 4·22.	Resulting Harvard Magnitude.
	1	2	3				
17 b Electra	0·31	0·62	0·41	0·45	0·12	3·77	3·82
16 g Celæno	1·01	1·15	1·28	1·15	0·09	5·37	5·23
19 e Taygeta	0·35	0·44	0·32	0·37	0·05	4·59	4·44
18 m	1·65	1·68	1·53	1·62	0·06	5·84	5·63
21 k Asterope	1·52	1·68	1·54	1·58	0·07	5·80	5·71
22 l	1·98	2·22	2·15	2·12	0·09	6·34	6·28
20 c Maia	0·19	0·50	0·34	0·34	0·10	3·88	3·98
10	2·87	2·78	3·16	2·94	0·15	7·16	6·96
19	2·68	2·64	2·29	2·54	0·16	6·76	6·69
22	2·65	2·59	2·28	2·51	0·15	6·73	6·74
25 η Alcyone	1·29	1·34	1·35	1·33	0·02	2·89	3·00
24	2·43	2·12	2·18	2·24	0·12	6·46	6·46
12	2·54	2·27	2·48	2·43	0·11	6·65	6·58
26 s	2·34	2·20	2·10	2·21	0·08	6·43	6·36
29	2·54	2·39	2·49	2·47	0·06	6·69	6·56

TABLE II.

Individual and Concluded Measures of the Magnitude of Eighteen Stars in the Pleiades made at Oxford. (Not observed by Prof. Pickering.)

Bessel's Desig-nation.	Diff. from Merope.	Observed Differences of Magnitude from Merope.			Mean difference of Mag.	Average Deviation.	Resulting Oxford Mag. if Merope =4·22.
		1	2	3			
1	3·0	2·95	2·99	3·10	3·01	0·06	7·23
4	3·0	3·38	3·36	3·28	3·34	0·04	7·56
6	4·0	5·14	5·00	4·93	5·02	0·08	9·24
7	3·0	3·17	3·36	3·20	3·24	0·08	7·46
8	3·5	3·16	3·11	3·30	3·19	0·07	7·41
9	3·5	3·68	3·67	3·59	3·65	0·04	7·87
13	3·5	3·94	3·70	4·07	3·90	0 14	8·12
24 p	2·5	1·66	1·55	1·65	1·62	0·05	5·84
15	3·5	4·30	4·15	4·41	4·29	0·09	8·51
17	3·0	2·36	2·47	2·47	2·43	0·05	6·65
18	3·0	3·78	3·68	3·96	3·81	0·10	8·03
21	3·5	3·74	4·02	3·93	3·90	0·10	8·12
23	3·5	3·19	3·24	3·22	3·22	0·02	7·44
30	3·5	3·55	3·76	3·46	3·59	0·11	7·81
31	3·0	2·41	2·49	2·64	2·51	0·08	6·73
33	3·5	3·06	2·98	2·89	2·98	0·06	7·20
35	4·0	5·29	5·39	5·38	5·35	0·04	9·57
36	4·0	4·50	4·60	4·65	4·58	0·06	8·80

The only remark necessary to make with regard to the above Tables is that the observations themselves are all differential. The magnitude assigned to *Merope*, which is the Oxford star of comparison, was 4·22, that being the value assigned to it by Professor Pickering. Argelander's value is 4·5. Had I assumed the magnitude of *Merope* as 4·16, instead of 4·22 with Professor Pickering (a course perfectly allowable), then the general agreement of the Oxford with the Harvard magnitudes would have been still more remarkably close.

Professor Pickering, in some valuable remarks which he has contributed to the *Astronomical Register* of February 1882, after kindly expressing the interest which he takes in the Oxford Observations, observes :—"In all forms of the method I have found the observed limit of visibility inconveniently variable, not only on different nights, and for different observers on the same night, but even in successive measures by the same observer at short intervals. The separate determinations included in a single series are sufficiently accordant ; but, on returning subsequently to the same object, a new result is often obtained, so that strictly differential measures cannot be made satisfactorily." In reply to this, I have looked carefully through a large mass of observations, made by the same observer on different nights and on the same night, and the result is, that the average deviations of all the several readings from the mean of the whole is sensibly the same as that which is recorded in the *Astronomical Register* as the result of the Harvard Observations. With regard to the fifteen stars in question observed by the two methods, the Harvard mean deviation is 0·13 magnitude ; the Oxford, 0·10 magnitude. On this head, too, I have with great interest consulted Professor Pickering's valuable photometric observations, in vol. xi. of the Harvard College Observations, where deviations of a very similar character and amount are exhibited. In fact there seems to me to be a limit to the accuracy to which stellar magnitudes are at present determinable, notwithstanding the skill of the observer or the ingenuity displayed in the construction of his instrument. Nevertheless, that limit is vastly beyond the amount of accuracy which has hitherto been secured by unaided comparisons with the eye.

As to the employment and results of different observers on the same or different nights, I have seen no differences such as in any degree to shake my confidence. I have, in fact, obtained on this head additional confirmation of what I have stated in the addendum to my paper in the *Monthly Notices*, November 1881, p. 11. Nevertheless, it is to be distinctly understood that, for the method of wedge-extinction, it is essential for accuracy that the meteorological circumstances under which any two stars are compared must be sensibly the same. Drifting clouds form the chief difficulty with which the observer has to contend, and I think it probable that it is this circumstance which explains Professor Pickering's remarks on occasional discordant results.

But meteorological effects on astronomical observations are not confined to photometry.

I have arranged a third Table, from which astronomers may form an opinion of the *advance* which is now being made in the accuracy and consistency of the estimates of the brightness of stars.

TABLE III.

Comparison of the Photometric Measures of the Magnitude of Fifteen Stars in the Pleiades made at Harvard and Oxford with the estimation of the Magnitude of the same stars made by Bessel, Argelander, and Wolf.

Name of Star.	Bessel.	Difference of Magnitude from Merope, Argelander.	Wolf.	Pickering.	Pritchard.
17 *b* Electra	−0·5	+0·2	−1·0	−0·40	−0·45
16 *g* Celæno	+0·5	+2·0	+0·5	+1·01	+1·15
19 *e* Taygeta	0·0	+0·5	0·0	+0·22	+0·37
18 *m*	+2·0	+1·8	+0·75	+1·41	+1·62
21 *k* Asterope	+2·5	+2·5	+1·0	+1·49	+1·58
22 *l*	+2·5	+2·5	+1·5	+2·06	+2·12
20 *c* Maia	0·0	+0·3	−1·0	−0·24	−0·34
10	+3·0	+3·5	+2·25	+2·74	+2·94
19	+3·0	+3·0	+2·75	+2·47	+2·54
22	+3·0	+3·0	+2·75	+2·52	+2·51
25 *η* Alcyone	−1·5	−1·3	−2·5	−1·22	−1·33
24	+3·0	+2·5	+2·0	+2·24	+2·24
12	+2·5	+3·0	+2·0	+2·36	+2·43
26 *s*	+2·5	+2·5	+2·0	+2·14	+2·21
29	+3·0	+3·3	+2·0	+2·34	+2·47

The minus sign implies that the particular star was brighter than *Merope.*

It is almost unnecessary to point out how closely the photometric measures made at Harvard and Oxford agree, in comparison with the estimations made by the careful and eminent astronomers who have given their attention to the same subject. Nevertheless, it is somewhat curious to remark that the *means* of the estimations made by Bessel, Argelander, and Wolf (widely as they individually differ from each other), agree in the main with the more accurate photometric measures.

Oxford University Observatory:
  1882, *March* 9.

*Spectroscopic Results for the Motions of Stars in the Line of Sight,
obtained at the Royal Observatory, Greenwich, in the year
1881. No. V.*

(Communicated by the Astronomer Royal.)

The results here given are in continuation of those printed in
the *Monthly Notices*, vol. xxxvi. p. 318, vol. xxxvii. p. 22, vol.
xxxviii. p. 493, and vol. xli. p. 109. The observations were made
with the "half-prism" spectroscope, one "half-prism" with a
dispersion of about 18°½ from A to H being used, except in a
few cases of bright stars, mentioned in the remarks, where a
train of two "half-prisms," with a dispersion of 83° from A to
H, was used. An eyepiece with a magnifying power of 14 was
employed throughout. Since the beginning of 1881 a convex
cylindrical lens, with its axis parallel to the length of the
spectrum, has been placed in the view-telescope within the focus,
instead of in front of the slit as before. Thus the angle of divergence of the rays which fall on the object-glass of the collimator
is not affected, as was formerly the case. A concave or Barlow
lens of 2 inches focus has been placed in the collimator between
the slit and the object-glass, so that the whole aperture of the
object-glass and of the prisms may be used, the ratio of aperture
to focal length being greater for the collimator of the spectroscope than for the refractor. In most of the observations a
diaphragm coated with Balmain's luminous paint has been used
in the micrometer-adapter to give a phosphorescent illumination
of the field.

Up to the present time the motions of 99 stars have been
spectroscopically determined, though the results in many cases
are still very uncertain.

The observations of the moon and of the sky spectrum have
been made as usual as a check on the general accuracy of the
results.

*Motions of Stars in the Line of Sight in Miles per Second, observed with the
Half-prism Spectroscope.*

(+ denotes Recession ; — Approach.)

The initials W. C. and M. are those of Mr. Christie and Mr. Maunder
respectively.

The weights are on a scale of ½ to 5 for each observation.

Stars marked with an asterisk have not been previously observed.

Date.	Obs.	Wt.	Line.	Earth's Motion in m. per sec.	Concluded Motion of Star. Meas.	Estimd.	Remarks.
					a Andromedæ.		
1881.							
Sept. 30	M	3	F	— 1·3	—23·0	—23·7	Star line dark, not very broad,
	M	3	F	— 1·3	—28·6	—32·0	though somewhat ill de-
							fined at the edges. Spec-
	M	3	F	— 1·3	—15·4	—15·4	trum steady.

Date.	Obs.	Wt.	Line.	Earth's Motion in m. per sec.	Concluded Motion of Star. Meas.	Estimd.	Remarks.
1881.							
Dec. 3	M	2	F	+ 14 7	− 37·5	− 37·2	Star-line dark, but broad, ill defined and nebulous.
	M	2	F	+ 14·7	− 57·2	− 50·6	
	M	2	F	+ 14·7	− 46·1	− 41 7	
7	M	3	F	+ 15·3	− 43·7	− 42·6	Definition fair. Direct comparison showed only a small displacement.
	M	3	F	+ 15·3	− 50 4	− 48·0	

β Cassiopeiæ.

Date.	Obs.	Wt.	Line.	Earth's Motion in m. per sec.	Concluded Motion of Star. Meas.	Estimd.	Remarks.
Dec. 7	M	½	F	+ 8 1	− 42·9	− 35·4	Star-line not very broad; but very difficult to bisect.
	M	½	F	+ 8·1	− 38·9	− 35·4	
	M	½	F	+ 8·1	− 25·7	− 26·3	

γ Pegasi.

Date.	Obs.	Wt.	Line.	Earth's Motion in m. per sec.	Concluded Motion of Star. Meas.	Estimd.	Remarks.
Dec. 3	M	2	F	+ 16·7	− 23·8	− 23·1	Star-line faint, broad and very diffused at the edges.
	M	2	F	+ 16·7	− 48·1	− 43·7	
	M	2	F	+ 16·7	− 43·5	− 39·2	

δ Andromedæ.

Date.	Obs.	Wt.	Line.	Earth's Motion in m. per sec.	Concluded Motion of Star. Meas.	Estimd.	Remarks.
Nov. 19	M	3	b_1	+ 10·0	− 72·0	− 68·5	Spectrum faint and tremulous, but definition good at times.
	M	3	b_1	+ 10·0	− 52·1	− 49·0	
22	M	1	b_1	+ 11·2	− 2·2	− 1·0	Definition variable.
	M	3	b_1	+ 11·2	− 30·3	− 31·6	
	M	2	b_1	+ 11·2	− 46·3	− 62·3	

β Ceti.

Date.	Obs.	Wt.	Line.	Earth's Motion in m. per sec.	Concluded Motion of Star. Meas.	Estimd.	Remarks.
Nov. 19	M	4	b_1	+ 14·7	− 45·5	− 53·7	Spectrum tremulous, but star-lines seen well at times.
	M	4	b_1	+ 14·7	− 47·4	− 53·7	
28	M	1	b_1	+ 16·1	− 28·6	− 28·0	Definition poor and star-lines faint.
	M	1	b_1	+ 16·1	− 14·2	− 7·8	
	M	1	b_1	+ 16·1	− 32·9	− 32·8	

γ Cassiopeiæ.

Date.	Obs.	Wt.	Line.	Earth's Motion in m. per sec.	Concluded Motion of Star. Meas.	Estimd.	Remarks.
Dec. 7	M	2	F	+ 6·9	− 20·5	− 20·5	The star-line, being bright instead of dark, was difficult to bisect.
	M	2	F	+ 6·9	− 18·5	− 17·8	
	M	2	F	+ 6·9	− 28·2	− 25·1	

β Andromedæ.

Date.	Obs.	Wt.	Line.	Earth's Motion in m. per sec.	Concluded Motion of Star. Meas.	Estimd.	Remarks.
Oct. 18	M	3	b_1	− 0·8	+ 44·8	+ 44·9	Spectrum bright. Star-lines dark and fairly well defined.
	M	3	b_1	− 0·8	+ 42·9	+ 44·9	
Nov. 19	M	2	b_1	+ 8·2	− 30·0	− 32·6	Image tremulous, definition variable.
	M	2	b_1	+ 8·2	− 56·1	− 56·9	
22	M	½	b_1	+ 9·0	− 57·7	− 60·1	Definition poor.

Date.	Obs.	Wt.	Line.	Earth's Motion in m. per sec.	Concluded Motion of Star. Meas.	Estimd.	Remarks.
1881.							
Nov. 22	M	¼	b_1	+ 9·0	− 5·5	+ 1·2	
	M	½	b_1	+ 9·0	+ 4·6	+ 11·4	
	M	½	b_1	+ 9·0	− 56·6	− 60·1	
				δ Cassiopeiæ.			
Dec. 7	M	2	F	+ 6·5	+ 16·9	+ 20·8	Star-line very broad. Spec-
	M	2	F	+ 6·5	+ 4·0	+ 4·4	trum faint.
				a Trianguli.			
Nov. 19	M	3	b_1	+ 6·9	+ 54·7	+ 58·1	Spectrum faint, but steady.
	M	3	b_1	+ 6·9	+ 56·6	+ 58·1	
				a Arietis.			
Nov. 19	M	5	b_1	+ 6·9	− 15·5	− 16·6	Definition very good.
	M	5	b_1	+ 6·9	− 7·7	− 6·9	
	M	5	b_1	+ 6·9	− 6·5	− 6·9	
				a Ceti.			
Nov. 28	M	2	b_4	+ 7·6	− 11·5	− 7·6	Definition poor.
	M	2	b_1	+ 7·6	− 12·3	− 7·6	
	M	2	b_1	+ 7·6	− 22·4	− 24·3	
				ρ Persei.			
Nov. 22	M	½	b_4	+ 2·3	− 37·8	− 55·5	The star-line, b_4, was not dis-
	M	½	b_4	+ 2·3	− 53·8	− 66·2	tinctly isolated.
	M	½	b_4	+ 2 3	− 22·2	− 23·6	
				ζ Persei.			
Oct. 18	M	1	b_1	− 10·6	− 14·3	− 11·4	Star-lines faint. Definition
	M	1	b_1	− 10 6	− 15·5	− 11·4	poor.
				ε Tauri.			
Nov. 22	M	2	h_1	− 1·9	+ 22·5	+ 27·5	Definition fair.
	M	2	h_1	− 1·9	+ 18·3	+ 27·5	
				Aldebaran.			
Mar. 28	M	2	b_1	+ 15·8	+ 78·6	+ 102·7	Star-lines dark, but diffi-
	M	2	b_1	+ 15·8	+ 71·9	+ 102·7	cult to isolate. Spectrum very tremulous. Definition
31	M	4	h_1	+ 15·3	+ 94·7	+ 62·3	poor. Star-lines dark. Spectrum
	M	4	h_1	+ 15·3	+ 78·7	+ 62·3	bright, but tremulous.
	M	4	h_1	+ 15·3	+ 85·3	+ 62·3	
April 7	M	2	h_1	+ 13·9	+ 26·0	+ 24·6	Star-lines ill defined. Spec-
	M	2	h_1	+ 13·9	+ 24·4	+ 24·0	trum very faint and trem- ulous. Two-prism train.

Date.	Obs.	Wt.	Line.	Earth's Motion in m. per sec.	Concluded Motion of Star. Meas.	Estimd.	Remarks.
1881.							
Oct. 18	M	3	b_1	+ 12·4	+ 14·0	+ 12·4	Star-lines dark and distinct.
	M	3	b_1	+ 12·4	+ 27·1	+ 21·2	Spectrum very bright.
	M	3	b_1	+ 12·4	+ 37·7	+ 34·4	
Nov. 22	M	3	b_1	− 2·3	+ 40·5	+ 53·4	Definition poor.
	M	3	b_1	− 2·3	+ 31·5	+ 36·3	
	M	3	b_1	− 2·3	+ 4·6	+ 2·3	
28	M	3	b_1	− 0·4	+ 26·1	+ 21·3	Definition poor.
	M	3	b_1	− 0·4	+ 9·7	+ 8·7	

Capella.

Date.	Obs.	Wt.	Line.	Earth's Motion in m. per sec.	Concluded Motion of Star. Meas.	Estimd.	Remarks.
Mar. 31	M	3	b_1	+ 15·8	+ 21·2	+ 23·0	Spectrum tremulous, but very bright. Definition fair.
	M	3	b_1	+ 15·8	+ 16·5	+ 23·0	
April 6	W.C.	2	b_1	+ 15·0		+ 51·0	Definition variable.
	W.C.	2	b_1	+ 15·0		+ 58·3	
7	M	5	b_1	+ 14·9	+ 37·1	+ 36·4	Spectrum bright. Definition fair. Two-prism train.
	M	5	b_1	+ 14·9	+ 37·3	+ 36·4	
	M	5	b_1	+ 14·9	+ 27·1	+ 27·9	
	M	5	b_4	+ 14·9	+ 28·2	+ 31·7	

Rigel.

Date.	Obs.	Wt.	Line.	Earth's Motion in m. per sec.	Concluded Motion of Star. Meas.	Estimd.	Remarks.
Feb. 28	M	½	F	+ 16·0	+ 12·4	+ 19·9	Definition very bad.
	M	½	F	+ 16·0	+ 15·1	+ 19·9	
Mar. 14	M	2	F	+ 15·7	+ 31·5	+ 21·8	Spectrum tremulous. Definition poor.
	M	2	F	+ 15·7	+ 32·4	+ 21·8	
Dec. 7	M	5	F	+ 0·3	+ 15·1	+ 13·3	Spectrum bright, but tremulous.
	M	5	F	+ 0·3	+ 8·3	+ 10·6	
	M	5	F	+ 0·3	+ 14·2	+ 13·3	

γ Orionis.

Date.	Obs.	Wt.	Line.	Earth's Motion in m. per sec.	Concluded Motion of Star. Meas.	Estimd.	Remarks.
Dec. 3	M	1	F	− 2·2	+ 30·0	+ 24·7	Definition poor.
	M	1	F	− 2·2	− 9·2	− 6·8	
	M	1	F	− 2·2	+ 1·6	+ 2·2	

β Tauri.

Date.	Obs.	Wt.	Line.	Earth's Motion in m. per sec.	Concluded Motion of Star. Meas.	Estimd.	Remarks.
April 9	M	4	F	+ 15·9	+ 5·1	+ 17·4	Spectrum steady. Definition good.
	M	4	F	+ 15·9	+ 26·6	+ 44·0	
	M	4	F	+ 15·9	+ 13·7	+ 34·1	
Dec. 3	M	1	F	− 2·9	− 19·9	− 19·6	Star-line dark, and, though broad, fairly well defined.
	M	1	F	− 2·9	− 14·7	− 12·1	
	M	1	F	− 2·9	− 12·2	− 8·4	

Date.	Obs.	Wt.	Line.	Earth's Motion in m. per sec.	Concluded Motion of Star. Meas.	Estimd.	Remarks.

a Orionis.

1881. Mar. 31	M	1	b_4	+ 17·1	+ 13·3	+ 22·5	Spectrum tremulous, but very
	M	1	b_4	+ 17·1	+ 23·8	+ 22·5	bright. Definition of star lines fair.
	M	1	b_1	+ 17·1	− 1·9	+ 2·3	
	M	3	b_1	+ 17·1	+ 35·9	+ 29·5	
April 7	M	3	b_1	+ 16·4	− 0·4	+ 5·0	Spectrum faint, but fairly
	M	3	b_1	+ 16·4	+ 6·1	+ 9·3	steady. Definition fair. Two-prism train.
	M	3	b_4	+ 16·4	+ 30·1	+ 24·4	
	M	3	b_4	+ 16·4	+ 6·5	+ 10·8	
Nov. 28	M	2	b_1	− 6·3	+ 26·6	+ 26·9	Definition poor.
	M	2	b_1	− 6·3	+ 53·5	+ 56·3	

β Aurigæ.

April 9	M	3	F	+ 15·8	+ 2·1	+ 9·2	Definition good, but star-
	M	3	F	+ 15·8	− 13·7	− 15·8	line not quite so well seen as in β Tauri.

γ Geminorum.

Feb. 28	M	½	F	+ 16·5	+ 11·0	+ 19·4	Definition bad. Measures
	M	½	F	+ 16·5	− 7·6	− 4·5	uncertain.
Mar. 14	M	1	F	+ 18·0	+ 8·5	+ 14·8	
	M	½	F	+ 18·0	+ 29·8	+ 25·8	

Sirius.

Feb. 16	M	½	F	+ 10·4	+ 9·6	+ 11·8	Spectrum very tremulous, and
	M	½	F	+ 10·4	+ 3·5	+ 11·8	definition very bad.
28	M	1	F	+ 12·2	+ 3·5	+ 11·8	Definition bad.
	M	1	F	+ 12·2	+ 14·6	+ 23·7	
Dec. 7	M		F	− 6·4	− 23·5	− 20·9	Spectrum very bright. Star-
	M		F	− 6·4	− 19·8	− 20·9	lines well seen. Direct comparison distinctly confirmed these measures.

Castor.

April 16	M	½	F	+ 17·7	− 49·1	− 51·1	Star-line very broad and dif-
	M	½	F	+ 17·7	+ 22·1	+ 24·1	fused at the edges.

Procyon.

Feb. 16	M	1	F	+ 10·2	+ 39·5	+ 43·1	Star-line faint. Spectrum
	M	1	F	+ 10·2	+ 18·8	+ 23·1	faint, and definition bad.
28	M	3	F	+ 13·0	+ 3·1	+ 2·0	Definition fair.
	M	3	F	+ 13·0	+ 1·8	+ 2·0	

Date.	Obs.	Wt.	Line.	Earth's Motion in m. per sec.	Concluded Motion of Star. Meas.	Estimd.	Remarks.
1881.							
Mar. 14	M	2	F	+15·6	+13·1	+17·2	
	M	1	F	+15·6	+31·0	+28·2	
April 9	M	4	F	+17·6	+45·9	+57·4	Definition good.
	M	4	F	+17·6	+29·6	+49·0	
	M	3	F	+17·6	+17·1	+6·8	Spectrum faint. Two-prism train.
	M	2	F	+17·6	+12·1	+0·7	

Pollux.

Mar. 25	M	2	b_1	+17·5	−71·7	−82·4	Spectrum very bright and fairly steady. Star-lines faint and ill defined.
	M	2	b_1	+17·5	−17·9	−17·5	
	M	1	b_1	+17·5	−77·5	−82·4	
	M	1	b_1	+17·5	−15·9	−17·5	
28	M	3	b_1	+17·7	−0·2	+11·9	Spectrum steady. Definition good.
	M	3	b_1	+17·7	+17·8	+21·8	
April 7	M	1	b_1	+18·1	−34·5	−39·5	Spectrum faint and tremulous. Star-lines faint. Definition poor. Two-prism train.
	M	1	h_4	+18·1	−2·9	+2·3	
	M	1	h_1	+18·1	−17·7	−18·1	
	M	2	b_1	+18·1	−10·0	−9·5	
	M	2	b_1	+18·1	−12·2	−13·8	

a Hydræ.

| Mar. 31 | | 1 | b_1 | +12·2 | −9·1 | −4·4 | Spectrum tremulous and faint. Star-lines faint. |
| | M | 1 | h_1 | +12·2 | −16·1 | −20·0 | |

ε Leonis.

| Mar. 25 | M | 2 | b_1 | +13·2 | +18·4 | +19·3 | Star-lines dark. Definition fair. |
| | M | 2 | b_1 | +13·2 | −3·9 | −0·2 | |

Regulus.

May 11	M	½	F	+17·9	−23·5	−22·9	Definition fair.
	M	3	F	+17·9	+1·8	+1·8	
	M	3	F	+17·9	+3·1	+6·8	

γ₁ Leonis.

| Mar. 25 | M | 1 | b_1 | +11·1 | +31·4 | +40·8 | Star-lines exceedingly difficult to see. |
| | M | 1 | b_1 | +11·1 | +11·9 | +14·9 | |

δ Leonis.

May 6	M	½	F	+16·0	+15·4	+11·4	Star-line faint and diffused.
	M	½	F	+16·0	+2·5	+2·3	
11	M	3	F	+16·6	+20·1	+20·4	Definition fair.
	M	3	F	+16·6	+16·1	+16·3	

Date.	Obs.	Wt.	Line.	Earth's Motion in m. per sec.	Concluded Motion of Star. Meas.	Estimd.	Remarks.

χ Ursæ Majoris.

Date.	Obs.	Wt.	Line.	Earth's	Meas.	Estimd.	Remarks.
1881. May 20	M	⅓	F	+ 13·4	+ 21·1	+ 18·8	Spectrum faint. Star-line diffused and faint.
	M	⅓	F	+ 13·4	− 20·5	− 19 9	

β Leonis.

Date.	Obs.	Wt.	Line.	Earth's	Meas.	Estimd.	Remarks.
May 4	M	2	F	+ 13·1	+ 10·9	+ 12·4	Star-line very diffused.
	M	2	F	+ 13·1	− 34·4	− 32·2	
	M	2	F	+ 13·1	− 4·5	− 5·5	
11	M	½	F	+ 15·5	− 28·1	− 27·8	Definition very poor.
	M	½	F	+ 15·5	− 39·8	− 40·2	

β Virginis.*

Date.	Obs.	Wt.	Line.	Earth's	Meas.	Estimd.	Remarks.
Mar. 25	M	⅓	b_1	+ 3·3	− 49·7	− 68·2	Spectrum faint.
	M	½	b_1	+ 3·3	− 35·6	− 46·6	

δ Ursæ Majoris.

Date.	Obs.	Wt.	Line.	Earth's	Meas.	Estimd.	Remarks.
May 20	M	1	F	+ 11·1	− 14·8	− 17·6	Star-line diffused.
	M	1	F	+ 11·1	− 20·4	− 17·6	

γ Virginis.

Date.	Obs.	Wt.	Line.	Earth's	Meas.	Estimd.	Remarks.
May 4	M	2	F	+ 10·6	+ 23·9	+ 27·6	Star-line diffused.
13	M	½	F	+ 12·7	+ 23·7	+ 17·1	Star-line dark. spectrum faint, and definition bad.
	M	½	F	+ 12·7	− 1·6	− 1·5	

a Canum Venaticorum.

Date.	Obs.	Wt.	Line.	Earth's	Meas.	Estimd.	Remarks.
May 11	M	2	F	+ 11·7	− 31·1	− 28·2	Definition fair.
	M	2	F	+ 11·7	− 41·9	− 41·3	

ε Virginis.

Date.	Obs.	Wt.	Line.	Earth's	Meas.	Estimd.	Remarks.
Mar. 31	M	½	b_1	+ 10·1	− 17·1	− 17·9	Star-lines very faint.
	M	⅓	b_1	+ 10·1	+ 3·9	+ 9·3	

Spica.

Date.	Obs.	Wt.	Line.	Earth's	Meas.	Estimd.	Remarks.
May 13	M	3	F	+ 9·3	− 21·6	− 20·5	Star-line very faint.
	M	3	F	+ 9·3	− 24·4	− 24·2	
	M	2	F	+ 9·3	− 35·5	− 31·6	
	M	2	F	+ 9·3	− 13·9	− 13·8	

ζ Ursæ Majoris.

Date.	Obs.	Wt.	Line.	Earth's	Meas.	Estimd.	Remarks.
May 20	M	1	F	+ 9·6	+ 24·6	+ 29·1	Star-line very diffused and difficult to bisect.
	M	1	F	+ 9·6	+ 12·9	+ 22·6	

η Ursæ Majoris.

Date.	Obs.	Wt.	Line.	Earth's	Meas.	Estimd.	Remarks.
May 20	M	1	F	+ 9·4	+ 15·9	+ 16·4	Star-line very diffused and difficult to bisect. Images steady.
	M	1	F	+ 9·4	+ 27·3	+ 29·3	

Date.	Obs.	Wt.	Line.	Earth's Motion in m. per sec.	Concluded Motion of Star. Meas.	Estimd.	Remarks.
1881.					*Arcturus.*		
June 22	M	3	b_1	+ 14·3	− 25·3	− 25·1	Definition variable, gener-
	M	3	b_1	+ 14·3	− 38·6	− 36·0	ally poor, but star-lines seen well at times. Two-prism train.
					ϵ_2 Boötis.		
June 22	M	1	b_1	+ 12·2	+ 2·6	+ 0·9	Definition variable.
	M	1	b_1	+ 12·2	+ 18·6	+ 14·0	
					*β Boötis.**		
June 22	M	1	b_1	+ 9·7	− 29·6	− 35·9	Definition variable.
	M	1	b_1	+ 9·7	− 54·5	− 49·0	
					β Libræ.		
May 6	M	1	F	− 0·4	− 32·0	− 30·0	Star-line dark, but diffused,
	M	1	F	− 0·4	− 37·8	− 30·0	and somewhat difficult to bisect.
					a Coronæ.		
May 13	M	½	F	+ 2·8	+ 12·3	+ 12·1	Spectrum faint. Definition
	M	½	F	+ 2·8	+ 16·9	+ 12·1	bad.
					γ Herculis.		
May 6	M	½	F	− 2·6	− 27·3	− 24·8	Observation interrupted by cloud.
					a Herculis.		
June 22	M	½	b_4	+ 4·2	+ 32·0	+ 28·0	Definition very poor.
	M	½	b_4	+ 4·2	+ 4·8	+ 8·7	
					a Ophiuchi.		
Sept. 30	M	1	F	+ 14·2	− 54·3	− 39·2	Spectrum bright and steady.
	M	1	F	+ 14·2	− 37·3	− 30·9	Star-line a broad faint nebulous haze.
					γ Draconis.		
June 22	M	½	b_4	+ 0·4	+ 18·7	+ 21·4	Definition bad.
	M	½	b_4	+ 0·4	+ 9·3	+ 16·0	
					a Lyræ.		
Sept. 14	M	1	F	+ 8·0	− 39·4	− 31·2	Spectrum tremulous. Defi-
	M	1	F	+ 8·0	− 26·5	− 19·6	nition very bad.
30	M	2	F	+ 8·6	− 31·7	− 31·1	Spectrum bright and fairly
	M	2	F	+ 8·6	− 18·5	− 21·1	steady. Star-line ill defined.
					γ Lyræ.		
Sept. 30	M	½	F	+ 10·3	− 49·2	− 47·7	Spectrum faint. Star-line
	M	½	F	+ 10·3	− 38·7	− 27·0	very ill defined.

Date.	Obs.	Wt.	Line.	Earth's Motion in m. per sec.	Concluded Motion of Star. Meas.	Estiml.	Remarks.
				ζ *Aquilæ.*			
1881. Sept. 19	M	2	F	+ 13·7	− 31·0	− 25·9	Images steady. Star-line
	M	2	F	+ 13·7	− 30·4	− 25·9	broad, diffused, and faint.
	M	2	F	+ 13·7	− 10·3	− 8·8	
				δ *Aquilæ.*			
Sept. 19	M	½	F	+ 15·0	+ 10·0	+ 4·6	Spectrum very faint. Mea-
	M	½	F	+ 15·0	− 85·3	− 76·2	sures worthless.
				β *Cygni.*			
Oct. 25	M	2	b_1	+ 12·2	− 7·5	− 3·7	Spectrum faint. Star-lines
	M	2	b_1	+ 12·2	− 39·5	− 40·5	exceedingly so.
	M	2	b_1	+ 12·2	+ 23·7	+ 30·3	
	M	2	b_1	+ 12·2	− 14·9	− 12·2	
				δ *Cygni.*			
Sept. 30	M	1	F	+ 6·3	− 0·7	+ 0·8	Spectrum faint, but steady.
	M	1	F	+ 6·3	+ 17·4	+ 10·4	Star-line da k, but ill de- fined at e dges.
				α *Aquilæ.*			
Sept. 30	M	1	F	+ 14·8	− 22·8	− 21·9	Spectrum tremulous. Star
	M	1	F	+ 14·8	− 13·6	− 14·8	line very ill defined.
	M	1	F	+ 14·8	− 37·9	− 39·8	
				α *Cygni.*			
Sept. 14	M	2	F	+ 2·9	+ 28·2	+ 20·3	Definition poor, evening
	M	1	F	+ 2·9	− 36·5	− 26·1	misty.
Oct. 18	M	1	b_1	+ 7·3	− 33·8	− 36·7	Definition poor, passing
	M	1	b_1	+ 7·3	− 69·3	− 66·1	cloud.
				ζ *Cygni.*			
Oct. 25	M	3	b_1	+ 11·8	+ 22·5	+ 30·7	Spectrum faint, but star-line fairly well seen.
Nov. 28	M	½	b_1	+ 13·6	− 61·9	− 63·6	Spectrum and star-lines very
	M	½	b_1	+ 13·6	+ 24·6	+ 28·1	faint. Definition very bad.
	M	½	b_1	+ 13·6	− 11·7	− 13·6	
	M	½	b_1	+ 13·6	− 40·5	− 41·4	
				α *Cephei.*			
Sept. 7	M	1	F	− 2·8	− 59·8	− 47·3	Star-line faint, broad, and
	M	2	F	− 2·8	− 24·7	− 39·0	diffused.

Date.	Obs.	Wt.	Line.	Earth's Motion in m. per sec.	Concluded Motion of Star. Meas.	Estimd.	Remarks.
				ζ Pegasi.			
1881. Sept. 19	M	1	F	+ 3·8	+ 6·7	+ 8·4	Star-line fairly sharp, but faint.
	M	1	F	+ 3·8	+ 28·6	+ 25·6	
	M	2	F	+ 3·8	− 14·0	− 16·0	
				η Pegasi.			
Nov. 22	M	½	b_1	+ 14·1	− 65·9	− 65·2	Star-lines very faint, and definition very bad.
	M	½	b_1	+ 14·1	− 0 5	− 3 9	
	M	½	b_1	+ 14·1	− 9·4	− 3·9	
	M	½	b_1	+ 14·1	− 22·3	− 24·3	
28	M	½	b_1	+ 14 8	+ 13·3	+ 13 0	Star-lines very faint. Definition very bad.
	M	½	b_1	+ 14·8	− 14·8	− 14·8	
				Fomalhaut.			
Sept. 19	M	½	F	+ 7·2	− 57·2	− 56·1	Very unsatisfactory.
				β Pegasi.			
Nov. 19	M	1	b_1	+ 14·0	− 0·8	− 4·3	Star-line faint and difficult to bisect.
	M	1	b_1	+ 14·0	− 8·2	− 4·3	
	M	1	b_1	+ 14 0	− 13·2	− 14·0	
22	M	½	b_1	+ 14 4	− 69·4	− 75 7	
	M	½	b_1	+ 14·4	− 3 5	+ 0 2	
	M	½	b_1	+ 14·4	+ 44·5	+ 36·7	
	M	½	b_4	+ 14·4	+ 12 1	+ 21·1	
	M	½	b_4	+ 14·4	+ 0 8	+ 12·2	
				a Pegasi.			
Sept. 30	M	3	F	+ 4·8	− 30·1	− 29·8	Spectrum steady, definition fair.
	M	3	F	+ 4·8	− 44 0	− 34·8	
Dec. 3	M	1	F	+ 17·6	− 12 3	− 13·1	Star-line dark but broad and diffused.
	M	3	F	+ 17 6	− 21·3	− 22 1	
7	M	4	F	+ 17 7	− 45·8	− 45·0	Star-line broad and ill defined at edges.
	M	4	F	+ 17·7	− 37·4	− 31·3	

Date.	Observer.	No. of Obs.	Line.	Motion Measured.	Remarks.

Moon.

1881, March 14	M	7	F	−0·6	
April 6	W.C.	3	b_1	−3·5	
7	M	3	b_2	−0·5	Two prism-train.
9	M	2	F	−0·5	Two prism-train.
9	M	4	F	+0·2	
May 6	M	3	F	−2·4	
11	M	5	F	+0·8	
13	M	2	F	+0·3	
Nov. 28	M	5	b_1	−0·7	
Dec. 3	M	5	F	−2·5	
7	M	5	F	−1·2	

Sky Spectrum.

1881, Feb. 28	M	5	F	+11·2	
March 26	M	6	b_1	+ 0·3	
29	M	6	b_1	+ 0·3	
April 1	M	5	b_1	− 4·6	
17	M	5	F	− 1·6	
May 5	M	5	F	+ 0·4	
June 23	M	3	b_1	− 1·1	
23	M	3	b_1	− 1·4	Two prism-train.
Sept. 20	M	5	F	+ 1·7	
Oct. 1	M	5	F	− 0·9	
19	M	5	b_1	+ 1·8	
Nov. 21	M	5	b_1	− 0·4	
23	M	5	b_1	0·0	

Observations of Occultations of Stars by the Moon, and of Phenomena of Jupiter's Satellites, made at the Royal Observatory, Greenwich, in the year 1881.

(Communicated by the Astronomer Royal.)

Occultations of Stars by the Moon.

Day of Observation.	Phenomenon.	Telescope.	Power.	Moon's Limb.	Mean Solar Time of Observation.	Observer.
					h m s	
1881, Jan. 5 (a)	Disapp. 19 Piscium	Altaz.	100	Dark	7 52 12·5	A. D.
12	Disapp. Piazzi V. 192	S. E. Eq.	320	,,	7 16 50·4	M.
March 8	Disapp. Piazzi V. 338	E. Eq.	70	,,	11 4 32·0	T.
16 (b)	Disapp. q Virginis	S. E. Eq.	130	Bright	12 1 0·9	M.
(c)	Reapp. q Virginis	,,	130	Dark	12 57 0·1	,,
May 4	Disapp. 5 Cancri	Altaz.	100	,,	9 26 33·2	A. D.
Sept. 3	Disapp. 33 Sagittarii	E. Eq.	140	,,	8 18 27·0	,,
Oct. 5	Disapp. κ Piscium	Altaz.	100	,,	12 9 5·6	L.
	Disapp. κ Piscium	E. Eq.	140	,,	12 9 7·0	J.
	Disapp. 9 Piscium	Altaz.	100	,,	12 13 47·9	L.
	Disapp. 9 Piscium	E. Eq.	140	,,	12 13 46·7	J.
Oct. 9	Reapp. 54 Arietis	Altaz.	100	,,	12 30 26·6	A. D.
Nov. 9 (d)	Reapp. 16 Geminorum	,,	100	,,	13 10 20·7	C.

(a) Instantaneous.
(b) The Moon's limb was smooth and regular at the point of disappearance. No projection. Disappearance not absolutely instantaneous.
(c) Very well seen; reappearance not absolutely instantaneous,　　(d) Much cloud about; observation doubtful.

Day of Observation.	Phenomenon.	Telescope.	Power.	Moon's Limb.	Mean Solar Time of Observation. h m	Observer.
1881, Nov. 12 (e)	Disapp. α Cancri	Altaz.	100	Bright	13 30 32.2	J. P.
	Reapp. α Cancri	,,	100	Dark	14 44 33.1	,,
29 (f)	Disapp. 16 Piscium	E. Eq.	310	,,	5 15 59.0	A. D.
	Disapp. 16 Dim	Altaz.	100	,,	5 16 2.0	J. P.
(g)	Disapp. 19 Piscium	E. Eq.	140	,,	11 27 21.7	A. D.
Dec. 30 (h)	Disapp. ρ² Arietis	Altaz.	100	,,	5 31 6.3	T.
	Disapp. ρ³ Arietis	E. Eq.	140	,,	5 31 6.1	A. P.

(e) Clouds passing.   (f) The star seemed very faint before the disappearance.
(g) Instantaneous.   (h) Image of star seemed disturbed on approaching the Moon's limb, but disappearance was instantaneous.

Phenomena of Jupiter's Satellites.

Day of Observation.	Satellite.	Phenomenon.	Telescope.	Powr.	Mean Solar Time of Observation. h m s	Mean Solar Time of N.A. h m s	Observer.
1881, Jan 31	II.	Ec. R. First seen	E. Eq.	140	7 3 29	7 4 11	A. D.
Feb. 15	III.	Ec. D. Last seen	,,	,,	8 18 59	8 19 41	W. C.
		,,	Altaz.	100	8 19 13		J. P.
Sept. 19 (a)	II.	Oc. R. Last contact	E. Eq.	140	10 44 27	10 38 0	L.
Oct. 6	III.	Tr. I. First contact	,,	,,	10 59 27	11 1 0	T.
		ion	,,	,,	11 3 11		
	I.	Ec. D. Last seen	,,	,,	11 29 11	11 28 40	,,
14 (b)	I.	Tr. I. First contact	,,	,,	11 22 45		,,
		Bisection	,,	,,	11 25 45	11 26 0	H. P.
		Last contact	,,	,,	11 29 44		,,

Day of Observation.	Satellite.	Phenomenon.	Telescope.	Power.	Mean Solar Time of Observation.	Mean Solar Time of N.A.	Observer.
					h m s	h m s	
1881, Oct. 14	I.	Tr. E. Last contact	E. Eq.	140	13 38 53	13 37 0	H. P.
29	I.	Ec. D. Last seen	,,	,,	11 40 35	11 40 41	J. P.
31	I.	Oc. R. First seen	,,	310	8 37 6	8 39 0	C.
		Last contact	,,	,,	8 38 53		
(c)	III.	Ec. D. Last seen	,,	,,	9 3 39	9 0 48	C.
(d)	III.	,,	S. E. Eq.	320	9 3 40		M.
Nov. 22	II.	Ec. R. First seen	E. Eq.	140	7 33 26	7 33 41	J.
	I.	Tr. I. First contact	,,	,,	8 54 8		
		Bisection	,,	,,	8 56 18	8 55 0	,,
		Last contact	,,	,,	8 58 22		
23 (e)	I.	Oc. D. First contact	Altaz.	100	6 3 30	6 5 0	C.
		Last contact	,,	,,	6 6 20		
28	I.	Ec. R. First seen	,,	,,	8 30 54	8 30 56	,,
	I.	Oc. D. First contact	E. Eq.	310	13 22 31	13 23 0	,,
		Bisection	,,	,,	13 24 10		
		Last contact	,,	,,	13 25 20		

(a) Observed through thin cloud; observation rough.   (b) Limb of Jupiter diffused, but occasionally very well defined.

(c) Observation very unsatisfactory. Clouds continually passing across the planet.

(d) This phenomenon was observed through a narrow slit—Jupiter being well outside the slit—in order to see if any faint light from the satellite could be discerned after its entry into the shadow.

(e) Limb of Jupiter very tremulous; observation difficult.

Day of Observation.	Satellite.	Phenomenon.	Telescope.	Power.	Mean Solar Time of Observation. h m s	Mean Solar Time of *N.d.* h m s	Observer.
1881, Nov. 29	II.	Oc. D. First contact	E. Eq.	140	6 37 22	6 45 0	A. D.
		Bisection	"	"	6 43 21		
		Last contact	"	"	6 47 35		
	II.	Fe. R. First seen	"	"	10 8 27	10 9 1	"
	I.	Tr. I. First contact	"	"	10 36 48		
		Bisection	"	"	10 40 42	10 39 0	"
		Last contact	"	"	10 43 12		
Dec. 8 (*f*)	II.	Tr. E. Bisection	"	"	5 58 42	6 0 0	T.
		Last contact	"	"	6 0 32		
13 (*g*)	III.	Oc. R. Last contact	"	100	7 43 5	7 42 0	H. C.
21	I.	Oc. D. First contact	"	140	13 3 36		
		Bisection	"	"	13 6 35	13 8 0	A. D.
		Last contact	"	"	13 9 20		
30 (*h*)	I.	Oc. D. Bisection	"	"	9 21 47	9 24 0	A. P.
		Last contact	"	"	9 24 42		

(*f*) Sky aky.   (*g*) Very faint; sky foggy.   (*h*) Limb of *Jupiter* diffused.

The clear aperture of the object-glass of the S.E. Equatoreal is 12¾ ins., of the E. Equatoreal 6·7 inches, and of the Altazimuth 3¾ inches.

The initials W. C., C., A. D., M., T., L., H. P., J. P., J., A. P., and H. C., are those of Mr. Christie, Mr. Criswick, Mr. Downing, Mr. Maunder, Mr. Eby, Mr. Lewis, Mr. H. Pead, Mr. Power, Mr. James, Mr. A. Pead, andMr. Cox.

Royal Observatory, Greenwich:
  1882, March 1.

Observations of Occultations of Stars by the Moon, and of Phenomena of Jupiter's Satellites, made at the Radcliffe Observatory, Oxford, in the year 1881. By E. J. Stone, Esq., M.A., F.R.S.

Occultations of Stars by the Moon.

Day of Observation.	Phenomenon.	Instrument.	Moon's Limb.	Greenwich Mean Solar Time of Observation. h m s	Observer.
1881, September 3 (a)	Disapp. of ξ² Sagittarii	10-foot Equatoreal	Dark	10 16 21·3	W.
"	"	42-inch telescope	"	10 16 21·5	R.
October 11 (b)	Reapp. of 105 Tauri	10-foot Equatoreal	"	17 6 35·7	R.
14 (c)	Reapp. of f Geminorum	42-inch telescope	"	13 56 52·8	R.
November 29 (d)	Disapp. of 16 Piscium	10-foot Equatoreal	"	5 14 39·8	W.
December 5 (e)	Disapp. of ι Tauri	42-inch telescope	Full	6 23 57·5	W.
(f)	Reapp. of ι Tauri	"	·	7 25 7·5	W.

Observers' Remarks.

(a) and (d) Instantaneous.   (b) Reappeared instantaneously.   (c) Reappeared instantaneously, but bad image.
(e) Moon partially eclipsed. Limb very sharp; observation good.   (f) Star's image diffused; so time noted may be a very little late.

Phenomena of Jupiter's Satellites.

Day of Observation.	Satellite.	Phenomenon.	Instrument.	Greenwich Mean Solar Time of Observation. h m s	Greenwich Mean Solar Time of N.A. h m s	Observer.
1881, August 24 (a)	III.	Tr. Ingr. Ext. contact	42-inch telescope	12 38 30	12 46	R.
		Bisection	,,	12 43 30±		
		Int. contact	,,	12 53 30		
September 19 (b)	II.	Occ. Reapp. First seen	,,	10 36 8	10 38	R.
		Bisection	,,	10 41 20±		
		Ext. contact	,,	10 45 19		
21 (c)	I.	Tr. Ingr. Ext. contact	,,	11 35 1	11 40	R.
		Bisection	,,	11 38 6		
		Int. contact	,,	11 42 20		
29 (d)	III.	Tr. Egr. Int. contact	,,	8 38 53	8 43	R.
		Bisection	,,	8 42 32		
		Ext. contact	,,	8 46 51		
29 (e)	I.	Ecl. Disapp. First dim. of brightness	,,	9 32 34	9 34 27	R.
		Disappearance	,,	9 34 53		
30	I.	Tr. Egr. Int. contact	,,	9 58 57	10 5	R.
		Bisection	,,	10 2 27		
		Ext. contact	,,	10 5 26		

Day of Observation	Satellite	Phenomenon	Instrument	Greenwich Mean Solar Time of Observation. h m s	Greenwich Mean Solar Time of N.A. h m s	Observer
October 3 (S)	II.	Ecl. Disapp. First diminution	42-inch telescope	10 54 11	10 55 7	R.
		Last seen	,,	10 55 53		
14 (g)	I.	Tr. Ingr. Ext. contact	10-foot Equatoreal	11 22 38	11 26	R.
		Bisection	,,	11 25 47		
		Int. contact	,,	11 30 37		
14 (h)	I.	Tr. Egr. Int. contact	,,	13 29 27	13 37	R.
		Bisection	,,	13 33 47		
		Ext. contact	,,	13 37 31		
15 (i)	I.	Ecl. Disapp. First diminution	42-inch telescope	7 49 43	7 51 33	W.
		Half brightness	,,	7 50 38		
		Disappearance	,,	7 51 34		
29 (k)	I.	Ecl. Disapp. First diminution	10-foot Equatoreal	11 39 1	11 40 41	W.
		Half brightness	,,	11 40 1		
		Disappearance	,,	11 41 9		
November 22 (l)	I.	Tr. Ingr. Ext. contact	,,	8 52 52	8 55	W.
		Bisection	,,	8 55 6		
		Internal contact	,,	8 57 34		
22 (l)	I.	Tr. Egr. Int. contact,	,,	11 4 18	11 7	W.
		Bisection	,,	11 6 24		
		Ext. contact	,,	11 8 32		

Day of Observation.	Satellite.	Phenomenon.	Instrument.	Greenwich Mean Solar Time of Observation. h m s	Greenwich Mean Solar Time of N.A. h m s	Observer.
1, November 23 (m)	I.	Ecl. Reapp. First appearance	10-foot Equatoreal	8 30 38	8 30 56	W.
29 (n)	II.	Ecl. Reapp. First seen	"	10 8 37		
		Half brightness	"	10 9 52	10 9 1	W.
		Full brightness	"	10 10 57		
December 7 (o)	I.	Occ. Disapp. First contact	"	9 31 12		
		Bisection	"	9 33 43	9 35	W.
		Last seen	"	9 36 5		

10-foot Equatoreal. Power = 180.   42-inch Telescope. Power = 100.

The initials W. and R. are those of Mr. Wickham and Mr. Robinson respectively.

Observers' Remarks.

(a) Cloudy at times.   (b) Observation fair.   (c) Good images.   (d) Observation good.   (e) Satisfactory observation.
(f) Satellite followed continuously until 10h 55m 50s, and at the time noted "last seen" (10h 55m 53s) it appeared for a moment, as the faintest possible point, and then vanished.   Good observation.
(g) Limb of planet tremulous at times.   (h) Limb of planet not well defined at times.
(i) Observation difficult, owing to bad image.   (k) Observation good.   (l) Images ill defined.
(m) Observation considered good, but images diffused.   (n) and (o) Observations good.

Note on the North Polar Distance of the Star Lacaille 4342. By David Gill, Esq., H.M. Astronomer at the Cape of Good Hope.

In his Paper on the North Polar Distances of the Cape Catalogue (*Monthly Notices*, November 1881), Mr. Downing, at page 20, directs my attention to the star Lacaille 4342, pointing out a discordance of 4″·69 between its N. P. D. as given in the Cape Catalogue for 1880 and that of the Melbourne Catalogue for 1870.

The discordance, however, is capable of easy explanation— viz. a typographical error in the Melbourne Catalogue.

The mean of the separate results, as published in the annual volumes of the Melbourne Observatory, reduced to 1870, the N.P.D. of the Star in question appears to be about

$$176° \ 16' \ 38''·8,$$

not

$$176° \ 16' \ 33''·8,$$

as printed.

The place of the Cape Catalogue reduced to 1870 is

$$176° \ 16' \ 38''·5,$$

a sufficiently satisfactory agreement.

Note on the Places of Three Stars in the Armagh Catalogue. By Rev. T. R. Robinson.

Dr. Robinson's attention has been called (by M. L. Schulhof, of Paris) to three stars in this Catalogue of which the places appeared to be affected with errors.

On re-examining the observations and reductions, he finds this to be the case.

A 1435 : the A.R. and N.P.D. given there belong to different stars—the A.R. to P. vi. 61 ; the N.P.D. to P. 55, and when properly reduced becomes N.P.D. 31° 20′ 54″·87.

A 1437 : N.P.D. should be 31° 30′ 20″·53.

A 2076 : N.P.D. should be 90° 46′ 31″·86.

A New Method of Bright-wire Illumination for Position Micrometers. By S. W. Burnham, Esq.

In all my double-star measures I have uniformly used a bright-wire illumination, that method having been found by experience to be much superior to any form of bright field, and particularly in the observation of very close pairs and very faint companions which could only be seen with difficulty without any illumination at all. The Fraunhofer plan (which seems to have been the best in general use), of attaching the lamp to a hollow arm forming an angle of about 45° with the axis of the telescope, was always very objectionable and inconvenient, from the fact that to obtain a proper light on the wires it was necessary to change the place of the lamp with every change of the wires in position-angle, so as to allow the light to strike the wires approximately at right angles. For some years past I have given much attention, and made many experiments, with a view to the improvement of the micrometer in this respect. This resulted, while at the Washington Observatory during the summer of 1881, in the execution of a plan for the "end" illumination of the Clark micrometer attached to the 15½-inch Refractor of that Observatory. That device was fully described and illustrated in the *English Mechanic* for September 16, 1881. This micrometer was regularly used for some months, and although the illuminating apparatus was somewhat roughly made, and intended only to demonstrate the practical success of this device, it was found to work so satisfactorily in every respect that the micrometer for the 12-inch Refractor of the Lick Observatory was constructed after the same plan, and was used by the writer for some weeks on Mount Hamilton, Cal., in the latter part of 1881. This illuminating apparatus is identical in principle with that previously described, but, as would be expected from the great skill and ingenuity of the firm of Alvan Clark & Sons, by whom it was made, the mechanical construction generally is vastly improved, and the convenience and usefulness of the micrometer increased thereby.

The accompanying illustration is engraved from a photograph of the Lick Observatory micrometer, the construction of which will be understood by a brief description.

The micrometer is of the usual form made by the Clarks, B being the graduated head of the micrometer-screw, and A another graduated head turning on the same axis for giving the whole revolutions of the screw; C is the head of a pinion attached to the plate under the micrometer-box, and gearing into the teeth of the rigid circular plate containing the position circle, for moving the wires in position angle; D is the head of another small pinion for sliding the eyepiece over the wires; E E' are heads of the bisecting-screw for moving the whole system of wires and the box S in a direction parallel to the micrometer-screw, and at right angles to the wires. The light from the lamp L is reflected by a mirror in N, and passes down that tube and through M, and then through a hole in the end of the box to the wires. A condensing lens is placed in N, for the purpose of concentrating the light on the wires. On the opposite side of the wires, towards the micrometer head, a small reflector is placed which reflects the light back, thereby symmetrically illuminating the wires on both sides. The lamp swings freely on its axis in the line of OT, but always maintains a vertical position, whatever may be the direction of the wires or the pointing of the telescope. The tube N, with the lamp and its

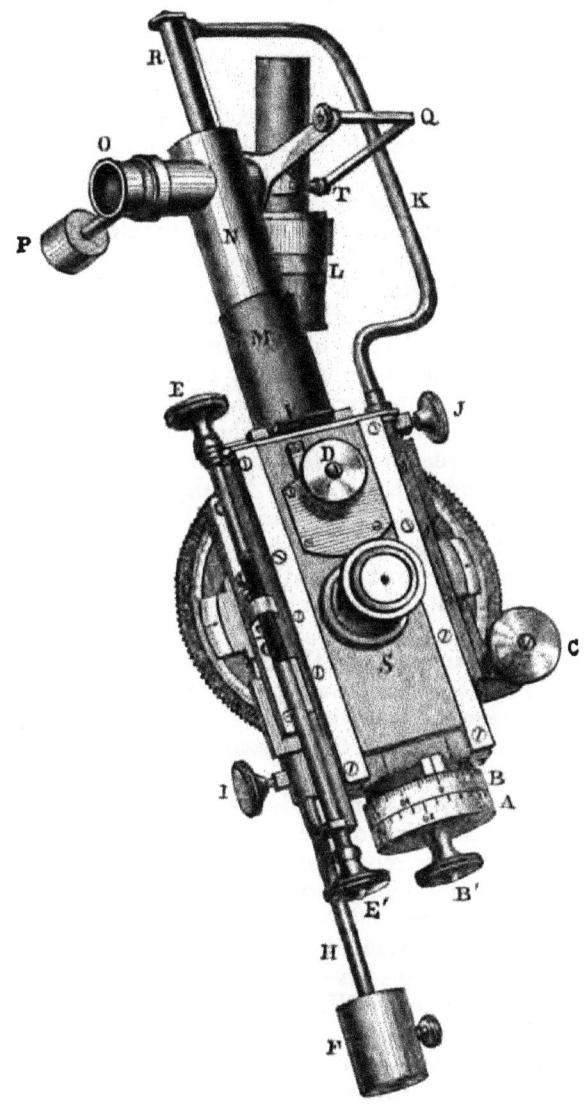

attachments, has an axle R supported by the fixed arm K. The bearings T, and axle of the lamp, are kept always horizontal by the weight of the counterpoise P. The tube M is fixed to the micrometer box, and projects loosely over N far enough to allow for the necessary movement of the box by the bisecting-screw E. The supporting arm K is attached by the set-screw J, not to the box, but to the plate underneath it, so that the weight moved by the bisecting-screw is not increased at all by the illuminating apparatus. Attached to the same plate, on the opposite side by a set-screw I, is the rod H, bent so as to be thrown forward out of the way of E' and B, with a weight F to balance the weight of the lamp attachments. The whole device can be instantly detached when desired, by loosening the screws I and J. In the tube M is a slot V, in which is placed a slip of red or other coloured glass, held in any desired place by a light spring pressing against it. All or any part of the light can be made to pass through the coloured medium. The mirror in N is attached to a tube which slides into the tube O. By turning this tube by the milled edge projecting at O, the inclination of the mirror may be varied to any extent, and the light reduced from the maximum amount until the wires become invisible. By turning the mirror 90° or more the light is entirely shut off. It will be seen that the lamp can revolve freely through the bent arm Q; and the whole moveable part of the device, lamp, arm Q, and counterpoise P, can turn through the supporting arm K, the lamp at all times remaining vertical, and in exactly the same position with respect to the wires. It might at first be supposed that the lamp, or some of the parts, would be in the way of the observer. I have never found it so in practice, and, although it is but a few seconds' work to either attach or detach it, I have very rarely removed it, whatever might have been the use of the telescope at the time.

It is important to preserve the relative positions of the micrometer-head, bisecting-screw, and pinion C, as here shown. No other arrangement will be as convenient. In every possible position of the micrometer, the necessary use of both hands at the same time will be found to be convenient and easy for the observer. Naturally the more delicate motions of the micrometer-screw and the pinion will be effected by the right hand, and the corresponding movement of the bisecting-screw by the left hand. When the micrometer box is anywhere near a horizontal position with respect to the observer (the wires at right angles to the line joining the eyes) C and E are used by the right and left hands respectively in measuring angles, and B' and E in measuring distances. When the box is more nearly vertical with respect to the observer, the head E' of the bisecting-screw will be worked by the left hand in each case. The convenience and practical value of this arrangement can only be appreciated by one who has used the old plans, and then tried this.

With respect to the practical working of the illumination, I will briefly say that it has proved a complete success in every respect. Any object that can be seen under any circumstances, however faint, can be well and accurately measured. There is no such thing as a star too faint for measurement, if it can be seen at all.   A very feeble light is sufficient to illuminate the wires perfectly for any object.   I believe far better results can be obtained by the use of bright wires in a large part of the most desirable and important double-star work, than is possible by the same observer using a bright field, and that sooner or later it will be generally used in all micrometrical observations.

Chicago:
  1882, *February* 25.

The Transit of Mercury, 1881, *November* 7, *observed in New South Wales.*

(*Communicated by H. C. Russell, Esq., Government Astronomer, Sydney.*)

In order to secure satisfactory observations of the Transit of *Mercury* the observers were divided into three parties: two, Mr. Lenehan, first assistant in the Observatory, and myself, observed at Sydney, using respectively 7¼-inch and 11½-inch Refractors, both stopped to 6 inches.   At Katoomba, a place on the Western Railway, 66 miles west of Sydney, and 3,400 feet above the sea level, Messrs. Hargrave and Bladen, assistants in the Observatory, observed ; the former using a 4¾-inch Troughton & Simms equatoreally-mounted Refractor, and the latter a 4¼-inch Cooke Equatoreal, both driven by clock-work.   At Bathurst, on the elevated table land, 2,300 feet above the sea and 134 miles by rail west of Sydney, the observers were Mr. Conder, chief of the Trigonometrical Branch of the Survey Department, and Mr. Brooks, on the Trigonometrical Survey staff ; the former used a 4¾-inch Schroeder Equatoreal, and the latter a 3¾-inch Troughton & Simms.

The morning was fine and clear, but the definition in Sydney was very bad ; on the high lands it was better.   At Egress, the definition in Sydney was steadier, but a thick cirrus haze covered the sky; the conditions were similar at the inland stations, only the definition got worse instead of better.

The times given are all in Sydney mean time, and corrected for clock errors, but not for the positions of the observers.   The following extracts from the observers' Reports will show the phenomena observed :—

Mr. Russell: " For *Ingress*, used 7¼-inch Merz Refractor, power 150.

" Air clear, but very unsteady.   8ʰ 21ᵐ 34ˢ·64, thought I saw notch in Sun's limb, but lost it.   8ʰ 21ᵐ 44ˢ·64, first contact certain.   8ʰ 23ᵐ 5ˢ, *Mercury* assumed a D shape; Sun's limb boiling, and *Mercury* seems to jump half its diameter.   8ʰ 23ᵐ 39ˢ·14, observed unsatisfactory contact—unsatisfactory, because of the vibration and tremulous definition, and that, for ten seconds before, contact of limbs was made and broken several times over. From this time I used 11½-inch Schroeder Refractor, stopped to 6 inches, polarising eyepiece, no coloured glass, powers 100 to 200, made of quartz. In moments of best definition, see clearly the black specks in the mottling on the Sun's surface. _Mercury_ intensely blue-black; no sign of satellite, and no halo of any kind visible round the planet; when definition gets bad, as it does periodically, _Mercury_ loses the blackness, and looks like a black ball shaded with white, the darkest part of the shading being near the planet's limb; thence, shading off to nothing before it reaches the centre, it extends half-way round the planet on the preceding side; I see no white spot in the centre. _Egress_: Definition steadier, but very thick haze. At $1^h 40^m 55^s$, Sun very unsteady, a shade or dark band connected the limbs, then broke; this was repeated several times in as many seconds, and then ceasing left a clear band of light between the limbs. Definition steady, band of light got gradually narrower, and broke at $1^h 40^m 26^s\cdot 65$; ten seconds later planet elongated. Clouds thick, obliged to turn polarising eyepiece to admit all the light it would; last contact observed at $1^h 42^m 8^s\cdot 95$. Nothing seen of the planet after this; haze too thick."

Mr. Lenehan : " _Ingress_ observed through 3-inch Refractor, hurriedly placed in position and moved by hand-screw. First indent, $8^h 22^m 8^s$. Planet moved steadily and clearly on towards internal contact, which was observed at $8^h 23^m 43^s$. Observations now made with 7¼-inch Merz, stopped to 6 inches. At $8^h 5^m$ an apparent halo round planet. $9^h 6^m$, halo still visible. $9^h 50^m$, disk of planet sharp, without marginal indistinctness or halo; no appearance of any satellite; clouds about. $10^h 50^m$, clear disk, halo not discernible. _Egress_: $1^h 10^m$, definition bad, with marginal indistinctness. First contact at Egress, $1^h 40^m 29^s\cdot 25$; no absolute certainty in the actual time, as the definition was not good and the wind high, causing vibration in the telescope, but I am satisfied with the time given. External contact occurred under same conditions at $1^h 42^m 4^s\cdot 25$."

Dr. Wright obtained time from the Observatory, and observed with his own 8½-inch Reflector : for Ingress stopped to 5 inches, power 80; for Egress stopped to 7½ inches. Browning's double-prism solar eyepiece used.

Mr. Morrice used his own 8½-inch Browning-With Reflector; obtained time from the Observatory, and observed in the suburbs of Sydney.

KATOOMBA.

Mr. Hargrave : "Weather all that could be wished. Used full aperture, 4¾ inches; power 100; darkest glass shade. Definition very good. Saw first contact at $8^h 22^m 0^s\cdot 66$, Sydney mean time. Planet three-quarters on, cusps slightly rounded, but it did not last until second contact, when the definition was very good and quite calm. At $8^h 23^m 41^s\cdot38$, there is an optical illusion like a spot of light in the centre of *Mercury*, which disappears on looking steadfastly at it; the limb of *Mercury* is a hard line, no colour or difference of light on the Sun at the planet's limb. $8^h 43^m$, cirrus clouds about. $9^h 21^m$, tried high power—definition bad, it is shaky with low one too; reduced aperture to 2 inches —much better definition. White spot still dancing about the middle of *Mercury*; blackness of planet more intense than that of Sunspot. $12^h 25^m$, white spot very persistent. $1^h 35^m$, light clouds passing. Observed internal contact at $1^h 40^m 10^s\cdot4$, both limbs moderately well defined, cusps sharp; last contact at $1^h 42^m 0^s\cdot09$."

Mr. Bladen: "Weather very clear. Used $4\frac{1}{4}$-inch Cooke Equatoreal, with clock-work, solar diagonal eyepiece, power 100; object-glass stopped to 2 inches. Definition very good indeed. First contact, slight indent, at $8^h 22^m 3^s\cdot34$; outline of planet as it crept on the Sun very clear and well defined. When planet was three-quarters on, a light cirrus cloud passed, making cusp indistinct; and at $8^h 23^m 24^s\cdot64$, I thought, from the irregular shape of *Mercury*, that internal contact would be complete but for the bad definition. At $8^h 23^m 42^s\cdot84$ a band of light was visible between the limbs, and from its breadth it would doubtless have been seen two or three seconds before with better definition.

"No sign of the part off the Sun's disk could be seen during the time *Mercury* was creeping on the Sun. *Mercury* on the Sun appears a perfect sphere and intensely black, without any halo or spots, and I saw no satellite. Definition so good for an hour after Ingress that I used the highest powers, and different apertures, but I could see nothing else to note.

"*Egress*: Internal contact, $1^h 40^m 44^s\cdot34$. I waited before taking this time until I was quite sure contact was complete, the bad definition making it difficult to decide. The planet did not seem to move steadily off the Sun, but in a series of jumps half a second in duration, which may have lasted ten or fifteen seconds; when I was quite certain that there was no trace of the planet, I noted the time, $1^h 42^m 23^s 64$."

BATHURST.

Mr. Conder: " I used $4\frac{3}{4}$-inch Schroeder telescope, equato-really mounted; solar eyepiece and power 90 were used. Tried various stops, and then used full aperture of objective. Having an electric chronograph, ticks of Sydney standard clock were recorded, beside those of my chronometer, a short time before and after transit, and errors of time were thus practically eliminated.

" In early morning weather very fine, but about time of Ingress clouds (cirrus) began to form, but definition was very good. My driving clock was not working well, so that specks of dust were made to move and caused apparent notches on the Sun's limb, and for a second or two I was in doubt, but as soon as I was certain I closed the chronograph key, and made the time $8^h 22^m 5^s$. Comparing mentally afterwards the distance between the cusps, I estimated this at six or seven-tenths of the planet's diameter; but as the estimate was not made at the time, it is not intended to be used. A few seconds before internal contact the telescope was accidentally touched, causing vibrations, which had scarcely ceased when the limbs of the planet and Sun appeared tangential; time by chronograph, $8^h 23^m 35^s\cdot75$. I felt in some doubt whether the cusps had really closed at this instant, and I continued to watch very carefully, and, within so small an interval as to be scarcely appreciable, an extremely fine line of light was noticed separating the planet from the Sun's limb.

"I watched the planet at intervals during its transit, the definition most of the time being magnificent. I failed to notice any peculiar appearances, except, perhaps, a very faint suspicion of a halo, or yellowish light surrounding the intensely black disk of the planet, and slightly brighter than the general illumination of the Sun's disk; this I attributed to an optical illusion.

"At Egress, clouds were so thick that I could not see the Egress through my dark glass, and I had not a second one.

"The position of the observing station, found by connection with the main triangulation, is—Lat. 33° 25' 45"·5 ; Long. 149° 33' 47"·9 East."

Mr. Brooks : " Telescope used was by Troughton & Simms, $3\frac{3}{4}$-inch aperture, and equatoreally mounted, stopped to 3 inches; solar reflecting eyepiece, and power 104, with lighter of the two coloured shades. The definition on Sunspots was very clear, and the whole of the Sun's surface had a faint mottled appearance, and the faculæ were at times very distinct. Three or four minutes before contact cirrus clouds covered the sky. I first caught sight of a decided notch in the Sun's limb at $8^h 22^m 3^s\cdot60$; the notch was quite decided before I was certain it was what I was looking for. On reflection, I think about one-tenth of the planet's disk must have been on the Sun at the time. At $8^h 23^m 36^s\cdot45$ I caught the first indication of a white line separating the Sun and planet's limbs. Just before this, the planet seemed to draw out slightly, as if unwilling to leave the Sun's limb, leaving in my mind a faint suspicion of a black drop. Three minutes after contact I saw a whitish spot on the south preceding quadrant, and an irregular white band of light inside the planet's disk. I suspect this was merely optical. About noon clouds were denser, causing the planet to lose its dead black appearance. Definition fairly steady. Ten minutes before Egress I changed the eyepiece for one marked 110, aperture and sunshade as before. At $1^h 40^m 26^s\cdot48$, I noted first internal contact, without seeing any indication whatever of the black drop; and at $1^h 42^m 8^s \cdot 59$ I saw the last contact; but owing to the passing clouds, there may be an uncertainty of, say, one second in first and about two seconds in last contact at Egress. Time by chronometer compared before and after transit with ticks from Sydney clock."

Observed Times of Transit of Mercury in New South Wales.

	First External Contact.	First Internal Contact.	Second Internal Contact.	Second External Contact.
	h m s	h m s	h m s	h m s
Russell	8 21 44·64	8 23 39·14	1 40 26·65	1 42 8·95
Lenehan	22 8·00	43·00	29·25	4·25
Wright	21 20·00	10·00	23·00	3·00
Morricp	22 15·00	46·00	16·00	16·00
Hargrave	22 0·66	41·38	10·40	0·09
Bladen	22 3·34	42·84	44·34	23·64
Conder	22 5·00	35·75	—	—
Brooks	22 3·60	· 36·45	26·48	8·59
Arithmetical Means }	8 21 57·53	8 23 36·82	1 40 25·16	1 42 9·217

Observations of the Transit of Mercury, 1881, *November* 7, *made at Honolulu, Sandwich Islands.* By C. H. Rockwell, Esq.

(Communicated by the Secretaries.)

I send herewith a Report of my movements in connection with observations made of the Transit of *Mercury*, at Honolulu, Sandwich Islands, on November 7, 1881.

The king, Kalakaua, offered me the free occupancy of the site from which the observations of the Transit of *Venus* were made in December 1874 by the English party under command of Capt. G. L. Tupman.

This civility on the part of His Majesty saved me the trouble and expense of building a pier of mason-work on which to mount my telescope, and, being on his own private grounds, I was enabled to occupy a quiet location whose geographical position had been determined with great accuracy.

The latitude and longitude given in this Report, and used in computing the times of contact, are the results obtained by Capt. Tupman.

I built a shed of rough boards over one of the piers left by the Transit of *Venus* party; there was no roof on the slope towards the South and West, the opening being covered at night with a part of an old sail, which could be easily put in position or withdrawn.

I set up the telescope in this temporary observatory on November 4, and during the 5th and 6th brought the instrument into proper adjustment.

The morning of the 7th was cloudy, and the prospects of success not very encouraging. The clouds passed away from the zenith, however, before the time of first contact, which was about 11.45 A M., local mean time.

Astronomically speaking, the day was not perfect, nor even first-rate; still it was very good indeed, and I consider my work, on the whole, as having been highly successful.

In making my observations I used the full aperture of the glass, 6½ inches, with a first surface solar prism, by Alvan Clark & Sons, and an eyepiece of power 60.

Twice during the day I used a power of 130, but for a few minutes only, as I found the increased size of the image was gained at the sacrifice of clearness of definition.

The intensely black disk of the planet was in strong contrast with a Sun-spot of considerable size, which was seen on the face of the Sun quite near to the point of second contact.

After having recorded the time of first and second contact, the telescope was moved in R. A. a little in advance of the Sun, and there clamped.

The time was then taken from the chronometer, when the advancing limb of the Sun, a large Sun-spot, the planet *Mercury*, and the following limb crossed the centre wire of the eyepiece: this being set at 90° to the line of the Sun's motion in the parallel, by allowing the Sun's limb to move along it.

I took about ninety of these combined observations, a copy of the record being appended to this Report.

At times during the continuance of the Transit the sky was free from clouds in the vicinity of the Sun, and the definition very satisfactory.

The spherical form of the planet was easily noticed; the edges of his disk sharp and clean, without any *fuzziness* or wavy outline to indicate the existence of an atmosphere.

I took advantage of these favourable moments to look carefully over the whole disk of the Sun, searching for anything which might look like an intra-Mercurial planet. I failed, however, to detect any such indications, although, from the clearness with which the *rice-grains* were shown, I am confident I could have detected the presence of any opaque body having one-tenth the diameter of *Mercury*. Nor did I at any time see any spots on the disk of the planet.

At the time of third and fourth contacts the Sun was only about twenty minutes above the horizon.

I could still see the planet distinctly, but the haze lying along the horizon caused a violent *boiling* of the edges of the Sun's disk, and the planet seemed to be dancing, or tossed about like a chip floating on the surface of broken water.

Three times I called to the recorder to *mark* the instant of third contact, supposing the planet to have reached the edge of the disk; again and then again I saw a strip of light in advance of the moving body, each time calling to *mark.*

The unfavourable atmospheric conditions at this phase would allow but little weight to be given to an opinion as to the existence of the *black-drop* phenomenon. I can say, however, quite positively, that no such appearance presented itself to me.

The whole disk of the planet seemed to be thrown towards the edge of the Sun, and then to rebound; but the advance and receding were well defined.

The form of the disk was distorted, but the edges were clean and sharp up to the moment when third contact was last called.

I attach but little value to the time given as fourth contact, it simply records the last glimpse which I had of the Transit of November 1881.

I give the computed and observed times of contacts below :—

Contacts.	Computed *Nautical Almanac.*	Computed *American Ephem.*	Observed G.M.T.
1st	10 16 29·79	10 16 23·01	10 17 27·25
2nd	10 18 12·62	10 18 05·84	10 18 25·75
3rd	15 35 21·77	15 34 39·66	15 35 36·25
4th	15 37 04·79	15 36 22·68	15 36 28·25

Transit of Mercury at Honolulu.

Lat. 21° 17' 56"·3 North; Long. 10ʰ 31ᵐ 27·3 West of Greenwich.

Chronometer: Negus, No. 1305. Correction + 38"·25.

First contact   10 16 49   by Chronometer.

Second contact 10 17 47·5   ,,

Sun's Advancing Limb.	Sun-spot.		Mercury (centre)		Sun's Following Limb.	
h m s	h m s	Diff. m s	h m s	Diff. m s	h m s	Diff. m s
10 19 58·5			10 21 56	1 57·5	10 22 12	2 13·5
22 59	10 24 12	1 13	24 56	57	25 14·5	15·5
26 44·5	27 57 5	13	28 39·5	55	28 59·5	15
29 38	30 51	13	31 31·5	53·5	31 52·5	14·5
32 23·5	33 37	13·5	34 17	53·5	34 39	15·5
35 00·5	36 13·5	13	36 53	52·5	37 15·5	15
37 45	38 57·5	12·5	39 35·5	50·5	39 59·5	14·5
40 22	41 34	12	42 11	49	42 36	14
42 49·5	44 01·5	12	44 38	48·5	45 04	14·5
45 31·5	46 44	12·5	47 20·5	49	47 46·5	15

Passing clouds.

Sun's Advancing Limb. h m s	Sun-spot. h m s	Diff. m s	Mercury (centre) h m s	Diff. m s	Sun's Following Limb. h m s	Diff. m s
11 05 29·5	11 06 42·5	1 13	11 07 10·5	1 41	11 07 44·5	2 15
08 16	09 29	13	09 56	40	10 31·5	15·5
10 52	12 05	13	12 31	39	13 07	15
13 37	14 49	12	15 15	38	15 52	15
16 15	17 28	13	17 52·5	37·5	18 31	16
18 52	20 04·5	12·5	20 28	36	21 07	15
21 28	22 40	12	23 03	35	23 43	15
24 12	25 24	12	25 46	34	26 27	15
26 40·5	27 53·5	13	28 14	33·5	28 57·5	17
29 20	30 32	12	30 52·5	32·5	31 35·5	15·5
38 18	39 30	12	39 47·5	29·5	40 33	15
41 14·5	42 25·5	11	42 43	29	43 29	14·5
44 40	45 52	12	46 08	28	46 55	15
47 18	48 29·5	11·5	48 44·5	26·5	49 32·5	14·5
50 43	51 54·5	11·5	52 08	25	52 58	15
53 29	54 40	11	54 52·5	23·5	55 43·5	14·5
56 24·5	57 36	11·5	57 47	22·5	58 39·5	15
59 07	12 00 18·5	11·5	12 00 29	22	12 01 22	15
12 02 12·5	03 24	11·5	03 33·5	21	04 27·5	15
04 54·5	06 06	11·5	06 14	19·5	07 09	14·5
07 45			09 04	19		
11 09·5	12 19·5	10	12 26	16·5	13 23	13·5
14 05	15 17	12	15 22·5	17·5	16 20·5	15·5
17 11	18 22	11	18 26	15	19 25·5	14·5
19 52	21 03	11	21 06	14	22 06·5	14·5
22 27	23 38	11	23 40	13	24 42	15
25 13·5	26 25	11·5	26 26	12·5	27 28·5	15
28 00	29 11	11	29 11	11	30 15	15

Mercury passes the Sun-spot.

Sun's Advancing Limb. h m s	Mercury (centre) h m s	Diff. m s	Sun-spot. h m s	Diff. m s	Sun's Following Limb. h m s	Diff. m s
12 37 19	12 38 26·5	1 07·5	12 38 29·5	1 10·5	12 39 34	2 15
39 53·5	41 00·5	07	41 05	11·5	42 09	15·5
42 24·5	43 30	05·5	43 35·5	11	44 39	14·5

Sun's Advancing Limb	Mercury (centre)	Diff.	Sun-spot.	Diff.	Sun's Following Limb.	Diff.
h m s	h m s	m s	h m s	m s	h m s	m s
12 45 08·5	12 46 14	1 05·5	12 46 20	1 11·5	12 47 24	2 15·5
49 22	50 25	03	50 33	11	51 37	15
51 56			53 06·5	10·5	54 11	15
54 42	55 43·5	01·5	55 53·5	11·5	56 57·5	15·5
57 14	58 14	00	58 25	11	59 29	15
59 48	1 00 47·5	0 59·5	1 00 59	11	1 02 03·5	15·5
1 02 38·5	03 37	58·5	03 49·5	11	04 54·5	16
05 30	06 27	57	06 40·5	10·5	07 45	15
08 06·5	09 03·5	57	09 17·5	11	10 21·5	15
11 11	12 06	55	12 12	11	13 26	15
14 12·5	15 06	53·5	15 23	10·5	16 27	14·5
16 57	17 49·5	52·5	18 07·5	10·5	19 12	15
19 43·5	20 35	51·5	20 54·5	11	21 58	14·5
22 57·5	23 48	50·5	24 08·5	11	25 12·5	15
25 47	26 37·5	50·5	26 58	11	28 02·5	15·5
28 43	29 31·5	48·5	29 53·5	10·5	30 58	15
31 45	32 33	48	32 56	11	34 00·5	15·5
35 30·5	36 17	46·5	36 41·5	11		
39 48·5						
40 14·5	40 58·5	44	41 25	10·5	42 30	15·5
42 42·5	43 26·5	43·5	43 54	11·5	44 58·5	16
45 42	46 23·5	41·5			47 56	14
48 43			Passing clouds.			
2 01 05	2 01 41	36				
15 44	16 13	29	2 16 52	08		
18 43	19 12	29	19 52·5	09 5 ·	2 20 57·5	14·5
22 42·5	23 11	28·5	23 51·5	10	24 56·5	14
25 21	25 48	27	26 30	09	27 36	15
28 26·5	28 52·5	26	29 36	09·5	30 41·5	15
31 01·5	31 27	25·5	32 11·5	10	33 16·5	15
33 31	33 55	24	34 40·5	09·5	35 45 5	14·5
36 30	36 53·5	23·5	37 40·5	10	38 45·5	15·5
39 27	39 49	22	40 36·5	09·5	41 42·5	15·5
42 04	42 24·5	20·5	43 14	10	44 19	15
45 06	45 26	20	46 16	10	47 21	15
47 42·5	48 01	18·5	48 51·5	09	49 57·5	15
50 13	50 31	18	51 23	10	52 29	16

Sun's Advancing Limb.	Mercury (centre)	Diff.	Sun-spot.	Diff.	Sun's Following Limb.	Diff.
h m s	h m s	m s	h m s	m s	h m s	m s
2 52 46	2 53 03	0 17	2 53 55	1 09	2 55 01	2 15
55 40·5	55 56·5	16	56 50	09·5	57 56	15·5
58 29			59 38	0·)	3 00 44	15
3 01 01·5	3 01 15	13·5	3 02 10·5	09	03 16	14·5
04 00·5	04 13	12·5	05 10	09·5	06 16	15 5
07 08·5	07 19 5	11	08 17 5	09	09 23·5	15
10 00	10 10	10	11 09	09	12 14·5	14·5
12 58·5	13 07	08·5	14 08	09 5	15 13·5	15
15 36	15 44	08	16 45	09	17 51	15
18 14	18 21	07	19 23	09	20 29	15
21 00	21 06	06	22 09·5	09·5	23 15·5	15·5
24 03	24 08	05	25 11·5	08 5	26 17·5	14·5

Third contact   3 34 58 by Chronometer.

Fourth contact   3 35 50   ..

Tarrytown, New York:
1881, *December.*

'*Notes on the Lunar Eclipse,* 1881, *December* 5.
By J. Rand Capron, Esq.

Though the Moon rose eclipsed at 3ʰ 50ᵐ 0ˢ G. M. T. it was, owing to slight banks of clouds on the horizon, not until 5ʰ 3ᵐ that it was seen from this locality. At that time it presented an interesting aspect, a small portion only being left uncovered. The opposite side of the disk to this glowed with a distinct, but not deep, red tint. This gradually melted into a golden tint, and finally into a pearly grey as it approached the bright part of the Moon. The uncovered part gradually got narrower, and at 5ʰ 10ᵐ (5ʰ 8ᵐ 0ˢ at Greenwich) was considered to be at its narrowest. No red patches or brilliant tints were seen, but the Moon's configuration was well made out through the shadow. In the 3¼-inch Cooke, power 50, the general appearance of the tinted globe, with its faint edging of partial light, was very charming. As the shadow passed off, the red tint was gradually lost, until when about two-thirds of the disk were uncovered it was not longer recognisable.

    Throughout the eclipse it was noticed that the lighted portion of the Moon was very much softened into the obscured part without definite outline; and also that there was difficulty in distinguishing any boundary between the umbra and the penumbra on the disk. The Moon was, however, low and the night misty. The occultation of *ι Tauri* was seen at about the time given for it, but the "boiling" of the Moon's limb rendered any time observation valueless. Five photographs were taken, with the 6" Equatoreal, of the gradual passing-off of the shadow, and I present with this paper prints from the enlargements. There should have been six, but one image was by mistake duplicated. The sixth place is filled by a photograph of the Moon showing the ordinary appearance of its bright edge.

Observatory, Guildown:
  1882, January 24.

Observations of Comet III. 1881, made at Windsor, New South Wales. By John Tebbutt, Esq.

This comet was very difficult to observe. I detected it while sweeping for comets with a 3¼-inch Refractor on the evening of September 17. It was then invisible to the unassisted eye, and presented the appearance of an oval nebula, with a very gradual condensation towards the centre. On the evening of the 19th its aspect was greatly changed. It was the only evening on which there was anything like a central condensation with tolerably defined limits. There were also rudiments of a tail. From this date the comet became rapidly fainter, and during the last few days of its visibility it could be observed only by looking obliquely into the telescope. At no time would it bear a field illumination sufficiently strong to exhibit the spider lines of the filar micrometer. I was therefore compelled to employ the ring micrometer on the 4½-inch telescope. The differential North Polar Distances for September 17, 18, 19, 30, and October 14, are unsatisfactory, owing to the near approach of the objects to the ring's centre. The mean places of the stars of comparison, where taken from the Washington Catalogues for 1872 and 1874, and the Sydney Catalogue for 1878, have been brought up to 1881·0 by means of the catalogue precessions. In all other cases the total precessions to 1881·0 have been derived from the annual precessions calculated for the mean dates by means of Peters's elements. Proper motion has been applied to the places of stars Nos. 11 and 14 only. For the parallax factors *p* denotes the corrections in seconds of time and arc for the Right Ascension and North Polar Distance respectively, and Δ the comet's distance from the Earth. The equatoreal horizontal parallax of the Sun has been assumed 8"·85.

I may add that with the help of M. Backlund's Ephemeris, received from St. Petersburg, I have sought for Encke's Comet,

but in vain. When the comet was sufficiently south, and removed from the morning twilight, it was considerably beyond
the grasp of my telescope.

Apparent Places of Comet III. 1881.

Windsor Mean Time 1881.	R.A.	Log (rΔ)	N.P.D.	Log (pΔ)	No. of Comps.	Comp. Star.
d h m s	h m s	+	° ′ ″	+		
Sept. 17 7 59 12	14 26 5·95	9·6994	104 44 41·9	0·6411	3	1
17 8 23 14	14 26 7 25	9 7052	—	—	7	2
17 8 42 22	14 26 8·50	9·7063	—	—	2	1
18 7 15 56	14 27 40·25	9·6797	105 36 3·0	0·6015	2	3
18 8 6 50	14 27 43·78	9·7041	—	—	3	3
19 7 26 20	14 29 13·20	9 6898	106 26 9·5	0·6075	7	4
20 7 27 2	14 30 40·05	9·6935	107 13 32·1	0·6055	8	5
26 7 53 31	14 37 55·65	9·7199	111 9 31·9	0·6309	7	6
29 7 43 28	14 40 53·28	9·7245	112 44 26·9	0·6222	5	7
29 8 13 41	14 40 54·14	9·7268	—	—	4	8
30 7 55 15	14 41 48·24	9·7282	113 13 39·8	0·6391	4	9
Oct. 1 7 49 25	14 42 41·92	9·7294	113 41 19·5	0·6337	5	10
2 7 38 3	14 43 34·70	9·7298	114 8 7·6	0·6206	7	11
3 7 55 41	14 44 24·82	9·7330	114 34 15·9	0 6467	5	12
4 7 33 38	14 45 13·82	9·7330	114 58 59·9	0·6191	2	13
5 7 44 51	14 46 3·28	9·7360	115 23 50·8	0·6374	5	14
8 7 33 9	14 48 22·99	9·7399	116 31 24·0	0 6297	3	15
12 7 53 59	14 51 19 80	9·7424	117 54 5·5	0 6732	3	16
12 7 53 59	14 51 19·85	9·7424	—	—	3	17
14 7 46 22	14 52 45·80	9·7453	118 32 31·3	0 6699	3	18
15 7 34 33	14 53 24·94	9·7481	118 50 42·6	0·6565	4	18

Mean Places of the Comparison Stars for 1881·0, *with the Reductions to the
Apparent Places for the Dates of Observation.*

Comp. Star.	Mean R.A.	Reduc-tion. +	Mean N.P.D.	Reduc-tion. +	Authority for Star's Mean Place.
	h m s	s	° ′ ″	″	
1	14 23 43·65	2·50	104 43 11·0	11·1	Lalande, 26453.
2	14 22 42·40	2·49	104 51 29 2	11·2	Lalande, 26425.
3	14 17 26·17	2·47	105 33 34·4	11·5	Arg. Oeltzen. S. Z. 13587 and 8.
4	14 27 28·95	2·52	106 17 42·5	11·2	Arg. Oeltzen. S. Z. 13727.
5	14 32 46·15	2·55	107 22 17·6	11·2	Arg. Oeltzen. S. Z. 13793, 4. 5, and 6.

Comp. Star.	Mean R.A.	Reduction. +	Mean N.P.D.	Reduction. +	Authority for Star's Mean Place.
	h m s	s	° ′ ″	″	
6	14 33 6·50	2·55	111 9 34·8	11·7	Arg. Oeltzen. S. Z. 13801.
7	14 35 34·22	2·55	112 54 56·5	11·7	Arg. Oeltzen. S. Z. 13848, 49 and 50.
8	14 39 16·99	2·57	112 38 50·1	11·5	Arg. Oeltzen. S. Z. 13909.
9	14 44 22·72	2·60	113 21 53·7	11·4	Arg. Oeltzen. S. Z. 13987, and 9.
10	14 47 0·80	2·60	113 29 11·3	11·2	Wash. Gen. Cat. 1860, 6118.
11	14 47 25·68	2·61	114 9 15·7	11·3	Wash. Gen. Cat. 1860, 6123.
12	14 37 44·49	2·56	114 32 36·4	11·6	Arg. Oeltzen. S. Z. 13885.
13	14 41 35·10	2·58	114 59 42·9	11·5	Wash. Gen. Cat. 1860, 6086.
14	14 40 48 09	2·57	115 35 17·1	11·6	Wash. Gen. Cat. 1860, 6082, and Sydney Cat. 1878, 170.
15	14 49 59·44	2·61	116 19 10·4	11·1	Wash. Gen. Cat. 1860, 6133.
16	14 46 43·46	2·60	117 55 3·4	11·1	Wash. Gen. Cat. 1860, 6117.
17	14 47 17·90	2·60	117 51 39·6	11·1	Wash. Gen. Cat. 1860, 6121.
18	14 50 7·83	2·62	118 40 30·8	11·0	Wash. Gen. Cat. 1860, 6135, Wash. Obs. 1872, 396, and Wash. Obs. 1874, 199.
18	14 50 7·83	2·62	118 40 30·8	10·9	,,

Windsor, N. S. Wales:
1881, December 22.

Comet b, 1881, *as seen from the Ship "Superb."* By D. W. Barker, Esq.

(Communicated by Capt. Henry Toynbee, R.N.)

Date	G.M.T.	Lat.	Long.	Bar.	Att. Ther.	Air Ther.	Remarks on Position, etc.
May 29	h m 7 17	36° 34' S.	2° 10' W.	30·268	56·2	55·5	First seen to S.W. of Canopus and Sirius. Fine bright nucleus comparable with β Columbæ, perhaps a shade brighter, round and well defined. Tail about 6° long, slightly inclined to S. of line joining nucleus and Canopus. Generally well defined. Distance from Sirius 26° 42' 45'.
June 2	5 53	40 21 S.	7 32 E.	30·582	52·5	48·0	Nucleus bright as α Columbæ. Tail faintly traceable to and slightly W. of o Columbæ, which was situated just on edge. Distance from Canopus 28° 51' 15" " Sirius 24 43 0 " Procyon 48 51 30
	5 48	41 4 S.	9 31 E.	30·745	53·5	49·0	Slightly brighter. Tail extending to o Columbæ. Showing well. Distance from Canopus 29° 54' 0" " Sirius 24 9 0 " Procyon 48 1 0
	5 33	41 12 S.	11 30 E.	30·663	52·5	50·0	Clearer and better defined. Tail longer, stretching to o Columbæ. Distance from Canopus 31° 5' 0" " Sirius 23 36 0

Chronometer time corrected for mean of measurements.

Distances measured by sextant, which was corrected for Index Error.

Note on a Term in the Perturbations of the Moon, due to the Action of Mars. By E. Neison, Esq.

When at work in the beginning of 1877 on the theory of the perturbations of the Moon arising from the disturbing action of the planets, I was lead to approximately calculate the value of a number of terms of long period. One of these came out with a coefficient of several seconds of arc, it being the term with the argument

$$l - 24\, l'' + 20\, l''',$$

where l, l'', l''', denote the mean longitudes of the Moon, Earth and Mars. It so happened that this term closely represented in both amount and period the correction required to bring theory and observation into accord during the long period 1765–1851, as indicated by the residuals given by Sir George Airy in his reduction of the Greenwich Observations (see *M. N.* Nov. 1873). This led me to communicate this provisional result to the Society, with the caution that, from "the nature of the new term, it is obvious that the value which has been deduced for it in the present Paper is at best merely provisional. It will require most careful revision, and the higher powers of the disturbing forces must be taken into account. . . . . The existence of this term is so strongly supported by observation, from the manner in which it removes the discordance between theory and observation, that I have thought it better to publish it in its present form rather than defer this for the considerable period which may elapse before I am able to verify these results" (*M. N.* vol. xxxviii. p. 53).

At that time I had not discovered how much these outstanding residuals required correction for existing errors in the theory before they properly represented the difference between observation and theory, or I should not have attached such a great weight to this agreement as a confirmation of the accuracy of my rough calculation.

The value of the coefficient of this term entirely depends on the exact value of the very small difference of the sum of two long series of very large quantities of different signs, so that if each term of these series be not calculated with the greatest exactness, the difference between the sums of these two series may be entirely incorrect. In my calculations I had employed quantities extending to five places, which seemed amply sufficient; but in the course of the following year I found that this was not so, and that, in consequence, the difference between the sums of these two long series was far too great, so that the value found for the coefficient of this new term of long period was far too large. This fact I mentioned in my paper on "Hansen's Terms of Long Period" (*M. N.* March 1878, p. 269).

Since then I have more accurately computed the value of this part of this term, and find its true value to be only a small fraction of a second, instead of several seconds, and the calculation forms a portion of the second part of my still unpublished Memoir on the " Perturbations of the Moon due to the Action of the Planets."

Lately I have received from M. Constantin Gogau an elaborate investigation of this portion of the complete value of this term—that is, the portion arising from the direct action of *Mars* on the Moon as disturbed by the Sun. Adopting the method employed by Delaunay for determining the value of the similar inequalities computed by Hansen, M. Gogau has investigated with great care the exact analytical expression for the coefficient of this term of long period as far as the third power of the eccentricities and inclinations of the Earth and *Mars*, and the second power of the ratio of the mean motions of the Earth and Moon. Then, by reducing these analytical results to numbers, he arrives at the conclusion that the value of the coefficient of this term is so small a fraction of a second of arc that the term must be considered quite insensible.

M. Gogau seems to have thought that my provisional calculation was intended for a complete investigation, for he points out that I have neglected a number of small quantities which ought to be included in a complete determination of the value of this coefficient. This is, of course, obvious; but the omission was intentional, for, as I state in my Paper, I omitted these terms because they could not yield any sensible portion of the coefficient. M. Gogau also lays much stress on the fact that I have omitted to take into account certain terms which are independent of m, the ratio of the mean motions of the Earth and Moon; but these terms, though they do not implicitly involve m, are none the less dependent on the disturbing force of the Sun, and cannot therefore be properly included in any investigation in which this disturbing force is neglected, as I explicitly stated was the case in the present instance. As I have pointed out elsewhere (*M. N.* March 1878) these terms depending on the solar disturbing force cannot be of any material importance in these inequalities of long period, but can only form a very small fraction of the value which is independent of them. Their consideration gives rise to an enormous increase in the labour of the investigation, but, as M. Gogau's results clearly show, their omission would not affect the conclusions in any material degree.

In so far as they extend, M. Gogau's results appear accurate. But they do not form a satisfactory determination of even this part of the complete coefficient, for they are not sufficiently extended. The terms depending on the fifth and seventh powers and products of the eccentricities of *Mars* and the Earth are capable of yielding far larger quantities than found to result from the third powers, and until it be shown that the mutual destruction of terms is as complete amongst the higher powers and products as it is amongst the lowest, the exact value of the coefficient cannot be held to have been satisfactorily determined. But M. Gogau has entirely neglected these terms, whereas in my investigation I retained them; hence his result is not sufficiently complete.

It is further to be noted that this elaborate investigation of M. Gogau, which reflects so much credit on its author, forms only the smaller portion of the complete work of determining the true value of this term of long period due to *Mars*, for, like Delaunay, he has entirely omitted to consider the difficult but most highly important effects arising from the direct action of the Sun on the perturbations in the motion of the Moon due to the direct action of *Mars*. But, as Hansen found, their indirect effect may be much more important than the direct effect of the planet. So until this additional portion of the complete value of the coefficient of this term has been computed, it cannot be said that the value of this term has been shown to be insensible, as M. Gogau believed he had done in his fine Memoir.

London :
   1882, *February* 15.

On the Correction to the Horizontal Diameter of the Moon, from the Observations made between 1851 *and* 1858. By James Campbell, Esq., and E. Neison, Esq.

In his Paper in the *Monthly Notices* for December 1881, p. 64, Mr. Stone states that he deduces from the Greenwich Observations for the years 1851–58 a very different value for the correction to the tabular semi-diameter to that given by us in our Paper as the correction deduced by one of us (Neison). It is necessary for us, therefore, to publish the details of this investigation made by Mr. Neison some years ago.

The determination of the correction to the semi-diameter may be divided into two portions—one, 1851–55, when it is necessary to correct the value of the semi-diameter given in the *Nautical Almanac* for the errors in Burckhardt's tables, by applying to the parallax and semi-diameter the corrections deduced by Adams, and given in the *Nautical Almanac* for 1856; and the other for the years 1856–58, where no such correction is required, as Adams's corrected parallax and semi-diameter were used in the *Nautical Almanac*.

Let

   S_a = Adams's original value of the mean semi-diameter of the Moon,

   δS_a = Correction to this value given by the Greenwich Observations,

   S'_h = Tabular horizontal semi-diameter of the Moon (apparent),

   S'_v = Tabular vertical semi-diameter of the Moon (apparent);

and put

$\Delta S'_a$ = Adams's correction to Burckhardt's tabular parallax,

$\Delta P'_a$ = Adams's correction to Burckhardt's tabular semi-diameter,

C_b = Observed correction to Burckhardt's horizontal semi-diameter,

C_a = Observed correction to Adams's horizontal semi-diameter.

Then

$$C_a = C_b + \Delta S'_a \times \frac{S'_h}{S'_v},$$

and

$$\$S_a = C_a \times \frac{S}{S'_h} = \left(C_b + \Delta S'_a \times \frac{S'_h}{S'_r}\right)\frac{S}{S'_h}$$

$$= C \times \frac{S}{S'_h} + 0.2725\,\Delta P'_a \times \frac{S}{S'_v};$$

where

S = Mean tabular semi-diameter = 932″5 approximately.

In both Burckhardt's tables and in Adams's original value, the semi-diameter is obtained by multiplying the parallax by the constant 0·2725. Burckhardt's value for the constant of parallax is (57′ 0″·5), so that his value for the mean semi-diameter is

$$(57' \ 0''5 \ ) \times 0.2725 = 15' \ 32''09.$$

The constant of parallax adopted by Adams is different, however, and is explicitly stated by him in the introduction to his supplementary tables (*Naut. Alm.*, 1856) to be (57′ 2″·325), hence the value of Adams's original mean semi-diameter is

$$(57' \ 2''325) \times 0.2725 = 15' \ 32''57,$$

or greater than Burckhardt's by 0″·48.

From the different volumes of the Greenwich Observations and the corresponding volumes of the *Nautical Almanac* are derived the quantities:—

No.	Date.	Approximate Hor. Semi-d.	Vert. Semi-d.	Mean semi-dia. Hor. semi-dia.	Mean semi-dia. Vert. semi-dia.	$\Delta P'_a$	Corr. applied in *N.A.* to Burck-hardt's Tab. Value.
		s	″			″	s
1	1851, Feb. 15	72·4	1·019	12 88	0·915	+0·8	+·00
	May 14	68·1	·964	13·70	0·968	+0·7	+·00
	June 13	68·5	·930	13·61	1·003	+3·0	+·00
4	July 12	67·4	·910	13·84	1·025	+4 1	+·00
5	Aug. 11	63·9	·893	14·58	1 044	+3·6	+·00
6	Sept. 9	62·1	·889	15·03	1·049	+1·8	+·00
7	1852, March 5	70·6	1·013	13·22	·921	+4·0	+·00

No.	Date.		Approximate Hor. Semi-d.	Vert. Semi-d.	Mean semi-dia. Hor. semi-dia.	Mean semi-dia. Vert. semi-dia.	ΔP′a	Corr. applied in N.A. to Burchhardt's Tab. Value.
			s	″			″	s
8	1852, May	3	70·8	1·004	13·17	·929	+ 2·5	+ ·00
9	Nov.	26	65·2	·902	14·31	1·034	+ 1·5	+ ·00
10	1853, Mar.	24	68·3	·995	13·65	·950	+ 2·6	+ ·19
11	May	22	74·2	1·013	12·57	·921	− 2·1	+ ·21
12	Sept.	16	65·1	·940	14·33	·993	+ 1·8	+ ·18
13	Oct.	16	62·6	·914	14·91	1·020	+ 2·4	+ ·17
14	1854, Feb.	12	66·2	·921	14·09	1·013	+ 0·4	+ ·18·
15	June	10	75·6	1·006	12·34	·927	− 1·6	+ ·22
16	Sept.	6	68·8	·996	13·56	·937	+ 2·8	+ ·19
17	1855, March	3	62·5	·905	14·93	1·030	− 1·0	+ ·17
18	July	28	76·9	1·004	12·13	·929	+ ·8	+ ·22
19	Sept.	25	69·1	1·009	13·50	·924	+ 1·5	+ ·19

The observations made by computers or other less experienced observers are marked with an asterisk, and, as stated on p. 443 of our paper, are rejected as liable to unknown error.

No.	Obs. Corr. to Naut. Alm.	Cb	Cb × $\frac{8}{\mathcal{B}'_h}$	0·2725 ΔP′a × $\frac{\mathcal{B}'}{\mathcal{B}'_v}$	δ8a
	s	s		″	″
	− ·17	− ·17	− 2·20	+ 0·20	− 2·00
	− ·18	− ·18	− 2·47	+ 0·19	− 2·28
	− ·18	− ·18	− 2·45	+ 0·82	− 1·63*
	− ·20	− ·20	− 2·77	+ 1·19	− 1·58
5	− ·17	− ·17	− 2·58	+ 1·03	− 1·55
6	− ·19	− ·19	− 2·86	+ 0·52	− 2·34
7	− ·23	− ·23	− 3·04	+ 1·00	− 2·04
8	− ·29	− ·29	− 3·82	+ ·63	− 3·19
9	− ·29	− ·29	− 4·15	+ ·43	− 3·72
10	− ·05	− ·24	− 3·27	+ ·67	− 2·60
11	− ·19	− ·40	− 5·03	− ·52	− 5·55
12	− ·07	− ·25	− 3·58	+ ·48	− 3·10
13	− ·15	− ·32	− 4·77	+ ·67	− 4·10
14	+ ·01	− ·17	− 2·39	+ ·11	− 2·28
15	+ ·16	− ·06	− ·74	− ·41	− 1·15*
16	+ ·01	− ·18	− 2·44	+ ·71	− 1·73
17	+ ·09	− ·08	− 1·19	− ·28	− 1·47*
18	+ ·10	− ·12	− 1·45	+ ·20	− 1·25
19	− ·05	− ·24	− 3·24	+ ·38	− 2·86

Hence we have

1851	6 obs. = $-1\overset{''}{\cdot}90.$	Excluding computers, 5 obs. = $-1\overset{''}{\cdot}95$
1852	? $-2\cdot97.$	3 $-2\cdot97$
1853	$-3\cdot84.$	$-3\cdot84$
1854	$-1\cdot72.$	$-2\cdot01$
1855	$-1\cdot86.$	$-2\cdot06$

For the year 1856 the value employed in the *Nautical Almanac* was derived from Adams's parallax multiplied by the constant 0·2725. The tables require no correction, and the corrections to the *Nautical Almanac* given by the Greenwich Observations yield directly the correction of the semi-diameter.

For this year, then, we have

Date.	S'_h	$\dfrac{S}{S'_h}$	C_a	$C'^1 \times \dfrac{S}{S'_h}$
1856, May 19	68·4	13·63	$-\cdot07$	$-\overset{''}{\cdot}95$
Aug. 15	72·2	12·93	$-\cdot01$	-13^*
Oct. 13	70·5	13·24	$-\cdot00$	$-\cdot00$

Hence

1856   3 obs. = $-0\overset{''}{\cdot}36.$ Excluding computers, 2 obs. $= -\overset{''}{\cdot}48.$

For the two years 1857–58 the value employed in the *Nautical Almanac* was derived from Adams's parallax multiplied by the constant 0·273114. Hence the mean diameter, instead of being $15' \, 32''\cdot57$, was

$$(57' \, 2''\cdot325) \times 0\cdot273114 = 15' \, 34'\cdot68,$$

or $2''\cdot11$ larger. For these years we have

Date.	S'_h	$\dfrac{S}{S'_h}$	C_a	$C_a \times \dfrac{S}{S'_h}$
1857, May 8	65·0	14·36	$-\cdot11$	$-1\overset{''}{\cdot}58$
Nov. 1	73·3	12·73	$+\cdot20$	$+2\cdot55^*$
Dec. 30	78·7	11·84	$+\cdot11$	$+1\cdot30$
1858, Feb. 27	66·0	14·13	$-\cdot02$	$-0\cdot28$
July 25	66·9	13·95	$+\cdot05$	$+0\cdot69$
Sept. 22	64·6	14·45	$+\cdot10$	$+1\cdot45$
Oct. 22	70·2	13·30	$+\cdot15$	$+2\cdot00$
Nov. 20	75·6	12·34	$+\cdot10$	$+1\cdot23$

Hence

1857	3 obs. = $+0\overset{'}{\cdot}76.$	Excluding computers, 2 obs. $= -0\cdot14$
1858	5 $+1\cdot02.$	5 $+1\cdot02$

To reduce all the results to the same value of the semi-diameter—namely, that commonly called Adams's value, and used in the *Nautical Almanac* for 1857–61—those for the years 1851–56 must be increased by $+2''\cdot11$. Hence the entire series become

1851	6 obs. =	$+0''21.$	Excluding computers 5 obs. =	$+0''16$
1852	2	$-0\cdot86.$	3	$-0\cdot86$
1853		$-1\cdot73.$		$-1\cdot73$
1854		$+0\cdot39.$		$+0\cdot10$
1855		$+0\cdot25.$		$+0\cdot05$
1856		$+1\cdot75.$		$+1\cdot63$
1857		$+0\cdot76.$		$-0\cdot14$
1858	5	$+1\cdot02.$	5	$+1\cdot02$

Multiplying each of these by the number of observations and taking the means, we have

$$30 \text{ obs.} = +0''21. \qquad\qquad 24 \text{ obs.} = -0''01$$

As explicitly stated by us, the observations of the computers were excluded; hence we have for the correction to the tabular semi-diameter $= -0''\cdot01$, or for the observed semi-diameter

$$(15' 34'68)-(-\cdot01) = 15' 34''69.$$

This result is exactly the same as that obtained by a second computation made by using the corrections to the actual diameter, instead of the semi-diameter, by which the necessity of dividing the corections by two was avoided.

The only other point raised by Mr. Stone which need be considered is, how far we are right in believing the large apparent value of the parallactic inequality obtained from the observations of the years 1846–50 to be probably due to the repeated changes in the observers during those years. Previous to 1846 Mr. Ellis seems to have made the greater number of observations, whilst subsequent to that year he made very few; and we point out that as Mr. Ellis seems to have made the correction to the diameter of the Moon nearly $3''$ greater than the other observers, the cessation of his observations must have tended to have raised the apparent value of the parallactic inequality.

This is questioned by Mr. Stone, who from *four* observations of the horizontal diameter of the Moon made by Mr. Ellis, as compared with twenty-three by other observers, shows that Mr. Ellis's value only exceeds those of the other observers by $1' \cdot94.$[*] But during the period covered by the observations quoted by Mr. Stone the instrument with which they were made was changed, and to obtain comparable results it is necessary to ex-

[*] These are taken direct from Mr. Stone's figures. There are two or three small errors in this portion of his paper.

clude the year 1851 and restrict ourselves to the consideration of the years prior to 1851, when the observations were
throughout made with the old Transit Instrument. Employing,
then, the twenty observations made between 1843 and 1850, we
find from *three* observations of Mr. Ellis, as compared with seventeen by other observers, that the mean of the former exceeds the
mean of the latter by 2"·62. This result practically accords with
those obtained by us ; but we demur to the idea that any trustworthy results can be obtained from the comparison of only
three or four observations of Mr. Ellis with other data. Our
own statement that the correction to the lunar diameter according to Mr. Ellis exceeded that according to the three other
observers, Messrs. Henry, Rogerson, and Dunkin, by nearly 3",
was founded on the direct comparison of all their observations,
made during the years 1843–50, taking Mr. Henry, who regularly observed throughout the entire period, as the standard
observer. This comparison yielded the following results :—

Correction to diameter.

Henry – Rogerson = + 0"·1.   From 174 obs. of the limb, and
                                        8 obs. of diameter.

Henry – Dunkin   + 0·2.   From 147 obs. of the limb.

Henry – Ellis   + 2·9.   From 325 obs. of the limb, and
                                        8 obs. of diameters.

As in the determination of the parallactic inequality, the
correction to the assumed diameter enters multiplied by a factor
whose mean value may be taken as 0·75, it follows that the
value found from Mr. Ellis's observations requires the correction
2"·9 × 0·75 = 2"·17 to bring it into accord with the value according to Mr. Henry's observations. Hence, owing to the cessation
of Mr. Ellis's observation in 1846, it follows that the value of
the parallactic inequality found for that year requires a correction of over 1" to bring it into harmony with the values found
for previous years. Similar corrections must also be applied to
the results for the other subsequent years.

This correction does not completely remove the discordance
between the results obtained from the observations made during
the short period 1846–50, and the long series prior to 1846.
But it weakens its weight enormously. Further examination
would probably reveal other causes; but the existence of this
discordance is not of sufficient importance to render its removal
worthy of the labour involved. To us it seems that this admitted
discordance of four or five years in a period of one hundred and
twenty cannot be held to be serious, when it is known to be coincident in period with the introduction of more inexperienced observers, and vanishes as soon as these observers become more
experienced, as in the period 1851–58.

MONTHLY NOTICES

ROYAL ASTRONOMICAL SOCIETY.

Vol. XLII.	April 14, 1882.	No. 6.

E. J. Stone, Esq., M.A., F.R.S., President, in the Chair.

Robert T. Pett, Esq., Royal Observatory, Cape of Good Hope,

was balloted for, and duly elected a Fellow of the Society.

Remarks on the Instructions for Observing the Transit of Venus formulated by the Paris International Conference. By Professor Simon Newcomb.

The recommendations of the Conference which met at Paris in October 1881 are entitled to such weight, both on account of the eminence of the astronomers who took part, and the advantages they naturally derived from mutual discussion, that I have hesitated in suggesting any possible improvement of their carefully-prepared instructions. This hesitation has been increased by the fact that my suggestions cannot now, as I should desire, be criticised by the Conference, to which I should have been glad to submit them. I submit them now to the Royal Astronomical Society with the hope that they will receive that critical discussion by all astronomers interested which alone can lead to a decision of all the questions involved.

The strongest reason for this course is, that as the entire Transit is visible on this continent, it is not improbable that the observations by astronomers in this country will exceed in number those made by the expeditions to be sent out by foreign Governments. The question whether American observers shall

z

adopt the Instructions of the Conference unchanged is therefore an important one, which can be decided only by a careful examination and discussion of the subject. The following is to be regarded simply as my individual contribution to this discussion.

The Articles recommended by the Conference are found in the *Comptes Rendus* of the French Academy for October 17, 1881, tome xciii. p. 569.

The first three Articles call for little important remark, as they refer principally to the precautions to be taken by the observer; and I am unable to suggest any decided emendation. It may, however, be remarked that the use of the silvered objective might be inconvenient in the event of the Sun's rays being so obscured by smoke or haze that a deeply-coloured shade would be unnecessary. It might then be found that the silvering darkened the Sun's light too much.

In respect to Article III. I feel some doubt whether the degree of illumination of the solar disk is best defined by the distance apart at which two spider lines can be seen projected on the Sun. If, indeed, the amount of contrast between the brilliancy of the Sun and that of the atmosphere immediately around the limb were always the same—if, in fact, all the observations were made through a very transparent atmosphere, such a method of securing uniformity would perhaps be the best. But, in the actual observation of contact, everything depends upon the observer seeing the gradual approach or separation of the sharp cusps around *Venus*, and estimating the moment at which the thread of light becomes visible or invisible. Now, the greater the brilliancy, so long as the comfort of the eye is not interfered with, the more definitely, it would seem, must such phases as these be discriminated. It therefore appears to me that a better rule for regulating the brilliancy of the solar disk would be to make it as bright as the eye could bear with entire ease and comfort. To increase the precision of the observation, the Dawes solar eyepiece, cutting off all that portion of the Sun distant from the point of contact, might have been recommended.

In Article IV. we find the times of internal contact defined as follows :—

" *A l'entrée* : le moment où l'on voit pour la dernière fois une discontinuité bien évidente et en même temps persistante dans l'illumination du bord apparent du Soleil, près du point de contact avec *Vénus*.

" *A la sortie* : le moment de la première apparition d'une discontinuité bien marquée et persistante dans l'illumination du bord apparent du Soleil près du point de contact."

I confess to some difficulty in forming a precise conception of what should be understood by the phrase " illumination of the apparent limb of the Sun " and how " discontinuity " in that illumination should be noted. It may be that on actually observing the approach of a contact, it would be sufficiently clear to the observer what was meant. At the same time I deem it proper to point out the source of the apparent difficulty. To do this, and to discuss the subsequent explanations made by the Conference, I shall have to begin with some general statements respecting the phenomena of contact, founded partly on numerous observations of an artificial Transit of *Venus* in 1874, partly on the general description of the observers of that Transit, and partly on the reason of the case.

In doing this I may refer in passing to the paper in the *Monthly Notices* for March 1877 (vol. xxxvii. p. 237).

The moment of true internal contact is that at which the sharp points of the cusps begin to meet around the dark body of *Venus* as it enters upon the solar disk. Now, if the limb of the Sun presented itself to the vision as a distinct mathematical line, a discontinuity in which could be observed with geometrical precision, there would be no difficulty in the interpretation. As a matter of fact, however, it needs little consideration to show that no such definiteness is possible. The imperfections of vision in the eye, the instrument, and the atmosphere, all combine to render it impossible. The result is that, as *Venus* is approaching internal contact, it is impossible to fix the precise point at which the supposed sharp cusps terminate. To fix such points is very much like trying to fix the point where the tail of a comet terminates. The cusps can be traced further according to the sharpness of definition, the steadiness of the atmosphere, the keenness of the eye, and the habit of the observer.

The result is that there is no one moment at which all observers, even under the same conditions, would agree that contact had occurred.

The ends of the cusps assume a certain haziness, and the distance to which each cusp can be traced depends upon the relative brilliancy of the Sun's limb and of the atmosphere around it.

Moreover, owing to the effect of atmospheric tremors, the very sharp cusps, just before internal contact, are broken up and diffused in various ways. The result is that internal contact does not show itself by the completion of what can be recognised as the limb of the Sun, but by the gradual combination of the waving and ill-defined lines of light which form the cusps. Confining the attention to the space immediately behind the planet, the discontinuity of the limb is resolved into continuity by a slow and gradual process, the region along which contact occurs being occupied by a waving mixture of light and darkness as it were. During a period after internal contact, which varies from a few seconds to a minute, according to the circumstances of vision, the minute portion of the Sun's disk visible behind the planet is broken, and darkened by the effect of irradiation, atmospheric undulation, &c. Now, during this gradual process from complete discontinuity of the Sun's limb to complete continuity, extending over a period of perhaps an entire minute, how is the observer to say at any moment whether there is or is not " une discontinuité bien évidente " ? Indeed, what should he consider a discontinuity ?

This difficulty the instructions seek to resolve by describing more fully what the observer is to look for under different supposed conditions. To analyse these several instructions we must premise some further considerations respecting the character of the phenomenon.

I remark, first, that the manner in which dark appearances, such as the black drop, are very often spoken of, does not seem to be that most conducive to entire precision on the part of the observer. The only phenomena actually seen are those of greater or less degree of illumination, and darkness can only be predicated of those portions of the field of vision where there is no visible illumination. Darkness, in a word, is not visible.

As an application of these considerations I may say that the asserted presence of a black drop can mean nothing more than that the sharp cusps of *Venus*, instead of appearing sharp, as in their true form, appear rounded at their termini. May it not be, therefore, that precision would be gained if the observer's attention were confined to the phenomena of illumination exhibited by the cusps, and the black drop, or other forms of distortion, were described only as a blunting, rounding, or diffusion of the cusps ?

Let us now take up the instructions in order :—

I. " Si les bords des deux astres viennent au contact géométrique sans déformation et sans obscurcissement du filet de lumière interposé, l'instant défini est celui même de ce contact."

This statement seems to me formulated with entire precision, and therefore does not call for further remark.

II. " S'il se produit une goutte noire ou ligament, bien net et aussi obscur que le corps même de la planète, les instants définis précédemment sont, à l'entrée celui de la rupture définitive, à la sortie celui de la première apparition du ligament."

I cannot but fear that in this case observers will note the time of ingress so much too late, and that of egress so much too early, as to destroy the value of their observations. All experience shows that there is no precise moment of rupture or formation of the ligament in such a case. Observers do, indeed, very often describe the formation as sudden, but, as is well known to all who have investigated the subject, the times they give for that sudden formation are so discordant that no weight can be assigned to it. In fact, as pointed out in the paper already referred to, if we suppose the planet approaching interior contact at egress, the band of light in front of it, as it grows thinner, appears darker to the vision from various optical causes too obvious to need enumeration. The amount of darkness is, however, uncertain, or, to speak with more precision, the brilliancy of the band as it fades away diminishes very gradually,.

until near the time of true contact, when the band is entirely lost to vision. Now, it would seem to be implied by the instructions that the moment to be noted was that of the first appearance of this darkening at egress, and, at ingress, that when it was about to disappear. To show the danger of such an interpretation I may advert to the observation of the Transit of 1874 made at Windsor, New South Wales, by Tebbutt, and published in the *Astronomische Nachrichten*, vol. lxxxv. s. 174.

It is evident from Mr. Tebbutt's description that he failed to observe either contact through waiting at ingress until almost every vestige of the ligament had disappeared, and at egress by endeavouring to catch its first formation.

III. " Si, les bords restant sans déformation, il se produit un obscurcissement du filet lumineux, sans que l'ombre devienne jamais aussi noire que le corps de la planète, l'observateur notera l'instant du contact géométrique. Il devra noter de plus l'instant de la formation ou de la disparition de l'ombre."

I only remark on the last sentence of this, that, as already explained, there can be no precise moment assigned at which the shade commences, nor do I clearly see how it can disappear otherwise than by becoming as dark as the body of the planet, a case supposed not to exist in the preceding sentence.

IV. " Si l'ombre interposée est d'abord ou devient aussi noire que le corps de la planète, l'instant défini précédemment est celui où cette égalité cesse ou celui où elle s'établit."

This definition of the time of contact concurs with the results of observations on the artificial Transit. But there seems room for question whether observers will in all cases be able to distinguish between this and the phenomenon described in II. It might be inferred, from the independence of the several definitions, that the Conference considered the phenomena of "l'ombre " and the " black drop " or " ligament " as distinct and separable. But there may be room for question whether this view can be sustained, and whether what is understood by the shadow is not simply the commencement or termination of what, when it becomes sufficiently marked, will be described as the black drop. Without, however, directly discussing this and the collateral question, I shall briefly indicate my views of the principles involved.

I. The great diversity of descriptions given by different observers arises as much from different ways of considering the phenomena as from differences among the actual phenomena. As a matter of fact, the character of the appearance is continually changing through the influence of atmospheric undulations, and one observer will catch and describe one appearance and one another. It is therefore essential to avoid merely subjective differences, that observers should have a clear idea of what they are to look for.

II. Assuming that the observations to be finally discussed will be made with sufficient optical power and with good

instruments all set to focus, there will be but two principal
sources of difference between actual phenomena. These are :—
  a. Different degrees of atmospheric tremors or blurring.
  b. Different degrees of contrast between the brilliancy of
the Sun's limb and that of the atmosphere immediately adjoining
it. The greater the irradiation produced by atmospheric tremors,
and the greater the contrast between the brilliancy of the Sun
and sky, the more decided may we expect to see that form of
distortion commonly described as the black drop. In order to
make all the observations comparable, we must seek out some
phenomenon which shall be independent of these various conditions. This seems difficult unless the observer has himself a
knowledge of the general theory of the subject so that he will
know how to make the best estimate under any conditions.

  III. Observations of the artificial Transit, made so as to
imitate as nearly as possible the appearance of the actual Transit,
seem to indicate that the only moment which remains nearly the
same under all circumstances is to be estimated by confining the
attention to the cusps of light and the general outline of the
Sun and planet near the time of contact. If the contrast
between the brilliancy of the Sun and sky is considerable, the
moment of true contact at ingress is that when light is about to
glimmer all the way across the dark space between the cusps.
The first glimmer may, however, be a fine line of light, a general
haziness, or a confused mixture of luminous filaments covering a
considerable breath, according to the distinctness of vision and
the steadiness of the atmosphere. Very generally there will be
a period of several seconds during which the light will glimmer
and disappear by turns. The middle of this period is that of
the true contact.

  IV. If, owing to haze, smoke, or clouds, the contrast between
the brilliancy of the Sun and of the sky is much less than usual,
it may be that the actual completion of the thread of light cannot
be noted for some seconds after true contact at ingress, but the
practised observer will still be able, by estimation, to say when
the cusps should have met had they been brilliant and therefore
sharply defined. The appearance to be noted is not that of supposed geometrical tangency, a moment which cannot be estimated with any precision whatever, and is substantially valueless,
but a moment defined by a general estimate for which no precise
directions can be given, and of which the eye itself is a good
judge. If in lieu of the first sentence of Article V. of the
instructions we should direct the observer to prolong the cusps
around *Venus* "*par la pensée*," and note the moment when the
cusps thus prolonged should meet, it would accord closely with
the view here indicated.

  V. I should regret to have exterior contacts neglected until
we have some better evidence than has yet been furnished that
they are not as accurate as interior contacts.

  VI. One disturbing circumstance which might frequently

interfere with the estimation of the contact seems to have been overlooked by the Conference. I refer to the bright line of light which encircles that portion of *Venus* outside the Sun's limb when a portion of her disk is projected upon the Sun. This line of light no doubt grows brighter as the moment of internal contact at ingress approaches, and there is great danger of its being mistaken for the Sun's limb about the moment of contact. It does not appear, however, that anything can be done except to caution the observer against this mistake. He must distinguish between this light and that of the true limb of the Sun as best he can.

VII. It is essential to the accuracy of observation that observers should practice on an artificial Transit. To get the best results from these instruments it is necessary that it should be so placed as to imitate as closely as possible the special phenomena of the actual Transit of *Venus*. These phenomena are atmospheric undulations, a softening of the outline of the Sun's disk and the partial illumination of the background.

I have said nothing on the subject of photography, because I cannot, in a small space, add anything of material value to the contents of my paper of 1872 published as Part I. of the papers prepared by the American Commission on the subject. It was there pointed out that only one practicable method had been proposed for an accurate determination of the solar parallax from photographs of the Transit, and I am not aware that the difficulties there pointed out were obviated in any other plan for photographing the Transit. As an example of what may be done under the most unfavourable circumstances, and on a first trial, by the horizontal meridian photoheliograph, reference may be had to Part I. of the American " Transit of Venus Observations."

Alleged Errors in the Time-Record of Observations of the Transit of Venus, at Hermitage, Rodriguez. By Lieut. W. Usborne Moore, R.N.

(*Communicated by the Astronomer Royal.*)

Page 397 of the account of the observations of the Transit of *Venus*, 1874, contains a note, which suggests four mistakes in time at Ingress and four at Egress, both groups of errors being in the same direction. I feel it my duty to record my respectful dissent from this note, feeling sure that no such mistakes were made, and that any calculations based on the assumption that a minute is to be added will be in error.

The Greenwich sidereal times at the principal phases observed at the three Rodriguez stations are recorded in the following table :—

Point Venus.

Phase observed.	Greenwich Sidereal Time.
	h m s
Ingress.	
Circular Contact	7 38 36·9
Change of colour of ligament	7 39 3·9
Disappearance of ligament .	7 39 27·2
Egress.	
Formation of ligament ...	11 8 39·7
Apparent Contact (fig. 2, p. 366)	11 9 57·3

Point Coton.

Phase observed.	Greenwich Sidereal Time.
	h m s
Ingress.	
Simple Internal Contact...	7 38 54·8
Total disappearance of black hazy drop	7 39 31·5
Egress.	
First appearance of brown haze .	11 8 42·2
First appearance of black hazy drop	11 8 50·8
Circular Contact	11 9 29·7

Hermitage.

Phase observed.	Greenwich Sidereal Time.
Phases from my Note-book.	h m s
Circular Contact ...	7 37 20·2
Broad ligament forming ...	7 37 31·6
Planet's disk apart from limb of Sun	7 38 12·6
Ligament assumed a more definite shape ...	7 39 23·6
Saw ligament snap ...	7 39 48·6
Link formed	11 7 46·4
Apparent Internal Circular Contact (fig. 3, p. 395) ...	11 8 37·5
Can see that portion of Venus outside the Sun's limb; a dark brown colour, with limb slightly illuminated (fig. 4, p. 395).	11 10 50·5

It is no part of my business to reconcile the times shown here to be at variance, but rather to demonstrate briefly the improbability of the mistakes imputed. I may, however, point out that the interval of 1^m 10^s without change of phenomenon, at Ingress, at Hermitage, is no more remarkable than that of 1^m 33^s during Egress at Suez. The observer remarks (p. 395) that the Sun was too strong for him, and from personal observation I can state that the "boiling," or agitation of the Sun's limb was excessive, owing to the close neighbourhood of a low cloud which was slowly moving, thus changing the conditions of refraction caused by the vapour at its edge. These circumstances may help to account for the interval.

Respecting Egress : it will be noticed that fig. 3, p. 395, illustrates a phase at least a minute previous to fig. 2, p. 366. It is clear that what was Internal Contact to the observer at Point Venus was not Internal Contact to the observer at Hermitage.

It may be considered as some evidence in favour of the correctness of the minutes at Egress that fig. 4, p. 395 (and fig. 145 of the Parliamentary Report), with accompanying description, assuredly indicate a phase not less than 2^m 13^s after Internal Contact.

It is right to mention that, to the best of my belief, the second and minute hands of the Parkinson and Frodsham chronometer corresponded very nearly ; that the second hand had unequal arms, and could not be observed at the wrong side of the dial; that I had been in the habit of using this and similar chronometers for over two years of surveying duty ; that during the week previous to the Transit I had taken time for over 200 observations; that comparisons were frequent; that between Ingress and Egress I took 45 observations of the Sun, timing myself: comparisons followed Egress ; then noon comparisons, afternoon Sun observations, and night observations of rockets. The watch, in short, was thoroughly familiar. Before Ingress the "minutes" especially were watched in order that the observer might be informed when the contact should occur by Nautical Almanac. The question arises—Is it probable that the watch was read wrong eight times, and only for the critical phenomena, and that the error of one minute had in every case the *same sign* ?

[The following is an exact copy of the original record of the observations in Lieut. Hoggan's note book. It will be remarked that in the case of one of the times recorded at Ingress, an alteration had been made in the figures.—W. H. M. Christie.]

Observations of the Transit of Venus, 1874, *at Hermitage, Rodriguez.*

Copy of Lieut. Hoggan's Note-book.

N.B. Until 6.10 clouds obscured the Sun.

About 6.10 first saw the planet on the Sun about ¼ on.

Great atmospheric disturbance. Planet and Sun's limb very
shaky, using lowest power in the telescope.
Sun obscured by thick clouds.
Cloud cleared away.

h m s	
6 21 40	Saw Sun and planet.
22 21·8	Planet a little more than ⅓ way on. Great atmospheric disturbance. 2nd power, yellow eyepiece. Venus boiling all round.
6 24 35·2	Cusps as sharp as can possibly be under circumstances. Outside the Sun (of *Venus*) brown colour. Inside the Sun, black.
6 26 30	Boiling seems to increase around the limb of *Venus* and the Sun. Cloud coming up.
6 27 30	A light cloud passing over. Shifted glass to light red.
6 28 0	Cusps still sharp.
6 28 44	Obscured.
6 33 0	Without shade. Cloud partially cleared away. Saw planet.
33 17·5	Cusps still sharp; not quite circular contact. Put on red-glass eyepiece (No. 2).
33 48	Just coming on to circular contact. ·
33 36·2	Not quite circular contact. (Nearly sure this is 56^s.—W. U. M.)
34 4·8	I conceive this to be circular contact.
34 16	Black drop forming, broad ligament—or disk of planet well on Sun.
34 57*	Only ligament, planet disk not joined to Sun. Immense boiling.
36 7·9	Ligament assumed a more definite shape.
36 17·4	Ligament getting perceptibly narrower.
36 32·8	Saw a ligament snap (fade away would be the proper term). I believe this to be internal contact.
36 43·2	Fancy saw a ligament about as broad as a needle snap. Immediately after planet appeared well on Sun's disk.
41 0	Sun obscured.

Remarks 15 minutes after :—
The atmospheric disturbance around the Sun's limb and the
planet's limb was such as to render it impossible for me,
although I watched very carefully, to distinguish the various
changes of the black drop with great accuracy.
This I did notice, that the ligament decreased in width
without increasing in length very perceptibly. I noticed no

---

* The figures have been altered in the observer's book. This appears to
have been written originally 35·37, and afterwards altered to 34·57.

brown haze whatever at internal contact. Near internal contact the Sun was as bright as could possibly be: so much so that, had there been time to shift, I should have used the darkest shade.

About ¼ hour after the planet was on the Sun's disk most of the atmospheric disturbance seemed to have ceased.

The weather before sunrise promised exceedingly well. I rose at two o'clock in the

[Here follow comparisons of chronometers.]

Egress.

	h	m	s	
10	3	18·2	B. d. forming.
	3	51·0	Light not quite interrupted.
	3	56·0	Link.
	4	47·0	Apparent internal circular contact.
	7	0·0	Can see *Venus's* limb outside Sun's limb. Portion of *Venus* inside the Sun appears black, that on outside appears brown with slight illumination of limb. Highest power used. Yellow glass.
About	33	45	Planet appeared to be just disappearing.
About	34	0	Telescope failed to work. Probable disappearance.

Remarks.

The observation of Egress was made under totally different circumstances from Ingress. There was no atmospheric disturbance whatever. The Sun's limb and *Venus's* limb were both beautifully defined, and the model was a most perfect representation of them.

On the best Mode of undertaking a Discussion of the Observations of Contact to be made at the approaching Transit of Venus. By David Gill, Esq., LL.D., Her Majesty's Astronomer, Cape of Good Hope.

In most discussions of previous Transits of *Venus* the difficulty has been felt that preconceived ideas on the part of those who have made the discussions may have unconsciously influenced the interpretation put upon the language of the observers.

This difficulty can only be obviated by so conducting matters that none of the persons concerned in the discussion can be aware what influence a particular interpretation will have upon the resulting parallax.

With this view I make the following proposals :—

(α) That an International Committee should be appointed whose secretaries or others appointed for the purpose shall receive the observations made by the various observers of the Transit.

(β) That the secretaries or others (who are to have no voice in the after discussion of the results) shall, as soon after the Transit as possible, prepare for publication the descriptions of the observations of contact made by the various observers, precisely in the language of the observers themselves (giving the original language and a translation in French), but without any statement of the names of the observers, of the locality or time of observation—merely bracketing together the successive observations made by the same observer, and dividing the observations into two lists—those of Ingress and those of Egress—and numbering the separate observations 1, 2, 3, &c., for identification. Such a list when published will afford all the data necessary for deciding the "phase" of contact observed, but would afford no information as to the effect which a particular interpretation of any given observation would have upon the resulting parallax.

(γ) On the publication of this list the Committee should be convened for the purpose of selecting all the observations which appear to answer to the given definition of "contact"—or, if necessary, arranging in corresponding groups the various kinds of contact observed. The result of this classification should then be published, but until it has been issued, no one, beyond the secretaries, should have access to the original observations.

(δ) The "phases" having been thus definitively "cast," the computation of the Greenwich mean time of each observation, and the equations which result from the observations, should be made under the direction of the Committee.

[It is here to be understood that the reduction of the time observations, and determinations of longitude have already been made by the observers themselves, or by the heads of the different national expeditions.]

(ε) The Committee should then proceed to combine the equations, dividing them into groups according to the phase observed and the class of telescope employed, and ascertain the relative weight of an equation of each group. The definitive value of the parallax would then be found from the combination of the results of each group, having regard to the weights of these results.

Such a discussion would rest on a broad and impartial basis and would command the acceptance of the scientific world.

Réponse à la communication faite à la Société Royale Astronomique par M. Marth, au sujet de mon appareil pour la détermination des flexions. Par M. Loewy, membre de l'Institut.

Monsieur Marth dans la communication * faite à la Société Royale Astronomique de Londres au mois de Décembre vient d'entretenir ce corps savant d'un appareil pour l'étude de la flexion proposé par lui en 1862, et il conclut à la similitude de son instrument avec le mien en ajoutant que les objections faites par moi contre le mode de construction et l'emploi de son appareil manquent de clarté, et sont pour lui d'une nature énigmatique.

Il m'a paru, en effet, superflu d'exposer les inconvénients des dispositions adoptées par M. Marth, parce que ces inconvénients sont tellement considérables et tellement évidents qu'ils doivent immédiatement frapper tout astronome au courant de ce problème si délicat.

Dans de telles conditions, il m'a paru bien préférable de ne pas soumettre à une critique plus détaillée l'appareil de M. Marth.

Je me vois à regret obligé de réfuter complètement toutes les allégations que M. Marth a produites sur cette question dans la séance précitée.

J'exposerai d'abord les raisons qui s'opposent à l'usage de l'appareil préconisé par M. Marth, et j'expliquerai ensuite toute la différence qui existe entre son instrument et le mien.

L'instrument de M. Marth se compose de deux objectifs et d'un miroir percé à son centre. Chaque objectif est lui-même formé de deux pièces (crown et flint) : c'est donc en réalité un appareil de cinq pièces.

Pendant la rotation de la lunette, ce système est soumis à des déplacements multiples : 1° l'appareil dans son ensemble éprouve divers mouvements de translation et de rotation ; 2° par suite du mode d'attache, quel qu'il soit, les diverses pièces donnent lieu à de nombreux mouvements indépendants les uns des autres : c'est ainsi entre-autres que le flint et le crown peuvent se déplacer l'un par rapport à l'autre, car pour ne pas déformer les images, on ne doit pas trop les serrer.

Les tiges, supports ou vis qui relient les deux objectifs et le miroir à la pièce fondamentale ou directement au cube fléchissent d'une façon quelconque sous le poids qu'ils portent pendant la rotation de la lunette et provoquent des déplacements relatifs.

M. Marth ne se préoccupe nullement des mouvements relatifs de ces diverses pièces; il ne considère que quelques-uns des mouvements d'ensemble dont il croit pouvoir déterminer l'effet par le retournement de son appareil ; c'est ainsi qu'après l'avoir enlevé et retourné M. Marth suppose que les déplacements de l'instru-

* M. Loewy refers to some remarks made by Mr. Marth at the December meeting, in the course of the discussion upon a paper by Mr. Stone on a new form of transit circle with a prismatic object-glass.

ment conservent les mêmes valeurs numériques (abstraction
faite du signe). Or aucune de ces deux hypothèses n'est admis.
sible, elles ne présentent même pas *a priori* le moindre degré
de probabilité, lorsqu'il s'agit d'évaluer des quantités du dernier
ordre de petitesse.

Lors des premières études avec mon appareil, d'une construction beaucoup plus simple que celui de M. Marth, M.
Perrotin, Directeur de l'observatoire de Nice, MM. Périgaud,
Renan, plusieurs autres astronomes de l'observatoire de Paris et
moi-même, après avoir complètement résolu le problème par la
détermination de toutes les inconnues, nous avons retourné
l'appareil pour opérer dans des conditions différentes, et malgré
toutes les précautions prises pour le fixer d'une manière identique
nous avons constaté que le mouvement de l'instrument dans la
position inverse était tout-à-fait différent du premier et qu'il
était impossible d'établir aucune relation entre les deux.

Nous avons aussi reconnu que le jeu des vis du réglage provo-
que des déplacements très-appréciables des images, et nous avons
été amené dans la seconde série d'expériences à les supprimer
presque toutes et à coller à l'arcanson celles que l'on avait con-
servées. Dans nos premières expériences, avec le grand cercle
méridien nous avons constaté un mouvement de rotation de l'est
à l'ouest. Le déplacement était *a priori* si peu probable qu'il
nous a fallu plusieurs semaines pour pouvoir trouver une cause
physique probable pour l'expliquer (voir mon mémoire publié
dans les Annales de l'Observatoire que j'ai adressé à la Société
Royale Astronomique).

Il est donc évident que ce qui se présente pour un appareil aussi
simple que le mien existe *a fortiori* pour la disposition beaucoup
plus compliquée de M. Marth. Lors qu'on veut atteindre la
précision la plus élevée, il n'est pas permis de recourir à des procédés hypothétiques, et l'on ne peut rien admettre qui ne soit
demontré par des expériences directes.

En résumé, pour étudier la variation d'une ligne de visée
unique, pendant la rotation de la lunette, M. Marth admet, contrairement à la réalité, l'invariabilité des cinq lignes de visées
différentes provenant des cinq pièces optiques composant son
instrument. Aucun de ces inconvénients n'existe dans mon
appareil, qui a été conçu dans le but de pouvoir déterminer avec
exactitude tous les mouvements susceptibles de se produire pendant la rotation de la lunette, sans avoir besoin de recourir *a
priori* à des lois hypothétiques tant sur la nature de plusieurs de
ces mouvements que sur l'action négligeable de quelques autres.

Mon appareil actuellement commandé pour plusieurs autres
observatoires se compose d'un seul disque de verre assez épais
pour pouvoir être collé à l'arcanson dans un barillet, de telle
sorte que les deux pièces forment un tout invariable, et tous les
déplacements de l'appareil ne peuvent alors provenir que de la
pesanteur de cette pièce unique attachée au cube de la lunette.
Une fois introduit dans le tube central sur la fibre neutre, cette

pièce y reste dans la même position sans qu'il soit besoin de l'enlever.

Par des dispositions prises d'avance, il est facile d'en doubler ou tripler la pesanteur par l'adjonction de poids supplémentaires symétriquement placés et autant que possible de même forme que l'appareil, et déterminer ainsi par un procédé nouveau les divers mouvements qui peuvent se produire pendant la rotation.

Toutes les méthodes proposées pour l'étude de la flexion ont échoué, comme celle de M. Marth, devant cette difficulté considérable de la détermination des mouvements propres des systèmes imaginés, et sont restés à l'état de simples conceptions théoriques, sans application possible.

Tous mes efforts ont précisément été dirigés vers ce double but : Trouver d'une part une méthode certaine pour la mesure des déplacements, et d'autre part imaginer une disposition d'appareil rendant possible l'emploi d'une rigoureuse méthode d'expérimentation.

J'ai la certitude aujourd'hui d'avoir complètement résolu le problème à ce double point de vue. Ce nouveau moyen d'évaluer les mouvements de l'appareil, comporte des vérifications nombreuses par la faculté même qu'a l'observateur de faire varier les poids.

Comme toutes les images sont produites par le seul disque de verre, on peut en outre déterminer ces mêmes mouvements d'une manière indépendante par trois opérations distinctes :

1° En comparant les résultats obtenus par transparence au moyen des traits tracés sur l'objectif.

2° Par les images réfléchies au moyen de la surface argentée du miroir.

3° Par la division insérée dans le tourillon.

Si l'on double et triple les poids de l'appareil, on arrive ainsi à trouver pour chaque déplacement six valeurs indépendantes, et par suite deux pour les flexions elles-mêmes.

Dans le cas présent, où il s'agit de dégager les observations les plus précises (positions des étoiles fondamentales) de certaines erreurs systématiques, il faut éviter d'introduire des corrections qui ne présentent pas les garanties d'exactitude les plus absolues, auraient pour effet contrairement au but poursuivi, d'affecter les résultats obtenus d'erreurs plus considérables que celles dont on cherchait l'élimination.

Or c'est en cela que réside le caractère dominant de mon appareil : à savoir qu'il offre des moyens de vérifications multiples et indépendants et présentant par cela même toutes les garanties d'exactitude qu'on est en droit d'exiger lorsqu'il s'agit d'évaluer des corrections à la fois si délicates et si importantes.

Beaucoup de mes confrères de l'Institut, plusieurs directeurs d'observatoires, et presque tous les astronomes de l'observatoire de Paris, ont fait des expériences avec mon appareil, et par des opérations dont la durée ne dépassait guère cinq à six minutes ont pu se convaincre de la précision de cet appareil.   La flexion

obtenue dans la position horizontale était identique à celle obtenue au moyen des collimateurs par les procédés physiques.
ordinaires.

En dehors des flexions que M. Marth croit à tort pouvoir
évaluer avec son appareil, mon instrument permet encore de
déterminer la flexion de l'axe central et la forme des tourillons.
On comprendra maintenant toute la différence qui existe entre
les deux appareils : avec le mien, on mesure avec la plus haute
précision et de plusieurs manières, non-seulement la flexion de
l'objectif et la flexion de l'oculaire, mais aussi la flexion absolue
de l'axe instrumentale et la forme des tourillons, ainsi que tous
les mouvements de l'appareil.

Avec l'autre, au contraire, en admettant, ce qui en réalité
est impossible, que l'on puisse évaluer les mouvements de
l'appareil, on n'obtiendrait que les deux premières données du
problème ; c'est-à-dire la flexion de l'objectif, et la flexion de
l'oculaire.

On some Results obtained from the Meridian Observations of Mars
at the Opposition of 1877. By E. J. Stone, M.A., F.R.S.

A very large number of meridian observations of *Mars* and
comparison stars near the planet were made during the Opposition
of 1877, at the principal Southern Observatories. The co-
operation of the Northern Observatories was not so general as.
could have been wished ; but a valuable series of corresponding
observations were secured at Leyden and Washington.

The results of the observations were forwarded, by arrangement, to the Washington Observatory for discussion. A value
of the solar parallax has been recently deduced from a discussion of these observations by Professor Eastman, an assistant
at the Washington Observatory, and the results have been published as Appendix III, "Washington Observations," 1877. The
value of the solar parallax thus found is

$$8\overset{''}{\cdot}953 \pm 0\overset{''}{\cdot}019.$$

This value has been deduced solely from a comparison of the
North Polar Distances of *Mars*, which were made within a few
hours of each other. The method has therefore the advantage
of practically avoiding the necessity for any correction of errors.
of tables ; but this advantage is dearly purchased by the loss.
of a large number of valuable observations, unless a very considerable number of corresponding observations should be found
available. The separate results obtained by Professor Eastman
were—

From Washington and Melbourne ...   ... $8''9712 \pm 0''0316$

Washington and Cape...   ...   ... $8·8960$ $0·0725$

Washington and Sydney   ...   ... $8·8846$ $0·0546$

Leyden and Melbourne   ...   ... $8·9693$ $0·0260.$

These separate results are accordant within their probable errors.

For some reasons, which are unassigned, no direct comparisons appear to have been made between the Leyden, the Cape, and Sydney Observations.

In addition to the Leyden and Washington observations, some observations of the Transits of *Mars* and the comparison stars were made over oblique systems of wires at Cambridge, United States; but it is very doubtful whether these observations, made by a method essentially different from that adopted at the other stations, can be properly combined with the other observations. These Cambridge observations, combined with the Melbourne observations, have given Professor Eastman a value of the solar parallax of

$$9''1382 \pm 0''0496.'$$

The results which would be obtained by a combination of these Cambridge observations with the corresponding observations made at the Cape and Sydney would not differ greatly from this value; but, for the reason given, it is hardly necessary to complete this part of the investigation.

There has unfortunately been a considerable delay in bringing forward the results obtained from these *Mars* observations; but I did not feel myself justified in discussing these observations before the publication of Professor Eastman's paper, and I should have been exceedingly glad if I could have regarded Professor Eastman's discussion as exhaustive. But of the thirty-eight observations of *Mars* which were made in 1877 at the Cape only nine are included in the discussion, and two of the results are rejected on account of supposed defects in the Cape observations; whilst of forty-four Sydney observations of *Mars* only eleven have been included in the discussion, and the results have in four cases been rejected as defective.

The rejection of the Cape observations of September 1 and September 3 is certainly, in my opinion, quite uncalled for; and I consider that the value of these observations cannot properly be estimated without a consideration of the residual errors of *all* the observations of *Mars* made at the different stations, and the consideration of what changes in these residuals would be introduced by any considerable alteration of the adopted value of the solar parallax. This work I have done, so far as the materials are at present available to me, and I hope to complete the work. But I have met with a difficulty which renders me unable, at

A A

present, to include the Sydney observations in the discussion. In the original scheme of observations it was recommended that the plan adopted by Dr. Winnecke in 1862 should be followed :—

" Two threads of equal size are inserted in the moveable declination system of the field of the telescope, the distance between them being 3″ or 4″ less than the minimum diameter of the planet during the proposed period of observation. The observation is effected by moving these threads until the two small segments of the planet outside of the threads are seen to be exactly equal. If the thread nearest the micrometer head is designated as thread *a* and the other as thread *b*, then the comparison stars should be observed by bisecting the first, fourth, fifth, and eighth stars with thread *a*, and the second, third, sixth, and seventh stars with thread *b*. This order should be reversed on alternate nights, but in all cases the thread used in bisecting ought to be carefully recorded for each star."

Now, on page 15 of Professor Eastman's paper it is stated as follows :—

" SYDNEY.

" The observations at Sydney were made with Simm's Transit Circle by Government Astronomer H. C. Russell and Assistant H. A. Lenehan, and according to the method proposed in the circular. They also made a separate observation of *Mars* by bisecting 'with the wire used for the Nadir observation, at the reading zero, *before* the planet came on to the Transit wires.' These extra observations differed so widely from the others that they have not been used."

The Sydney observations of *Mars* and stars are given on pages 31 to 36 of the Sydney Observations, 1877–78. The only information respecting these observations which is available to me at present is on page 6 of the Introduction, where it is stated :—

" The declination micrometer has only one wire, and is fitted with a head counter similar to that on the collimation micrometer. For the purpose of observing *Mars* a second wire was put in 19″·68 from the zero wire, but as it was found to be somewhat in the way when taking the Nadir readings it was removed."

Now, the following is a specimen of a day's work as given in the Sydney volume 1877–1878.

1877, *September* 2.

Name.	R.A.	N.P.D
70 Aquarii	22 42 5·69	101 11 55·56
74 Aquarii	—	102 15 51·66
B.A.C. 8004		103 43 24·65
Mars	—	101 56 1·60 *a*
Mars	22 12 21·37	101 56 13·49 *β*
B.A.C. 8199	23 25 53·62	102 12 57·59
B.A.C. 8266	23 40 59·33	102 35 2·97

No intimation is here given of the stars which are bisected by the wires *a*, *b*, respectively. And unless great care is taken to balance the observations of stars made with each wire there is always a suspicion of the possibility of systematic error in this method. It is not stated whether the observations of *Mars* are or are not corrected for defective illumination. And there appears to me a considerable difficulty in determining which of the two observations of *Mars* (*a*) or (*β*) is really that made by the method of equal segments. Professor Eastman has taken the observation (*a*) with N.P.D.$=101°$ $56'$ $1''\cdot60$ as that made by the method of equal segments, and *β* with N.P.D.$=$ $101°$ $56'$ $13''\cdot49$ as that made by simple bisection; and the observations *β* have been rejected from the discordance of the results.

I presume that Professor Eastman has had direct information on the point from Mr. Russell, and that the course which he has followed is correct. But I must confess that I should hardly have come to this conclusion from the mere observations, as recorded in the Sydney volume 1877–1878.

It would almost appear from these observations, as recorded, that the stars have been observed on, or reduced to, the principal wire, and that (*a*) represents the attempt to fix the centre of *Mars* by " bisecting with the wire in ordinary use, before the planet came to the transit wires." In this case the reading *β*, or $101°$ $56'$ $13''\cdot49$, would represent the principal observations of *Mars* made by the method of equal segments ; but, supposing it recorded as observed, it would require a correction of $9''\cdot84$, which corresponds to half the distance between the parallel wires *a* and *b*. This would make the two results *a* and *β* agree fairly with each other. On the other hand, if Professor Eastman is right in his selection of the principal observations—and I presume that such must be the case—then, undoubtedly, there is something wrong in the observations as recorded, and some confusion in reducing the observations to a common wire, for it is quite impossible that a constant error of about $10''$ can have been made in bisecting such a planet as *Mars* even by a direct attempt to find the centre.

I have worked up these Sydney observations; but it appeared to me not desirable to give them until the points mentioned have been cleared up.

The results contained in the Table have been obtained as follows :—

I have first computed star-corrections for all the observations of the comparison stars selected for observation with *Mars*. With these star-corrections, mean North Polar Distances have been formed. The mean of the separate results for each star for the Cape, Leyden, Melbourne, and Washington, have then been separately found, and the mean of these four separate results adopted as the North Polar Distance of the star for the purposes of this discussion.

In the determination of a value of the solar parallax from a discussion of observations like those under consideration, the tabular places of the stars are entirely eliminated when the same stars are observed at all the stations. The course now adopted allows of the exhibition of the results on days when some of the comparison stars are wanting, without the introduction of any important systematic errors ; and, if desired, the errors can be exhibited as corrections to the final result. The mean of the differences between the mean North Polar Distances of the comparison stars and the adopted mean North Polar Distances is then found for each day at each station, and these mean differences are considered as index corrections applicable to the North Polar Distance of *Mars* on the same day for the station. The tabular North Polar Distances have been, for the present purpose, directly interpolated from the *Nautical Almanac*, and the value 8″·95 has been adopted as the value of the horizontal equatoreal solar parallax.

The "error of the tables" is the excess of the observed North Polar Distance over the tabular North Polar Distance.

These errors have been laid down as ordinates, of which the corresponding times are the abscissæ ; and a curve has been swept amongst the points laid down. The corrections for "errors of tables" are the corrections read off from this curve. The residual errors are the excesses of the tabular errors over those read off from the curve.

The residual errors, therefore, represent the real errors on the assumption of the accuracy of the mean curve of errors. It will be seen that the residual errors are small for all the stations. And there appears no strongly-marked preponderance of positive or negative signs between the results for the Northern and Southern Observatories over any considerable portion of the curve. This is the only important point so far as a determination of the Solar Parallax is concerned.

The mean residual errors are very small indeed for all the stations.

Day	Washington			Leyden			Melbourne			Cape			Day
	Errors of Tables.	Corr. for Errors of Tables.	Residual Errors.	Errors of Tables.	Corr. for Errors of Tables.	Residual Errors.	Errors of Tables.	Corr. for Errors of Tables.	Residual Errors.	Errors of Tables.	Corr. for Errors of Tables.	Residual Errors.	
July 20				2·74	2·52	+0·22							July 20
21				2·10	2·56	−0·46	2·65	2·54	+0·11				21
22				2·45	2·69	−0·24	3·26	2·60	+0·66				22
23							2·69	2·66	+0·03				23
24													24
25													25
26													26
27				2·81	3·04	−0·23	2·78	2·87	−0·09				27
28							3·66	2·93	+0·73				28
29	3·52	3·03	+0·49				2·69	2·96	−0·27				29
30	3·50	3·06	+0·44				3·30	3·03	+0·27				30
31													31
Aug. 1										3·00	3·13	−0·13	Aug. 1
2										3·46	3·16	+0·30	2
3							2·83	3·17	−0·34	2·21	3·20	−0·99	3
4													4
5							3·56	3·26	+0·30				5
6	3·26	3·35	−0·09	3·73	3·33	+0·40							6
7							2·97	33·3	−0·36	3·75	3·35	+0·40	7

Day	Washington Errors of Tables	Washington Corr. for Errors of Tables	Washington Residual Errors	Leyden Errors of Tables	Leyden Corr. for Errors of Tables	Leyden Residual Errors	Melbourne Errors of Tables	Melbourne Corr. for Errors of Tables	Melbourne Residual Errors	Cape Errors of Tables	Cape Corr. for Errors of Tables	Cape Residual Errors
Aug. 8				2·92	3·39	−0·47	2·92	3·37	−0·45			
9							4·27	3·42	+0·85	3·76	3·43	+0·33
10							3·10	3·44	−0·34			
11							2·89	3·47	−0·58			
12							3·07	3·52	−0·45			
13				3·39	3·62	−0·23	4·35	3·55	+0·80	2·96	3·56	−0·60
14							3·50	3·57	−0·07	4·02	3·58	+0·44
15							4·26	3·60	+0·66	3·87	3·62	+0·25
16				3·89	3·66	+0·23	4·31	3·63	+0·68			
17	3·61	3·70	−0·09				3·60	3·65	−0·05			
18										3·32	3·68	−0·36
19												
20	1·20	3·75	−2·55	3·74	3·74	0·00	2·63	3·73	−1·10	3·24	3·73	−0·49
21				3·63	3·76	−0·13						
22												
23							3·84	3·76	+0·08			
24							3·49	3·77	−0·28			
25							4·54	3·79	+0·75			
26							4·68	3·80	+0·88			

Day.	Washington.			Leyden.			Melbourne.			Cape.		
	Errors of Tables.	Corr. for Errors of Tables.	Residual Errors.	Errors of Tables.	Corr. for Errors of Tables.	Residual Errors.	Errors of Tables.	Corr. for Errors of Tables.	Residual Errors.	Errors of Tables.	Corr. for Errors of Tables.	Residual Errors.
Aug. 27	3·49	3·80	−0·31				3·93	3·80	+0·13			
28	3·73	3·79	−0·06				2·72	3·80	−1·08			
29							3·21	3·79	−0·58	2·86	3·78	−0·92
30				3·78	3·77	+0·01	4·10	3·78	+0·32	3·67	3·77	−0·10
31										3·44	3·76	−0·32
Sept. 1	5·41	3·73	+1·68	4·13	3·74	+0·39	3·47	3·75	−0·28	2·96	3·74	−0·78
2				3·51	3·72	−0·21	4·33	3·73	+0·60			
3	5·49	3·71	+1·78				3·80	3·71	+0·09	3·87	3·72	+0·15
4				3·32	3·70	−0·38	2·60	3·69	−1·09	3·95	3·70	+0·25
5							2·60	3·68	−1·08			
6							4·10	3·66	+0·44	3·83	3·66	+0·17
7												
8												
9												
10										2·85	3·57	−0·72
11										3·35	3·54	−0·19
12										4·07	3·52	+0·55
13							4·14	3·50	+0·64	3·46	3·49	−0·03
14							1·80	3·46	−1·66			

Day	Washington			Leyden			Melbourne			Cape			Day
	Errors of Tables	Corr. for Errors of Tables	Residual Errors	Errors of Tables	Corr. for Errors of Tables	Residual Errors	Errors of Tables	Corr. for Errors of Tables	Residual Errors	Errors of Tables	Corr. for Errors of Tables	Residual Errors	
Sept. 15	3·23	3·37	−0·14				3·48	3·43	+0·05	4·12	3·40	+0·72	Sept. 15
16							3·62	3·38	+0·24				16
17				2·59	3·33	−0·74	2·93	3·34	−0·41	3·50	3·33	+0·17	17
18							3·57	3·31	+0·26	3·76	3·29	+0·47	18
19										2·60	3·24	−0·64	19
20				3·09	3·15	−0·06				3·46	3·20	+0·26	20
21	2·74	3·13	−0·39										21
22							2·77	3·13	−0·36				22
23													23
24	2·82	2·99	−0·17	2·60	2·95	−0·35	3·09	3·05	+0·04	2·53	3·03	−0·50	24
25							3·30	2·97	+0·33	3·05	2·95	•10	25
26	3·15	2·86	+0·29	2·56	2·83	−0·27	2·76	2·92	−0·16	3·04	2·89	+0·15	26
27							2·47	2·86	−0·39	2·51	2·83	−0·32	27
28							2·04	2·79	−0·75	2·32	76	−0·44	28
29										2·86	2·70	•16	29
30													30
Oct. 1	2·96	2·54	+0·42	2·29	2·56	−0·27	1·73	2·58	−0·85	2·06	2·56	−0·50	Oct. 1
2	2·14	2·47	−0·33	2·09	2·50	−0·41	2·92	2·52	+0·40	2·07	2·50	−0·43	2
3				2·25	2·44	−0·19	2·39	2·46	−0·07	2·61	2·44	+0·17	3

Day	Washington			Leyden			Melbourne			Cape			Day
	Errors of Tables	Corr. for Errors of Tables	Residual Errors	Errors of Tables	Corr. for Errors of Tables	Residual Errors	Errors of Tables	Corr. for Errors of Tables	Residual Errors	Errors of Tables	Corr. for Errors of Tables	Residual Errors	
Oct. 4				2·01	2·38	−0·37	2·12	2·40	−0·28	1·14	2·38	−1·24	Oct. 4
5							2·27	2·34	−0·07	1·52	2·32	−0·80	5
6	1·62	2·20	−0·58	2·20	2·22	−0·02	2·94	2·24	+0·70	3·20	2·22	+0·98	6
7							1·58	2·18	−0·60				7
8							3·02	2·13	+0·89				8
9	2·23	2·01	+0·22	3·03	2·03	+1·00	2·31	2·06	+0·25				9
10							1·78	2·00	−0·22				10
11							2·72	1·94	+0·78				11
12	1·52	1·83	−0·31				1·21	1·87	−0·66				12
13													13
14	1·58	1·71	−0·13	2·63	1·73	+0·90	2·10	1·76	+0·34				14
15	1·77	1·65	+0·12	2·14	1·67	+0·47	1·71	1·71	0·00				15
16							2·51	1·63	+0·88				16
17	2·34	1·55	+0·79										17
18	1·82	1·50	+0·32				1·32	1·54	−0·22				18
19													19
20													20
21				1·94	1·41	+0·53	1·28	1·43	−0·15				21
Oct. 22							1·76	1·33	+0·43				Oct. 22

The results given by this Table appear to me important. The Cape observations began on August 1, and were continued to October 6, and embrace the period when the horizontal equatoreal parallax of *Mars* was 20″.

If we take the corresponding residuals for all four stations during the period August 1 to October 6, we obtain the following results :—

Mean Results.	*Southern Stations.*		*Northern Stations.*	
	Cape 38 obs.	Melb. 46 obs.	Leyden 19 obs.	Wash. 14 obs.
Residual Error ...	$-0''118$	$-0''085$	$-0''162$	$-0''039$
Error of Tables ...	$+3·112$	$+3·261$	$+3·038$	$+3·203$
Correction applied .	$-3·230$	$-3·346$	$-3·199$	$-3·242$
Factor of Parallax .	$+0·90$	$+1·07$	$-2·19$	$-1·89$

If we include the observations during the whole period July 20 to October 22, we obtain the following results :—

Mean Results.	*Southern Station.*	*Northern Stations.*	
	Melb. 65 obs.	Leyden 27 obs.	Wash. 22 obs.
Residual Error	$-0''011$	$-0''033$	$+0''064$
Error of Tables	$+2·990$	$+2·873$	$+2·870$
Correction applied ...	$-3·001$	$-2·905$	$-2·806$

In both cases these observations admit of no important alteration of the adopted value of the solar parallax, 8″·95 ; and the agreement between the results affords no grounds for suspicion of a difference of personality in the contacts with the upper and lower limbs of *Mars*.

As the observations of *Mars* at the Cape and Melbourne were made by the method of tangents to the limbs, whilst those at Leyden and Washington were made by the method of equal segments, it may be desirable to point out that the results obtained by both methods lead, as might be expected, to sensibly identical results. In 1862, the Pulkova and Cape observations were made by the method of equal segments ; but the Greenwich and Melbourne observations were made by the method of tangents. The Pulkova and Cape observations gave 8″·96 for the value of the solar parallax, whilst Greenwich and Cape gave 8″·92, and Greenwich and Melbourne gave 8″·94.

Unless it can be proved that there is some cause which affects differently the observations of *Mars* at the Northern and Southern stations, these observations and the corresponding results obtained in 1862 would appear to render it impossible that any value of the solar parallax as small as 8″·75 can be the true value.   The weight of the result given by these observations is certainly very great.

I know of only one cause which has yet been assigned as a possible explanation of the large value given by these meridian observations—viz. chromatic dispersion by the atmosphere. There is, however, no difficulty in proving mathematically that the illumination of the upper limb of a planet is increased and not decreased relatively to the lower limb from this cause. This will be clear enough if anyone will merely consider that the red light from the lower limb cannot be strengthened by the yellow, green, and blue lights of greater refrangibility, whilst the red light from the upper limb will be supplemented by light of greater refrangibility from portions of the planet at slightly greater zenith distances.   Under any conditions of illumination, therefore, this cause cannot make the observer cut deeper into the upper limb than into the lower.   Atmospheric dispersion might, therefore, be appealed to with some slight plausibility as a cause of the small value which has been deduced from the Heliometer measures, but most certainly it is not directly the cause of the large value found from the meridian observations.

Note on Mr. Neison's Paper on the Corrections to Adams's Semi-Diameter of the Moon.   By E. J. Stone, M.A., F.R.S.

I have felt great reluctance to continue any longer in our *Notices* a discussion upon these questions of the diameter of the Moon.   The question is not, however, simply whether I am right, or Mr. Neison right, upon a point of combined mathematics and arithmetic, but whether the parallactic inequality deduced from a discussion of a valuable series of observations affords, or does not afford, evidence in favour of a certain supposed long inequality in the Moon's longitude.   If that long inequality does exist with a coefficient $1''\cdot2$ it is of the utmost importance that its existence should be recognised, and that its changes should be allowed for in our Lunar Theory ; but if the inequality does not exist with any such coefficient, it is of even greater importance that its variations should not be allowed to vitiate our results. I have, since the last meeting, re-examined my work, and I find it correct.   But before the publication of Mr. Neison's separate results I was unable to offer any possible explanation of the systematic discordances between the results obtained by us. The error into which Mr. Neison has fallen is now, however, clear to me.

The following are the formulæ from which my results have been obtained :—

Let

B = Burckhardt's parallax in seconds of arc.

δA = The correction to B found directly from Prof. Adams's formulæ.

S = Tabular semi-diameter adopted in any year.

x = The mean correction required by Adams's semi-diameter with the factor 0·273114.

Let

$$\frac{\text{Observed diameter} - \text{Tabular diameter}}{\text{Tabular diameter.}} = n.$$

Then

$$n + 1 = \frac{(B + \delta A) \, 0\cdot273114 + x}{S},$$

or

$$x = \left(n + 1 - \frac{B}{S}\, 0\cdot273114\right) S - \delta A \cdot 0\cdot273114.$$

Now, in

1853, 1854, 1855,

$$S = B\,0\cdot2725 \left(1 + \frac{1}{360}\right);$$

$$x = \left\{n + 1 - \frac{0\cdot273114}{0\cdot2725\left(1 + \frac{1}{360}\right)}\right\} S - \delta A\, 0\cdot273114,$$

$$x = (n + 0\cdot0005233)\, S - \delta A\, 0\cdot273114.$$

If we refer to the " Tables containing corrections to be applied to the values of the Moon's equatoreal horizontal parallax given in the *Nautical Almanac* 1840–1856, in order to make them agree with those calculated from Mr. Adams's tables " (*Appendix N.A.* 1856), we can extract numbers which will give us δA. But, on reference to the *Nautical Almanac* 1853, 1854, and 1855, it will be found that the *N.A.* Parallax = B $\left(1 + \frac{1}{1200}\right)$.

It follows, therefore, that if δN is the correction found in the *Appendix N.A.* 1856,

$$B + \delta A = B \left(1 + \frac{1}{1200}\right) + \delta N,$$

or

$$\delta A = \frac{B}{1200} + \delta N;$$

therefore

$$x = (n + 0\cdot0005233)\, S - \delta N \cdot 0\cdot273114 - \frac{B}{1200}\, 0\cdot273114.$$

Now, if we neglect the term $\frac{B}{1200}\, 0\cdot273114$, or assume that the corrections given in the *Appendix N.A.* 1856, are corrections to Burckhardt's Parallax, and not to the *N.A.* Parallax, we shall substantially obtain Mr. Neison's results; but these results are clearly wrong to an extent

$$= \frac{B}{1200} \, 0.273114 = \frac{3420}{1200} \times 0.273114 = 0''.78$$

very approximately.

To show that the corrections given in the *Appendix N.A.* 1856, are corrections to the adopted parallaxes of the *N.A.*, and not to Burckhardt's tables, it is sufficient to extract the following corrections :—

1852	Dec. 30 =	$+2''.8$	} Corrections to Burckhardt
1852	31 =	$+2.6$	
1853	Jan. 1 =	-0.4	} Corrections to Burckhardt $(1 + \frac{1}{1200})$
1853	2 =	-0.3	

If we apply the correction $0''.78$ to Mr. Neison's results, and compare them with mine given in the *Notices* for December 1881, we obtain as follows. I have changed the signs, so that the quantities given in these tables are the *corrections* to be applied to Adams's semi-diameter to make it agree, in the mean, with the observed horizontal semi-diameters :—

	Neison's corrections to Adams'.		Stone's corrections to Adams'		Neison's results corrected.
	Obs.	$''.$	Obs.	$''$	$''$
1851	6	-0.21	7	$-0''.43$	$-0''.21$
1852	3	$+0.86$	3	$+0.71$	$+0.86$
1853	4	$+1.73$	4	$+0.97$	$+0.95$
1854	3	-0.39	3	-1.17	-1.17
1855	3	-0.25	3	-0.94	-1.03
1856	3	-1.75	3	-1.77	-1.75
1857	3	-0.76	3	-0.76	-0.76
1858	5	-1.02	5	-1.02	-1.02

I have no doubt about the substantial accuracy of my results, but the discordances between the outstanding results are now unimportant. These observations make Adams's semi-diameter too large by $0''.51$, but Messrs. Campbell and Neison have assumed, in their work, that this semi-diameter is too small by $0''.18$ for the same group of observations. The rejection by Mr. Neison of the diameters observed by the computers at Greenwich, whilst their observations at Quadratures are retained, is certainly unsatisfactory. With respect to the second point, I have nothing more to remark than that my work is accurate, and that the corrections to the diameter deduced by me refer to the period 1843–1851 under discussion; and the results certainly do not bear out the inferences which Messrs. Campbell and Neison have made from them. The authors appear to have overlooked the fact that if personal systematic differences exist in the observations of the limbs, it is only the unequal distribution amongst the observers of the observed diameters and observations at Quadratures which affect the final result.

On the Inclination of the Ring of Saturn to its Orbit, deduced from Washington Observations. By Professor Edward S. Holden.

(Communicated by the Secretaries.)

The observations given in the following pages were made by Professor A. Hall and myself during the years 1877, 1878, and 1879, which were favourable for the purpose.

The instruments employed were the 26-inch Clark Refractor, ortomethe filar mier, and eyepieces magnifying from 400 to 800 times. The resulting position-angles of the major axis of the ring are given in the fifth column of the table. The reductions were made by the following formulæ, where p is the angle of the northern half of the minor axis of the Ring and the Declination-circle through *Saturn's* centre.

The formulæ are Bessel's, and are given in Engelmann's Bessel's *Abhandlungen*, vol. i. p. 321.

$$\cos a \cdot \cos A = \cos p,$$
$$\cos a \cdot \sin A = \sin p \cdot \sin \delta,$$
$$\sin a = \sin p \cdot \cos \delta,$$
$$\tan B = \frac{\tan a}{\cos (n - a + A)},$$
$$\tan C = \tan (n - a + A) \frac{\cos B}{\cos (B + i)},$$
$$\tan i' = \tan (B + i)' \frac{\cos C}{\cos (n'' - n' + C)}.$$

In these formulæ a, δ, are the geocentric coordinates of *Saturn*; i is the inclination of *Saturn's* orbit to the Earth's equator, and n the longitude of the node of *Saturn's* orbit on the same plane; n' is the node of the Earth's equator on *Saturn's* orbit, and n'' is the node of the Ring-plane on *Saturn's* orbit, i' being the (required) inclination of these planes.

I have assumed Bessel's values of n, n', n'', and i, since the introduction of Le Verrier's or Hill's values for ♌ and i will not change the results for i'.

The following table gives each observation and the resulting value of i'.

Table giving Washington Observations of Saturn's Ring, and the Computed Values of the Inclinations of the Plane of the Ring to Saturn's Orbit.

No.	Date.		Date.	Wash. M.T.	p	Obs.	Hall. $i'=$			Holden. $i'=$		
				h m	°		°	′	″	°	′	″
1	June	18	1877·46	15 8	94·82	H.	26	42	35			
⁝		19	·47	15 20	4·80	H.	26	44	39			
		22	·47	14 50	5·18	H.	27	4	34			
4		23	·48	14 57	4·76	H.	26	39	23			
5		25	·49	14 46	4·80	H.	26	41	55			
6	July	10	·52	14 50	5·20	Hn.				27	5	14
7		11	·53	14 6	4·68	Hn.				26	40	52
8		11	·53	14 7	4·58	Hn.				26	28	27
9		13	·53	14 0	5·03	Hn.				26	55	19
10	Aug.	3	·60	12 13	5·3	Hn.				27	7	43
11		10	·61	13 38	4·75	H.	26	32	49			
12		11	·61	11 36	4·73	H.	26	59	1			
13		15	·63	12 35	5·01	H.	26	46	55			
14		16	·63	12 54	4·98	H.	26	44	46			
15		17	·63	13 8	4·93	H.	26	41	24			
16		18	·64	12 12	4·98	H.	26	44	10			
17		27	·66	13 34	4·93	H.	26	38	0			
18		28	·66	13 13	4·78	H.	26	28	40			
19	Sept.	23	·73	10 44	5·13	H.	26	40	12			
20		29	·75	9 40	5·08	H.	26	35	9			
21	Oct.	1	·75	10 9	5·18	H.	26	40	36			
22		⁝	·75	10 28	5·00	H.	26	29	22			
23		5	·76	7 17	5·27	Hn.				26	44	46
24		9	·78	9 24	5·75	Hn.				27	12	30
25		13	·79	10 47	5·77	Hn.				27	11	35
26		13	·79	9 19	5·18	H.	27	6	40			
27		14	·79	8 25	5·30	H.	26	47	38			
28		14	·79	9 59	5·56	Hn.				26	58	44
29		15	·79	11 12	5·68	Hn.				27	6	45
30		16	·79	10 39	5·71	Hn.				27	8	20
31		17	·80	7 20	5·09	H.	26	30	40			
32		18	·80	6 56	5·26	R.	26	40	40			
33		18	·80	9 49	5·66	Hn.				27	4	49
34		23	·81	7 5	5·21	H.	26	36	33			
.35		24	·81	7 35	5·29	H.	26	41	25			

No.	Date.		Date.	Wash. M.T.	p	Obs.	Hall. t'=	Holden. t'=
				h m	°		° ′ ″	° ′ ″
36	Dec.	2	1877·93	6 26	5·10	H.	26 28 38	
37		7	·94	6 48	5·40	H.	26 47 18	
38		8	·94	6 9	5·15	H.	26 32 30	
39		9	·94	6 18	5·27	H.	26 39 56	
40		15	·96	7 2	4·80	H.	26 12 56	
41		19	·97	6 52	5·20	H.	26 37 56	
42		31	·99	6 23	5·55	H.	27 2 44	
43	Jan.	3	1878·01	6 12	5·50	H.	27 0 44	
44		5	·01	6 18	4·95	H.	26 28 27	
45		12	·03	6 4	5·17	H.	26 44 30	
46		16	·04	6 6	4·90	H.	26 30 2	
47		17	·05	6 10	4·97	H.	26 34 40	
48		19	·05	5 51	4·83	H.	26 27 10	
49	Sept.	9	1878·69	10 30	4·5	Hn.		27 15 52
50		17	·71	10 40	4·1	Hn.		26 52 41
51		19	·72	10 50	3·9	Hn.		26 40 55
52	Oct.	1	·75	11 30	4·5	Hn.		27 11 8
53		8	·77	11 0	4·4	Hn.		26 59 19
54		14	·79	9 30	3·9	Hn.		26 29 14
55		21	·81	8 30	4·6	Hn.		27 6 7
56		25	·82	9 18	4·5	Hn.		26 59 56
57		26	·82	9 37	4·2	Hn.		26 41 55
58	Nov.	5	·85	10 44	4·5	Hn.		26 59 35
59		13	·87	8 33	3·60	H.	26 12 52	
60		13	·87	9 42	4·2	Hn.		26 36 56
61		14	·87	7 50	3·90	H.	26 18 50	
62		14	·87	8 53	4·3	Hn.		26 42 55
63		15	·88	7 30	3·85	H.	26 15 45	
64		18	·89	9 13	4·23	H.	26 38 29	
65		20	·89	8 23	4·2	Hn.		26 36 41
66	Oct.	16	1879·80	11 5	3·4	Hn.		27 11 30
67		20	·81	9 48	3·3	Hn.		27 8 34
68		23	·81	12 0	3·2	Hn.		27 4 40
69		24	·82	8 48	3·3	Hn.		27 11 35
70		25	·82	9 47	3·2	Hn.		27 6 7
71		26	·82	8 38	3·4	Hn.		27 19 15
72		27	·82	12 5	3·2	Hn.		27 7 45
73		28	·83	10 0	3·1	Hn.		27 2 11

No.	Date.	Date.	Wash. M.T.	p	Obs.	Hall. $i'=$	Holden. $i'=$
			h m	°		° ' "	° '
74	30	·83	10 6	3·3	Hn.		26 59 41
75	31	·83	10 21	3·1	Hn.		26 46 55
76	Nov. 11	·87	9 49	3·5	Hn.		27 18 40
77	22	1879·89	8 36	93·3	Hn.		26 55 2

From the 40 observations of Professor Hall we have for

1877·948 = 1877, Dec. 11,

$$i' = 26° \; 38' \; 47'' \pm 1'·4.$$

The probable error of a single observation is $+8'·6$.
From the 37 observations made by me, we have for

1878 771 = 1878, Oct. 8,

$$i' = 26° \; 57' \; 2'' \pm 1'·7.$$

The probable error of a single observation is $\pm 10'·3$.
Bessel's 22 observations give for

1818·726 = 1818, Sept. 20,

$$i' = 27° \; 0' \; 9'' \pm 5'·2.$$

The probable error of a single observation was $\pm 24'·1$.
I have not combined the results obtained by Professor Hall
and myself, since the observations of 1877, October 13 and 14,
and of 1878, Nov. 13 and 14, show that they are not strictly
comparable. My result agrees with Bessel's within the limits
of the probable errors. A comparison of the results of Professor
Hall and of Bessel would indicate a diminution of the inclination
since 1818.

Washburn Observatory,
  University of Wisconsin, Madison:
    1882, January.

Postscript.—Since submitting the above to Professor Hall, I
have found two observations of the position-angle of *Saturn's*
Ring made by Dr. Auwers in 1861 with Königsberg Heliometer
in the publications of that Observatory for 1865, page 132.
They are

1861, April 15; $p = 83° \; 14'$ (2),
23; $-82 \; 39$ (2).

These, when reduced as above, give

$$1861\cdot29 \quad i' = 26 \ 48\cdot6$$
$$1861\cdot31 \quad i' = 27 \ 56\cdot0 \ \text{or}$$
$$1861\cdot30 \quad i' = 26 \ 52\cdot3 \ (2 \ \text{nights.})$$

1882, *March.*

Conjunctions of the Interior Satellites of Saturn.
By Professor Asaph Hall.

Date.	Wash. M. T.	Satellite.	Conjunction.	Remarks.
	h m			
1881, Sept. 19	13 35·8	Enceladus	Superior	
Oct. 22	10 39·3	„	„	,
Nov. 14	9 45·1	Dione	Inferior	Poor
15	10 50·4	Tethys	„	Good
16	9 29·7	„	Superior	„
28	10 12·8	Enceladus	„	Faint
Dec. 8	10 27·9	Rhea		Good
10	9 42·5	Dione	„	
1882, Jan. 11	7 25·6	Rhea	Inferior	
27	7 14·1	Dione	∷	

Note.—The above observations were made by setting the wire of the micrometer at the zero of position, so as to observe the time of the conjunction in Right Ascension. On November 16, *Mimas* was noticed to be very bright, and near the following elongation at $9^{\text{h}} 37^{\text{m}}$ m. t.

U.S. Naval Observatory,
Washington:
    1882, *March* 6.

Errata in the First Melbourne General Catalogue of Stars.
By R. L. J. Ellery, Esq., F.R.S.

In Mr. Downing's paper on the N.P.D.'s of the Cape Catalogue for 1880, which appeared in the *Monthly Notices* of the R. A. S. for November 1881, he refers to a discrepancy between the N.P.D. of Lacaille 4342 as given in the Cape and Melbourne Catalogues, and suggests that it might be partly due to proper motion. The difference, however, is explained by a typographical error in the Melbourne Catalogue, where 3 has been printed instead of 8, so that the seconds ought to be 38·83 instead of 33·83.

Our observations, extending over a period of fifteen years, show no appreciable motion in the P.D. of this star.

I append a list of the errata of the Melbourne First General Catalogue detected up to the present time.

No.	Place.	*For*	*Read*
74	d	6·2865	1·2865
80	a'	6·6287	9·6287
194	P′	·0034	·0008
347	N.P.D.	13·81	5·02
358	Name	Majoris	Minoris
511	N.P.D.	33·83	38·83
552	P′	·2099	·1485
	p'	·033	·007
796	N.P.D.	10·10	40·10
	P	5·833	5·834
	a	9·0054	9·0056
		9·1710	9·1712
	c	0·7659	0·7660
	d	8·9732	8·9734
	m	(−0·25)	(+0·01)
	a'	9·7822	9·7823
895	P′	·8356	·7589
910	R.A.	42·95	42·97
	M.	−0·004	0·000
	N.P.D.	58·70	57·48
	m	+0·23	0·00
947	N.P.D.	30′	20′
1067	P	102·365	102·372

The Observatory, Melbourne:
1882, *January* 31.

The Variable Star β Ursæ Minoris. By T. E. Espin, Esq., B.A.

This star has been frequently suspected of variation. Herschel, Struve, and Heis have each observed alterations in its light. The star was seen of unusual brilliancy on the evening of November 5, 1881, and observations were at once commenced which soon showed signs of a period of about ten days. The observations were discontinued of necessity at the end of December, but were resumed again in March 1882. The total number of nights on which this star has been observed is 42, namely :—

1881, November	... ...	13	1882, March	15
December	... ...	7	April	7

The star was near a maximum or minimum on the following nights :—

Near Maximum.			Near Minimum.		
	d	h		d	h
1881, Nov.	5	0	1881, Nov.	9	11
	15	6		20	6
	25	13		29	6
1882, March	3	8	Dec.	21	6
	13	8	1882, March	5	8
April	3	11	April	6	10

The star was also observed to be at a maximum on the evening of March 11, 1879. The following elements have been found to represent the observations very satisfactorily :—

$$\text{Period } 10\overset{d}{\cdot}6747$$

Variation from 2·2 to 2·8

$$\text{Epoch of Maximum, 1882, April } 4\overset{d}{\cdot}10.$$

The star's magnitude, for the greater part of its period, is 2·5 or 2·6. The minimum takes place two or three days after the maximum. Observations were also made by Mr. T. Read in November and December, which were found to agree with my own. The stars used as comparison stars were :—

a Ursæ Minoris	...	...	Mag.	2·4
γ „ „	...	...		3·2

The determinations were made with an opera-glass, the stars being placed slightly out of focus. The colour of β Ursæ Minoris is usually a bright yellow; it probably becomes redder at minimum.

Orbits of Meteor-streams, deduced from Observations made during the years 1871–1880 in Hungary. By M. R. de Kövesligethy.

(Communicated by Dr. N. de Konkoly.)

For some years past observations of meteors have been made in Hungary, especially at O-Gyalla, Selmec, Zágráb, Szathmár-Némethy, Hódmezö-Vásárhely and Gyulafehérvár. There was thus a considerable amount of material, about 5,000 meteors having been observed. The number of those, however, which were employed in the determination of the radiant points was only 1,088, the greater part of the meteors proving to be scattered, or to have been incorrectly observed. Since 91 radiant points have been deduced, each radiant is derived on an average from 12 meteors.

The calculations have been made with the formulæ published by Dr. Charles Schræder in one of the *O-Gyalla Annales*, which do not differ very much from the approximate method given by Klinkerfues.

Let $\odot$ be the ephemeridical, $\odot'$ the corrected longitude of the Sun, R the radius vector of the Earth. By assumption of parabolical velocity, that is by neglecting $\dfrac{1}{a}$ in the expression

$$V = \kappa \sqrt{\frac{2}{R} - \frac{1}{a}},$$

where

$$\log \kappa = 8 \cdot 2356,$$

we have the following relations :

$$\cos \eta = \cos b \sin (l - \odot'),$$

where

$$\odot' = \odot + \Delta \odot,$$

and

$$\sin \theta = m \sin \eta,$$

where

$$m = \sqrt{1 - \frac{R}{2}}.$$

$\Delta \odot$ and $\log m$ are to be taken out of the latter following table. For the inclination we have

$$\tan i = \frac{\sin b}{\sin \eta} \tan (\eta - \theta),$$

and for the anomaly

$$\tan \frac{v}{2} = \frac{\cos i \tan (\eta - \theta)}{\tan (l - \odot') \tan \eta}.$$

Finally,

$$\pi - \Omega = v, \quad \text{and } \Omega = \begin{cases} \odot \\ 180 + \odot \end{cases} \text{if } b \text{ is} \begin{cases} \text{negative} \\ \text{positive} \end{cases}.$$

The perihelion distance is derived by

$$q = R \cos^2 \frac{v}{2}.$$

As a control I used

$$\cos \frac{v}{2} = \frac{\cos (\eta - \theta)}{\cos i}.$$

The motion is direct, if

$$\eta - \theta < 90°,$$

retrograde, if

$$\eta - \theta > 90°.$$

TABLE.

With the argument $\odot$ *we get* $\Delta \odot$ *and log m.*

$\odot$	$\Delta\odot$	log. m	$\odot$	$\Delta\odot$	log. m
0°	−0·93°	9·8502	190°	+0·95°	9·8495
10	0·97	8495	200	0·95	8501
20	0·95	8489	210	0·92	8507
30	0·92	8483	220	0·87	8512
40	0·85	8477	230	0·77	8517
50	0·75	8472	240	0·65	8522
60	0·63	8467	250	0·52	8525
70	0·50	8463	260	0·35	8528
80	0·35	8460	270	0·18	8529
90	0·18	8459	280	0·02	8530
100	0·02	8458	290	−0·17	8530
110	+0·15	8459	300	0·33	8529
120	0·32	8460	310	0·48	8526
130	0·47	8463	320	0·62	8522
140	0·62	8467	330	0·75	8518
150	0·73	8471	340	0·85	8513
160	0·83	8476	350	0·88	8508
170	0·90	8482	360	−0·90	9·8502
180	+0·93	9·8488			

The perihelia have been excluded as being of no importance for comparison.

Dr. Edmund Weiss has given a table of the comets which nearly approach the Earth's orbit. I have changed this table only by adding the elements of these comets, which are based on the Olbers-Galle Catalogue. The second column gives the comet and node in which the approximation happens, the third the time of this approximation, the fourth the distance from the Earth's orbit, and the fifth the radiant point in longitude and latitude. The data are reduced to the mean equinox of 1850.

Catalogue of Comets which approach near to the Earth's Orbit.

No	Comet	Time		R−r	l	b	ι	ϖ	Ω	q	Motion.
1	1792, II. ☍☌	Jan.	5	−0·066	182·2	27·9	49·12	135·87	283·23	0·9668	R.
2	1840, I. ☌		20	+0·036	141·1	−45·5	53·08	192·20	119·95	0·6185	D.
3	1718, ☌		29	−0·042	217·6	−18·2	31·13	121·65	127·92	1·0254	R.
4	1857, I, ☍	Feb.	2	−0·028	258·3	46·3	87·93	74·73	313·15	0·7725	D.
5	1092, ☌		5	−0·012	110·1	−56·9	88·92	156·33	125·67	0·9281	D.
6	1854, IV. ☍		13	+0·015	320·7	55·0	40·90	94·40	324·47	0·7987	D.
7	1858, IV. ☍		13	+0·045	272·2	35·2	80·03	226·10	324·97	0·5543	R.
8	1862, IV. ☍	Mar.	16	+0·013	247·9	22·9	42·47	125·18	355·77	0·8032	R.
9	1683, ☌		16	−0·052	224·0	−34·5	83·78	86 52	173·28	0·5533	R.
10	1763, ☍		18	−0·026	322·4	37·7	72·57	8·95	356·28	0·4983	D.
11	1861, I. ☍	April	20	+0·002	270·6	57·0	79·75	243·37	29·92	0·9207	D.
12	1790, III. ☍		24	−0·063	328·5	33·1	63·58	274·95	35·23	0·7910	R.
13	1863, II. ☌	June	2	−0·054	338·9	−40·3	67·37	47·25	251·27	1·0682	R.
14	1684, ☌		22	−0·010	40·1	−65·8	65·80	238·87	268·25	0·9602	D.

No.	Comet.	Time.	R−r	l	b	i	π	Ω	φ	Motion.
15	1850, I. ☍	24	−0·065	1·9	9·5	68·18	272·42	92·88	1·0815	D.
16	1864, II. ☍	27	+0·047	13·5	1·1	1·87	304·22	95·20	0·9993	R.
17	1737, II. ☍	July 29	+0·025	129·2	58·8	39·23	262·60	123·88	0·8670	D.
18	1852, II. ☍	Aug. 10	−0·013	33·6	−27·7	49·18	278·70	317·48	0·9129	D.
19	1862, II. ☍	19	−0·027	486	−4·6	7·90	299·33	326·53	0·9813	R.
20	1854, III. ☍	Sept. 10	−0·018	45·8	−33·7	71·32	273·68	347·65	0·6481	R.
21	1790, I. ☍	16	−0·053	104·7	15·2	29·73	58·40	172·83	0·7473	R.
22	1763, ☍	20	+0·029	33·2	−39·0	72·57	84·95	356·28	0·4983	D.
23	1864, IV. ☍	Oct. 16	−0·044	185·6	50·1	48·87	321·70	203·22	0·7709	D.
24	1779, ☍	19	+0·022	24·5	−42·3	32·52	87·23	25·07	0·7132	D.
25	1849, I. ☍	29	−0·027	147·2	55·1	85·05	63·23	215·20	0·9597	D.
26	Bila ☍	Nov. 28	+0·011	38·7	30·6	12·55	109·13	245·85	0·8606	D.
27	1819, IV ☍	Dec. 9	+0·086	327·9	−35·2	9·02	67·32	77·23	0·8926	D.
28	1680, ☍	26	+0·050	128·6	3·4	60·67	262·82	272·15	0·0062	D.

I consider now the influence of neglecting $\frac{1}{a}$ on the elements. From

$$V = \kappa \sqrt{\frac{2}{R} - \frac{1}{a}}$$

we find by differentiation

$$dV = \frac{\kappa^2}{2\,a^2 V}\,da\,;$$

κ^2 being small, and a^2 large, the error in V will be but small. For the November stream, $a = 10\cdot340$; for the August stream, $a = 22\cdot355$. If we put, for instance, $da = 12\cdot015$, we get $dV = 0\cdot00015$; we can say, therefore, that, by neglecting the quantity $\frac{1}{a}$ the velocity is hardly altered.

The approximate method follows from the strict one by putting $L = \odot - 90^\circ$, and $R = 1$, where L is defined by the equation

$$\sin(\odot - L) = \frac{0\cdot9998}{\sqrt{2\,R - R^2}}.$$

It easily follows that

$$dL = \frac{0\cdot9998\,(1 - R)\,dR}{(2\,R - R^2)\,\sqrt{2\,R - R^2} - 0\cdot9996}\,;$$

therefore, in the most unfavourable case, $dL = 0\cdot0144$. The expression for the inclination is

$$\tan \imath = \frac{\tan \beta}{\sin(\odot - \lambda)},$$

where

$$\tan \beta = \frac{\tan b}{\sin(l - L)}\,\sin(\lambda - L)\,;$$

l and b are the latitude and longitude of the radiant point, and λ, β, the longitude and latitude of the end point of the tangent drawn to the orbit of the meteor. By substitution and differentiation we have

$$di = \frac{\tan b\,.\,\sin(\odot - \lambda)\,\sin(l - \lambda)\,dL}{\tan^2 b\,\sin^2(\lambda - L) + \sin^2(l - L)\,\sin^2(\odot - \lambda)}\,;$$

and we may put

$$di = \frac{\tan b\,\sin(\odot - \lambda)\,\sin(l - \lambda)\,dL}{\tan^2 b\,\cos^2(\odot - \lambda) + \cos^2(\odot - l)\,\sin^2(\odot - \lambda)}.$$

If we choose for ⊙, λ and l the most unfavourable possible values, we get, in the case when $b=89°·17$, an error of $di=1·0000$. In the two extreme cases of our calculation, the least latitude is $-0·66$, and the greatest$=81·12$, giving respectively $di=-0·0002$ and $+0·0922$.

We determine still the error in the anomaly, which is also identical with the error of the longitude of perihelion.

By differentiation of the formula

$$p = \frac{V^2 R^2 \sin^2 \sigma}{\kappa^2}$$

with respect to V, regarding σ as independent from L, an omission of very slight importance, we get

$$dp = \frac{2 VR^2 \sin^2 \sigma}{\kappa^2} \, dV,$$

and thus in the worst case, $dp=0·0240$.

From the formula for the anomaly

$$e \cos v = \frac{p}{R} - 1$$

we find, regarding p as variable,

$$dv = \frac{-dp}{R\sqrt{e^2 - \left(\frac{p}{R} - 1\right)^2}};$$

so that the greatest value of $dv : da = -0·1580$. The longitude of the node is determined only by the longitude of the Sun; the supposition $\frac{1}{a}=0$ has no influence on it.

These few data show sufficiently that the calculation with the approximate formulæ introduces errors which are less than the errors of observation.

In a series of observations there frequently occur meteors which might with equal justice be reckoned as belonging to different radiant points. It is therefore of the greatest importance to note that meteors belonging to different streams are essentially different in colour, in the character of their trains, and in their apparent velocity. It is to be remarked that these differences form the characteristics of the meteor-streams. The Perseids are, for instance, yellow, the trains disappear rapidly, and their brightness continually increases. The following elements are arranged according to the longitude of the radiant point, which allows a better comparison and view. Our purpose is, not to record as many radiant points as possible, but to reduce the many observations to the best determined ones : as, however, among the radiant points from which I have calculated there may possibly still be some uncertainties, I give the following table of identical radiant points and meteor-streams :—

Identical Radiant Points and Meteor Orbits.

Time.		Central No.	No. of Meteors.	Time.		Central No.	No. of Meteors.
1875, July	27·48	2	10	1874, Aug.	9·95	6	11
1878,	29·44	3	9	1874,	9·95	5	8
				1875,	10·96	7	9
1872, Aug.	12·01	9	11		11·95	3	23
1875,	10·96	10	10	1876,	10·92	4	11
1874,	9·95	7	10				
				1875, Aug.	10·96	5	11
1872, Aug.	7·42	1	9	1874,	9·95	·3	50
1875,	9·99	2	6				
				1876, July	27·98	4	14
1879, July	26·96	1	10		27·98	1	17
1875,	26·96	1	18	1873,	26·48	1	8
1874, Aug.	12·45	10	7	1874, Aug.	9·95	2	9
1872,	10·98	6	9	1876,	13·40	2	6
				1876, Aug.	10·92		15
1872, Aug.	10·98	5	8	1875,	8·93	1	9
1877,	12·93	2	19				
				1880, Aug.	9·44	7	22
1880, July	29·99	9	13	1875,	10·96	6	14
1878,	26·43	1	10	1872,	10·98	4	8
				1874,	9·95	4	25
1875, July	28·45	2	23	1879,	11·94	3	12
1876,	27·98	2	9	1875,	11·95	2	12
1873, July	27·99	2	13	1874, Aug.	9·95	1	14
1880,	27·95	11	13	1877,	13·44	1	12
				1872,	12·42	11	9
1874, Aug.	12·45	9	9	1876,	10·92	1	9
1875,	12·41	12	6				
	10·96	8	8	1872, Aug.	10·98	8	12
	12·96	9	14		9·97	3	16
1871,	9·97	2	7				
				1872, Aug.	12·42	13	9
1878, Apr.	20·94	1	6	1879,	12·46	5	13
1874,	20 63	1	6				
				1876, Aug.	10·92	6.	16
				1872,	12 01	10	22

Time	Central No.	No. of Meteors	l	b	l	ϖ	Ω	q	Motion	η	η – θ	Station
1880, July 9·42	10	7	358·93	44·13	15·43	297·41	287·43	1·0093	D.	47·13	16·20	Selmec.
1875, 27·48	2	10	358·03	22·06	56·48	45·15	303·95	0·4091	R.	138·33	110·52	„
1876, 27·98	3	13	357·99	48·04	84·85	247·91	305·17	0·7823	D.	122·00	85·48	O·Gyalla.
1878, 29·44	3	9	357·15	18·16	53·24	60·27	306·03	0·2992	R.	137·36	108·97	„
1872, Aug. 12·01	9	11	356·60	72·17	55·93	295·30	319·50	0·9683	D.	100·49	56·79	„
1880, 9·44	8	8	353·65	37·98	74·50	227·19	317·07	0·5074	D.	117·60	79·10	„
1875, 10·96	10	10	352·72	56·63	63·28	265·99	317·85	0·8195	D.	108·05	66·15	„
1874, 9·95	7	10	352·46	57·36	63·33	266·64	317·10	0·8292	D.	107·96	66·03	„
1875, July 31·98	7	9	347·82	33·20	82·69	209·05	308·27	0·4250	D.	121·86	85·27	„
1872, Aug. 10·98	6	9	344·24	66·19	54·10	282·67	318·53	0·9172	D.	99·87	56·08	„
1875, July 28·45	2	23	335·40	60·02	58·97	258·60	304·92	0·8582	D.	104·50	61·71	„
1873, 29·9	3	10	330·26	15·65	44·79	172·57	305·37	0·1627	D.	113·53	73·49	„
1875, 31·98	5	6	328·62	45·60	52·81	235·71	308·27	0·6595	D.	103·79	60·83	„
1872, Aug. 7·42	1	9	326·10	59·10	46·78	268·66	315·08	0·8564	D.	95·36	51·00	„
1875, July 31·98	6	9	325·71	32·50	45·50	214·37	308·27	0·4729	D.	104·27	61·42	„
Aug. 9·99	2	6	322·50	48·95	40·77	257·76	316·88	0·7670	D.	93·31	48·78	„
1880, 9·44	6	7	320·62	22·43	23·13	230·07	317·07	0·5334	D.	92·75	48·20	„
1879, July 26·96	1	10	319·22	22·96	36·20	196·80	303·52	0·3617	D.	104·08	61·21	„
1875, 26·96	1	18	316·51	12·96	20·96	190·28	303·48	0·3077	D.	102·33	59·07	Zágráb.

Time	Central No.	No. of Meteors	l	b	b	■	Ω	g	Motion	η	η−θ	Station
1874, Aug. 12·45	10	7	319·96	70·02	41·23	292·74	319·52	0·9584	D.	87·55	42·97	O-Gyalla.
1872, 10·98	5	8	306·45	47·09	29·54	270·33	318·53	0·8444	D.	81·41	37·42	"
1875, July 28·45	1	9	305·10	73·58	44·17	281·68	304·92	0·9740	D.	89·94	45·37	"
1877, Aug. 12·93	2	19	303·47	32·10	19·33	266·71	320·17	0·8080	D.	75·40	32·57	Zágráb.
1876, July 27·98	2	9	302·87	66·54	41·91	272·99	305·17	0·9372	D.	88·93	44·37	O-Gyalla.
1875, 31·98	4	7	296·83	15·23	10·85	241·05	308·27	0·7036	D.	78·54	35·09	"
1878, 29·44	1	9	293·65	53·12	32·51	263·07	306·03	0·8760	D.	82·36	38·30	"
1873, 27·99	2	13	277·21	40·82	20·38	266·37	305·37	0·9018	D.	·68·79	27·92	"
1880, 29·99	9	13	270·00	77·55	37·96	294·82	307·10	1·0035	D.	84·46	38·38	Fehérvár.
1878, 26·43	1	10	269·52	73·06	35·98	295·39	303·17	1·0108	D.	80·62	36·82	Selmec.
1877, Aug. 12·93	1	18	269·13	23·76	8·72	296·07	320·17	0·9687	D.	44·13	14·84	Zágráb.
1875, 10·96	9	14	268·25	74·05	34·02	306·17	317·85	1·0028	D.	77·82	34·46	O-Gyalla.
1871, 9·97	2	7	267·27	71·64	32·47	303·83	316·85	1·0001	D.	76·00	33·04	"
1874, April 20·63	2	7	267·11	56·54	78·58	172·14	210·28	0·8980	D.	117·80	79·21	"
1880, July 27·95	11	13	260·17	47·08	19·69	280·38	305·18	0·9689	D.	60·98	23·13	Selmec.
1874, Aug. 12·45	9	9	250·44	81·12	37·60	315·74	319·52	1·0120	D.	81·68	37·64	O-Gyalla.
1878, Apr. 21·44	1	10	245·83	45·93	69·66	138·30	211·10	0·6513	D.	113·89	73·76	Vásárhely.
1875, Aug. 12·41	12	6	244·74	70·88	29·66	314·12	319·18	1·0110	D.	71·56	29·76	O-Gyalla.
10·96	8	8	220·73	73·85	31·52	320·17	317·85	1·0130	D.	74·00	31·54	"

Time.	Central No.	No. of Meteors.	l	b		*	Ω	q	Motion.	ψ	γ-θ	Station.
1876, Aug. 13·93	1	10	178·75	63·82	30·24	343·71	321·35	0·9742	D.	74·74	32·07	O-Gyalla.
1878, Apr. 20·94	1	6	174·98	64·25	30·72	187·70	210·60	0·9654	D.	75·68	32·58	„
1874, 20·63	1	6	168·91	71·69	33·77	198·62	210·28	0·9948	D.	78·24	34·58	„
1880, Nov. 28·61	12	7	113·26	14·99	56·76	197·23	66·73	0·1728	R.	134·05	103·27	Selmec.
1871, Aug. 9·97	1	9	106·60	62·82	58·25	357·97	316·85	0·8884	D.	103·54	60·48	O-Gyalla.
1874, 9·95	6	11	81·79	42·37	85·35	254·04	317·10	0·7360	R.	127·71	93·96	„
1876, 10·92	5	15	79·65	53·91	82·76	358·89	318·47	0·8922	D.	120·47	83·21	Vásárhely.
1874, 9·95	5	8	75·80	45·68	85·10	267·80	317·10	0·8370	R.	128·04	94·45	O-Gyalla.
1875, 10·96	7	9	72·15	45·07	81·52	274·59	317·85	0·8758	R.	130·28	97·88	„
1875, 11·95	3	23	64·20	32·45	57·41	281·40	318·80	0·9090	R.	144·65	120·66	Szathmár.
1876, 10·92	4	11	63·33	40·49	70·13	288·07	318·47	0·9434	R.	137·49	109·15	Vásárhely.
1875, 8·93	1	9	62·31	57·24	84·43	335·73	315·83	0·9837	D.	121·36	84·51	O-Gyalla.
1875, 10·96	5	11	61·67	46·68	79·40	293·79	317·85	0·9682	R.	131·90	100·37	„
1874, 9·95	3	50	60·59	47·53	80·57	339·92	317·10	0·9734	R.	131·16	99·24	„
1880, 9·44	7	22	60·37	36·13	62·64	287·07	317·07	0·9450	R.	141·99	116·36	„
1875, 10·96	6	14	59·78	32·26	55·90	288·93	317·85	0·9502	R.	146·00	122·87	„
1872, 10·98	4	8	59·23	39·60	67·43	296·35	318·53	0·9758	R.	129·36	112·13	„
1874, 9·95	4	25	56·32	39·00	66·14	297·84	317·10	0·9848	R.	140·21	113·50	„
1879, 11·94	3	12	55·08	36·29	61·09	305·11	318·73	0·9988	R.	143·32	118·51	„

Time.	Cen-tral No.	No. of Meteors		b		ϖ	Ω		Motion.	η	η–θ	Station.
1875, Aug. 11·95	2	12	54·04	32·10	54·29	306·84	318 80	1·0023	R.	147·60	125·49	Szathmár.
10·96	11	15	52 72	66·93	72·71	321·33	317·85	1·0125	D.	113·00	72·72	O-Gyalla.
10·96	4	15	51·43	39·89	85·00	29·17	317·85	0·6690	R.	127·78	94·06	,,
1876, 10·92	3	9	49·80	48·40	80·09	317·17	318·47	1·0133	R.	131·60	99·91	Vásárhely.
1874, 9·95	2	9	47·61	33·48	56·28	317·28	317·10	1·0132	R.	146·52	123·72	O-Gyalla.
1876, 13·40	2	6	46·96	32·40	54·78	332·51	320·87	1·0025	R.	147·30	125·00	Vásárhely.
July 27·98	4	14	46·28	66·62	72·64	296·41	305·17	1 102	D.	112·95	72·69	O-Gyalla.
27·98	1	17	43·38	57·19	86·17	315·15	305·17	1·0078	D.	122·48	86·18	,,
189, Aug. 13·92	4	22	42·58	28·72	49·36	344·39	320·67	0·9700	R.	150·10	129·60	,,
1873, July 26·48	1	8	39·84	57·63	85·73	311·03	303·45	1·0106	D	122·17	85·74	,,
1876, Ag. 13·40	1	8	37·77	49·43	83·69	343·73	320·87	0·9730	R.	129 18	96·18	Vásárhely.
1875, July 28·45	3	7	37·58	74·28	63·20	306 06	304·92	1·0152	D.	105·70	63·20	O-Gyalla.
1872, Aug. 10·98	7	26	35·08	66·84	71·85	307·27	318·53	1·0035	D.	112·43	71·94	,,
1874, 9·95	1	14	31·96	35·64	62·95	354·30	317·10	0·9102	R.	141·48	115·54	,,
1877, 13·44	1	12	30·17	42·60	76·42	2·27	320·63	0·8845	R.	133·39	102·68	,,
1879, July 26·96	2	8	29·49	42·67	71·31	315·56	303·52	1·0138	R.	137·14	108·63	,,
1872, Aug. 12·42	11	9	28·79	38·68	70·78	7·02	319 88	0·8510	R.	13649	796	,,
1874, 9·95	8	28	26·68	47·06	82·96	86·16	317·10	0·9130	R.	129 50	96·68	,,
1876. 11·40	3	8	25·63	52·36	87·24	283·77	320·87	0·9102	D.	123·33	87·38	Vásárhely.

Tim.	Cen-tral No.	No. of dirs.	l	b	i	*	Ω	q	Motion.	π	π−θ	Station.
876, Aug. 10·92	1	9	24·30	35·97	68·48	14·93	318·47	0·7865	R.	137·30	108·86	Vásárhely.
1872, 10·98	8	12	20·15	53·87	83·69	279·79	318·53	0·9018	D.	121·05	84·05	O-Gyalla.
1875, 11·95	1	22	20·13	8·92	20·93	54·90	318·80	0·4527	R.	149·52	128·64	Szathmár.
1872, 9·97	3	16	19·99	53·27	84·86	278·92	317·58	0·9024	D.	121·81	85·15	O-Gyalla.[1]
1872, Nov. 28·90	1	10	18·50	66·14	27·23	55·08	66·98	0·9758	D.	69·70	27·82	„
1872, Aug. 12·42	13	9	18·45	43·16	84·65	18·66	319·88	0·7686	R.	128·18	94·66	„
1875, July 27·48	1	12	18·44	57·36	84·30	284·15	303·95	0·9850	D.	121·25	84·39	Selmec.
1872, Oct. 24·32	1	13	17·47	64·87	36·42	1·04	31·18	0·9264	D.	83·85	39·01	O-Gyalla.
1872, Aug. 12·42	12	6	16·87	64·38	69·50	231·62	319·88	0·9540	D.	111·10	70·14	„
1879, 12·46	5	13	11·87	39·73	85·92	32·28	319·32	0·6550	R.	127·27	93·28	Fehérvár.
1876, 10·92	2	8	7·34	-0·66	2·66	359·35	138·47	0·1236	R.	138·29	110·42	Vásárhely.
1875, 9·99	3	20	7·08	59·87	70·87	277·76	316·88	0·9000	D.	112·48	72·01	O-Gyalla.
1876, 10·92	6	16	6·19	41·74	86·37	243·99	318·47	0·6423	D.	123·14	87·11	Vásárhely.
1872, 7·42	2	7	5·98	50·07	80·95	259·48	315·08	0·7930	D.	119·63	81·00	O-Gyalla.
1872, 12·01	10	22	1·46	48·55	74·22	254·42	319·50	0·7198	D.	115·93	76·74	„

The radiant 2, 1878, July 29·44, had such an unfavourable situation that the orbit could not be strictly calculated, it has therefore been excluded.

In the comparison of meteor orbits with those of known comets, I found but two which accorded.

	1872, Aug. 7·42 } 1875, Aug. 9·99 }	Comet 1854, IV. Lesser.	1874, April 20·63 Central No. 2.	Comet 1861, I. Oppolzer.
l	324·30	320·7	267·11	270·6
b	54·02	55·0	56·54	57·0
i	43·77	40·90	78·58	79·75
π	263·21	94·40	172·14	243·37
Ω	315·98	324·47	210·28	209·92
q	0·8117	0·7987	0·8980	0 9207
	D	D	D	D

The cause of it may be partly that all radiant points have northern latitude—that is, that all meteor-streams have been observed in the descending node.

Finally, I must pay my best thanks to Dr. de Konkoly and Dr. Weiss, who were so kind as to aid me with assistance and advice.

Vienna:
  1882, *February.*

Observations of the Companion of Sirius, made at the U.S. Naval Observatory, Washington. By Professor Asaph Hall.

Date.	Sid. Time.	p	s	Weight.	Obs.
	h	°	''		
1882, Feb. 27	8·9	42·7	9·93	2	F.
Mar. 2	8·0	42·2	9·88	2	F.
4	6·7	41·4	9·91	3	F.
11	7·0	42·6	9·88	3	F.
13	6·8	42·3	10·12	2	F.
14	7·1	42·3	10·01	3	F.
22	7·6	41·9	clouds	2	H
23	6·5	42·1	9·53	3	H.

Date.	Sid. Time.	*p*	*s*	Weight.	Obs.
	h	°	ʺ		
1882, Mar. 24	7·2	43·9	9·79	2	H.
25	6·8	42·2	· 9·59	3	H.
28	7·2	42·7	9·70	2	H.
29	7·1	41·9	9·61	3	H.
31	7·5	42·7	9·79	2	H.

Results.

	p	*s*	Observers.
1882·183	42·25	9·955	E. Frisby.
1882·235	42·49	9·668	A. Hall.

MONTHLY NOTICES

OF THE

ROYAL ASTRONOMICAL SOCIETY.

| VOL. XLII. | MAY 12, 1882. | No. 7. |

E. J. Stone, Esq., M.A., F.R.S., President, in the Chair.

Henry George Hollingworth, Esq., F.R.G.S., 319 Vauxhall Bridge Road, S.W.;
Samuel Okell, Esq., Bowden, near Manchester;
William Barrott Roué, Esq., M.B., 165 Whiteladies Road, Clifton, Bristol; and
Hesketh Goddard Williamson, Esq., Shrigley Road, Bollington, near Macclesfield,

were balloted for and duly elected Fellows of the Society.

Discussion of the Observations of γ Draconis, made with the Greenwich Reflex Zenith Tube during the years 1857 to 1875 inclusive. By A. M. W. Downing, Esq., M.A.

In the twenty-ninth volume of the *Memoirs* of this Society there is a paper by Mr. Main containing a discussion of the observations made with the Reflex Zenith Tube during the eight years 1852-1859. The results of this discussion were not altogether satisfactory. The value of the constant of aberration which was deduced was $20''\cdot335$, or more than a tenth of a second smaller than Struve's value, which has been universally received by astronomers with such great confidence. Whilst the annual parallax of γ Draconis resulting from the observations was $-0''\cdot242$, a quantity too large to allow us to consider it accidental, and which probably indicated (the author considered) the existence of some periodic error, whether arising from the

D D

instrument or from some deformation of the atmosphere, which affects the observations.

As the Reflex Zenith Tube has been in continuous use up to the present time (with the exception of some interruption during the years 1871 and 1872, when the experiments with the Water Telescope were in progress), I have felt that it was desirable that another attempt should be made to utilise the observations for a determination of the constant of aberration; and, as it seemed unlikely, owing to the discouraging circumstances mentioned above, that any person not connected with the Greenwich Observatory would undertake such a laborious task, I have myself entered on the investigation, the results of which I have now the honour to present to the Society.

I have begun my investigation with the year 1857, as previously to that date the instrument was located in a different and (apparently) most unsatisfactory position, and the mercury trough was supported in a different manner to that finally adopted, so that the star's image was subject to great unsteadiness, and it was, I believe, impossible to take an observation in the daytime. It appeared to me, therefore, that it would be better to begin my discussion from the time when the instrument was brought into its present position, and the suspension of the mercury trough arranged in the manner which has remained unaltered up to the present time. And I have concluded my investigation with the year 1875, thus embracing the period of a complete revolution of the Moon's Node, so that a value of the constant of nutation may also be determined from these observations, and the number of observations made during this interval of time being very great, it seemed probable that as good a value of the constant of aberration could be obtained from this series as the instrument was capable of giving even from a longer term of years.

The observations of apparent zenith distance have been extracted from the volumes of Greenwich Observations for the different years, those observations only being used which were made in both positions of the instrument. The observations were then combined in convenient groups, each group not extending over more than twelve or, at the most, fourteen days, it being assumed that the mean of the observed Z. D.'s corresponds to the mean of the times of observation. We have thus a series of observed apparent zenith distances, which are to be compared with the assumed mean zenith distance by the application (to the latter) of corrections for precession and proper motion, aberration, annual parallax, and nutation.

The assumed mean Z. D. North is $102''\cdot37 + w$ for the epoch 1866·0.

The precession for this epoch applicable to mean Z. D. North of γ *Draconis* is $-0'''\cdot569$, and the assumed proper motion in Z. D. North is $-(0'''\cdot028 + \delta p)$; so that the annual variation for 1866 is $-(0'''\cdot597 + \delta p)$. If the mean Z. D. North for 1866·0 be assumed to be $102''\cdot37$, the following are the corresponding values for the different years :—

Date.	Mean Z. D. North.	Date.	Mean Z. D. North.
1857·0	107·83	1867·0	101·78
58·0	107·21	68·0	101·18
59·0	106·60	69·0	100·59
60·0	105·99	70·0	100·00
61·0	105·38	71·0	99·41
62·0	104·78	72·0	98·83
63·0	104·17	73·0	98·24
64·0	103·57	74·0	97·66
65·0	102·97	1875·0	97·08
1866·0	102·37		

And the correction for annual variation applicable at any date is the annual variation for the year multiplied by the fraction of the year counted from the instant when the Sun's mean longitude is 280°.

The assumed value of the constant of aberration is $20''\cdot4+x$, and therefore the correction applicable to mean Z. D. North is generally

$$-(20''\cdot4+x)\{\cos a \cos \Delta \sin \odot + (\sin \omega \sin \Delta - \cos \omega \sin a \cos \Delta)\cos \odot\}.$$

Assuming that, in the case of γ *Draconis*, for 1860,

$$a = 268° \, 20' \, 15'', \qquad \Delta = 38° \, 29' \, 36'', \qquad \omega = 23° \, 27' \, 27'',$$

and for 1870

$$a = 268° \, 23' \, 45'', \qquad \Delta = 38° \, 29' \, 42'', \qquad \omega = 23° \, 27' \, 22'',$$

the above expression becomes, for 1860,

$$-\cdot96580 \cos(1° \, 21' + \odot) \times (20''\cdot4+x),$$

and for 1870,

$$-\cdot96574 \cos(1° \, 18' + \odot) \times (20''\cdot4+x).$$

From these formulæ it is easy to find the correction for any epoch between 1857 and 1875 corresponding to any value of ⊙, and it evidently consists of two parts—a numerical correction to the assumed mean Z. D. North, corresponding to the value $20''\cdot4$ of the constant of aberration, and the correction x to that value multiplied by a numerical coefficient.

If y be the assumed value of the annual parallax of γ *Draconis*, the corresponding correction to the assumed mean Z. D. North is found immediately from the above expressions to be, for 1860,

$$+ \cdot 96580 \, R \, \sin (\overset{\circ}{1} \, 2\overset{'}{1} + \odot) \times y,$$

and for 1870,

$$+ \cdot 96574 \, R \, \sin (\overset{\circ}{1} \, 1\overset{'}{8} + \odot) \times y,$$

where R is the Earth's radius vector.

The expression for nutation is taken from Peters' *Numerus Constans Nutationis*, p. 74, omitting terms which are quite insensible in the case of γ *Draconis*. Assuming the constant to be, for 1866,

$$9'' \cdot 2237 \left(1 + \frac{z}{10} \right),$$

the correction applicable to mean Z. D. North is, for 1860,

$$\left(1 + \frac{z}{10} \right) \left\{ \begin{array}{l} -6'' \cdot 8669 \cos \alpha \sin \Omega + 9'' \cdot 2236 \sin \alpha \cos \Omega + 0'' \cdot 0822 \cos \alpha \sin 2\Omega \\ \cdot 8672 \cdot 2237 \end{array} \right.$$

$$\left. - 0'' \cdot 0896 \sin \alpha \cos 2\Omega - 0'' \cdot 0813 \cos \alpha \sin 2 \mathbb{C} + 0'' \cdot 0886 \sin \alpha \cos 2 \mathbb{C} \right\}$$

$$+ \left(1 - 2 \cdot 162 \times \frac{z}{10} \right) \left\{ -0'' \cdot 5053 \cos \alpha \sin 2\odot + 0'' \cdot 5508 \sin \alpha \cos 2\odot \right\}$$

where the figures below the line are the corresponding values of the coefficients for 1870. For γ *Draconis* this expression is easily reduced to the following:—

$$\left(1 + \frac{z}{10} \right) \left\{ \begin{array}{l} -9'' \cdot 2219 \cos (\Omega + 1^{\circ} \, 14') + 0'' \cdot 090 \cos (2\Omega + 1^{\circ} \, 32') \\ \cdot 2221 \phantom{\cos (\Omega + 1^{\circ} \, 1} 12 \phantom{) + 0'' \cdot 090 \cos (2\Omega + 1^{\circ} \, 3} 29 \end{array} \right.$$

$$\left. \begin{array}{r} -0'' \cdot 089 \cos (2 \mathbb{C} + 1^{\circ} \, 32') \\ 29 \end{array} \right\}$$

$$- \left(1 + \frac{z}{10} - 0'' \cdot 316 \, z \right) \times 0'' \cdot 551 \cos (2\odot + 1^{\circ} \, 31'),$$
$$\phantom{- \left(1 + \frac{z}{10} - 0'' \cdot 316 \, z \right) \times 0'' \cdot 551 \cos (2\odot + 1^{\circ} \,} 28$$

where the figures on the line are the values for 1860, those below the line the corresponding values for 1870. From this expression has been found the correction applicable to the assumed mean Z. D. North, at the mean date of each normal group, and it evidently, like the correction for aberration, consists of a numerical correction, corresponding to the value of the constant $9'' \cdot 2237$, and the correction z multiplied by a numerical coefficient. The values of $\odot$, $\mathbb{C}$ and Ω have been taken from the *Nautical Almanacs* of the different years.

In this manner have been formed the equations of condition, corresponding to each normal group of observed Z. D.'s, the equations containing the five unknown quantities—w, x, y, z, and δp. I have not attempted to determine the value of δp, but merely exhibit its effect on the values of the other quantities.

From the 1,044 observations made during the period under

consideration, I have formed 266 equations of condition as is shown in the following tabular statement.

Means of Observed Z. D.'s of γ Draconis from Groups.

No.	Extent of Group.		Mean Day and Hour.		Mean of Observed Z. D.'s.	No. of Obs.
			d h		″	
1	1857, Jan.	1	Jan.	1·23	94·67	1
	Feb.	9–16	Feb. 13	6	82·35	5
	Mar.	12–17	Mar. 15	18	78·46	4
.	April	14–20	Apr. 17	12	80·86	6
5		29–May 14	May 7	11	85·36	7
6	May	25–June 8	31	10	92·47	6
7	June	10–18	June 14	6	97·08	7
8		23–30	26	8	100·43	6
9	July	7–18	July 12	21	106·00	9
10		20–Aug. 5	27	10	110·01	8
11	Aug.	12–27	Aug. 21	5	115·11	8
12	Sept.	3–9	Sept. 6	2	116·57	5
13		16–Oct. 1	23	18	117·88	8
14	Oct.	5–14	Oct. 8	13	116·91	2
15		26–30	28	3	114·60	
16	Nov.	11	Nov. 11	3	111·14	
17		27–Dec. 4	29	17	105·76	3
18	Dec.	10–11	Dec. 10	13	102·24	2
19		31	31	23	95·63	–
20	1858, Jan.	6–13	Jan. 10	23	91·71	3
21		25–31	28	4	86·42	4
22	Feb.	18–25	Feb. 22	12	80·12	6
23	Mar.	8–10	Mar. 9	19	78·59	3
24		21–26	23	18	78·67	4
25	April	13–21	Apr. 18	10	80·34	
26	May	4–13	May 8	11	85·35	7
27		18–29	22	19	89·06	5
28	June	3–19	June 13	2	95·79	11
29		22–July 8	29	8	101·52	6
30	July	12–21	July 16	10	106·68	7
31		26–Aug. 7	Aug. 1	22	110·82	9
32	Aug.	10–19	13	8	113·07	5
33		26–Sept. 8	31	7	116·14	5
34	Sept.	13–24	Sept. 17	9	117·34	8
35	Oct.	5–12	Oct. 7	21	116·89	3

No.	Extent of Group.	Mean Day and Hour.	Mean of Observed Z. D.'s.	No. of Obs.
		d h	"	
36	1858, Oct. 30–Nov. 11	Nov. 4 11	112·59	
37	Nov. 18	Nov. 18 2	109·20	
38	Dec. 1–3	Dec. 2 1	105·84	
39	20–23	21 12	98·38	2
40	1859, Jan. 23–Feb. 6	Jan. 31 22	85·53	3
41	Feb. 17–27	Feb. 23 20	80·06	5
42	Mar. 7–18	Mar. 11 4	79·32	5
43	31–April 10	April 5 17	80·23	4
44	April 22	22 16	82·29	1
45	May 5–14	May 10 15	87·18	5
46	23–June 7	29 14	93·11	3
47	June 16–July 1	June 25 0	102·05	6
48	July 4–19	July 11 9	106·44	12
49	28–Aug. 5	Aug. 2 9	112·15	4
50	Aug. 16–26	21 18	115·36	
51	30–Sept. 6	Sept. 2 13	117·55	
52	Sept. 18–24	21 6	118·65	
53	Oct. 3–14	Oct. 7 21	117·41	
54	22–Nov. 3	27 20	114·74	
55	Nov. 10–21	Nov. 16 20	110·75	4
56	23–Dec. 10	Dec. 1 1	106·55	5
57	Dec. 26	26 0	98·52	1
58	1860, Jan. 2–11	Jan. 7 11	94·18	
59	22–23	23 10	89·40	2
60	Feb. 19–29	Feb. 25 0	81·36	6
61	Mar. 4–5	Mar. 5 7	80·96	2
62	14–21	17 18	80·40	
63	April 1	April 1 17	80·35	1
64	May 9–24	May 19 22	91·83	6
65	June 4–6	June 5 13	96·58	3
66	13–20	16 18	99·83	4
67	July 4–14	July 10 18	107·65	3
68	20–21	20 22	110·20	2
69	30–Aug. 14	Aug. 7 4	114·48	5
70	Sept. 4–14	Sept. 10 6	119·57	5
71	26–Oct. 3	29 5	118·65	3
72	Oct. 9–22	Oct. 16 22	118·28	
73	30–Nov. 9	Nov. 3 3	115·44	

No.	Extent of Group.	Mean Day and Hour.		Mean of Observed Z. D.'s.	No. of Obs.
		d	h		
74	1860, Nov. 20–23	Nov. 21	14	110·80	
75	Dec. 19–28	Dec. 24	15	100·25	
76	1861, Jan. 8–9	Jan. 9	10	94·90	2
77	Feb. 11–26	Feb. 20	20	83·75	3
78	Mar. 20–21	Mar. 21	6	81·61	2
79	May 16–22	May 19	4	93·47	5
80	June 8–19	June 14	12	101·45	7
81	22–27	25	0	104·48	4
82	July 5–16	July 11	10	109·12	5
83	23–Aug. 2	29	3	114·86	4
84	Aug. 6–15	Aug. 12	3	117·73	4
85	19–30	25	1	120·43	10
86	Sept. 3–10	Sept. 6	15	121·39	3
87	30–Oct. 8	Oct. 4	5	122·87	2
88	Oct. 23–Nov. 1	27	12	119·06	6
89	Dec. 2–11	Dec. 7	5	108·11	2
90	1862, Jan. 9	Jan. 9	22	96·66	
91	Mar. 18	Mar. 18	18	84·69	
92	April 15–24	April 20	4	85·81	
93	May 20	May 20	14	95·84	
94	June 3	June 3	13	99·74	
95	25–30	28	0	108·22	
96	July 12	July 12	13	113·08	1
97	19–28	23	20	116·77	5
98	Aug. 1–14	Aug. 6	14	119·20	5
99	19–Sept. 1	26	7	123·66	5
100	Sept. 10–19	Sept. 15	18	126·22	4
101	Oct. 1–16	Oct. 8	0	124·22	5
102	1863, Jan. 6	Jan. 6	23	101·54	1
103	30–Feb. 3	Feb. 1	21	92·97	2
104	Feb. 12–17	14	20	90·77	3
105	Mar. 2	Mar. 2	19	87·71	1
106	13	13	19	86·32	
107	May 13	May 13	15	95·87	
108	25–26	26	2	99·52	
109	June 1–4	June 3	1	102·92	
110	15–30	23	19	109·30	7
111	July 2–18	July 11	23	115·12	11

No.	Extent of Group.	Mean Day and Hour.		Mean of Observed Z. D.'s.	No. of Obs.
		d	h	"	
112	1863, July 23–Aug. 1	28	17	119.74	6
113	Aug. 8–15	Aug. 12	4	122.22	5
114	17–24	20	8	124.50	
115	Sept. 17–29	Sept. 23	9	127.73	7
116	Oct. 9–23	Oct. 16	12	125.98	3
117	Nov. 25–Dec. 4	Nov. 29	7	116.44	4
118	1864, Jan. 27–31	Jan. 29	13	95.98	3
119	Feb. 9	Feb. 9	20	96.42	1
120	Mar. 17–23	Mar. 19	10	89.72	
121	May 13–23	May 18	8	100.35	
122	30–June 7	June 4	7	105.36	
123	June 10–16	13	13	108.07	
124	30–July 5	July 2	17	115.36	4
125	July 14–23	18	20	119.38	5
126	29–Aug. 8	Aug. 4	3	123.44	4
127	Aug. 10–17	13	8	125.33	6
128	26–Sept. 10	Sept. 4	23	127.66	3
129	Sept. 22–28	25	6	129.17	6
130	Oct. 1–7	Oct. 4	5	128.93	2
131	20–Nov. 3	26	4	126.22	
132	Nov. 25–Dec. 9	Dec. 2	6	116.85	
133	1865, Jan. 12	Jan. 12	22	100.87	
134	Feb. 19	Feb. 19	20	93.49	
135	Mar. 19	Mar. 19	18	90.89	
136	May 12	May 12	15	99.25	
137	24–31	27	6	105.08	-
138	June 8–12	June 10	7	109.23	4
139	19–28	22	12	112.73	5
140	July 3–15	July 9	19	118.16	6
141	19–28	24	0	121.97	5
142	Aug. 1–11	Aug. 5	3	124.66	4
143	18–30	25	0	129.47	3
144	Sept. 2–8	Sept. 5	15	129.57	6
145	14–29	22	1	130.83	9
146	Oct. 2–12	Oct. 5	19	129.95	7
147	1866, Jan. 14–18	Jan. 16	22	102.77	
148	Feb. 18–23	Feb. 21	8	94.33	
149	Mar. 1–15	Mar. 6	19	91.92	4

No. 148.　An observation taken on 1866, Feb. 13, has been rejected.

No.	Extent of Group.	Mean Day and Hour.		Mean of Observed Z. D.'s.	No. of Obs.
		d	h		
150	1866, May 3-4	May 4	3	99·09	2
151	16-28	21	17	103·61	8
152	June 8	June 8	13	109·09	1
153	19-28	24	12	114·14	5
154	July 4-14	July 10	11	118·71	5
155	18-20	19	10	122·19	2
156	Aug. 3-9	Aug. 5	17	125·73	3
157	14-25	19	13	128·17	5
158	Sept. 10	Sept. 10	7	129·79	1
159	27	27	6	128·45	1
160	Oct. 8-16	Oct. 12	5	129·13	3
161	Oct. 26-31	28	16	127·81	2
162	Nov. 30	Nov. 30	3	119·60	
163	1867, Feb. 6	Feb. 6	21	96·13	
164	Mar. 15	Mar. 15	18	91·05	
165	May 29-30	May 30	1	104·11	2
166	June 10-17	June 13	13	109·09	5
167	22-28	26	2	113·94	5
168	July 6-16	July 10	23	118·01	6
169	24-Aug. 2	29	18	122·49	3
170	Aug. 6-20	Aug. 13	12	126·44	7
171	24-Sept. 5	30	17	128·60	5
172	Sept. 16-Oct. 1	Sept. 23	10	130·55	6
173	Oct. 17	Oct. 17	4	129·66	1
174	Nov. 6	Nov. 6	3	125·77	
175	1868, Feb. 5	Feb. 5	21	94·58	
176	27	27	19	92·18	
177	Mar. 5-19	Mar. 11	13	89·84	
178	May 14-20	May 18	11	99·68	4
179	23-28	26	9	101·03	5
180	June 1-11	June 6	5	104·63	6
181	15-19	17	13	107·61	5
182	22-29	25	16	110·95	7
183	July 7-10	July 9	0	115·89	4
184	13-17	15	10	117·05	5
185	20-27	23	20	119·68	5
186	31-Aug. 12	Aug. 4	21	122·42	6
187	Sept. 1-11	Sept. 5	16	127·30	8
188	21-23	22	6	127·40	3

No.	Extent of Group.	Mean Day and Hour.		Mean of Observed Z. D.'s.	No. of Obs.
		d	h	"	
189	1868, Oct. 7–12	Oct. 9	14	126·49	
190	29	29	3	123·32	
191	1869, Jan. 21–24	Jan. 23	10	95·62	
192	Feb. 5–18	Feb. 12	9	90·75	
193	Mar. 12	Mar. 12	19	86·30	
194	April 1	April 1	18	86·24	
195	29–May 12	May 5	15	92·70	
196	May 20–31	27	5	100·27	
197	June 3–18	June 11	6	104·32	
198	25–28	26	20	109·56	3
199	July 10–24	July 18	3	115·18	7
200	27–Aug. 7	Aug. 3	3	119·38	
201	Aug. 11–24	17	15	122·45	
202	26–Sept. 3	29	2	124·69	
203	Oct. 6–12	Oct. 9	5	124·17	
204	Nov. 15–18	Nov. 16	14	115·65	
205	Dec. 1	Dec. 1	1	114·54	
206	1870, Jan. 12–27	Jan. 20	10	94·43	
207	Feb. 24	Feb. 24	20	85·74	
208	Mar. 13	Mar. 13	18	84·59	1
209	May 16–26	May 21	2	94·24	8
210	June 1–10	June 7	1	100·43	4
211	14–24	19	9	103·30	8
212	29–July 7	July 2	11	108·16	4
213	July 12–28	20	1	112·82	13
214	Aug. 6–19	Aug. 13	c	118·05	9
215	25–Sept. 8	Sept. 1	2	120·73	5
216	Sept. 21–Oct. 2	26	10	121·52	10
217	Oct. 10–11	Oct. 10	17	120·47	2
218	Nov. 1–2	Nov. 1	15	117·15	
219	12	12	2	114·82	
220	1871, Feb. 16	Feb. 16	20	84·92	
221	May 22–26	May 24	14	92·57	
222	June 5	June 5	12	95·70	
223	July 4–8	July 6	11	105·29	
224	15–19	17	18	109·12	
225	29–Aug. 2	31	15	112·13	4
226	Aug. 7–15	Aug. 11	8	114·02	8
227	1872, Jan. 17–18	Jan. 18	10	88·47	2

No.	Extent of Group.	Mean Day and Hour.	Mean of Observed Z. D.'s.	No. of Obs.
		d h	"	
228	1872, Feb. 12	Feb. 12.21	81·61	
229	Aug. 15	Aug. 15 8	112·03	
230	Dec. 6	Dec. 6 1	100·21	1
231	25–28	27 15	91·50	3
232	1873, Jan. 26–27	Jan. 27 10	82·35	2
233	May 13–22	May 17 14	83·25	3
234	June 16–25	June 20 20	93·75	3
235	July 1	July 1 11	98·50	1
236	18–28	23 1	103·45	8
237	31–Aug. 9	Aug. 5 0	107·38	8.
238	Aug. 13–26	18 14	108·60	8
239	30–Sept. 11	Sept. 5 23	109·68	3
240	Sept. 18–27	24 1	111·06	5
241	Oct. 17	Oct. 17 4	108·69	1
242	Nov. 15	Nov. 15 3	102·97	
243	25	25 2	102·24	1
244	1874, May 15–29	May 22 9	80·59	5
245	June 1–13	June 8 0	85·52	9
246	22–24	23 12	89·98	2
247	July 1–7	July 4 7	94·28	6
248	13–23	17 13	98·03	8
249	28–30	29 9	101·63	3
250	Aug. 6–13	Aug. 9 9	103·51	4
251	17–21	19 8	105·87	
252	28–Sept. 4	31 13	107·41	4
253	Sept. 10–15	Sept. 13 6	107·51	3
254	24–26	25 6	108·67	2
255	Oct. 8–12	Oct. 10 5	108·43	
256	20–28	24 4	104·90	
257	Nov. 20	Nov. 20 2	99·96	1
258	1875, May 20–24	May 22 14	79·73	3
259	31–June 1	June 1 1	81·92	2
260	June 8–19	14 22	86·08	5
261	22–26	24 12	90·88	3
262	July 1–12	July 7 5	95·49	4
263	26–Aug. 4	30 4	101·72	5
264	Aug. 14–25	Aug. 20 18	106·24	5
265	30–Sept. 9	Sept. 3 23	106·97	6
266	Sept. 14–25	17 16	108·07	5

Equations of Condition.

No.	Equations.	Weights.	Residuals.
1	$w - 0.224x - 0.924y - 1.010z + 8.9958p + 0.05 = 0$	1	$+0.29$
2	$w - .804 - .528 - 0.863 + 8.879 - 0.28 = 0$	5	-0.06
3	$w - .965 - .051 - .783 + 8.795 + 0.02 = 0$	4	$+0.19$
4	$w - .843 + .473 - .840 + 8.705 + 0.24 = 0$	6	$+0.33$
5	$w - .638 + .733 - 0.888 + 8.651 + 0.19 = 0$	7	$+0.24$
6	$w - .304 + .930 - 1.014 + 8.585 + 0.14 = 0$	6	$+0.14$
7	$w - .086 + .977 - 1.036 + 8.547 + 0.05 = 0$	7	$+0.04$
8	$w + .108 + .976 - 1.035 + 8.514 + 0.65 = 0$	6	$+0.63$
9	$w + .365 + .908 - 1.008 + 8.461 + 0.13 = 0$	9	$+0.12$
10	$w + .568 + .793 - 0.958 + 8.429 + 0.06 = 0$	8	$+0.06$
11	$w + .835 + .490 - .865 + 8.361 - 0.11 = 0$	8	-0.06
12	$w + .934 + .248 - .818 + 8.318 + 0.21 = 0$	5	$+0.29$
13	$w + .965 - .042 - .787 + 8.269 - 0.42 = 0$	8	-0.29
14	$w + .924 - .282 - .803 + 8.229 - 0.21 = 0$	3	-0.05
15	$w + .776 - .570 - .880 + 8.175 - 0.79 = 0$	3	-0.60
16	$w + .613 - .738 - .940 + 8.137 - 0.42 = 0$	1	-0.21
17	$w + .341 - .891 - 0.995 + 8.086 - 0.21 = 0$	3	$+0.02$
18	$w + .162 - .937 - 1.026 + 8.056 - 0.32 = 0$	2	-0.09
19	$w - .204 - .928 - 1.012 + 7.998 - 1.09 = 0$	1	-0.85
20	$w - .367 - .878 - 0.985 + 7.971 - 0.60 = 0$	3	-0.36
21	$w - .619 - .730 - .922 + 7.924 - 0.76 = 0$	4	-0.53
22	$w - .879 - .397 - .815 + 7.854 - 0.18 = 0$	6	$+0.03$
23	$w - .953 - .153 - .778 + 7.811 - 0.36 = 0$	3	-0.18
24	$w - .963 + .079 - .773 + 7.774 - 0.69 = 0$	4	-0.54
25	$w - .837 + .483 - .820 + 7.704 + 0.50 = 0$	4	$+0.59$
26	$w - .629 + .740 - .910 + 7.649 - 0.06 = 0$	7	-0.02
27	$w - .436 + .874 - .960 + 7.610 + 0.41 = 0$	5	$+0.42$
28	$w - .108 + .974 - .988 + 7.551 + 0.75 = 0$	11	$+0.74$
29	$w + .151 + .970 - .990 + 7.507 + 0.26 = 0$	6	-0.25
30	$w + .414 + .887 - .956 + 7.460 + 0.21 = 0$	7	$+0.20$
31	$w + .635 + .739 - .888 + 7.415 + 0.38 = 0$	9	$+0.40$
32	$w + .762 + .601 + .844 + 7.383 + 0.47 = 0$	5	$+0.51$
33	$w + .904 + .343 - .765 + 7.334 + 0.17 = 0$	5	$+0.25$
34	$w + .964 + .069 - .730 + 7.289 + 0.07 = 0$	8	$+0.18$
35	$w + .928 - .267 - .750 + 7.232 - 0.21 = 0$	3	-0.05
36	$w + .699 - .660 - .834 + 7.156 - 0.10 = 0$	3	$+0.11$
37	$w + .521 - .803 - .888 + 7.119 - 0.09 = 0$	1	$+0.14$

No.	Equations.	Weights.	Residuals.
38	$w + 0\cdot307x - 0\cdot902y - 0\cdot925s + 7\cdot0818p - 0\cdot82 = 0$	2	$-0\cdot59$
39	$w - \cdot021 - \cdot951 - \cdot935 + 7\cdot027 + 0\cdot20 = 0$	2	$+0\cdot45$
40	$w - \cdot664 - \cdot691 - \cdot811 + 6\cdot914 - 0\cdot49 = 0$	3	$-0\cdot24$
41	$w - \cdot886 - \cdot379 - \cdot713 + 6\cdot851 + 0\cdot08 = 0$	5	$+0\cdot30$
42	$w - \cdot956 - \cdot135 - \cdot670 + 6\cdot809 - 0\cdot71 = 0$	5	$-0\cdot52$
43	$w - \cdot922 + \cdot286 - \cdot679 + 6\cdot739 - 0\cdot86 = 0$	4	$-0\cdot72$
44	$w - \cdot802 + \cdot540 - \cdot716 + 6\cdot692 - 0\cdot11 = 0$	1	$-0\cdot01$
45	$w - \cdot605 + \cdot760 - \cdot795 + 6\cdot643 - 0\cdot76 = 0$	5	$-0\cdot72$
46	$w - \cdot339 + \cdot918 - \cdot846 + 6\cdot591 - 0\cdot85 = 0$	3	$-0\cdot83$
47	$w + \cdot079 + \cdot979 - \cdot868 + 6\cdot519 - 1\cdot09 = 0$	6	$-1\cdot09$
48	$w + \cdot335 + \cdot920 - \cdot820 + 6\cdot474 - 0\cdot17 = 0$	12	$-0\cdot16$
49	$w + \cdot638 + \cdot735 - \cdot751 + 6\cdot413 - 0\cdot15 = 0$	4	$-0\cdot11$
50	$w + \cdot835 + \cdot490 - \cdot652 + 6\cdot361 + 0\cdot54 = 0$	7	$+0\cdot61$
51	$w + \cdot915 + \cdot311 - \cdot612 + 6\cdot329 - 0\cdot17 = 0$	4	$-0\cdot07$
52	$w + \cdot966 + \cdot008 - \cdot574 + 6\cdot277 - 0\cdot27 = 0$	3	$-0\cdot16$
53	$w + \cdot933 - \cdot248 - \cdot591 + 6\cdot232 + 0\cdot32 = 0$	3	$+0\cdot43$
54	$w + \cdot784 - \cdot560 - \cdot633 + 6\cdot178 + 0\cdot43 = 0$	3	$+0\cdot64$
55	$w + \cdot542 - \cdot789 - \cdot711 + 6\cdot123 - 0\cdot13 = 0$	4	$+0\cdot11$
56	$w + \cdot327 - \cdot896 - \cdot755 + 6\cdot083 - 0\cdot08 = 0$	5	$+0\cdot17$
57	$w - \cdot094 - \cdot946 - \cdot758 + 6\cdot015 - 0\cdot31 = 0$	1	$-0\cdot04$
58	$w - \cdot303 - \cdot901 - \cdot726 + 5\cdot981 - 0\cdot22 = 0$	$1\frac{1}{2}$	$+0\cdot06$
59	$w - \cdot548 - \cdot782 - \cdot672 + 5\cdot937 - 0\cdot69 = 0$	2	$-0\cdot42$
60	$w - \cdot892 - \cdot365 - \cdot533 + 5\cdot848 - 0\cdot19 = 0$	6	$-0\cdot03$
61	$w - \cdot940 - \cdot219 - \cdot498 + 5\cdot823 - 0\cdot78 = 0$	2	$-0\cdot55$
62	$w - \cdot966 - \cdot013 - \cdot476 + 5\cdot789 - 0\cdot74 = 0$	4	$-0\cdot54$
63	$w - \cdot937 + \cdot234 - \cdot480 + 5\cdot748 + 0\cdot01 = 0$	1	$+0\cdot18$
64	$w - \cdot470 + \cdot854 - \cdot609 + 5\cdot616 - 0\cdot87 = 0$	6	$-0\cdot81$
65	$w - \cdot220 + \cdot955 - \cdot639 + 5\cdot571 - 0\cdot18 = 0$	3	$-0\cdot14$
66	$w - \cdot042 + \cdot980 - \cdot647 + 5\cdot539 + 0\cdot29 = 0$	4	$+0\cdot32$
67	$w + \cdot337 + \cdot921 - \cdot615 + 5\cdot474 + 0\cdot09 = 0$	3	$+0\cdot12$
68	$w + \cdot485 + \cdot847 - \cdot578 + 5\cdot445 + 0\cdot46 = 0$	2	$+0\cdot51$
69	$w + \cdot702 + \cdot671 - \cdot501 + 5\cdot399 + 0\cdot41 = 0$	5	$+0\cdot48$
70	$w + \cdot950 + \cdot176 - \cdot352 + 5\cdot306 + 0\cdot24 = 0$	5	$+0\cdot39$
71	$w + \cdot956 - \cdot136 - \cdot354 + 5\cdot254 + 1\cdot18 = 0$	3	$+1\cdot37$
72	$w + \cdot871 - \cdot416 - \cdot365 + 5\cdot205 + 0\cdot23 = 0$	4	$+0\cdot46$
73	$w + \cdot709 - \cdot650 - \cdot414 + 5\cdot158 + 0\cdot18 = 0$	4	$+0\cdot37$
74	$w + \cdot462 - \cdot837 - \cdot491 + 5\cdot108 + 0\cdot09 = 0$	2	$+0\cdot37$
75	$w - \cdot084 - \cdot946 - \cdot510 + 5\cdot017 + 0\cdot05 = 0$	3	$+0\cdot35$
76	$w - \cdot348 - \cdot887 - \cdot465 + 4\cdot974 + 0\cdot05 = 0$	2	$+0\cdot35$

No.	Equations.	Weights.	Residuals.
77	$w - 0\cdot869x - 0\cdot418y - 0\cdot267z + 4\cdot8588p + 0\cdot12 = 0$	3	$+0\cdot30$
78	$w - \cdot965 + \cdot042 - \cdot204 + 4\cdot780 + 0\cdot16 = 0$	2	$+ \cdot39$
79	$w - \cdot484 + \cdot845 - \cdot348 + 4\cdot619 - 0\cdot83 = 0$	5	$-0\cdot74$
80	$w - \cdot081 + \cdot978 - \cdot387 + 4\cdot547 - 0\cdot16 = 0$	7	$-0\cdot09$
81	$w + \cdot087 + \cdot978 - \cdot368 + 4\cdot518 + 0\cdot42 = 0$	4	$+0\cdot48$
82	$w + \cdot344 + \cdot918 - \cdot336 + 4\cdot473 + 0\cdot92 = 0$	5	$+0\cdot99$
83	$w + \cdot591 + \cdot775 - \cdot260 + 4\cdot424 + 0\cdot11 = 0$	4	$+0\cdot20$
84	$w + \cdot752 + \cdot613 - \cdot192 + 4\cdot386 + 0\cdot38 = 0$	4	$+0\cdot49$
85	$w + \cdot865 + \cdot433 - \cdot136 + 4\cdot351 - 0\cdot18 = 0$	10	$-0\cdot04$
86	$w + \cdot936 + \cdot239 - \cdot096 + 4\cdot316 + 0\cdot20 = 0$	3	$+0\cdot37$
87	$w + \cdot942 - \cdot215 - \cdot068 + 4\cdot241 - 0\cdot99 = 0$	2	$-0\cdot75$
88	$w + \cdot782 - \cdot562 - \cdot110 + 4\cdot177 + 0\cdot09 = 0$	6	$+0\cdot37$
89	$w + \cdot218 - \cdot927 - \cdot225 + 4\cdot066 + 0\cdot43 = 0$	3	$+0\cdot75$
90	$w - \cdot351 - \cdot885 - \cdot175 + 3\cdot973 + 0\cdot81 = 0$	1	$+1\cdot15$
91	$w - \cdot986 - \cdot004 + \cdot089 + 3\cdot789 - 0\cdot59 = 0$	1	$-0\cdot32$
92	$w - \cdot822 + \cdot510 + \cdot060 + 3\cdot699 + 1\cdot80 = 0$	2	$+1\cdot99$
93	$w - \cdot467 + \cdot856 - \cdot043 + 3\cdot616 - 0\cdot36 = 0$	1	$- 0\cdot23$
94	$w - \cdot259 + \cdot944 - \cdot067 + 3\cdot578 + 0\cdot32 = 0$	1	$+0\cdot42$
95	$w + \cdot131 + \cdot973 - \cdot051 + 3\cdot510 + 0\cdot12 = 0$	2	$+0\cdot22$
96	$w + \cdot356 + \cdot912 - \cdot017 + 3\cdot471 - 0\cdot24 = 0$	1	$-0\cdot13$
97	$w + \cdot518 + \cdot829 + \cdot036 + 3\cdot440 - 0\cdot63 = 0$	5	$-0\cdot51$
98	$w + \cdot690 + \cdot685 + \cdot101 + 3\cdot401 + 0\cdot30 = 0$	5	$+0\cdot44$
99	$w + \cdot873 + \cdot418 + \cdot173 + 3\cdot348 - 0\cdot82 = 0$	5	$-0\cdot64$
100	$w + \cdot961 + \cdot093 + \cdot250 + 3\cdot292 - 1\cdot45 = 0$	4	$- 1\cdot22$
101	$w + \cdot927 - \cdot270 + \cdot240 + 3\cdot231 - 0\cdot08 = 0$	5	$+0\cdot20$
102	$w - \cdot300 - \cdot902 + \cdot124 + 2\cdot982 - 0\cdot83 = 0$	1	$-0\cdot45$
103	$w - \cdot676 - \cdot680 + \cdot243 + 2\cdot911 - 0\cdot14 = 0$	2	$+0\cdot23$
104	$w - \cdot815 - \cdot512 + \cdot308 + 2\cdot875 - 0\cdot92 = 0$	3	$-0\cdot56$
105	$w - \cdot926 - \cdot271 + \cdot365 + 2\cdot832 - 0\cdot34 = 0$	1	$0\cdot00$
106	$w - \cdot961 - \cdot091 + \cdot400 + 2\cdot801 + 0\cdot41 = 0$	1	$+0\cdot73$
107	$w - \cdot566 + \cdot792 + \cdot282 + 2\cdot636 - 0\cdot13 = 0$	1	$+0\cdot05$
108	$w - \cdot391 + \cdot900 + \cdot249 + 2\cdot601 + 0\cdot06 = 0$	2	$+0\cdot22$
109	$w - \cdot271 + \cdot941 + \cdot251 + 2\cdot579 - 0\cdot56 = 0$	4	$-0\cdot41$
110	$w + \cdot060 + \cdot980 + \cdot230 + 2\cdot522 - 0\cdot05 = 0$	7	$+0\cdot08$
111	$w + \cdot345 + \cdot917 + \cdot289 + 2\cdot473 - 0\cdot03 = 0$	11	$+0\cdot11$
112	$w + \cdot580 + \cdot784 + \cdot357 + 2\cdot428 + 0\cdot60 = 0$	6	$+0\cdot76$
113	$w + \cdot748 + \cdot619 + \cdot442 + 2\cdot387 + 0\cdot95 = 0$	5	$+1\cdot14$
114	$w + \cdot824 + \cdot509 + \cdot460 + 2\cdot364 - 0\cdot05 = 0$	4	$+0\cdot16$
115	$w + \cdot966 - \cdot029 + \cdot548 + 2\cdot272 - 0\cdot56 = 0$	7	$-0\cdot27$

No.	Equations.	Weights.	Residuals.
116	$w+0·879x-0·398y+0·535s+2·2088p-0''17=0$	3	$+0''16$
117	$w+ ·354 - ·885 + ·400 +2·088 -0·34=0$	4	$-0·05$
118	$w- ·631 - ·720 + ·497 +1·921 -0·06=0$	3	$+0·34$
119	$w- ·764 - ·581 + ·550 +1·890 -(3·40)=0$	0	$(-3·00)$
120	$w- ·966 + ·016 + ·677 +1·785 -0·95=0$	3	$-0·61$
121	$w- ·492 + ·840 + ·540 +1·621 -0·83=0$	4	$-0·64$
122	$w- ·240 + ·948 + ·516 +1·574 -0·26=0$	4	$-0·08$
123	$w- ·092 + ·976 + ·493 +1·549 -0·03=0$	3	$+0·14$
124	$w+ ·213 + ·859 + ·531 +1·496 -0·85=0$	4	$-0·68$
125	$w+ ·456 + ·865 + ·577 +1·452 -0·08=0$	5	$+0·11$
126	$w+ ·638 + ·717 + ·636 +1·407 -0·34=0$	4	$-0·10$
127	$w+ ·767 + ·594 + ·693 +1·382 +0·03=0$	6	$+0·26$
128	$w+ ·930 + ·261 + ·766 +1·320 +0·66=0$	3	$+0·93$
129	$w+ ·963 - ·071 + ·802 +1·265 -0·10=0$	6	$+0·21$
130	$w+ ·941 - ·219 + ·798 +1·240 -0·21=0$	2	$+0·13$
131	$w+ ·793 - ·547 + ·735 +1·170 -0·26=0$	3	$+0·12$
132	$w+ ·296 - ·906 + ·632 +1·079 -0·09=0$	5	$+0·33$
133	$w- ·402 - ·864 + ·665 +0·964 +1·66=0$	$\frac{1}{2}$	$+2·10$
134	$w- ·861 - ·433 + ·838 +0·861 -0·76=0$	1	$-0·33$
135	$w- ·965 + ·018 + ·904 +0·784 -0·45=0$	1	$-0·08$
136	$w- ·573 + ·786 + ·777 +0·637 +0·07=0$	1	$+0·31$
137	$w- ·366 + ·906 + ·732 +0·597 -1·28=0$	3	$-1·06$
138	$w- ·149 + ·970 + ·710 +0·559 -0·82=0$	4	$-0·62$
139	$w+ ·047 + ·981 + ·707 +0·525 -0·28=0$	5	$-0·04$
140	$w+ ·319 + ·926 + ·734 +0·477 -0·29=0$	6	$-0·09$
141	$w+ ·523 + ·823 + ·778 +0·439 -0·14=0$	5	$+0·08$
142	$w+ ·676 + ·698 + ·838 +0.405 +0·19=0$	4	$+0·42$
143	$w+ ·865 + ·434 + ·907 +0·351 -1·29=0$	3	$-1·02$
144	$w+ ·932 + ·255 + ·941 +0·320 -0·08=0$	6	$+0·22$
145	$w+ ·966 - ·015 + ·968 +0·275 -0·66=0$	9	$-0·31$
146	$w+ ·936 - ·239 + ·960 +0·236 -0·29=0$	7	$+0·07$
147	$w- ·460 - ·835 + ·824 -0·045 -0·59=0$	2	$-0·13$
148	$w- ·870 - ·415 +0·974 -0·142 -1·11=0$	2	$-0·68$
149	$w- ·944 - ·203 +1·010 -0·178 -0·48=0$	4	$-0·08$
150	$w- ·681 + ·690 +0·923 -0·339 -1·61=0$	2	$-1·34$
151	$w- ·450 + ·868 + ·834 -0·386 -1·30=0$	8	$-0·76$
152	$w- ·180 + ·963 + ·802 -0·436 -1·09=0$	1	$-0·88$
153	$w+ ·075 + ·979 + ·807 -0·480 -0·74=0$	5	$-0·54$
154	$w+ ·325 + ·924 + ·838 -0·523 -0·28=0$	5	$-0·07$

No.	Equations	Weights	Residuals
155	$w + 0{\cdot}457x + 0{\cdot}864y + 0{\cdot}851z - 0{\cdot}548\,p - 1{\cdot}33 = 0$	2	$-1{\cdot}11$
156	$w + {\cdot}679 + {\cdot}695 + {\cdot}930 - 0{\cdot}594 \quad -0{\cdot}50 = 0$	3	$-0{\cdot}26$
157	$w + {\cdot}820 + {\cdot}509 + 0{\cdot}980 - 0{\cdot}633 \quad -0{\cdot}35 = 0$	5	$-0{\cdot}08$
158	$w + {\cdot}948 + {\cdot}182 + 1{\cdot}029 - 0{\cdot}693 \quad +0{\cdot}20 = 0$	1	$+0{\cdot}51$
159	$w + {\cdot}962 - {\cdot}096 + 1{\cdot}047 - 0{\cdot}740 \quad +1{\cdot}84 = 0$	⅓	$+2{\cdot}20$
160	$w + {\cdot}905 - {\cdot}336 + 1{\cdot}027 - 0{\cdot}780 \quad +0{\cdot}17 = 0$	3	$+0{\cdot}55$
161	$w + {\cdot}773 - {\cdot}575 + 0{\cdot}974 - 0{\cdot}825 \quad -0{\cdot}96 = 0$	2	$-0{\cdot}55$
162	$w + {\cdot}337 - {\cdot}892 + {\cdot}834 - 0{\cdot}914 \quad -1{\cdot}26 = 0$	1	$-0{\cdot}91$
163	$w - {\cdot}734 - {\cdot}619 + 0{\cdot}938 - 1{\cdot}102 \quad -0{\cdot}26 = 0$	1	$+0{\cdot}19$
164	$w - {\cdot}964 - {\cdot}059 + 1{\cdot}043 - 1{\cdot}204 \quad -0{\cdot}30 = 0$	1	$+0{\cdot}09$
165	$w - {\cdot}332 + {\cdot}920 + 0{\cdot}824 - 1{\cdot}410 \quad +0{\cdot}20 = 0$	2	$+0{\cdot}42$
166	$w - {\cdot}105 + {\cdot}975 + {\cdot}802 - 1{\cdot}449 \quad -0{\cdot}00 = 0$	5	$+0{\cdot}31$
167	$w + {\cdot}099 + {\cdot}977 + {\cdot}791 - 1{\cdot}484 \quad -0{\cdot}86 = 0$	5	$-0{\cdot}66$
168	$w + {\cdot}329 + {\cdot}923 + {\cdot}821 - 1{\cdot}524 \quad -0{\cdot}26 = 0$	6	$-0{\cdot}05$
169	$w + {\cdot}593 + {\cdot}774 + {\cdot}883 - 1{\cdot}577 \quad -0{\cdot}66 = 0$	3	$-0{\cdot}43$
170	$w + {\cdot}761 + {\cdot}601 + {\cdot}935 - 1{\cdot}617 \quad -0{\cdot}54 = 0$	7	$-0{\cdot}29$
171	$w + {\cdot}899 + {\cdot}352 + 0{\cdot}979 + 1{\cdot}663 \quad -0{\cdot}34 = 0$	5	$-0{\cdot}05$
172	$w + {\cdot}965 - {\cdot}030 + 1{\cdot}018 - 1{\cdot}729 \quad -1{\cdot}07 = 0$	6	$-0{\cdot}73$
173	$w + {\cdot}874 - {\cdot}408 + 0{\cdot}976 - 1{\cdot}794 \quad -1{\cdot}89 = 0$	⅓	$-1{\cdot}50$
174	$w + {\cdot}682 - {\cdot}677 + {\cdot}882 - 1{\cdot}848 \quad -1{\cdot}91 = 0$	1	$-1{\cdot}49$
175	$w - {\cdot}720 - {\cdot}635 + {\cdot}875 - 2{\cdot}099 \quad +0{\cdot}63 = 0$	1	$+1{\cdot}07$
176	$w - {\cdot}909 - {\cdot}323 + {\cdot}936 - 2{\cdot}159 \quad -1{\cdot}61 = 0$	1	$-1{\cdot}20$
177	$w - {\cdot}959 - {\cdot}115 + {\cdot}958 - 2{\cdot}194 \quad -0{\cdot}43 = 0$	4	$-0{\cdot}05$
178	$w - {\cdot}490 + {\cdot}842 + {\cdot}755 - 2{\cdot}380 \quad -0{\cdot}35 = 0$	4	$-0{\cdot}12$
179	$w - {\cdot}376 + {\cdot}903 + {\cdot}746 - 2{\cdot}402 \quad +0{\cdot}90 = 0$	5	$+1{\cdot}11$
180	$w - {\cdot}209 + {\cdot}955 + {\cdot}713 - 2{\cdot}431 \quad +0{\cdot}72 = 0$	6	$+0{\cdot}92$
181	$w - {\cdot}029 + {\cdot}980 + {\cdot}690 - 2{\cdot}462 \quad +1{\cdot}40 = 0$	5	$+1{\cdot}59$
182	$w + {\cdot}102 + {\cdot}975 + {\cdot}677 - 2{\cdot}483 \quad +0{\cdot}51 = 0$	7	$+0{\cdot}70$
183	$w + {\cdot}311 + {\cdot}930 + {\cdot}694 - 2{\cdot}521 \quad -0{\cdot}34 = 0$	4	$-0{\cdot}15$
184	$w + {\cdot}406 + {\cdot}888 + {\cdot}710 - 2{\cdot}539 \quad +0{\cdot}35 = 0$	5	$+0{\cdot}54$
185 ?	$w + {\cdot}525 + {\cdot}824 + {\cdot}730 - 2{\cdot}562 \quad -0{\cdot}08 = 0$	5	$+0{\cdot}12$
186	$w + {\cdot}675 + {\cdot}700 + {\cdot}773 - 2{\cdot}595 \quad +0{\cdot}01 = 0$	6	$+0{\cdot}23$
187	$w + {\cdot}934 + {\cdot}250 + {\cdot}866 - 2{\cdot}682 \quad -0{\cdot}31 = 0$	8	$-0{\cdot}02$
188	$w + {\cdot}966 - {\cdot}024 + {\cdot}888 - 2{\cdot}727 \quad +0{\cdot}19 = 0$	3	$+0{\cdot}51$
189	$w + {\cdot}917 - {\cdot}303 + {\cdot}853 - 2{\cdot}775 \quad +0{\cdot}06 = 0$	4	$+0{\cdot}42$
190	$w + {\cdot}763 - {\cdot}588 + {\cdot}855 - 2{\cdot}828 \quad +0{\cdot}02 = 0$	1	$+0{\cdot}42$
191	$w - {\cdot}558 - {\cdot}775 + {\cdot}659 - 3{\cdot}064 \quad +0{\cdot}74 = 0$	2	$+1{\cdot}17$
192	$w - {\cdot}801 - {\cdot}527 + {\cdot}716 - 3{\cdot}119 \quad +0{\cdot}09 = 0$	2	$+0{\cdot}50$
193	$w - {\cdot}961 - {\cdot}100 + {\cdot}772 - 3{\cdot}197 \quad +0{\cdot}61 = 0$	1	$+0{\cdot}97$

No.	Equations.	Weights.	Residuals.
194	$w - 0.938x + 0.231y + 0.768z - 3.2518p + 1.19 = 0$	1	$+ 1.49$
195	$w - .660 + .712 + .619 - 3.343 + 0.49 = 0$	2	$+ 0.72$
196	$w - .367 + .905 + .544 - 3.403 - 0.75 = 0$	3	$- 0.56$
197	$w - .117 + .974 + .501 - 3.445 + 0.31 = 0$	7	$+ 0.48$
198	$w + .117 + .975 + .382 - 3.487 - 1.36 = 0$	3	$- 1.21$
199	$w + .444 + .871 + .516 - 3.547 + 0.31 = 0$	7	$+ 0.48$
200	$w + .653 + .721 + .568 - 3.589 + 0.04 = 0$	4	$+ 0.24$
201	$w + .805 + .540 + .615 - 3.629 - 0.29 = 0$	7	$- 0.07$
202	$w + .892 + .370 + .641 - 3.661 - 1.08 = 0$	4	$- 0.83$
203	$w + .920 - .294 + .631 - 3.773 - 0.39 = 0$	2	$- 0.06$
204	$w + .540 - .791 + .460 - 3.878 + 0.54 = 0$	2	$+ 0.92$
205	$w + .319 - .899 + .395 - 3.918 - (2.70) = 0$	0	$(- 2.30)$
206	$w - .512 - .805 + .458 - 4.056 - 0.25 = 0$	2	$+ .016$
207	$w - .894 - .360 + .514 - 4.153 - 0.16 = 0$	1	$+ 0.20$
208	$w - .961 - .086 + .529 - 4.199 - 0.74 = 0$	1	$- 0.41$
209	$w - .460 + .861 + .290 - 4.386 - 0.04 = 0$	8	$+ 0.13$
210	$w - .204 + .958 + .226 - 4.433 - 1.01 = 0$	4	$- 0.87$
211	$w - .007 + .982 + .206 - 4.466 + 0.09 = 0$	8	$+ 0.22$
212	$w + .218 + .956 + .209 - 4.501 - 0.35 = 0$	4	$- 0.22$
213	$w + .466 + .858 + .237 - 4.550 - 0.31 = 0$	13	$- 0.16$
214	$w + .759 + .607 + .291 - 4.616 - 0.40 = 0$	9	$- 0.23$
215	$w + .909 + .328 + .366 - 4.668 - 0.24 = 0$	5	$- 0.02$
216	$w + .962 - .082 + .364 - 4.737 - 0.40 = 0$	10	$- 0.13$
217	$w + .913 - .313 + .332 - 4.777 - 0.38 = 0$	2	$- 0.08$
218	$w + .731 - .626 + .211 - 4.837 - 0.65 = 0$	2	$- 0.32$
219	$w + .604 - .746 + .229 - 4.865 - 0.69 = 0$	1	$- 0.34$
220	$w - .832 - .484 + .194 - 5.130 - 1.45 = 0$	½	$- 1.11$
221	$w - .413 + .883 - .013 - 5.393 - 0.87 = 0$	2	$- 0.74$
222	$w - .216 + .955 - .061 - 5.428 - 0.01 = 0$	1	$+ 0.10$
223	$w + .260 + .946 - .098 - 5.513 - 0.34 = 0$	4	$- 0.24$
224	$w + .430 + .878 - .067 - 5.544 - 0.85 = 0$	3	$- 0.74$
225	$w + .617 + .754 - .033 - 5.581 - 0.41 = 0$	4	$- 0.29$
226	$w + .739 + .630 + .009 - 5.611 - 0.07 = 0$	8	$+ 0.07$
227	$w - .476 - .827 - .217 - 6.049 - 0.61 = 0$	2	$- 0.28$
228	$w - .794 - .543 - .142 - 6.119 - 1.05 = 0$	1	$- 0.74$
229	$w + .786 + .567 - .280 - 6.624 - 0.86 = 0$	1	$- 0.71$
230	$w + .233 - .923 - .521 - 6.932 - 0.87 = 0$	1	$- 0.58$
231	$w - .128 - .942 - .537 - 6.991 + 0.41 = 0$	3	$+ 0.70$
232	$w - .613 - .755 - .471 - 7.075 - 0.99 = 0$	2	$- 0.70$

No.	Equations.	Weights.	Residual.
233	$w - 0{\cdot}505x + 0{\cdot}833y - 0{\cdot}578z - 7{\cdot}3778p - 0{\cdot}''56 = 0$	3	$-0{\cdot}''49$
234	$w + {\cdot}022 + {\cdot}980 - {\cdot}670 - 7{\cdot}471 \quad -0{\cdot}32 = 0$	3	$-0{\cdot}30$
235	$w + {\cdot}190 + {\cdot}963 - {\cdot}695 - 7{\cdot}500 \quad -1{\cdot}85 = 0$	1	$-1{\cdot}83$
236	$w + {\cdot}511 + {\cdot}832 - {\cdot}621 - 7{\cdot}559 \quad -0{\cdot}61 = 0$	8	$-0{\cdot}57$
237	$w + {\cdot}675 + {\cdot}699 - {\cdot}577 - 7{\cdot}595 \quad -1{\cdot}41 = 0$	8	$-1{\cdot}35$
238	$w + {\cdot}814 + {\cdot}526 - {\cdot}534 - 7{\cdot}631 \quad -0{\cdot}17 = 0$	8	$-0{\cdot}08$
239	$w + {\cdot}934 + {\cdot}247 - {\cdot}508 - 7{\cdot}682 \quad +0{\cdot}62 = 0$	3	$+0{\cdot}74$
240	$w + {\cdot}964 - {\cdot}048 - {\cdot}498 - 7{\cdot}731 \quad -0{\cdot}28 = 0$	5	$-0{\cdot}12$
241	$w + {\cdot}871 - {\cdot}416 - {\cdot}558 - 7{\cdot}796 \quad +0{\cdot}14 = 0$	1	$+0{\cdot}34$
242	$w + {\cdot}559 - {\cdot}777 - {\cdot}691 - 7{\cdot}874 \quad -0{\cdot}22 = 0$	1	$+0{\cdot}02$
243	$w + {\cdot}415 - {\cdot}861 - {\cdot}754 - 7{\cdot}902 \quad -2{\cdot}27 = 0$	1	$-2{\cdot}04$
244	$w - {\cdot}441 + {\cdot}870 - {\cdot}818 - 8{\cdot}390 \quad +0{\cdot}65 = 0$	5	$+0{\cdot}67$
245	$w - {\cdot}188 + {\cdot}962 - {\cdot}871 - 8{\cdot}435 \quad +1{\cdot}04 = 0$	9	$+1{\cdot}05$
246	$w + {\cdot}060 + {\cdot}980 - {\cdot}889 - 8{\cdot}478 \quad +1{\cdot}46 = 0$	2	$+1{\cdot}46$
247	$w + {\cdot}231 + {\cdot}954 - {\cdot}882 - 8{\cdot}507 \quad +0{\cdot}54 = 0$	6	$+0{\cdot}54$
248	$w + {\cdot}431 + {\cdot}878 - {\cdot}852 - 8{\cdot}546 \quad +0{\cdot}65 = 0$	8	$+0{\cdot}66$
249	$w + {\cdot}592 + {\cdot}775 - {\cdot}820 - 8{\cdot}576 \quad +0{\cdot}14 = 0$	3	$+0{\cdot}16$
250	$w + {\cdot}721 + {\cdot}650 - {\cdot}763 - 8{\cdot}607 \quad +0{\cdot}71 = 0$	4	$+0{\cdot}75$
251	$w + {\cdot}818 + {\cdot}520 - {\cdot}734 - 8{\cdot}634 \quad +0{\cdot}04 = 0$	4	$+0{\cdot}10$
252	$w + {\cdot}905 + {\cdot}338 - {\cdot}704 - 8{\cdot}667 \quad +0{\cdot}00 = 0$	4	$+0{\cdot}09$
253	$w + {\cdot}957 + {\cdot}134 - {\cdot}689 - 8{\cdot}701 \quad +0{\cdot}70 = 0$	3	$+0{\cdot}81$
254	$w + {\cdot}964 - {\cdot}063 - {\cdot}692 - 8{\cdot}735 \quad -0{\cdot}43 = 0$	2	$-0{\cdot}28$
255	$w + {\cdot}916 - {\cdot}306 - {\cdot}719 - 8{\cdot}776 \quad -1{\cdot}13 = 0$	2	$-0{\cdot}96$
256	$w + {\cdot}815 - {\cdot}514 - {\cdot}766 - 8{\cdot}813 \quad +0{\cdot}49 = 0$	1½	$+0{\cdot}68$
257	$w + {\cdot}491 - {\cdot}821 - {\cdot}886 - 8{\cdot}887 \quad -0{\cdot}55 = 0$	1	$-0{\cdot}32$
258	$w - {\cdot}441 + {\cdot}869 - {\cdot}936 - 9{\cdot}389 \quad -0{\cdot}25 = 0$	3	$-0{\cdot}23$
259	$w - {\cdot}301 + {\cdot}930 - {\cdot}978 - 9{\cdot}415 \quad +0{\cdot}30 = 0$	2	$+0{\cdot}30$
260	$w - {\cdot}082 + {\cdot}978 - 0{\cdot}999 - 9{\cdot}453 \quad +0{\cdot}80 = 0$	5	$+0{\cdot}79$
261	$w + {\cdot}056 + {\cdot}979 - 1{\cdot}010 - 9{\cdot}480 \quad -0{\cdot}98 = 0$	3	$-0{\cdot}99$
262	$w + {\cdot}273 + {\cdot}942 - 0{\cdot}996 - 9{\cdot}514 \quad -1{\cdot}65 = 0$	5	$-1{\cdot}66$
263	$w + {\cdot}599 + {\cdot}769 - {\cdot}913 - 9{\cdot}577 \quad -1{\cdot}45 = 0$	5	$-1{\cdot}44$
264	$w + {\cdot}828 + {\cdot}503 - {\cdot}847 - 9{\cdot}636 \quad -1{\cdot}89 = 0$	5	$-1{\cdot}84$
265	$w + {\cdot}922 + {\cdot}287 - {\cdot}804 - 9{\cdot}675 \quad -0{\cdot}91 = 0$	6	$-0{\cdot}83$
266	$w + {\cdot}964 + {\cdot}065 - {\cdot}786 - 9{\cdot}712 \quad -1{\cdot}33 = 0$	5	$-1{\cdot}22$

Out of the entire number, three observations have been rejected, viz. those made on 1864, February 9, 1866, February 13, and 1869, December · These give very discordant results, and in the first two instances the readings of one of the micrometers have been altered, which throws a certain amount of suspicion on the observations. In the third case the bisection of the star's image with the micrometer wire was made when the star was very near the edge of the field. The rejection of these observations reduces the effective number of the equations of condition to 264.

As a general rule, the weight assigned to an equation of condition is the same as the number of observations contained in the group of observed Z. D.'s corresponding to the equation; but in some cases half weight has been given to an observation when it was made under very unfavourable circumstances.

Applying the method of least squares to the 264 equations of condition, I find the normal equations to be

$$+ \quad 1038w + 264\cdot445x + 419\cdot123y - \quad 93\cdot978z + \quad 902\cdot6068p - 172\cdot81 = 0$$
$$+ 264\cdot445 \quad + 446\cdot723 \quad + \quad 98\cdot288 \quad + \quad 31\cdot426 \quad - \quad 448\cdot785 \quad - \quad 43\cdot39 = 0$$
$$+ 419\cdot123 \quad + \quad 98\cdot288 \quad + 526\cdot628 \quad - \quad 28\cdot858 \quad - \quad 275\cdot246 \quad - \quad 24\cdot18 = 0$$
$$- \quad 93\cdot978 \quad + \quad 31\cdot426 \quad - \quad 28\cdot858 \quad + 512\cdot619 \quad - 1310\cdot491 \quad - \quad 42\cdot84 = 0$$

Solving these, I find the following values of the unknown quantities :—

$$w = +0\overset{''}{\cdot}236 - 0\cdot8878p, \quad \text{Weight (putting } \delta p = 0\text{) } 608\cdot580$$
$$x = -0\cdot022 + 1\cdot448 \qquad \text{,,} \qquad \text{,,} \qquad 373\cdot019$$
$$y = -0\cdot131 - 0\cdot358 \qquad \text{,,} \qquad 356\cdot990$$
$$z = +0\cdot121 + 2\cdot285 \qquad \text{,,} \qquad 495\cdot730$$

From these values we have

Mean Z. D. North of γ *Draconis* for 1866·0 =	$102\overset{''}{\cdot}606$	$-0\cdot887$ δp
Constant of Aberration	= $20\cdot378$	$+1\cdot448$
Annual Parallax of γ *Draconis* ...	= $-0\cdot131$	$-0\cdot358$
Constant of Nutation for 1866 ...	= $9\cdot3353$	$+2\cdot1076$

The residuals given above have been formed by assuming δp=0, and from them I find sum of

$$\text{Weight} \times \text{Square of residual} = 343\cdot917$$

(the sum of weight × square of absolute term is 391·145), and therefore the probable error of an observation of weight unity is ±0 ˊˊ776, and combining this with the weights of w, x, y and z given above we have

Probable error of	$w = \pm 0\overset{''}{\cdot}031$
,,	$x = \pm 0\cdot040$
,,	$y = \pm 0\cdot041$
,,	$z = \pm 0\cdot035$

Therefore, finally, assuming $\delta p = 0$, we have

Mean Z. D. North of γ *Draconis*, 1866·0 = $102''·606 \pm 0''·03I$

Constant of Aberration   ...   ...   = $20·378 \pm 0·040$

Annual Parallax of γ *Draconis*   ...   = $- 0·13I \pm 0·04I$

Constant of Nutation for 1866   ...   = $9·3353 \pm 0·0323$

It will be remarked that the parallax of γ *Draconis* again comes out a negative quantity, and as it is three times greater than its probable error we cannot regard it as accidental. It is, however, somewhat smaller than the quantity found by Main, whilst at the same time the constant of aberration is larger, so that the results of the present discussion are rather more satisfactory. It must, however, be assumed that there is a periodic error of some kind which affects the determination of annual parallax (perhaps arising from the effect of temperature on the instrument), and which therefore tends to throw a certain amount of suspicion on the value of the constant of aberration here deduced.

The above value of the constant of nutation agrees closely with that found by Main from the observations of γ *Draconis*, made with the 25-foot Zenith Tube.

As mentioned above, the probable error of an observation is $\pm 0''·776$. The corresponding quantity for the Transit Circle is, in the case of a star near the zenith, $\pm 0''·47$. The larger quantity found for the Reflex Zenith Tube may partly be accounted for by the difficulty of obtaining a perfectly satisfactory image by reflexion from mercury; and partly by the remarkable circumstance that, in this latitude, the sky is seldom quite dark when a star of 18^h Right Ascension is passing the meridian. It may also be remarked that considerable difficulty is experienced in observing γ *Draconis* with the Reflex Zenith Tube during the daytime, and that really comparatively few observations are made when the aberration has its greatest effect—*i.e.* when the star passes the meridian at six o'clock (mean time) in the morning or evening, so that the series of observations here discussed is not so favourable for a determination of the constant of aberration as might *à priori*.be imagined.

Royal Observatory, Greenwich :
    1882, *May* 11.

Remarks on an Apparatus for determining those Errors of astronomical Observations which are caused by the Flexures of an Instrument and by Defects in the Shape of its Pivots. By A. Marth, Esq.

At the meeting of last December, in the course of the discussion on Mr. Stone's paper on a new form of Transit-Circle, I stated that in the spring of 1862 I had published * a proposal of a new method for determining the flexure- and pivot-errors, which was specially applicable in the case of an instrument with a prism in the centre—a case to which Mr. Stone had alluded. I mentioned that in 1869 an apparatus such as I proposed had been constructed by Messrs. T. Cooke & Sons, of York, for a Transit-Circle with telescope of the ordinary form, and that the instrument was still at their works. I further remarked that, in December 1878, there had appeared in the "Comptes Rendus" of the Paris Academy a paper by M. Loewy, in which he raised some objections to my proposal which had been a great puzzle to me; that, while in his own proposal M. Loewy had adopted the principle of my method, he had made alterations in its application about which astronomers would form their own judgment by-and-by. I mentioned also that Messrs. Cooke would probably bring or send up their apparatus for inspection at one of our meetings, so that those who take an interest in the matter—and especially Mr. Stone—might criticise it and compare it with M. Loewy's, of which a description had lately appeared.

The apparatus is to be shown at the May meeting, and I will, therefore, now offer some explanatory remarks. But before doing so, it will be desirable or even essential, on account of some extraordinary assertions which M. Loewy makes in the April number of the *Monthly Notices*, that I should first submit to the critical consideration of readers a preliminary question referring to elementary geometrical optics.

If a distant object is looked at through a refracting telescope, how is the place of the object's image in the field of view affected by some slight shifting of the object-glass?

The question must have presented itself to every thinking observer who uses an astronomical instrument intelligently, so that there should be no difficulty in getting an answer, even if the text-books are silent about it. I have not been aware that there was or could be a doubt about the correctness of the answer, that the effect of the shifting depends entirely upon the shifting of the *optical centre* of the object-glass (if by optical centre is understood that point where the straight lines cross which join the points of an infinitely distant object with the

---

* *Astr. Nachrichten*, No. 1361 in Vol. LIII. "Vorschlag eines neuen Verfahrens, die von der Biegung eines Instruments und von den Unregelmässigkeiten seiner Zapfen erzeugten astronomischen Beobachtungsfehler zu bestimmen."

corresponding points of its image), and that any slight shifting of the object-glass round its optical centre does *not* affect the line of vision or the direction in which the object appears. I cannot find, at present, any flaw in the reasoning which leads to this conclusion. But if the answer is really wrong, I shall be much obliged to any reader who will enlighten me (and probably others) about what is the truth in the matter, and I shall *then* most readily acknowledge that I have been all along under a delusion respecting an elementary question of practical optics.

The exact position of the optical centre and other cardinal points in any system of lenses on a common axis may be determined either from the optical elements of the system, in case these are known, or by means of special experiments, the theory having been cleared up by the dioptrical investigations of Gauss * and Bessel † published nearly simultaneously at the beginning of 1841.

As some readers may be interested to get distinct notions of the actual forms of two celebrated object-glasses, of which the optical elements are known, they will perhaps avail themselves of the following data for the graphical representation of sections of these glasses passing through the axis of the lenses. The x abscissæ, expressed in English inches, are reckoned along the axis from the centre of the outer surface of the crown lens, the y ordinates perpendicular to the axis.

Object-glass of the Koenigsberg Heliometer.

The elements of its construction are given on p. 101 of Bessel's "Astr. Untersuchungen."

	$y = 0$	$\pm 1\cdot07$	$\pm 2\cdot13$	$\pm 3\cdot20$
Crown lens	$x = 0\cdot0$	$\cdot008$	$\cdot031$	$\cdot069$
	$= 0\cdot234$	First principal point.		
	$= 0\cdot533$	$\cdot514$	$\cdot456$	$\cdot360$
Flint lens	$= 0\cdot533$	$\cdot514$	$\cdot458$	$\cdot363$
	$= 0\cdot556$	Second principal point = optical centre.		
	$= 0\cdot888$	$\cdot883$	$\cdot866$	$\cdot839$

Focal length $100^{in}\cdot50$, effective aperture $6^{in}\cdot23$.

Object-glass of the great Washington Equatoreal.

According to Prof. Holden's "Investigation of the Objective and Micrometers of the Twenty-six Inch Equatoreal" (Appendix

---

* Gauss, "Dioptrische Untersuchungen," *Werke*, Bd. V. The optical centre in the sense of the term as used above, is identical with the point called by Gauss "zweiter Hauptpunkt."

† Bessel, "Ueber die Grundformeln der Dioptrik," *Astr. Nachr.* No. 415. Cf. in the *Briefwechsel zwischen Gauss und Bessel* the letters No. 179 and 180, exchanged in January 1841. The further investigations of Bessel must be studied in the "Besondere Untersuchung des Heliometers . . . . ," the second paper of his *Astron. Untersuchungen.*

I. of Wash. Obs. for 1877), the elements of its construction are approximately (using the notation employed by Bessel)—

$$
\begin{aligned}
&\text{in.}\\
r &= +\quad 161\text{·}39 \\
d &= \qquad 1\text{·}884 \\
\rho &= -\quad 161\text{·}39 \\
\end{aligned}
\left.\right\} n = 1\text{·}5172
$$

$$
e = \qquad 0\text{·}029
$$

$$
\begin{aligned}
r' &= -\quad 162\text{·}07 \\
d' &= \qquad 0\text{·}958 \\
\rho' &= -19466\text{·}
\end{aligned}
\left.\right\} n' = 1\text{·}6280
$$

Hence the coordinates for graphical representation of a section of the object-glass:

$y=$	0	±3	±6	±9	±12	±13	±13·6
$x=$	−0·344	First principal point.					
Crown lens	= 0·0	0·028	0·112	0·251	0·447	0·524	0·574
	= 0·669	Second principal point = optical centre.					
	= 1·884	1·856	1·772	1·633	1·437	1·360	1·310
Flint lens	= 1·913	1·885	1·802	1·663	1·468	1·391	1·341
	= 2·871	2·871	2·870	2·869	2·867	2·867	2·866

The focal length, deduced from the assumed elements by means of the catenary fraction employed by Bessel, is 389·12 inches; Prof. Holden, using a modification of Gauss's formulæ, finds it = 389·66 inches.

While the optical centre of the object-glass of the Koenigsberg Heliometer is a little within the flint lens very near to the point where the two lenses touch, that of the great Washington object-glass is within the crown lens, at about one-third of its thickness from the outer surface.

With the exception of a few cases, the elements of the construction of the object-glasses employed are not known to observers. The exact places of the optical centres cannot, on that account, be computed, and their practical determination is only feasible where the means are available for making proper experiments for the purpose. For ordinary observations a knowledge of the position of the optical centre may, indeed, be dispensed with; it may be sufficient to know that there is such a point through which the line of vision passes. In applying my method for determining flexure, in which a marked point on the object-glass is observed, it will be desirable to know the place of the optical centre at least approximately, so that the effect of any shifting of the object-glass may be duly estimated.

I will now proceed to explain how my old proposal of an apparatus for determining the astronomical errors which affect the line of vision has been carried out in Messrs. Cooke's apparatus which is to be shown at the meeting. The Transit-Circle to which the apparatus belongs has an object-glass of 7 inches aperture and 88 inches focal length. The pivots of the axis have apertures of $2\frac{1}{4}$ inches diameter in order to allow the determination of the errors arising from flexure of the axis and shape of the pivots. The axis of the instrument is cast in one piece; the middle part is 13 inches across and has in the middle of its two exposed sides apertures in the form of squares of 4 inches, with rounded-off corners. Usually these apertures are closed by covers, which are held in their places by press-clamps attached to the instrument. When the flexure-apparatus is to be used, these covers (which together have the same weight as the apparatus, about 14 lbs.) are removed and the apparatus is inserted so that its stout end-pieces or end-plates fill the apertures, and are held there so that the clamped parts are flush with the sides of the instrument. The stout end-plates fit very closely, but some small portions of the metal are to be taken off on two sides of the apparatus in order to allow the insertion of wedges so as to guard against any possibility of shake. The connection of the apparatus with the central part of the instrument is very strong and yet very simple; in planning this connection it was made a condition that the apparatus should be fit to be placed in the right position without troublesome preparations, so that it might be ready for use at any spare time in the regular course of observations. The middle part of the apparatus contains a plane-parallel glass between two achromatic object-glasses of $2\frac{1}{2}$ inches aperture and of nearly half the focal length of the chief object-glass. One half of the plane glass is to be silvered on both sides, so that the silvered half may serve as a mirror. The silvered portion may be the middle or the outer part; or a half-circle or the whole may be silvered, according to the observer's judgment. The frame which holds the mirror may be taken out and inserted in the reversed position; of course, proper care must be taken to fasten it so as to guard against any possibility of shake. The two small object-glasses must be so placed that the mark on the great object-glass is in the focal plane of the one, while the focus of the other is in the plane of the webs at the eye-end. In order to enable the observer to effect exact focussing, the cells of the small object-glasses can be slowly moved in the direction of their axes by being turned round with the help of wheel-work, the last wheels of which are outside the end-plates of the apparatus, within the handles, where they may be turned from the eye-end. After focussing, the cells may be further fastened, if need be ; but, as an inspection of the apparatus will show, Messrs. Cooke's contrivance is so well executed that the precaution will probably be found superfluous.

When the apparatus and the illumination are got ready, there are to be seen in the field of view: the webs, their image reflected by the mirror, and the image of the cross-lines by which the chief object-glass is marked. The relative positions of these images (in the direction of the transit-observations as well as of the circle-observations) are then to be measured in different directions of the telescope. The measured changes in the relative positions of the images are the effect of the flexure of the telescope, yet mixed up with some possible changes in the position of the mirror relative to the parts of the instrument where the apparatus is fastened, and also with some possible changes in the relative position of the optical centres of the two small object-glasses. Moreover, the line of vision is thus referred to the actual axis of rotation of the apparatus, the direction of which is liable to change on account of flexure of the axis of the instrument, and of possible irregularities in the pivots and their supports. For the determination of the united effect of these latter sources of error, the flexure apparatus carries at the side facing one of the pivots a plane mirror. Further, on an entirely independent support, outside of the instrument, an auxiliary telescope with micrometer, &c., is to be firmly mounted, so that it points through the aperture in the pivot of the Transit-Circle to the side-mirror of the flexure-apparatus. By bringing the (horizontal and vertical) webs of the auxiliary micrometer to coincidence with their image reflected from the side mirror, and making the observation for many readings of the Transit-Circle, the coordinates of the curve become known which the normal to the mirror describes on the sphere, and from them the effect of the flexure of the axis and of the pivot-errors upon the line of vision of the Transit-Circle telescope is found.

In order to examine and to eliminate the effects of any possible changes in the position of the central mirror, &c., and of other possible sources of error, the flexure-apparatus is to be reversed as well in regard to the axis of the telescope as in regard to the axis of rotation, and the observations are to be made in the four positions of the apparatus. I will not dilate, at present, further on this question. As my remarks are intended for critical readers who are capable of forming a judgment of their own, and who are sufficiently interested in the subject not to mind the trouble of looking over my old paper of 1862, I should be obliged for their critical examination of what I have stated there on this point, so that we may get at the truth. They will perhaps also form their own judgment on the curious objections and assertions which M. Loewy makes in his paper in the last number of the *Monthly Notices*, and on which I shall have something to say in the next number.

At the December meeting I stated that my old proposal was especially applicable in the case of an instrument with a prism in the centre, as, in fact, I had mentioned long ago in the introductory remarks of my old paper. The advantages of such an instrument are obvious : one of the most important, where the highest attainable accuracy is required, is the absence of a very serious source of error—namely, the presence of the observer's body underneath or close to the instrument. The question whether it is preferable to break the line of vision by a prism or by a piece of silvered glass will be settled by considerations of the available facilities for renewing the silvered surface. If a prism is preferred, I may perhaps mention an old suggestion of mine to make use of an elongated prism, of which only the middle part need be brought to perfection for reflection, while the ends are to serve for the firmer mounting of the prism. But I will not enter here into considerations of the proper construction of the middle part of the axis of the instrument for securing strength and the firm mounting of the prism and of the flexure-apparatus. I will merely indicate the simplified conditions which the latter has to fulfil. The side-mirror, of course, disappears. As the apparatus is mounted near the prism at the side of the eye-end, the lenses and the mirror turn, with the telescope, round their own axes, and reversion with regard to the line of vision becomes superfluous. The small object-glasses, the apertures of which ought not to be less than the side of the prism, are now of unequal focal length, since, for the sake of using a prism of moderate size, the distance to the webs in the eye-end will probably be made considerably shorter than that to the mark on the great object-glass. The flexure-apparatus will perhaps be best so arranged that both the small object-glasses may be taken out and the (fully) silvered mirror put in place of that farthest from the eye-end. When the eye-end micrometer is removed and the auxiliary telescope employed, the curve described by the normal to the plane of the mirror is referred to a fixed direction. If, then, the small object-glass nearest the eye-end is inserted, and the eye-end micrometer reinserted and adjusted, the observed coincidences of the webs with their reflected image will refer the positions of the webs to the normal of the mirror. If, further, the mirror is replaced by the other object-glass, the changes of the image of the mark in reference to a zero point of the net of webs will show the effect of the flexure of the telescope-tube and of the bending of the prism, this effect multiplied by the proportion of the focal lengths of the two object-glasses and added to the changes of the zero point. These experiments may then be varied, &c. But the hints here given will perhaps be sufficient for thinking readers who take an interest in the problem, to induce them to consider it in all its bearings, and I must leave it to those astronomers who have means and opportunities to get ideas carried out in practice, whether they will avail themselves of these suggestions.

On the Spectrum of Comet a, 1882 (*Wells*), *observed at the Royal Observatory, Greenwich.*

(*Communicated by the Astronomer Royal.*)

The spectrum of this object was first examined on the evening of April 22, with the Single-prism spectroscope. No bright bands could be detected, but a continuous spectrum was traced from about λ 5900 to λ 4900. This spectrum was very fairly bright from the star-like nucleus, and a faint spectrum could also be perceived some little distance down the tail.

The comet was observed again with the same instrument on April 24, when the continuous spectrum was traced from about λ 6000 to λ 4400. This spectrum was not equally bright throughout, there being two ill-defined maxima in the green and greenish-blue, but no bright bands could be made out.

On May 11 the "Half-prism" spectroscope was employed, with one half-prism, reversed as in the observations of prominences, so as to give great purity of spectrum. No bright bands could be seen, nor any irregularities of brightness or break of continuity detected in the continuous spectrum which was traced from about λ 6000 to about λ 4700. The continuous spectrum seemed remarkably bright from the nucleus, considering the comparative faintness of the comet, and the high dispersion used, viz. 5° from A to H, the magnifying power being 28. The observations were made by Mr. Maunder throughout.

Royal Observatory, Greenwich:
1882, *May* 12.

Elements of the Orbit of Comet Wells. By A. Graham, Esq.

(*Communicated by Professor Adams.*)

From the following observations of the Comet Wells, made at this Observatory with the Northumberland Equatoreal and square-bar Micrometer, under favourable conditions, after carefully correcting for aberration and parallax, I have obtained parabolic elements of the orbit, which may be of some value in tracing its course until better elements are forthcoming.

Observed places:—

G.M.T.	R.A.	N. Declination.
	h m s	° ′ ″
1882, April 5·42729	18 26 49·06	44 0 58·3
14·42002	18 51 32·80	51 32 15·1
22·42595	19 24 55·81	59 21 17·6
·45297	25 4·09	22 59·9

Elements of crbit :—

$$_{,}T = 1882, \text{ June } 10\cdot58295, \text{ G.M.T.}$$

$$\left.\begin{array}{l} \pi = 53° \; 53' \; 48''2 \\ \Omega = 205 \quad 2 \quad 0\cdot8 \end{array}\right\} \text{Mean Equinox 1882·0.}$$

$$i = 73 \; 52 \; 54\cdot6$$

$$\log q = 8\cdot788203. \quad \text{Motion direct.}$$

Comparison with middle place (C—O) :—

$$\Delta L \cos l = 3''4 \qquad \Delta l = -0''3.$$

Heliocentric equatoreal coordinates, referred to apparent equinox and obliquity, April 1 :—

$$x = [9\cdot9607526] \; r \sin (126° \; 15' \; 11''2 + v)$$

$$y = [9\cdot8607709] \; r \sin (61 \; 12 \; 38\cdot4 + v)$$

$$z = [9\cdot9026397] \; r \sin (196 \; 41 \; 38\cdot6 + v).$$

where r = radius vector, and v = true anomaly of the comet.

Satisfactory comparisons with the star numbered 6093 in the Radcliffe Catalogue for 1845 were obtained on May 10. These give for the correction of the Ephemeris deduced from the above elements

$$\Delta \alpha \; . \; \cos \delta = +4^{s} \cdot 36 ; \qquad \Delta \delta = -1'' \cdot 2.$$

Cambridge Observatory:
   1882, May 11.

An Examination of the Roorkee Observations of the Transit of Venus, 1874, *December* 8. By Col. J. F. Tennant, R.E., F.R.S.

It was proposed by Sir George Airy that the observations of this transit should be reduced, so that the results of observations should be shown in the form of equations of condition, their further discussion being left to be taken up by some person in whom there should be general confidence after the whole had been so reduced. In accordance with these views I was instructed by the Government of India to reduce my observations in this manner, and the results of these were published in 1877 in a Report which has, I believe, been largely distributed through the Royal Observatory.

Since that time a Parliamentary Paper was issued in which the contact observations made by the observers whose expenses were defrayed from the Parliamentary Grants were compared and a value of the solar parallax deduced. Colonel Tupman then, in a Paper published in the *Monthly Notices*, vol. xxxviii., discussed these, with the addition of a number of observations by British observers, among which were those I had made at

Roorkee. As a result, he ascribed a considerable error to my observed Contact at Ingress, which he considered was late observed, and rejected it in his deduction of parallax, on which, indeed, it would have produced little effect. Disagreeing in his conclusion, I recently wrote to him, pointing out that the observation did not stand alone, and that the micrometer observations confirmed its accuracy. Colonel Tupman admits this, but I think that an examination of my results may be interesting to the Society, especially as I saw what I believed to be geometric contacts free from all halo, black drop, or other peculiarity. I am aware of no details of other micrometer observations than my own, and I much regret that they have not been reduced and published, as the discussion of them would, I feel sure, have led to confidence in them.

Comparison of Cusp Distances with Observed Contacts.

At page 40 of my Report the following values are given of the time of contact as deduced from the cusp measures, the third being corrected by half a minute :—

$$
\begin{array}{ll}
\text{h\ m\ s} & \\
12\ 42\ 44.6 - 22.461\ \Delta\sigma + 0.011\ \Delta S \\
42.0 - 14.215 & +0.009 \\
40.0 - \ 9.738 & +0.006 \\
39.3 - \ 3.643 & +0.003
\end{array}
$$

It will not much matter what value is assigned to ΔS, the correction to the solar semi-diameter. I will assume it as $-1''.85$, as found at Greenwich, and then I find the most probable value of $\Delta\sigma$ is $0''.29$, and of the time of contact, T, it is $12^h\ 42^m\ 37^s.7$. The several observations give $38^s.2$, $37^s.7$, $37^s.1$, $38^s.1$, as the values of the seconds for contact. 1 have weighted the observations equally in this deduction, and it is evident that it cannot matter how the weights are assigned to the equations in determining $\Delta\sigma$. If the above value of $\Delta\sigma$ be substituted in the weighted mean value of the time at page 41, the result will be $38^s.1$ for the time of contact.

It is evident that the values of the contact time from the cusp measures at Egress (p. 42) will not admit of this treatment. They are

$$
T_2 = \left\{
\begin{array}{ll}
\text{h\ m\ s} & \\
16\ 28\ 28.7 + \ 9.561\ \Delta\sigma - 0.006\ \Delta S \\
34.9 + 12.722 & -0.008 \\
26.8 + 18.573 & -0.008
\end{array}
\right.
$$

If, however, we substitute the values of ΔS and $\Delta\sigma$ first used in them, we have for the seconds of contact the three values $31^s.6$, $38^s.7$, $32^s.3$, whose arithmetical mean is 34.1. With these assumptions regarding the diameters the weighted mean equation (p. 43) gives for contact $16^h\ 28^m\ 34^s.0$.

The observed contact at Ingress was at $12^h 42^m 30^s$, and at Egress at $16^h 28^m 40^s$, the former being eight seconds earlier, while the latter is six seconds later, than the values from cusps. These intervals are such as would be due to a larger semi-diameter of the Sun for visible contacts than for those deduced from the cusps of about a quarter of a second of space.

Comparison of Observed Contacts with Measures of the Distance of the Planet from the Limb.

In making these comparisons it is necessary to know the error of the tabular places of the planet and Sun relatively to each other, and also that of the assumed parallax. I assume here that the mean solar parallax is $8''\cdot73$, so that the value of $\Delta\pi$ in my equations is $0''\cdot15$. Also the longitude of Madras having been determined by the officers of the Great Trigonometrical Survey of India, by the telegraphic method, I shall use their corrections to the assumed longitude of my Report, thus : $\Delta L = +2^s\cdot15$.

On the last page are given the following values for the correction of the relative places of the Sun and *Venus*, in which I introduce the above values—

$$\lceil\Delta A.R. = +6\overset{''}{\cdot}502 - 0\cdot071\,\Delta L - 0\cdot989\cdot\,\Delta\pi = +6\overset{''}{\cdot}50,$$

$$\Delta N.P.D. = -2\cdot247 - 0\cdot017 \quad -2\cdot629 \quad = -1\cdot89.$$

These have been deduced from Colonel Campbell's observations with an Altazimuth. I also learn from Colonel Tupman that a discussion of the measures of the American photographs has given the following values :—

$$\Delta R.A. = +4\overset{''}{\cdot}83 \text{ ; and } \Delta N.P.D. = -2\overset{''}{\cdot}89.$$

From the equations of condition in my Report, I get, after substituting the values of ΔL and $\Delta\pi$,

At Ingress $\begin{cases} \text{From contact} \quad \Delta S - \Delta\sigma \quad = -7\overset{''}{\cdot}56 + 0\cdot6418\,\Delta A.R. \\ \qquad\qquad\qquad\qquad\qquad -0\cdot7183\,\Delta N.P.D. \\ \text{,,} \quad \text{measures} \quad \Delta S - 0\cdot833\Delta\sigma = -6\cdot41 + 0\cdot5865\,\Delta A.R. \\ \qquad\qquad\qquad\qquad\qquad -0\cdot7710\,\Delta N.P.D. \end{cases}$

At Egress $\begin{cases} \text{,,} \quad \text{measures} \quad \Delta S - \Delta\sigma \quad = -2\cdot70 - 0\cdot1744\,\Delta A.R. \\ \qquad\qquad\qquad\qquad\qquad -0\cdot9817\,\Delta N.P.D. \\ \text{,,} \quad \text{contact} \quad \Delta S - \Delta\sigma \quad = -2\cdot81 - 0\cdot2557\,\Delta A.R. \\ \qquad\qquad\qquad\qquad\qquad -0\cdot9610\,\Delta N.P.D. \end{cases}$

Confining our attention at present to the contact equations, I substituted first the Roorkee corrections to the tables and then the American, calling the results R and A. Thus

		R.	A.
At Ingress	$\Delta S - \Delta\sigma = $	$-2\overset{''}{\cdot}03$	$-2\overset{''}{\cdot}37,$
Egress		$-2\cdot65$	$-1\cdot26.$

These values can be made identical by values of the correc-tions between the Roorkee and American results.
Assuming

$$\Delta \text{ A.R.} = +5''\!\cdot\!90 ; \qquad \Delta \text{ N.P.D.} = -2\overset{''}{\cdot}25,$$

we have

$$\Delta S - \Delta \sigma = -2\overset{''}{\cdot}15$$

in both cases.

If ΔS be $-1'''\!\cdot\!85$, then $\Delta \sigma$ would be $+0'''\!\cdot\!30$, much as de-duced from the cusps; but it must be remembered that if ΔS here be $-1'''\!\cdot\!85$, then, as I have shown before, the cusp times require for that class of observation $\Delta S = -2''\!\cdot\!10$.

If the last values of the corrections of tables be now substi-tuted in the equations from measures, we have

$$\Delta S - 0\!\cdot\!833 \, \Delta \sigma \text{ (Ingress)} = -1\overset{''}{\cdot}22,$$
$$\Delta S - \Delta \sigma \qquad \text{(Egress)} = -1\!\cdot\!51.$$

If we assume $\Delta \sigma = +0\!\cdot\!29$, as found from cusp measures, we shall have the mean value of $\Delta S - \Delta \sigma = 1'''\!\cdot\!39$, which is three-quarters of a second greater than from the contacts, and a change in $\Delta \sigma$ would not sensibly alter this.

Thus the cusp measures point to a value of $\Delta \sigma$ of $+0'''\!\cdot\!29$, and the contact times then derived from them require that the centres of Sun and planet should be a quarter of a second nearer than at the moment of visible contact. On the other hand, the measures of the distance of limbs of Sun and *Venus* show that the solar limb seemed to extend three-quarters of a second beyond the place where the visible internal contact was seen.

With these adjustments of the diameters and errors of the tabular places which fall between those determined at Roorkee and by the American parties the Roorkee contacts and cusp equations of condition are satisfied, and the residual errors in the distances from the limb are very small, the discordance be-tween the Ingress and Egress measures being only about $0''\!\cdot\!36$.

Calcutta:
1882, *March* 25.

Note on the Illumination of Micrometers. By A. A. Common, Esq.

The satisfactory illumination of the wires of the micrometer of the astronomical telescope has hitherto been a difficult and troublesome affair. Certainly it is not that much light is wanted, but what little is needed should be in the right place and in one constant direction.

I have tried many different ways experimentally, but in practice always the only thing that answered well was the oil

The observed contact at Ingress was   well described b
Egress at 16ʰ 28ᵐ 40ˢ, the former b·   *Notices.* The fir
while the latter is six seconds later   *or* four years ago, wa
These intervals are such as w·   re brought this to sho
diameter of the Sun for visibl   *om* the lamp comes on t
from the cusps of about a qu·   rer's eye). This plan wa
   nt of my 3-foot Reflector i
Comparison of Observed C·   ttached to the position circle
the J   intensity and colour of the light
   th an improved micrometer makes

In making these
error of the tabular ř·   wire was tried; and this, from the
other, and also th·   anywhere and requires, therefore, no
that the mean sol·   the wire when heated high enough to give
my equations i   burn on the slightest increase of current.
been determin·   tried, but the experiment of one shock
Survey of In   small coil used was sufficient.
corrections t   the small incandescent carbon lamps that
+2ˢ·15.   commercially, and can be readily obtained.
On th   small as five-candle power, and as this is
rection   any purpose, and they can be used at any re-
introdr   der this, and will last a long time, they seem to
   things for the work. These lamps are now very
   and their advantages for this special work are very
   only thing that may be a trouble is the production
   current of electricity. Where a large number could be
t   an observatory, the use of a dynamo-machine driven
   would be best, but in ordinary cases, where the great
   is only with the micrometer, a small battery of, say,
   bichromate cells will do well, and by a cord worked by the
   or any other means, the light can be produced or varied by
   ing the plates in the solutions. The wires may be led any-
   and the connection with the micrometer made either by
   ible wire or by insulated rings and contact springs, in a
   that will naturally be apparent to one.
   From the trials I have made I consider it well worth trying
   anyone who is not satisfied with the old oil lamp.

1882, *May* 12.

On the Solar Spots of April and May 1882. By the
Rev. F. Howlett.

The great solar spot which, according to the observations of
the Astronomer Royal and Mr. Whipple of the Kew Observatory,
first made its appearance on the visible disk in the Sun's
southern hemisphere on April 13, 1882, was in every way
worthy of the Sun-spot maximum of the present epoch, accompanied as it was with remarkable magnetic storms, which com-

the 14th, but which more especially manifested
I was informed at Kew, on the night of the 16th
on the 17th, after which they diminished until
the 20th of the same month, when they were
tive.

Earth currents considerably interfered at times with
ic communication, especially on the 17th aforesaid, not
this country, but at Brussels, and doubtless many other
of the world.

At Alton I was informed by the Post Office authorities that
the dial plate of their telegraphic apparatus had to be turned
round many degrees before the needles were free to oscillate.
As regards the magnitude of the largest spot, on April 19, at
3^h 30^m P.M. this had a length of about $2'$ $15''$, with a mean
breadth of about $1'$ $15''$, and which, making due allowance for
irregularities of contour, implies a superficial area of not less than
2,050 millions of square miles.

Another large pair of spots, which preceded the former one
just described, were each about $65''$ in mean length and breadth,
possessing, therefore, a joint area of about 855 million square
miles, whilst the remaining small spots contained about 730
millions; making a total disruption of the solar photosphere of
not less than 3,644 millions of square miles—a condition of things
which I have not seen equalled since the still more wonderful
outbursts of February 1870, when a group was on the disk
subtending $6'$ of arc, or of September 21 of the same year,
when the total area occupied by the various groups amounted
at one time to not less than 5,000 million square miles. I possess
a superb photograph of the disk of that date 2 feet in diameter.

I would observe also that the forms assumed by the great
spot of April of the present year not a little resembled those
of the memorable outburst of August and September 1859,
which were distinguished also by very remarkable and repeated
magnetic storms. The principal spot of April of the present
year was characterised, as then also, by a very peculiar whirling
or cyclonic-looking disposition, especially during the earlier
stages of its development, and was accompanied, as was the case
in August 1859, by a large pair of associated spots on another
part of the disk, which, however, *followed* the principal group
in 1859, but *preceded* it in April 1882.

Some fine and interesting spots have been also visible during
the present month of May, one of which, making due allowance
for the time required by a synodic revolution of a spot, would
appear to be the diminished representative of that of April 13;
but of this I am not sure.

On May 8th a fine and conspicuously round spot was to be
seen nearly in the middle of its passage across the Sun's northern
hemisphere, and having a mean diameter of about $46''$. It
is not often that neatly-rounded spots of such magnitude are
seen on the disk.

Finally, during the whole nearly of April and May of the present year faculæ of unwonted extent and brilliancy have been visible on the disk, being plainly traceable sometimes to a distance of from five to six minutes of arc from the limb, instead of from about three to four minutes only, which latter are about the limits wherein they are more generally to be seen. Is it not therefore probable that simultaneously with greatly increased solar activity, as indicated by the spots, the faculæ are also upheaving the solar photosphere in mountainous billows of unusual altitude?

The Great Sun-spots of April. By Henry Pratt, Esq.

Having secured a series of photographs of the grand display of spots on the Sun's southern hemisphere, I send copies of some of them with this, although not as good as I desired.

By measures taken from the negative of April 18, 2 P.M., it appears that the total length of the group containing the grandest spot extended to a distance of about 3', while the diagonals of the principal spot (which formed a rude square) equalled 2' 7", or about 59,000 miles, and 1' 57", about 52,000 miles, the average length of its sides being 1' 35", about 43,000 miles. The total length of the other group reached 2' 52", about 76,000 miles.

From a comparison between the negative taken on April 20, 9.20 A.M., and that last referred to, it appears that the great spot had not only considerably altered in form (so that the Maltese cross-like markings had disappeared), but that its length had diminished, whilst its breadth had remained nearly constant. The total length of the last-described group had increased, while an enormous development had taken place between the two principal spots, which had previously been separated by a space almost free from penumbra.

18 *Preston St., Brighton.*

On the Nebula near Merope. By T. W. Backhouse, Esq.

Mr. Burnham having expressed a doubt as to the existence of this nebula, I looked for it on February 13, about 10.15 P.M., with my 4¼-in. Refractor, and a power of 38. I had sometimes looked for it before, but never carefully, and had never seen it. In order that I might be unbiassed, I avoided looking at the descriptions of it in the *Monthly Notices* previous to looking for it. I did not even know which star was *Merope*. I soon found the nebula, though to see it well it was necessary to have

Merope out of the field. It was very faint, and the only definite feature was the *f* edge, which, however, was quite plain. I carefully noticed its position among the stars, and afterwards looked at the *Monthly Notices*, vol. xl., and found that its position was almost exactly the same as in Tempel's drawing, p. 622. Taking the squares in that drawing to be 15' in length, the two brightish stars in it at about 5' and 10' off *Merope* were upon the *f* edge ; while the bright star about 16' off *Merope* was outside it, rather further off it than in Tempel's sketch, only I suspected that the nebula came up to that star, but more faintly.

Mr. Common's drawing, page 376, appears to be very rough, or perhaps he has confounded two stars, for his principal nebula does not touch the nebula as I saw it: indeed, the sky is perfectly dark where he draws it.

On February 18 I looked for the *n* edge of the nebula, but could not detect it with any certainty, owing to the brightness of the neighbouring stars ; *Electra* at least being in the field at the same time.

Sunderland :
    1882, April 25.

Determination of the Orbit of η Cassiopeiæ. By J. B. Coit, Esq.

(*Communicated by the Astronomer Royal.*)

I have recently completed a determination of an orbit for this beautiful binary, and a comparison of the same with all the observations at my command.

Before giving my results I would offer a few suggestions concerning the work, which, to my mind, is most needed in the line of double-star observations.

There is at present a manifest tendency upon the part of many to encourage the search for new pairs, to the neglect of the old binaries of Herschel and Struve. It is undoubtedly desirable that our catalogue of multiple stars should be extended as far as possible, in order that we may investigate the law of their distribution. At the same time, however, it is of primary importance that the orbits of several binaries be determined with all possible precision, for by this means alone can we hope to discover the laws which govern their motions, and the forces that rule in these sidereal systems. To accomplish this, the computer needs multiplied observations, with all the data that can enable him to properly determine the relative weight of each measurement.

I suggest, then, that all the systematic labour possible be centred upon a definite list of binaries, representing fairly the different classes, as to period, eccentricity, &c., as already approximately determined. If this work were faithfully done, though the present generation might see but little of the fruits of the labour, the future computer, with the data furnished by our concentrated efforts, would be enabled to determine the orbits with all needed definiteness.

The following is the orbit determined for η *Cassiopeiæ* :—

$$P = 167\cdot4 \text{ years.}$$
$$T = 1904\cdot0 \text{ A.D.}$$
$$\lambda = 233\overset{o}{\cdot}1$$
$$\gamma = 52\cdot09$$
$$\Omega = 41\cdot02$$
$$\epsilon = 0\cdot622$$
$$a = 8''\cdot702$$

This would give a mean distance 56·5 times our mean distance from the Sun, and a mass 6·4 times that of our Sun, if we accept the parallax as given by O. Struve, viz. 0''·154.

The annexed table exhibits the comparison of this orbit with the observations at hand. The computation is carried only to the first decimal place, as this seemed to be the degree of accuracy warranted by the discordant character of the observations, which is revealed as soon as they are arranged in chronological order. By tabulating the comparison with different observers many interesting results might be discovered, which we have not the space here to note. Without assuming for the above orbit absolute accuracy, it is safe to affirm that some of the observations are affected by a systematic error.

In the following table θ_c indicates the angle as computed by this orbit; θ_o the angle given by the observer, reduced to 1881·0, for those dates where the correction would be appreciable. $\theta_c - \theta_o$ presents the resulting residuals. Similarly for the distances ρ_c and ρ_o.

Mr. L. E. Holden's Observatory,
Cleveland, Ohio:
1882, *April* 12.

Observer.	No.	Epoch.	θ_c	θ_o	$\theta_c - \theta_o$	ρ_c	ρ_o	$\rho_c - \rho_o$
						''	''	''
W. Herschel	1	1779·80	—	—		11·1	11·1	0·0
,,	2	1780·50	—	—		11·1	11·5	−0·4
	3	1782·40	57·6	61·0	−3·4	11·1	—	
,,	4	1803·10	69·5	70 8	−1·3	11·6	—	
Struve	5	20·16	80·0	81·2	−1·2	10·9	10·7	+0·2
Hers. & South	6	21·90	81·0	82·9	−1·9	10·8	8·8	+2·0
,,	7	25·78	83·7	83·2	+0·5	10·5	9·9	+0·6
Struve	8	27·21	84·6	85·7	−1·1	10·4	10·2	+0·2
Hers. & South	9	28·90	85·9	86·5	−0·6	10·3	12·0	−1·7

Observer.	No.	Epoch.	θ_c	θ_o	$\theta_c \div \theta_o$	ρ_c	ρ_o	$\rho_c - \rho_o$
Bessel	10	1830·75	87·2	86·3	+0·9	10·1	10·1	0·0
Smyth	11	30·91	87·3	87·9	−0·6	10·1	9·8	+0·3
,,	12	31·92	88·2	88·3	−0·1	10·0	9·9	+0·1
Struve	13	32·05	88·3	87·7	+0·6	10·0	9·8	+0·2
Dawes	14	32·87	89·0	88·7	+0·3	9·9	9·7	+0·2
Smyth	15	33·74	89·6	89·0	+0·6	9·9	9·9	0·0
Bessel	16	34·76	90·4	89·7	+0·7	9·8	9·8	0·0
Smyth	17	35·20	90·8	91·0	−0·2	9·7	9·7	0·0
Struve	18	35·26	90·9	91·3	−0·4	9·7	9·5	+0·2
,,	19	36·74	92·0	92·2	−0·2	9·6	9·4	+0·2
Smyth	20	36·81	92·1	92·1	0·0	9·6	9·4	+0·2
Encke	21	37·62	92·7	92·6	+0·1	9·5	9·6	−0·1
Galle	22	38·68	93·7	92·7	+1·0	9·4	9·5	−0·1
Kaiser	23	40·43	95·2	95·9	−0·7	9·2	9·0	+0·2
G. O.	24	40·44	95·2	96·5	−1·3	9·2	9·0	+0·2
O. Struve	25	41·34	96·1	98·2	−2·1	9·2	9·2	0·0
Mädler	26	41·57	96·3	96·5	−0·2	9·1	9·2	−0·1
Dawes	27	41·80	96·5	95·8	+0·7	9·1	9·3	−0·2
Mädler	28	42·39	97·0	98·4	−1·4	9·0	8·7	·+0·3
Smyth	29	43·19	97·8	95·9	+1·9	9·0	9·4	−0·4
Mädler	30	44·56	99·1	100·2	−1·1	8·8	8·6	+0·2
..	31	45·39	100·0	102·0	−2·0	8·7	8·5	+0·2
,,	32	46·67	101·3	101·2	+0·1	8·6	8·6	0·0
Smyth	33	46·73	101·3	101·6	−0·3	8·6	8·5	+0·1
D. O.	34	47·08	101·8	104·1	−2·3	8·6	8·8	−0·2
O. Struve	35	47·40	102·0	101·8	+0·2	8·5	8·5	0·0
Mädler	36	47·42	102·0	102·8	−0·8	8·5	8·3	+0·2
Miller	37	47·60	102·2	101·5	+0·7	8·5	8·6	−0·1
Mitchel	38	47·63	102·2	101·4	+0·8	8·5	8·6	−0·1
O. Struve	39	49·66	104·5	105·0	−0·5	8·3	8·3	0·0
Mädler	40	50·80	105·8	106·6	−0·8	8·2	8·0	+0·2
Jacob	41	50·87	105·9	105·6	+0·3	8·2	8·2	0·0
Mädler	42	51·76	107·0	107·0	0·0	8·1	7·7	+0·4
O. Struve	43	51·84	107·0	108·1	−1·1	8·1	8·0	+0·1
Jacob	44	51·88	107·1	106·5	+0·6	8·1	8·0	+0·1
Miller	45	51·90	107·1	107·0	+0·1	8·1	8·1	0·0
Mädler	46	52·67	108·0	108·7	−0·7	8·0	7·6	+0·4
Jacob	47	52·75	108·0	108·0	0·0	8·0	8·0	0·0
..	48	53·13	108·5	109·1	−0·6	7·9	7·9	0·0

Observer.	No.	Epoch.	θ_c	θ_o	$\theta_c - \theta_o$	ρ_c	ρ_o	$\rho_c - \rho_o$
Mädler	49	1853·29	108·8	110·2	−1·4	7·9	7·6	+0·3
„	50	53·90	109·5	112·8	−3·3	7·8	7·5	−0·3
Peters	51	53·93	109·5	109·7	−0·2	7·8	8·2	−0·4
Powell	52	53·94	109·5	109·5	0·0	7·8	7·6	+0·2
Jacob	53	53·98	109·6	109·8	−0·2	7·8	8·0	−0·2
Dawes	54	54·00	109·6	109·6	0·0	7·8	7·9	−0·1
Smyth	55	54·17	109·7	110·6	−0·9	7·8	7·7	+0·1
O. Struve	56	54·56	110·3	112·0	−1·7	7·8	8·0	−0·2
Dembowski	57	54·77	110·6	111·2	−0·6	7·8	7·9	−0·1
Mädler	58	54·80	110·6	111·8	−1·2	7·7	7·6	+0·1
Powell	59	54·94	110·8	111·5	−0·7	7·7	7·5	+0·2
Morton	60	54·95	110·8	111·0	−0·2	7·7	8·1	−0·4
Dembowski	61	55·08	111·0	112·8	−1·8	7·7	7·8	−0·1
Winnecke	62	55·25	111·2	110·9	+0·3	7·7	7·9	−0·2
Mädler	63	55·51	111·6	112·0	−0·4	7·7	7·6	+0·1
Secchi	64	55·79	111·9	112·2	−0·3	7·6	7·9	−0·3
Mädler	65	55·87	112·0	111·0	+1·0	7·6	7·8	−0·2
Powell	66	55·92	112·1	112·5	−0·4	7·6	7·6	0·0
Morton	67	55·96	112·1	112·4	−0·3	7·6	7·8	−0·2
Luther	68	56·57	112·9	117·5	−4·6	7·5	8·3	−0·8
Dembowski	69	56·58	113·0	114·2	−1·2	7·5	7·4	+0·1
„	70	57·11	.113·7	114·5	−0·8	7·5	7·3	+0·2
Secchi	71	57·15	113·8	112·8	+1·0	7·5	7·9	−0·4
O. Struve	72	57·22	113·8	114·1	−0·3	7·5	7·6	−0·1
Dembowski	73	57·82	114·6	115·9	−1·3	7·4	7·2	+0·2
„	74	58·46	115·6	115·8	−0·2	7·4	7·3	+0·1
Mädler	75	58·52	115·7	114·3	+1·4	7·4	7·1	+0·3
„	76	59·26	116·7	115·7	+1·0	7·3	7·0	+0·3
Powell	77	59·72	117·4	116·6	+0·8	7·3	7·0	+0·3
Morton	78	59·94	117·7	117·3	+0·4	7·2	7·1	+0·1
O. Struve	79	60·68	118·8	119·8	−1·0	7·1	7·2	−0·1
Powell	80	60·97	119·3	118·3	+1·0	7·1	7·0	+0·1
Auwers	81	61·58	120·2	119·8	+0·4	7·1	7·4	−0·3
Main	82	61·82	120·6	118·1	+2·5	7·0	6·4	+0·6
Mädler	83	61·91	120·7	119·0	+1·7	7·0	7·0	0·0
Powell	84	61·95	120·7	120·6	+0·1	7·0	6·7	+0·3
Dembowski	85	62·74	122·0	121·3	+0·7	6·9	7·0	−0·1
Romberg	86	62·86	122·2	119·1	+3·1	6·9	7·0	−0·1
·,	87	62·90	122·3	121·7	+0·6	6·9	7·3	−0·4
	88	63·04	122·5	122·9	−0·4	6·9	6·9	0·0

Observer.	No.	Epoch.	θ_c	θ_0	$\theta_c-\theta_0$	ρ_c	ρ_0	$\rho_c-\rho_0$
Romberg	89	1863·06	122·5	122·5	0·0	6·9	6·9	0·0
,,	90	63·12	122·6	121·0	+1·6	6·9	7·0	−0·1
Luther,	91	63·18	122·7	123·6	−0·9	6·9	7·1	−0·2
Dembowski	92	63·48	123·2	122·6	+0·6	6·8	6·9	−0·1
..	93	64·10	124·3	124·5	−0·2	6·8	6·8	0·0
	94	64·71	125·2	124·6	+0·6	6·7	6·8	−0·1
,,	95	65·51	126·7	126·3	+0·4	6·6	6·7	−0·1
Knott	96	65·69	126·9	125·3	+1·6	6·6	6·7	−0·1
..	97	65·69	126·9	126·7	+0·2	6·6	6·7	−0·1
,,	98	65·70	126·9	125·2	+1·7	6·6	6·8	−0·2
Talmage	99	65·73	126·9	123·9	+3·0	6·6	6·4	+0·2
O. Struve	100	66·22	128·1	132·6	−4·5	6·6	6·4	+0·2
Talmage	101	66·63	128·7	124·6	+4·1	6·5	6·4	+0·1
Secchi	102	66·86	129·0	127·7	+1·3	6·5	6·8	−0·3
Dembowski	103	67·16	129·7	129·3	+0·4	6·5	6·6	−0·1
Main	104	67·65	130·4	129·9	+0·5	6·4	6·3	+0.1
Dunér	105	68·37	131·7	131·8	−0·1	6·4	6·3	+0·1
O. Struve	106	68·53	132·1	132·9	−0·8	6·4	6·4	0·0
Dembowski	107	68·55	132·1	132·4	−0·3	6·4	6·3	+0·1
Brünnow	108	68·84	132·8	131·5	+1·3	6·3	6·3	0·0
Talmage	109	68·89	132·8	124·3	+8·5	6·3	6·2	+0·1
Main	110	69·67	134·3	132·4	+1·9	6·3	6·1	+0·2
Dembowski	111	69·68	134·3	134·1	+0·2	6·3	6·2	+0·1
Talmage	112	69·72	134·3	124·8	+9·5	6·3	6·6	−0·3
Dunér	113	69·93	134·7	135·2	−0·5	6·2	6·1	+0·1
O. Struve	114	70·18	135·3	136·2	−0·9	6·2	6·3	−0·1
Dembowski	115	70·52	136·0	135·4	+0·6	6·2	6·2	0·0
Gledhill	116	70·65	136·3	135·7	+0·6	6·2	6·1	+0·1
..	117	70·70	136·3	138·8	−2·5	6·2	6·1	+0·1
,,	118	70·80	136·5	136·0	+0·5	6·1	6·0	+0·1
Dembowski	119	71·56	138·1	137·5	+0·6	6·1	6·1	0·0
Gledhill	120	71·60	138·3	137·6	+0·7	6·1	6·1	0·0
,,	121	71·80	138·7	138·3	+0·4	6·1	6·0	+0·1
Wilson & Seab.	122	71·93	139·0	140·9	−1·9	6·0		—
,,	123	72·01	139·2	—		6·0	6·0	0·0
O. Struve	124	72·18	139·5	140·8	−1·3	6·0	5·9	−0·1
Dunér	125	72·50	140·1	140·5	−0·4	6·0	6·6	−0·6
Dembowski	126	72·62	140·3	139·1	+1·2	6·0	6·0	0·0
Knott	127	72·65	140·6	138·0	+2·6	6·0	6·1	−0·1
..	128	72·65	140·6	137·7	+2·9	6·0	6·1	−0·1

Observer.	No.	Epoch.	θ_c	θ_o	$\theta_c - \theta_o$	ρ_c	ρ_o	$\rho_c - \rho_o$
Knott	129	1872·66	140·6	138·0	+2·6	6·0	6·0	0·0
„	130	72·66	140·6	137·7	+2·9	6·0	6·1	−0·1
Main	131	72·77	140·8	143·9	−3·1	6·0	5·9	+0·1
Talmage	132	72·86	140·9	124·4	+16·5	6·0	6·3	−0·3
Wilson & Seab.	133	73·06	141·4	142·3	−0·9	6·0		—
Gledhill	134	73·51	142·4	143·1	−0·7	5·9	6·1	−0·2
O. Struve	135	73·53	142·4	144·6	−2·2	5·9	5·7	+0·2
Dembowski	136	73·65	142·6	140·7	+1·9	5·9	5·8	+0·1
Gledhill	137	73·73	142·9	143·8	−0·9	5·9	5·8	+0·1
„	138	73·81	143·0	144·3	−1·3	5·9		—
Wilson & Seab.	139	73·83	143·0	144·7	−1·7	5·9	6·2	−0·3
„	140	73·83	143·0		—	5·9	6·6	−0·7
„	141	73·83	143·0		—	5·9	6·2	−0·3
Talmage	142	73·86	143·0	141·2	+1·8	5·9	5·7	+0·2
Nobile	143	73·98	143·5	143·6	−0·1	5·9		—
Dunér	144	74·22	144·0	144·9	−0·9	5·9	5·7	+0·2
Dembowski	145	74·63	144·8	142·5	+2·3	5·8	5·8	0·0
Wilson & Seab.	146	74·90	145·6	146·0	−0·4	5·8	5·8	0·0
O. Struve	147	75·15	146·1	148·6	−2·5	5·8	5·6	+0·2
Dunér	148	75·51	146·9	146·7	+0·2	5·8	5·7	+0·1
Dembowski	149	75·60	147·2	146·2	+1·0	5·8	5·7	+0·1
Gledhill	150	75·69	147·2	147·5	−0·3	5·7		—
Main	151	75·78	147·3	146·1	+1·2	5·7	5·8	−0·1
Doberck	152	75·93	147·8	147·8	0·0	5·7		—
Gledhill	153	76·37	148·7	147·9	+0·8	5·7	5·6	+0·1
Talmage	154	76·86	150·0	149·3	+0·7	5·7	4·7	+1·0
Dembowski	155	76·94	150·1	149·9	+0·2	5·7	5·6	+0·1
Gledhill	156	77·41	151·4	149·9	+1·5	5·6		—
Doberck	157	77·76	152·3	150·2	+2·1	5·6	5·7	−0·1
Wilson & Seab.	158	77·95	152·8	153·5	−0·7	5·6	5·3	+0·3
Gledhill	159	78·67	154·4	153·3	+1·1	5·6		—
Hall	160	78·86	154·9	154·5	+0·4	5·5	5·3	+0·2
..	161	78·97	155·3	154·7	+0·6	5·5	5·4	+0·1
	162	78·97	155·3	155·5	−0·2	5·5	5·5	0·0
	163	79·06	155·6	156·2	−0·6	5·5	5·3	+0·2
	164	79·06	155·6	160·8	−5·2	5·5	5·2	+0·3
	165	79·06	155·6	156·0	−0·4	5·5	5·4	+0·1
	166	1879·08	155·6	160·2	−4·6	5·5	5·3	+0·2

MONTHLY NOTICES

OF THE

ROYAL ASTRONOMICAL SOCIETY.

Vol. XLII.	June 9, 1882.	No. 8.

E. J. Stone, Esq., M.A., F.R.S., President, in the Chair.

Franklen George Evans, Esq., Tynant House, near Cardiff, was balloted for and duly elected a Fellow of the Society.

The Meridian Photometer. By Professor Edward C. Pickering.

This instrument is designed for the photometric observation, on or near the meridian, of objects not fainter than stars of the tenth magnitude. A smaller instrument constructed upon the same general principles has been used during the past three years in comparing the brightness of all stars visible in this latitude to the naked eye. This work is now nearly completed. The new instrument will be largely used in determining the ratios of light corresponding to the scale of magnitudes employed in the construction, and in the recent revision, of the "Bonner Durchmusterung." For this purpose, measurements will be made of the light of the stars observed during this revision at each of two co-operating Observatories. These stars are situated near the borders of the zones into which the northern heavens were divided in order to distribute the work of determining the positions of the stars.

One of the annexed photographs * exhibits the whole of the new instrument; while the other represents, on a larger scale, the end of it farthest from the eyepiece. The two plane mirrors at this end of the photometer are connected with tubes placed horizontally east and west, and capable of revolution in the

---

* The photographs are in the library of the Society.

G G

plane of the meridian. The inclination of each mirror to the axis of its tube may also be varied. By means of connecting-rods and cords these movements of both kinds may be effected by the observer without leaving his place at the eyepiece, and also by an assistant stationed near the mirrors. One of the mirrors is ordinarily directed to λ *Ursæ Minoris*, with which the other stars are compared. The second mirror is connected with a longer tube, and projects beyond the first, so that it may be directed to any star within a few minutes of its meridian passage.

The light from the two mirrors is received respectively by two object-glasses of about four inches aperture and sixty inches focal length, placed within the instrument, and not shown in the photographs. Both the pencils of rays, before reaching the eyepiece, pass through a double-image prism, which can be reversed by the observer, in order to eliminate the error due to the partial polarisation of the light from the mirrors. The ordinary image received from one mirror and object-glass is compared with the extraordinary image from the other mirror and object-glass by means of a Nicol connected with the eyepiece. The double-image prism is then reversed, and the comparison repeated with the other pair of images, which are thus brought nearly to the place in the field previously occupied by the first pair. The pair not in use at the time is excluded from the field by an eye-stop.

A curtain, not shown in the photographs, protects the observer from the light of the gas-burners which illuminate the circles of the instrument. The assistant directs the mirror to the successive stars to be observed, and records the readings of the photometer-circle near the eyepiece.

The two images compared in an observation with this instrument are placed under the same conditions, and it may therefore be hoped that the results obtained will be free from systematic error. Among causes likely to produce such error in photometric observations may be named those stated in the list below. Care has been taken to guard against them either in the construction of the instrument or in the methods of observation and of reduction which have been adopted.

1. Variation in light of the star or other object used as the standard of comparison.

2. Difference in the background of the respective images, due to twilight, moonlight, or other causes.

3. Varying atmospheric absorption, when the same object is observed at different altitudes.

4. Difference in reflecting-power of two mirrors.

5. Partial polarisation produced by reflection from mirrors.

6. Difference in the appearance of the images, due to unequal magnifying power, or to unlike sources of light.

7. Difference in the size or position of the two pencils of rays forming the images.

8. Difference in the relative positions in the field of the two images in different observations.

Harvard College Observatory:
1882, May 8.

On Photographs of the Nebula in Orion, and of its Spectrum. By Henry Draper, Esq., M.D.

(*Communicated by A. C. Ranyard, Esq.*)

At the meeting of the Royal Astronomical Society in January 1881 I presented a photograph of the nebula in *Orion*, which was taken on September 30, 1880, with my Clark telescope of 11 inches aperture and an exposure of 51 minutes. This was the first photograph ever taken of a nebula. It comprised the brightest parts in the neighbourhood of the trapezium, and showed the condensed masses well. In March 1881 a number of photographs of the nebula were taken, the best being on March 11 with an exposure of 104 minutes. By comparison with the former picture this made a marked advance, and minute stars down to the 14·7 magnitude of Pogson's scale were shown. An account of it was read before the French Academy of Sciences, and printed in the "Comptes Rendus," April 18, 1881.

On March 14, 1882, the negative was made from which the enlarged photolithographs herewith presented were produced. The instrument used was the Clark photographic refractor of 11 inches aperture, mounted on the equatoreal stand, and driven by the clock I had constructed. The exposure was from $7^h 8^m$ to $9^h 25^m$—that is, 137 minutes. Gelatino-bromide plates were employed. The night was clear, but cold and windy. The mean temperature was 27° F.; the wind NNW. and in gusts, the strongest pressure being 5 pounds per square foot, about 9 o'clock; the whole travel of the wind during the exposure was 35 miles. The variation in the force of the wind is one reason why the stars show some ellipticity under this magnifying power. The gusts of course displaced the telescope somewhat, though the mounting is firm and the clockwork strong.

In the photograph the larger stars are much over-exposed, the proper time to make a good picture of the trapezium being about 2 minutes. The twinkling of these stars is therefore recorded on the sensitive plate, and gives to them an excess of size. If a photograph should be taken on a steady night, the stars of the trapezium would be easily separated, and in the original negative of this picture, in a strong light, the separation can be seen. The variation in size of the stellar images gives an idea of the relative magnitude of the stars, though that estimate requires correction for the colour of the stars. It must be remembered that no one enlargement can do justice to the original negative: various exposures, various intensities of light, and various points of view are necessary for a complete examination.

On comparing this photograph with the well-known drawings of the same object by Lord Rosse, Bond, Struve, Lassell, Trouvelot, Secchi, and others, it will be noticed that most of the small stars, about which there is no dispute, are represented, and the outlying streamers are well indicated. It is worthy of remark also that in this photograph with an exposure of only 137 minutes I have depicted stars almost the *minimum visible* in this telescope, and it is not unreasonable to hope that by still further prolonging the exposure and by still further study of photographic processes, stars and details entirely invisible to the eye may be secured. With this object in view, I have under consideration a new form of mounting which will permit of continuous exposures of six hours.

During the month of March 1882 I also made four photographs of the spectrum of the nebula in *Orion*, which are described in the May number of the "American Journal of Science." Two of these were obtained with the slit spectroscope I usually employ for photographing spectra of the stars, and they show two lines in the ultra-violet plainly, besides the traces of two others. The first-mentioned two are hydrogen γ λ 4340 and hydrogen δ λ 4101; the others are too faint to give a good estimate of the wave-length.

The other spectrum photographs, taken without a slit, show that two of the condensed masses preceding the trapezium give a continuous spectrum, and therefore contain either gas under pressure, or liquid or solid matter.

271 *Madison Avenue, New York :*
    1882, *May.*

Some Remarks on Mr. Newcomb's Paper " On the Instructions for Observing the Transit of Venus formulated by the Paris International Conference." By E. J. Stone, Esq., M.A., F.R.S.

It is greatly to be regretted that the step which was taken by the English Committee in the October of 1881 of issuing a draft of Instructions to Observers for consideration and discussion did not lead to any direct interchange of ideas between the American and European astronomers. I believe that an interchange of ideas, such as that contemplated, would have led to the adoption of " Instructions " which were identical in all essential points, and I hope that this object may yet be secured. The remarks made in the *Notices* for April by Mr. Newcomb, on the Instructions issued by the International Committee on the Transit of *Venus*, are interesting, and require consideration; but unfortunately they are stated to be simply "an individual contribution to the discussion," and the time for discussion is fast passing away and that for action is approaching. It appears to me, however, that there are really but few important points upon which Mr. Newcomb's views differ essentially from those which were attempted to be embodied in the International Instructions, and that something like common action is still possible. All astronomers have their favourite way of describing the "real contact," but mutual concessions on unimportant points are necessary if combined action is desired. I have, in this note, thrown together a few remarks which have occurred to me whilst reading Mr. Newcomb's paper. I hope these remarks will, at least, be considered before any Instructions are issued to the American observers which may prevent their observations of contacts from being ultimately combined in one general discussion with those observed by the French, German, and English observers. A chain is not stronger than its weakest link, and the value of the large number of observations of contact at "Ingress retarded" and "Egress accelerated," which will undoubtedly be secured, if the weather is favourable, by the American observers, will depend very much indeed upon the number of contacts of "Ingress accelerated" and "Egress retarded" which may be available for direct comparison with them, and for most of these the American astronomers will be dependent upon the observers of other nations.

The most important point upon which Mr. Newcomb's views appear to differ from mine is the question of the brightness of the field of view in which the contacts should be observed. Mr. Newcomb considers "that the observer should have the solar disk as bright as the eye could bear with entire ease and comfort." I cannot think this desirable. If the sky was covered with thin cloud or haze at one of the stations, the difference would be excessive between the brightness of the field available at that station and the brightness which would be adopted at stations where the sky was perfectly clear. There are also very considerable differences in the degree of illumination of the field of view which different observers can bear with ease and comfort, and all the phenomena of irradiation are presented in an aggravated form when the brightness of the field is excessive. The complications introduced by the phenomena of irradiation are perhaps the difficulties chiefly to be feared in attempts to determine the value of the solar parallax from a discussion of internal contacts. I think, therefore, that we are not justified in adopting a brightness of the field of view which must increase to the utmost these difficulties.

But the changes in the illumination of the Sun's limb, which take place near the point of contact, can only be recognised by contrast ; it is therefore essential that the brightness of the field should be sufficient to allow these differences of illumination to be readily distinguished, and the adoption of a field of view with a very feeble illumination is not, therefore, desirable. It will be found, on consideration, that difficulties quite as great as those dependent on the irradiation phenomena can be introduced by unduly diminishing the brightness of the field. Pairs of very fine spider webs at distances corresponding to a second of arc in the focus of a positive eyepiece appear to afford a delicate test of a suitable field. But this test requires great care in its application, and I am inclined to think that a rougher but better practical test is afforded as follows :—

Let an observer mark the part of his wedge where he can just observe the Sun's limb in a perfectly clear sky without discomfort. This degree of brightness is, I consider, excessive. Let him also mark the part of the wedge at which, under the same circumstances, he could just distinctly and clearly observe the Sun's limb. This field is the faintest which that observer could adopt, and is, in my opinion, decidedly *too* faint. If the observer adopts the mean portion of the wedge as giving, under a clear sky, a standard brightness of the field of view, he cannot err much either in excess or defect. There will of course be slight differences of opinion amongst observers of what constitutes the *best* degree of illumination of the field, but all the skilled observers who have examined the point at Oxford agree that the field of view thus found is an agreeable one to observe in, and that the brightness is sufficient to allow the rice-grains and the minute details of solar spots to be well seen. Slight changes in the portion of the wedge recommended for use are unimportant, but it is desirable that excessive differences in the brightness should be avoided, and that some approximation to uniformity in this respect should be secured. The brightness of the field fixed as described, on a clear day, would have to be learnt by practice as a habit of the eye, and adopted, as closely as possible, by shifting the wedge to meet the varying conditions of the atmosphere at the time of observation. This is the recommendation which I should make with regard to the brightness of the field of view in which the contact should be observed. It secures approximate uniformity, and it excludes extremes of either excessive brightness or undue faintness in the illumination of the field of view.

I could not join in Mr. Newcomb's recommendation of the use of Dawes' solar eyepiece. This eyepiece is valuable for the examination of detached portions of the solar disk ; but the field of view is exceedingly limited, and when clouds were passing there would be practical difficulties in keeping the point of contact exactly in the centre of the field, whilst the effects of the stop, near the edge of the field, would be most injurious in the contact observations.

I must confess that I cannot understand the difficulties which Mr. Newcomb experiences in forming a precise conception of what should be understood by the phrase " illumination of the apparent limb of the Sun " and " how discontinuity in that illumination should be noted." The "illumination of the *apparent* limb of the Sun " means the illumination of the limb of the Sun as seen—the visible limb; and the discontinuity of the illumination can be noted as follows:—Suppose a, c, b to be three small portions of the visible, or apparent, limb of the Sun, c being the portion near which the internal contact takes place. Before contact at Egress, or after contact at Ingress, there will be no sensible difference between the illumination of the portions a, c, b, but near the contact the illumination of the Sun's limb at c will be less than the illumination either at a or b. This difference between the illumination at c and at a and b constitutes a discontinuity in the illumination of the Sun's limb near the point of contact, which every observer must see, if the sky is clear, near the internal contacts. The observers are therefore simply asked to give at Ingress the last time at which they are *certain* of the existence of such a discontinuity as independent of mere atmospheric tremor; and at Egress, the first time at which they are certain of such a discontinuity. In the English Instructions the observers will be asked to give also the times of deepest shade, when the black drop, ligament, haze, or shadow ceases to be as dark as the outer edge of the planet *Venus* at Ingress, or first becomes as dark as the outer edge of that planet at Egress. I hope and believe that the observers will experience no difficulty in understanding these instructions, and that they will observe the contacts as defined.

The contacts as defined in the International Instructions are the same as those which Mr. Newcomb would, *apparently*, wish to have observed. I say *apparently*, for in many parts of Mr. Newcomb's paper it would appear that he prefers to leave the observer without any precise definition of the contacts to which his attention should be directed—a course of proceeding which, in my opinion, could lead to nothing but a complete failure: but if Mr. Newcomb simply recommends the observers to give at Ingress "the time when light is about to glimmer all the way across the dark space between cusps," the recommendation would do no harm. The time which should then be given by the observers would be the time when there was last seen a distinct and persistent discontinuity in the illumination of the Sun's limb at the point of contact. The chief difference between the definitions appears to be that Mr. Newcomb prefers to direct the attention of the observer to the light of the cusps which is encroaching upon the "*dark space*" between them, whilst in the definition adopted by the International Conference the attention of the observer is directed to the disappearance of the *dark space* between the cusps.

But if Mr. Newcomb's definition were adopted it would be necessary to introduce an exception, and to caution the observers from taking the time when the light of the aureole, penumbra, or "sunlight through the atmosphere of *Venus*" began to glimmer across the dark space between the cusps. Without this caution the observed contacts would be very uncertain and the times recorded would be earlier at Ingress than those which Mr. Newcomb requires by about a minute of time. Mr. Newcomb's definition and that of which it is a modification and improvement, given in the *Notices* for March 1877—viz. "the time at which true sunlight is first seen all the way around the following limb of the planet"—is open to the objection that it is of a negative character. The first time at which bright sunlight is seen may be caught up, through clouds, long after all touch between the limbs has ceased. Such observations when recorded as contacts destroy the value of a large number of good observations, unless rejected, and this is always an unsatisfactory proceeding, and one sometimes open to misconceptions. If properly understood and strictly followed, the definition of Mr. Newcomb should lead to identical results with that adopted at the Paris Conference; but I cannot regard it as an improvement. To the recommendations to leave each observer from a knowledge of the general theory of the subject to observe any contact which he considers best under existing conditions, and to the proposals of IV., page 280, I object *in toto*.

I have no wish to indulge in mere verbal criticisms, but it is desirable to point out that Mr. Newcomb's views of the nature of the internal contacts have been very much based on model-practice, and that, without great care, this model-practice is misleading. In the case of the model we have the Sun's limb represented by one hard edge and the planet represented by a disk. There are therefore complicated phenomena of diffraction beyond the hard edge which represents the Sun's limb, and of interference between the limbs of *Venus* and the Sun, which vary as the contact approaches, and which have nothing strictly analogous to them in the case of the actual transit; whilst, on the other hand, the light which is refracted through the atmosphere of *Venus* in the real transit introduces complications to which there is nothing strictly analogous in the models in general use. Whilst therefore model-practice may be useful to observers, in preparing them for the slowness with which the contacts are established and in exhibiting changes somewhat similar in general character to those which will be presented in the real transits, the observers require to be carefully cautioned against expecting to see the *same* succession of phases identically reproduced in their model-practice and in the real transit. And here I must point out that when Mr. Newcomb speaks of the "true internal contact" and of the "sharp cusps of *Venus*, instead of appearing sharp as in *their true form*, appear rounded at their termini by the black drop or other forms of distortion," the language used is misleading. There is no "true internal contact," neither are there cusps, except as we see them projected on the Sun's disk; and the phenomena as seen are the *real phenomena*. No doubt, if there was no such thing as the diffraction of light, if the telescopic image of a bright point was a point, if there was no dispersion of light through the atmosphere of *Venus*, then the definition of the true internal contact would be easy, and the sharp cusps of *Venus* might then be seen without any blunting, rounding, or diffusion of the cusps. But we must accept the laws of nature as we find them, and no good, but much harm, will result from our ignoring the existence of causes which prevent the appearances which would be found to exist under the supposed laws of formal *geometrical* optics from being realised in the actual transit. We have already suffered too much from this error.

I am unable to follow Mr. Newcomb's remarks respecting Mr. Tebbutt's observations in 1874. There is no doubt that Mr. Tebbutt's " contact " at Ingress was observed rather later than the general "average contact," but there are later times still given for contact; and the error of Mr. Tebbutt's observation, as measured by the difference between the angular separation of the centres of *Venus* and the Sun at the time given and the mean angular separation, is only about two-tenths of a second of arc. The observation is therefore neither worthless nor very wild. At Egress the time given by Mr. Tebbutt corresponds to an angular separation of the centres which is nearly the mean angular separation of all the observers.

In conclusion, I would remark that if some approximation to uniformity in the instrumental equipments of the observers be secured—if good telescopes of nearly the same, and not too small, apertures be employed; if magnifying powers of not much less than 150 be used; if the observations of contact are made in fields of view of which the degree of illumination is not greatly in excess or defect; and if, above all, the attention of the observers is directed to a contact which is distinctive enough to be recognised : then, in my opinion, we may be certain of a substantial success, unless clouds should intervene and spoil the observations at several of the most important stations. To diminish this risk as much as possible, we have increased the number of stations ; more we cannot do to insure and deserve success. But, whilst I feel confident of success if all the observers attempt to observe the same thing under somewhat similar conditions, if each observer is to observe any kind of contact which appears good in his own estimation, without reference to what is being done by other observers, then there must be as many values of the solar parallax deducible from attempts to combine these discordant observations as there are different kinds of contact observed at the opposed stations of accelerated and retarded effects of parallax.

In the paper of March 1877, *Monthly Notices* (vol. xxxvii.), to which Mr. Newcomb refers as explaining his views of the internal contact, the attention of the observers is directed to observing the first time at Ingress and last time at Egress when direct sunlight is seen all around *Venus*. Now, if such a definition should be adopted by the American observers, whilst the observers at the Cape, Madagascar, Australia, and New Zealand record the times when they last see "the illumination of the

Sun's limb near the point of contact as dark, or nearly as dark, as the outer edge of the planet," then an attempt to obtain a value of the Sun's parallax from a discussion of these observations, on the assumption that the contacts observed refer to the same angular separation of the limbs, will most certainly lead to a value of the solar parallax which is in excess of the true value. If the contacts observed were reversed at the different stations, then the result that would be obtained would be too small. I call attention to the point now, before the transit has taken place, in order, if possible, to prevent the serious confusion and doubt which may arise if any large number of observers are directed to observe a contact of an essentially different character from that which will be observed by the general body of the observers. If this caution is not attended to, it will be possible to obtain different values of the solar parallax by different combinations of the contacts; but of course no one entrusted with the calculations should attempt to combine in one discussion discordant material. It would, however, be far more satisfactory if the collection of this discordant material could be prevented rather than it should be rejected after the transit as worthless for the objects in view.

Curves showing the Changes in the Adopted Diameter of the Moon as given by the Observations in the Greenwich Lunar Reductions 1750–1830. By E. J. Stone, M.A., F.R.S.

The results upon which the curves are based have been extracted from "The Greenwich Lunar Reductions," vol. ii., Section iii., Comparison of Moon's Observed and Tabular Place, pages [1] to [293]. It is from a discussion of the results given in this section that the coefficients of the parallactic inequality were deduced by Sir G. B. Airy, which show an inequality with a period of about forty-six years. The observations have been divided into groups extending over periods of about nine years, identical with those adopted by Sir G. B. Airy. If E_1 and E_2 denote the excess of observed longitudes over tabular longitudes as deduced respectively from the observations of the first and second limbs, then E_1 and E_2 have been extracted directly from Section iii. for all the days on which both limbs were observed in Right Ascension and a North Polar Distance is available for the deduction of the observed longitude. The mean results are given in the following table. The results, divided by 2, will be found essentially the same as those which I have already given in the *Notices* for Dec. 1881, but in the former paper the results for the last year of each period were not included. In the present table the results of the last year of each group are included. The general form of the curve is not, however, altered by this difference in the grouping :—

12
13
14
15
16
17
18
19
1820
21
22
23
24
25
26
27
28
29
1830
31
32
33
34
35

The Dotted curve shows Corrected values.

Spottiswoode :

TABLE A.

Group.	Mean Year.	No. of Obs.	Mean Value of $E_1 - E_2$	Second Set.
1	1750–1759	1754·7	32	+ 0·145	+ 0·145
2	1755–1764	1758·8	28	+ 0·709	+ 0·222
3	1760–1768	1763·4	18	+ 0·546	− 0·248
4	1765–1773	1769·5	16	+ 0·533	+ 0·533
5	1769–1778	1773·2	18	+ 1·184	+ 1·184
6	1774–1782	1777·8	16	+ 1·664	+ 1·664
7	1779–1787	1783·1	16	+ 4·069	+ 3·326
8	1783–1791	1787·4	18	+ 4·053	+ 3·397
9	1788–1796	1792·7	23	+ 2·179	+ 2·179
10	1792–1801	1796·7	23	+ 2·086	+ 2·086
11	1797–1805	1801·3	15	+ 2·053	+ 2·053
12	1802–1810	1805·3	13	+ 0·756	+ 0·756
13	1806–1815	1809·3	12	− 0·027	− 0·027
14	1811–1819	1814·0	8	− 1·008	− 0·768
15	1816–1824	1820·9	9	+ 0·314	+ 0·954
16	1820–1829	1825·0	17	+ 0·526	+ 1·166
17	1825–1833	1828·0	16	+ 1·083	+ 1·723
18	1830–1838	1834·7	11	+ 1·721	+ 2·361

The errors in the table may of course be affected by the existence of systematic errors in the North Polar Distance, for these observations are involved in the determination of the observed longitudes.   There are two very large errors—

| 1763 | Sept. 21 | 14·05 |
| 1784 | 28 | 15·21 |

The rejection of these two results, and the correction of the results after 1816, August 7, for the slight difference between the adopted diameters before and after that period, lead to the second set of results.

The curves have been laid down from the above results with the mean values of $(E_1 - E_2)$ as ordinates, and the corresponding mean times as abscissæ.

On account of the importance of these results, from the use which has recently been made of these old Lunar Reductions, I have not only computed them myself, but I have also had them recomputed independently.   I have no doubt whatever about the substantial accuracy of these results.   But if correct, they *prove* conclusively that in these old observations, *as reduced*, we cannot pass from observations of the centre as deduced from the first limb of the Moon to observations of the centre as deduced from the second limb of the Moon without the large systematic and variable discordances which are indicated in the table and by the curves.

Addendum.

In the discussion between myself and Mr. Neison, we are concerned with the observations as reduced in Section iii., and to these I have confined my attention : but if the direct comparison between the times of passage of the diameter over the meridian as observed in Right Ascension did not indicate any such discordances as those found from the longitudes, then we should certainly be able to find the source of the error which has vitated the results as reduced. I have therefore computed the corrections to the provisionally adopted diameter which follow directly from the observations in R.A., as given on pages lxiv and lxv of the Introduction, vol. i. I have, however, found two editions of this Table : one marked k, the other marked $*k$. My private copy is marked $*k$. The Observatory copy is marked k. The copy at the Royal Society is marked k. The copy at the Royal Astronomical Society $*k$. There is no doubt that the copy $*k$ is the more correct, and my results are derived from that copy. I have had the work done in duplicate, and I have examined it and it is substantially correct. There is only one unit in such inquiries in which the errors can be properly expressed, and that is in seconds of arc and the decimal parts of such a second. The errors in my table are therefore given in such units.

The following are the results :—

TABLE B

Group.	No. of Obs.	Corrections to Provisional Diameter.	Excess of Adopted Diameter over Observed.
	1750-1759	34	$+4\overset{''}{\cdot}686$	$-1\overset{''}{\cdot}012$
	1755-1764	29	$+4\cdot068$	$-0\cdot394$
	1760-1768	19	$+3\cdot762$	$-0\cdot088$
	1765-1773	18	$+4\cdot310$	$-0\cdot636$
5	1769-1778	22	$+4\cdot198$	$-0\cdot524$
6	1774-1782	18	$+3\cdot150$	$+0\cdot524$
7	1779-1787	16	$+2\cdot420$	$+1\cdot254$
8	1783-1791	18	$+2\cdot266$	$+1\cdot408$
9	1788-1796	24	$+2\cdot590$	$+1\cdot084$
10	1792-1801	24	$+3\cdot162$	$+0\cdot512$
11	1797-1805	15	$+3\cdot038$	$+0\cdot636$
12	1802-1810	13	$+3\cdot276$	$+0\cdot398$
13	1806-1815	12	$+4\cdot440$	$-0\cdot766$
14	1811-1819	8	$+5\cdot004$	$-1\cdot330$
15	1816-1824	9	$+2\cdot798$	$+0\cdot876$
16	1820-1829	17	$+2\cdot654$	$+1\cdot020$

The correction to provisional diameter to find adopted diameter is 3″·674. The results of this table are but slightly affected by any errors in the N.P.D. The corrections are certainly less in amount, but they follow the same general law as those given in Table A.

The deduced corrections to the adopted diameter do not sensibly agree amongst themselves for groups of observations extending over nine years. The errors are systematic. The diameters for a long period about 1783 are about a second of arc less, whilst those about 1765 are about half a second, and those about 1811 about a second of arc, greater than the mean value. It is certain, therefore, that if such corrections should be applied to coefficients of the parallactic inequality deduced from the series of observations, the result must be to strengthen the impression of the existence of a periodic term with a coefficient of about a second of arc and with a period of about forty-six years. And this is just what Mr. Neison found. (See *Notices*, 1880, May, pages 402–405.) Before corrections, such as those indicated, were applied, there appeared to be some indications of a periodic term in the coefficients of parallactic inequality with a coefficient of about a second of arc and a period of about thirty years ; but after the application of corrections deduced from the observed durations, a periodic term was supposed to be found with a period of forty-six years.

My position with respect to the discussion is simply as follows : (1.) That I have shown that these old observations, as reduced, are affected with such systematic errors connected with the diameter that they cannot be applied with safety to the determination of values of the coefficients of the parallactic inequality until the source of these errors has been determined and the effects removed from the results given in Section iii. (2.) That the so-called forty-six-year period is certainly mainly due to the same sources of error which directly affect the diameters. (3) That the introduction of an empirical periodic term with a coefficient of about a second of arc and a period of forty-six years as a *real term* in the expression for the tabular longitude of the Moon is a serious mistake, unless it can be justified by showing that the existence of such a term is a necessary consequence of the theory of gravitation. At present the only evidence in its favour is that such a term would approximately account for the appearance of some systematic discordance in the coefficients of the parallactic inequality as deduced from the results of Section iii. after correction for supposed errors of diameter. But it has been shown that these discordances are largely due to the errors of the diameters themselves; and the introduction of such a term in the longitude would not account for the existence of these errors in the diameter.

Until it has been shown that the results which I have given are substantially in error it will be useless for me to continue any discussion on these questions. They have already absorbed

much valuable time which I could ill spare, but the introduction
of inequalities of long period into the expression of the Moon's
longitude which are not necessary is a serious matter, and I
have thought it desirable at all events to call attention to what
I believe is a case in point.

---

*On a supposed Periodical Term in the Values found for the Co-
efficient of the Parallactic Inequality.* By E. Neison, Esq.

When Sir G. Airy determined the correction to the value of
the parallactic inequality, he contented himself with applying a constant correction for error of semi-diameter for each
of the two periods 1751–1815, and 1815–1839. But in the
investigation undertaken by Mr. Campbell and myself, this was
thought insufficient, and it was judged better to apply to each
group of observations the correction to the tabular semi-diameter
derived from the observations made during the same period.
And it was shown that the values of the parallactic inequality
when thus corrected strongly indicated the existence of a
periodical term with a period of about forty-five years.

It is now objected that these corrections to the semi-diameter
of the Moon exhibit a similar periodical term, and the conclusion
is drawn that the term shown by the values of the parallactic
inequality may have been introduced through the application of
these corrections to the semi-diameter, and so be fictitious.

It became necessary to investigate this point.

To establish this objection it is not merely sufficient to show
that there exists a periodical inequality in the correction applied
for errors of semi-diameter, not even if it be shown that this
inequality has a similar period; but it is imperative to show that
in the correction applied for error of semi-diameter, there exists
an inequality of the same value, period, and epoch as that found
in the values of the parallactic inequality, and that when this
inequality is eliminated from the corrections applied for errors
of semi-diameter, the inequality also disappears from the values
found for the parallactic inequality. Nothing less than this will
serve to establish the conclusion it has been sought to draw.

From Sir G. Airy's reduction of the Greenwich observations
(" Reduction of the Greenwich Lunar Observations," vol. i. pages
lxiv–lxv) are derived the following corrections to the tabular
semi-diameter of the Moon, divided into the same groups as those
used for determining the correction to the parallactic inequality,
and for convenience they have been expressed in terms of the
same unit as employed by Sir George Airy:—

No.	Group.	No. of Obs.	Corr.	No.	Group.	No. of Obs.	Corr.
i.	1750–1759	34	+·684	v.	1769–1778	22	+·608
ii.	1755–1764	29	+·592	vi.	1774–1782	19	+·392
iii.	1760–1768	19	+·560	vii.	1779–1787	17	+·280
iv.	1765–1773	18	+·648	viii.	1783–1791	18	+·300

No.	Group.	No. of Obs.	Corr.	No.	Group.	No. of Obs.	Corr.
ix.	1788–1796	24	+ ·356	xv.	1816–1824	9	+ ·408
x.	1792–1801	24	+ ·472	xvi.	1820–1829	17	+ ·400
xi.	1797–1805	15	+ ·464	xvii.	1825–1833	16	+ ·376
xii.	1802–1808	13	+ ·472	xviii.	1830–1838	11	+ ·408
xiii.	1806–1815	12	+ ·628	xix.	1834–1842	14	+ ·580
xiv.	1811–1819	8	+ ·740	xx.	1839–1847	20	+ ·492

The periodical term in the values of the parallactic inequality was taken as having the approximate value

$$B \cos b = +0\text{·}300 \cos \{8° \times [T - 1825\text{·}5]\}.$$

It was then assumed that a similar term existed in the above correction. In this manner each of the above values gave an equation of the form

$$S' + \cos b \times S'' = \text{correction to semi-diameter.}$$

This furnished twenty equations of condition, from which by the method of least squares there was derived the values

$$S' = +·490 \pm ·015; \qquad S'' = -·078 \pm ·022.$$

It was obvious, therefore, that the much greater value derived from the parallactic inequality could not have been introduced through the corrections to the semi-diameter. To render this unquestionable, the effect of this small inequality was altogether eliminated from the above corrections, by applying as the correction to the semi-diameter the quantity

$$\{\text{Correction to semi-diameter} - S'' \times \cos b\}.$$

The resulting values for the parallactic inequality furnished twenty equations of condition from which to determine B by the method of least squares. The result was

$$B = +·210 \pm ·040.$$

Thus, even after every trace of such an inequality had been eliminated from the values of the correction to the semi-diameter, the values of the parallactic inequality gave a coefficient three-fourths as large as before.

In the previous investigation, it was not thought necessary to attempt to attain any minute accuracy, as it was considered sufficient to adopt the first approximate determination of the value and period of this term. But from the first it was pointed out that by slightly altering the period and epochs of maximum, more accurate values could be obtained (see *Monthly Notices*, vol. xl. page 405). The previous investigation showed that the yearly motion of the argument should be slightly diminished, and instead of 8° it would be better to take $7\frac{1}{2}°$. It was assumed, therefore, that the argument was

$$b' = 7\tfrac{1}{2}° \times [T - 1826\cdot0],$$

and each value of the correction to the tabular semi-diameter was equated to an expression of the form

$$S' + sS'' + \sin b'S''' + \cos b'S''''.$$

The second term sS'' was introduced to allow for the effect of any difference which might have been produced in the value of the semi-diameter from the change of instruments in 1816, it being supposed zero after that date. The resulting twenty equations were solved by the method of least squares and gave the following values :—

$$S' = +0\cdot528$$
$$S'' = -0\cdot017$$
$$S''' = -0\cdot109$$
$$S'''' = -0\cdot088$$

The periodical term had the form, therefore,

$$= -0\cdot140 \cos \{b' - 51°\cdot1\}$$
$$= -0\cdot140 \cos \{7\tfrac{1}{2}° \times [T - 1832\cdot9]\}.$$

It is obvious that this term is in amount far smaller and in epoch very different from that existing in the values of the parallactic inequality.

The effects of this assumed term were now eliminated as before from the correction to the semi-diameter applied to the parallactic inequality. The resulting twenty values of the parallactic inequality were then equated to expressions of the form

$$A + a + B' \sin b' + B'' \cos b',$$

and the resulting equations solved as before. The result was the values

$$A = +0\cdot599$$
$$a = -0\cdot585$$
$$B' = -0\cdot114$$
$$B'' = +0\cdot200$$

Hence the periodical term existing in the values of the parallactic inequality, after carefully eliminating any portion introduced through the correction to semi-diameter, is

$$= +0\cdot230 \cos \{b + 29°\cdot6\}$$
$$= +0\cdot230 \cos \{7\tfrac{1}{2}° \times [T - 1822\cdot1]\}.$$

This result completely confirms that previously obtained.

It may be remarked that the comparison with observation shows that this adopted period is somewhat too small, and that the observations would be best satisfied by the term

$$= +0\cdot240 \cos \{7\tfrac{3}{4}° \times [T - 1824\cdot0]\}.$$

Converting into seconds of arc, this becomes

$$= + 0''96 \; \cos \{7\tfrac{3}{4}° \times [\text{T} - 1824\cdot0]\}.$$

It remains now to see how far this value agrees with the results of the Greenwich observations since 1851. We have (*Monthly Notices*, vol. xli. page 262)

$$
\begin{aligned}
1851\text{-}1858 \quad &(\text{P}) = -123''55 \\
1862\text{-}1869 \quad &= -125\cdot09 \\
1870\text{-}1876 \quad &= -125\cdot13
\end{aligned}
$$

Then applying the correction for this term, these become

$$
\begin{aligned}
1855\cdot0 \quad (\text{P}) &= -123''55 - \cdot55\text{B} = -124''08 \\
1866\,0 \quad &= -125\cdot09 + \cdot79\text{B} = -124\cdot33 \quad (\text{A}) \\
1873\cdot5 \quad &= -125\cdot13 + \cdot94\text{B} = -124\cdot23
\end{aligned}
$$

This term, therefore, still brings these three values into thorough accord.

If in obtaining the value for 1855·0, my own results be throughout adopted instead of using the Greenwich value for the correction to the semi-diameter—and this would perhaps be more satisfactory—then the result for this period must be increased by 0·''05 bringing it up to −124''·13 in still better accord.

If the effects of any assumed term in the correction to the semi-diameter be neglected, the value of the term in the parallactic inequality becomes

$$+ 1''12 \; \cos \{7\tfrac{3}{4}° \times [\text{T} - 1824\cdot5]\},$$

and the three groups become

$$
\begin{aligned}
1855\cdot0 \quad (\text{P}) &= -123''55 - \cdot49\text{B} = -124''10 \\
1866\cdot0 \quad &= -125\cdot09 + \cdot83\text{B} = -124\cdot16 \quad (\text{B}) \\
1873\cdot5 \quad &= -125\cdot13 + \cdot91\text{B} = -124\cdot11
\end{aligned}
$$

These results may be regarded as showing quite clearly that this periodical term exists in the values of the parallactic inequality itself, and are not introduced by the corrections which have been applied for the errors of tabular semi-diameter.

London :
    1882, *March* 25.

Addendum, 1882, June 1.—Since the preceding was written it has been shown by Mr. Stone (*Monthly Notices*, page 303) that the correction to Adams's value of the semi-diameter of the Moon, deduced by me from the observations of the years 1853-54-55, requires to be increased by +0''·78. As these comprise exactly a third of the total number of observations, the mean from the whole must be increased by +0''·26· Hence it

Après cette simple constatation de faits je demanderai aussi s'il est possible d'admettre que l'*Indus*, à peu près aussi large, et certainement aussi noir que la *mer de Kaiser* en 1881–82, peut n'être que la limite de deux régions inégalement sombres ou diversement colorées, imparfaitement vues ou imparfaitement distinguées l'une de l'autre ?

Measures of the Companion to Sirius. By S. W. Burnham, Esq.

During the present season I have made the following measures of the companion to *Sirius*:—

1881·835	42·6	9·72
81·840	43·9	9·56
81·843	43·6	9·20
81·859	43·1	8·87
81·865	43·2	9·62
82·085	47·4	9·48
82·088	44·4	9·64
82·102	43·6	9·11
82·104	43·4	9·35
82·145	41·7	9·25
82 148	43·0	9·36
Mean = 1881·99	43·6	9·38 11 nights.

The measures taken in 1881, the first five of the series, were made with the 12-inch Refractor of the Lick Observatory at Mount Hamilton. The other measures were made with the 18½-inch Refractor of the Dearborn Observatory.

Chicago :
1882, *May* 3.

On the Variable Star U Cephei. By George Knott, LL.B., B.A.

I succeeded in getting observations of four minima of U *Cephei* in March and April last, the results of which I give below, together with a comparison of the observed times with those deduced by carrying on Dr. Schmidt's Ephemeris (*A. N.* 2382) by the aid of his period 2·49277 days. In accordance with the practice generally adopted in the case of variable stars, the times are reduced to Paris Mean Time, by adding 9ᵐ·3 to the G. M. T. of observation, which was read off from the light-curve to the nearest minute. The observations show that the period is subject to irregularities—a result quite in accordance with those obtained by other observers.

Min. Observed.	Min. Calc.	C – O.	Obs. Mag.
d h m	h m	m	
1882, March 18 12 30·3	12 9·3	−21·0	9·5
April 7 11 14·3	10 45·9	−28·4	9·45
22 10 14·3	9 43·5	−30·8	9·5
27 9 54 3	9 22·7	−31 6	9·45

As is well known, the minima of this variable occur in sets, alternate minima being observable at intervals of about five days. The intermediate minima are observed when the next set comes round. For convenience I will call these A and B sets. From an examination of my observations I have a strong suspicion that there is a difference of about three-tenths of a magnitude between the minimum magnitude reached in the A and B sets respectively. I have now observed two sets of each, and in the following table give the magnitudes observed by me in the minima of each set respectively:—

A₁ Set.	B₁ Set.	A₂ Set.	B₂ Set.
m	m	m	m
9·1	9·4	9·2	9·5
9·1	9·4	9·2	9·45
9·2	9·4	9·2	9·5
9·1	9·4	9·25	9·45
	9·4		
	9·4		

The dates of the several sets are : *—

A₁	1880, Oct. 23	—	1881, Jan. 6
B₁	1881, Mar. 29	—	May 8
A₂	Sept. 27	—	Dec. 21
B₂	1882, Mar. 18	—	1882, April 27

The differences, though not great, are so fairly consistent as to lead me to think that they can hardly be due to accident merely. If further observation should confirm my suspicion, it would appear that we must regard the period as a double one.

Professor Pickering in his " Photometric Measurements of the Variable Stars β *Persei* and D.M. 81° 25 " (the latter being U *Cephei*) remarks of a neighbouring star D. M. 81° 18 that it " is either variable or its light in grades is erroneously given by M. Glasenapp." My own observations show that this star is certainly variable to the extent of some six-tenths of a magnitude, but I am unable as yet to see my way to any suggestion as to its probable period.

Knowles Lodge, Cuckfield :
    1882, *June* 8.

* The dates given are those of the first and last minimum observed in each set.

Ephemeris for Physical Observations

Greenwich Noon. 1882.	Angle of Position of ♃'s Axis.	Latitude of Earth above ♃'s Equator.	Sun above ♃'s Equator.	Annual Parallax A—L.	O—L.	Long. of ♃'s Central Meridian.
Aug. 2	357·538	+2·551	+2·680	−8·445	52·319	108·08
7	357·973	2·537	2·669	8·946	51·381	139·53
12	358·392	2·524	2·657	9·411	50·479	171·03
17	358·792	2·511	2·645	9·836	49·618	202·59
22	359·173	2·498	2·633	10·216	48·802	234·21
27	359·531	2·485	2 621	10·549	48·033	265·88
Sept. 1	359·866	+2·473	+2·609	−10·832	47·315	297·62
6	0·175	2·462	2·597	11·060	46·652	329·42
11	0·457	2·451	2·584	11·229	46·047	1·29
16	0·711	2·441	2·572	11·337	45·504	33·23
21	0·933	2·431	2·559	11·380	45·028	65·24
26	1·123	2·423	2·546	11·353	44·621	97·31
Oct. 1	1·278	+2·415	+2·533	−11·252	44·288	129·46
6	1·398	2·409	2·520	11·075	44·031	161·69
11	1·481	2·403	2·506	10·819	43·854	193·98
16	1·525	2·399	2·493	10·482	43·758	225·34
21	1·531	2·395	2·479	10·062	43·746	258·78
26	1·498	2·392	2·466	9·558	43·817	291·28
31	1·426	2·390	2·452	8·972	43·972	323·85
Nov. 5	1·316	+2·389	+2·438	−8·303	44·209	356·48
10	1·168	2·389	2·424	7·554	44·526	29·17
15	0·985	2·389	2·409	6·730	44·920	61·91
20	0·769	2·389	2·395	5·837	45·382	94·69
25	0·524	2·389	2·381	4·882	45·907	127·49
30	0·255	2·389	2·366	3·874	46·485	160·31
Dec. 5	359·965	+2·388	+2·351	−2·822	47·108	193·14
10	359·659	2·387	2·336	1·738	47·762	225·97
15	359·345	2·384	2·321	−0·634	48·436	258·77
20	359·027	2·381	2·306	+0·476	49·117	291·55
25	358·713	2·377	2 291	1·578	49·791	324·28
30	358·408	2·372	2·276	2·661	50·446	356·94

of Jupiter, 1882–83.   By A. Marth, Esq.

Corr. for Phase.	Equat. Diam.	Phase pr. limb.	Polar Diam.	Diff. of Limbs in R.A.	Phase pr. limb.	d.	w.	Greenwich Noon. 1882.	
°	″	″	″	s	s	°	°		
+0·31	34·28	0·19	32·10	2·479	0·013	8·44	268·86	Aug.	2
·35	34·63	·21	32·43	2·505	·015	8·94	268·88		7
·38	35·01	·24	32·78	2·533	·017	9·40	268·91		12
·42	35·41	·26	33·16	2·563	·019	9·83	268·93		17
·45	35·84	·28	33·56	2·594	·021	10·21	268·94		22
·48	36·30	·31	33·99	2·627	·022	10·54	268·96		27
+0·51	36·78	0·33	34·44	2·663	0·024	10·82	268·98	Sept.	1
·55	37·19	·35	34·92	2·701	·025	11·05	268·99		6
·55	37·83	·36	35·43	2·740	·026	11·22	269·01		11
·56	38·39	·37	35·96	2·781	·027	11·33	269·03		16
·56	38·98	·38	36·50	2·823	·028	11·37	269·05		21
·56	39·58	·39	37·06	2·866	·028	11·34	269·07		26
+0·55	40·20	0·39	37·64	2·911	0·028	11·24	269·10	Oct.	1
·53	40·83	·38	38·24	2·957	·028	11·06	269·13		6
·51	41·48	·37	38·84	3·003	·027	10·81	269·17		11
·48	42·12	·35	39·44	3·050	·025	10·47	269·21		16
·44	42·76	·33	40·04	3·097	·024	10·05	269·26		21
·40	43·39	·30	40·63	3·143	·022	9·55	269·32		26
·35	44·01	·27	41·21	3·187	·020	8·96	269·38		31
+0·30	44·60	0·23	41·76	3·230	0·017	8·29	269·46	Nov.	5
·25	45·15	·20	42·28	3·271	·014	7·55	269·55		10
·20	45·66	·16	42·76	3·308	·011	6·72	269·66		15
·15	46·12	·12	43·19	3·341	·009	5·83	269·81		20
·11	46·51	·08	43·55	3·370	·006	4·88	270·01		25
·07	46·84	·05	43·85	3·393	·004	3·87	270·27		30
+0·04	47·08	0·03	44·08	3·411	0·002	2·82	270·76	Dec.	5
+ ·01	47·24	·01	44·24	3·423	·001	1·74	271·73		10
·00	47·31	·00	44·31	3·428	·000	0·62	276·12		15
·00	47·29	foll. l	44·29	3·426	foll. l	0·48	80·33		20
− ·01	47·18	·01	44·18	3·418	·001	1·58	86·70		25
·03	46·99	·03	44·00	3·403	·002	2 66	87·83		30

Greenwich Noon. 1883.	Angle of Position of ♃'s Axis.	Latitude of Earth above ♃'s Equator.	Sun	Annual Parallax A – L.	O – L.	Long. of ♃'s Central Meridian.
	°	°	°	°	°	°
Jan. 4	358·117	+ 2·366	+ 2·260	+ 3·713	51·069	29·54
9	357·847	2·359	2·245	4·721	51·650	62·06
14	357·603	2 350	2·229	5·677	52·177	94·48
19	357·387	2·341	2 213	6·566	52·641	126·82
24	357·204	2·330	2·197	7·385	53·034	159·05
29	357·057	2·319	2·181	8·131	53·352	191·18
Feb. 3	356·947	+ 2·307	+ 2·165	+ 8·796	53·591	223·20
8	356·874	2·294	2·149	9·378	53·748	255·12
13	356·840	2·281	2·133	9·875	53·820	286·94
18	356·846	2 267	2·116	10·288	53·807	318·65
23	356·890	2·253	2·100	10 617	53·711	350·27
28	356·972	2·238	2·083	10·864	53·534	21·80
Mar. 5	357·091	+ 2·223	+ 2·066	+ 11·031	53·277	53·25
10	357·245	2·207	2·049	11·122	52 544	84·61
15	357·433	2·191	2·032	11·139	52·537	115·90
20	357 654	2·174	2 015	11·086	52·061	147·12
25	357·906	2·157	1·998	10·967	51·520	178·28
30	358·187	2·139	1·981	10·787	50 917	209 38
Apr. 4	358·495	+ 2 121	+ 1·964	+ 10·549	50·256	240·43
9	358 829	2·102	1·946	10·256	49·541	271·43
14	359·186	2 082	1·929	9·912	48·776	302 40
19	359·565	2·062	1·911	9 522	47·964	333·34
24	359·965	2 041	1·893	9·089	47·110	4·25
29	0·383	2·019	1·875	8·617	46·218	35·13
May 4	0 817	+ 1·996	+ 1·857	+ 8 110	45·290	66·co
9	1·267	1·973	1·839	7·569	44·330	96·85
14	1·730	1·949	1·821	6 997	43·339	127·69
19	2·204	1·924	1·813	6·397	42·318	158·53

Corr. for Phase.	Equat. Diam.	Phase pr. limb.	Polar Diam.	Diff. of Limbs in R.A.	Phase pr. limb.	d.	w.	Greenwich Noon. 1883.	
°	″	″	″	s	s	°	°		
− 0·06	46·71	0·05	43·74	3·382	0·004	3·71	88·30	Jan.	4
·10	46·35	·08	43·40	3·356	·006	4·72	88·62		9
·14	45·92	·11	43·00	3·325	·008	5·67	88·82		14
·19	45·43	·15	42·54	3·289	·011	6·56	88·94		19
·24	44·88	·19	42·03	3·249	·013	7·38	89·05		24
·29	44·30	·22	41·48	3·207	·016	8·12	89·14		29
−0·34	43·68	0·26	40·90	3·162	·019	8·79	89·20	Feb.	3
·38	43·03	·29	40·30	3·115	·021	9·37	89·25		8
·42	42·37	·31	39·68	3·068	·023	9·87	89·28		13
·46	41·71	·33	39·05	3·020	·024	10·28	89·31		18
·49	41·04	·35	38·42	2·972	·025	10·61	89·33		23
·51	40·37	·36	37·80	2·925	·026	10·85	89·35		28
−0·53	39·72	0·37	37·19	2·878	·027	11·02	89·35	Mar.	5
·54	39·08	·37	36·59	2·832	·027	11·11	89·35		10
·54	38·46	·36	36·01	2·787	·026	11·13	89·34		15
·53	37·86	·35	35·45	2·745	·026	11·08	89·34		20
·52	37·28	·34	34·91	2·704	·025	10·96	89·33		25
·51	36·73	·32	34·39	2·665	·024	10·78	89·31		30
−0·49	36·20	0·31	33·90	2·627	·022	10·54	89·29	Apr.	4
·46	35·71	·29	33·43	2·592	·021	10·25	89·27		9
·43	35·24	·26	32·99	2·559	·019	9·91	89·24		14
·39	34·79	·24	32·58	2·527	·017	9·52	89·21		19
·36	34·38	·22	32·20	2·498	·016	9·08	89·17		24
·32	34·00	·19	31·84	2·470	·014	8·61	89·14		29
−0·29	33·64	0·17	31·51	2·444	0·012	8·11	89·10	May	4
·25	33·32	·15	31·20	2·421	·011	7·57	89·06		9
·21	33·02	·12	30·92	2·399	·009	6·99	89·01		14
−0·18	32·75	0·10	30·67	2·379	0·007	6·39	88·96		19

The " Annual Parallax," $\Lambda - L$, is the difference of the Jovicentric longitudes of the Sun and the Earth, reckoned in the plane of *Jupiter's* equator; the angle $O - L$ is the difference of longitudes of *Jupiter's* vernal equinoctial point O and of the point of his equator which is in opposition to the Earth, or $180° + L - O$ is the Jovicentric longitude of the Earth reckoned from O.

The daily rate of rotation, on which the "Longitude of ♃'s Central Meridian" depends, is that adopted in last year's Ephemeris—namely, $870°·42$—the corresponding period being $9^h 55^m 34^s·47$. As the motion of the great reddish spot is not uniform, but slackening, it will be better for the present purpose not to make any alteration. The successive values of the Long. of C. Mer. differ, for an interval of five days, by twelve rotations and some thirty degrees, so that, for instance, the first difference is $4351°·45$ and the last $4350°·84$, which must be borne in mind in interpolating. If the "Corr. for Phase" is added to the "Longitude of ♃'s Central Meridian," or of the meridian directed to the Earth, the longitude of the meridian is found which bisects the illuminated disk of *Jupiter*. A list of the Greenwich mean times, when the adopted First (or Zero) Meridian passes the middle of the disk, will be given further on.

The assumed value of *Jupiter's* equatorial diameter is $37''·60$ at the distance $5·20273$. The assumed proportion of the polar axis to the equatorial diameter is $0·9363$. As during the present apparition the axis of *Jupiter* is little inclined to the circle of declination, the difference of limbs in declination does not sensibly differ from the polar diameter, and the defect of illumination or the phase in declination is insensible. The last columns give the values of the auxiliary angles d and w required in the computations, as explained in vol. xl. p. 490 ff.

The inclinations γ and the ascending nodes Γ of the orbits of the four satellites in reference to the plane of *Jupiter's* equator are the following, the nodes being reckoned from O, the point of the vernal equinox of *Jupiter's* northern hemisphere :—

1882.	Sat. I.		Sat. II.		Sat. III.		Sat. IV.	
	γ_1	Γ_1	γ_2	Γ_2	γ_3	Γ_3	γ_4	Γ_4
Aug. 2	0·0112	342·2	0·4783	344·04	0·1688	259·97	0·3228	331·06
Oct. 1	·0111	340·5	·4789	342·10	·1677	259·46	·3219	331·06
Nov. 30	·0110	338·8	·4794	340·16	·1667	258·94	·3210	331·05
1883. Jan. 29	·0110	337·0	·4799	338·22	·1658	258·53	·3201	331·01
Mar. 30	·0109	335·1	·4804	336·28	·1649	257·92	·3193	330·96
May 29	·0109	333·0	·4809	334·33	·1641	257·42	·3187	330·87

If these values of Γ are added to the elongations $O - L$ of the point O from superior conjunction, given before, the angles $\Gamma + O - L$ are the elongations of the ascending nodes of the orbits from superior conjunction; and the latitudes of the satellites above the plane of *Jupiter's* equator are easily found, if their elongations are known.

The following is a list of the Greenwich mean times, when the assumed First Meridian of *Jupiter* passes the middle of the illuminated disk :—

1882.	h	m	1882.	h	m	1882.	h	m	1882.	h	m
Aug. 2	16	52·0	Aug. 18	0	11·0	Sept. 2	7	29·2	Sept. 17	14	46·5
3	2	47·6		10	6·7		17	24·8	18	0	42·0
	12	43·3		20	2·3	3	3	20·4		10	37·6
	22	39·0	19	5	57·9		13	16·1		20	33·2
4	8	34·6		15	53·6		23	.11·7	19	6	28·8
	18	30·3	20	1	49·2	4	9	7·3		16	24·3
5	4	25·9		11	44·8		19	2·9	20	2	19·9
	14	21·6		21	40·5	5	4	58·5		12	15·5
6	0	17·2	21	7	36·1		14	54·1		22	11·1
	10	12·9		7	31·7	6	0	49·7	21	8	6·7
	20	8·6	22	3	27·4		10	45·3		18	2·3
7	6	4·2		13	23·0		20	40·9	22	3	57·9
	15	59·9		23	18·6	7	6	36·5		13	53·5
8	1	55·5	23	9	14·3		16	32·1		23	49·0
	11	51·2		19	9·9	8	2	27·7	23	9	44·6
	21	46·8	24	5	5·5		12	23·3		19	40·2
9	7	42·5		15	1·2		22	18·9	24	5	35·8
	17	38·1	25	0	56·8	9	8	14·6		15	31·4
10	3	33·8		10	52·4		18	10·2	25	1	26·9
	13	29·4		20	48·1	10	4	5·8		11	22·5
	23	25·1	26	6	43·7		14	1·4		21	18·1
11	9	20·7		16	39·3		23	57·0	26	7	13·7
	19	16·4	27	2	34·9	11	9	52·6		17	9·2
12	5	12·0		12	30·6		19	48·2	27	3	4 8
	15	7·7		22	26·2	12	5	43·8		13	0·4
13	1	3·3	28	8	21·8		15	39·3		22	55·9
	10	59·0		18	17·4	13	1	34·9	28	8	51·5
	20	54·6	29	4	13·0		11	30·5		18	47·1
14	6	50·3		14	8·7		21	26·1	29	4	42·6
	16	45·9	30	0	4·3	14	7	21·7		14	38·2
15	2	41·5		9	59·9		17	17·3	30	0	33·8
	12	37·2		19	55·5	15	3	12·9		10	29·3
	22	32·8	31	5	51·1		13	8·5		20	24·9
16	8	28·5		15	46·7		23	4·1	Oct. 1	6	20·5
	18	24·1	Sept. 1	1	42·4	16	8	59·7		16	16·0
17	4	19·7		11	38·0		18	55·3	2	2	11·6
	14	15·4		21	33·6	17	4	50·9		12	7·2

1882.	h	m	1882.	h	m	1882.	h	m	1882.	h	m
Oct. 2	22	2·7	Oct. 19	1	9·1	Nov. 4	4	14·3	Nov. 20	7	18·6
3	7	58·3		11	4·6	14		9·8		17	14·1
	17	53·8		21	0·1	5	0	5·3	21	3	9·6
4	3	49·4	20	6	55·7		10	0·8		13	5·1
	13	45 0		16	51·2		19	56·3		23	0·5
	23	40 5	21	2	46 7	6	5	51·8	22	8	56·0
5	9	36·1		12	42·2		15	47·3		18	51·5
	19	31·6		22	37·8	7	1	42·8	23	4	47·0
6	5	27·2	22	8	33·3		11	38·3		14	42·5
	15	22·7		18	28 8		21	33·8	24	0	38 0
7	1	18·3	23	4	24·4	8	7	29·3		10	33·5
	11	13·9		14	19·9		17	24·8		20	28·9
	21	9·4	24	0	15·4	9	3	20·3	25	6	24·4
8	7	5·0		10	10·9		13	15·8		16	19·9
	17	0·5		20	6·5		23	11·3	26	2	15·4
9	2	56·1	25	6	2·0	10	9	6·8		12	10·9
	12	51·6		15	57·5		19	2·3		22	6·3
	22	47·2	26	1	53·0	11	4	57·8	27	8	1·8
10	8	42·7		11	48·5		14	53·3		17	57·3
	18	38·3		21	44·1	12	0	48·8	28	3	52·8
11	4	33·8	27	7	39·6		10	44·3		13	48·3
	14	29·3		17	35·1		20	39·8		23	43·8
12	0	24·9	28	3	30·6	13	6	35·3	29	9	39·2
	10	20·4		13	26·1		16	30·8		19	34·7
	20	16·0		23	21·6	14	2	26·3	30	5	30·2
13	6	11·5	29	9	17·2		12	21·8		15	25·7
	16	7·1		19	12·7		22	17·3	Dec. 1	1	21·2
14	2	2 6	30	5	8·2	15	8	12·8		11	16·6
	11	58·2		15	3·7		18	8 2		21	12·1
	21	53·7	31	0	59·2	16	4	3·7	2	7	7·6
15	7	49·2		10	54·7		13	59·2		17	3·1
	17	44·8		20	50·2		23	54·7	3	2	58·5
16	3	40·3	Nov. 1	6	45·8	17	9	50·2		12	54·0
	13	35·9		16	41·3		19	45·7		22	49·5
	23	31·4	2	2	36·8	18	5	41·2	4	8	45·1
17	9	26·9		12	32·3		15	36·7		18	40·5
	19	22·5		22	47·8	19	1	32·2	5	4	35·9
18	5	18·0	3	8	23·3		11	27·6		14	31·4
	15	13·5		18	18·8		21	23·1	6	0	26·9

1882.	h	m	1882.	h	m	1883.	h	m	1883.	h	m
Dec. 6	10	22·4	Dec. 22	13	26·1	Jan. 7	6	35·4	Jan. 23	9	41·7
	20	17·9		23	21·6		16	30·9		19	37·3
7	6	13·3	23	9	17·1	8	2	26·5	24	5	32·8
	16	8·8		19	12·6		12	22·0		15	28·4
8	2	4·3	24	5	8·1		22	17·5	25	1	24·0
	11	59·8		15	3·6	9	8	13·0		11	19·6
	21	55·2	25	0	59·1		18	8·6		21	15·1
9	7	50·7		10	54·6	10	4	4·1	26	7	10·7
	17	46·2		20	50·1		13	59·6		17	6·3
10	3	41·7	26	6	45·6		23	55·2	27	3	1·9
	13	37·2		16	41·1	11	9	50·7		12	57·4
	23	32·6	27	2	36·6		19	46·2		22	53·0
11	9	28·1		12	32·1	12	5	41·8	28	8	48·6
	19	23·6		22	27·6		15	37·3		18	44·2
12	5	19·1	28	8	23·1	13	1	32·8	29	4	39·8
	15	14·6		18	18·6		11	28·4		14	35·4
13	1	10·0	29	4	14·1		21	23·9	30	0	31·0
	11	5·5		14	9·6	14	7	19·5		10	26·5
	21	1·0	30	0	5·1		17	15·0		20	22·1
14	6	56·5		10	0·6	15	3	10·6	31	6	17·7
	16	52·0		19	56·1		13	6·1		16	13·3
15	2	47·4	31	5	51·6		23	1·7	Feb. 1	2	8·9
	12	42·9		15	47·1	16	8	57·2		12	4·5
	22	38·4	1883.				18	52·8		22	0·1
16	8	33·9	Jan. 1	1	42·6	17	4	48·3	2	7	55·0
	18	29·4		11	38·2		14	43·9		17	51·3
17	4	24·9		21	33·7	18	0	39·4	3	3	46·9
	14	20·3	2	7	29·2		10	35·0		13	42·5
18	0	15·8		17	24·7		20	30·5		23	38·1
	10	11·3	3	3	20·2	19	6	26·1	4	9	33·7
	20	6·8		13	15·2		16	21·6		19	29·3
19	6	2·3		23	11·2	20	2	17·2	5	5	24·9
	15	57·7	4	9	6·8		12	12·7		15	20·5
20	1	53·2		19	2·3		22	8·3	6	1	16·1
	11	48·7	5	4	57·8	21	8	3·9		11	11·7
	21	44·2		14	53·3		17	59·4		21	7·3
21	7	39·7	6	0	48·8	22	3	55·0	7	7	2·9
	17	35·2		10	44·4		13	50·6		16	58·5
22	3	30·7		20	39·9		23	46·1	8	2	54·2

1883.	h	m	1883.	h	m	1883.	h	m	1883.	h	m
Feb. 8	12	49·8	Feb. 24	15	59·5	Mar. 12	19	10·7	Mar. 28	22	22·9
	22	45·4	25	1	55·2	13	5	6·4	29	8	18·7
9	8	41·0		11	50·8		15	2·0		18	14·4
	18	36·6		21	46·5	14	0	57·7	30	4	10·1
10	4	32·2	26	7	42·1		10	53·4		14	5·8
	14	27·8		17	37·8		20	49·1	31	0	1·5
11	0	23·5	27	3	33·4	15	6	44·8		9	57·2
	10	19·1		13	29·1		16	40·5		19	52·9
	20	14·7		23	24·8	16	2	36·2	Apr. 1	5	48·7
12	6	10·3	28	9	20·4		12	31·9		15	44·4
	16	5·9		19	16·1		22	27·6	2	1	40·1
13	2	1·6	Mar. 1	5	11·8	17	8	23·3		11	35·8
	11	57 2		15	7·4		18	19·0		21	31·5
	21	52·8	2	1	3·1	18	4	14·6	3	7	27·2
14	7	48·5		10	58·8		14	10·3		17	23·0
	17	44·1		20	54·4	19	0	6·0	4	3	18·7
15	3	39·7	3	6	50·1		10	1·7		13	14·4
	13	35·4		16	45·8		19	57·4		23	10·1
	23	31·0	4	2	41·4	20	5	53·1	5	9	5·8
16	9	26·6		12	37·1		15	48·8		19	1·5
	19	22·3		22	32·8	21	1	44·5	6	4	57·3
17	5	17·9	5	8	28·4		11	40·2		14	53·0
	15	13·5		18	24·1		21	35·9	7	0	48·7
18	1	9·2	6	4	19·8	22	7	31·6		10	44·4
	11	4·8		14	15·5		17	27·3		20	40·2
	21	0·5	7	0	11·1	23	3	23·0	8	6	35·9
19	6	56·1		10	6·8		13	18·7		16	31·6
	16	51·7		20	2·5		23	14·4	9	2	27·3
20	2	47·4	8	5	58·2	24	9	10·2		12	23·0
	12	43·0		15	53·9		19	5·9		22	18·8
	22	38·7	9	1	49·5	25	5	1·6	10	8	14·5
21	8	34·3		11	45·2		14	57·3		18	10·2
	18	30·0		21	40·9	26	0	53·0	11	4	5·9
22	4	25·6	10	7	36·6		10	48·7		14	1·7
	14	21·3		17	32·3		20	44·4		23	57·4
23	0	16·9	11	3	27·9	27	6	40·1	12	9	53·1
	10	12·6		13	23·6		16	35·8		19	48·9
	20	8·2		23	19·3	28	2	31·5	13	5	44·6
24	6	3·9	12	9	15·0		12	27·2		15	40·3

1883.	h	m	1883.	h	m	1883.	h	m	1883.	h	m
Apr. 14	1	36·0	Apr. 22	18	6·4	May 1	10	36·8	May 10	3	7·4
	11	31·7	23	4	2·1		20	32·6		13	3·1
	21	27·5		13	57·8	2	6	28·3		22	58·8
15	7	23·2		23	53·6		16	24·0	11	8	54·6
	17	18·9	24	9	49·3	3	2	19·8		18	50·3
16	3	14·7		19	45·0		12	15·5	12	4	46·1
	13	10·4	25	5	40·8		22	11·3		14	41·8
	23	6·1		15	36·5	4	8	7·0	13	0	37·6
17	9	1 8	26	1	32·2		18	2·7		10	33·3
	18	57·6		11	28·0	5	3	58·5		20	29·0
18	4	53·3		21	23·7		13	54·2	14	6	24·8
	14	49·0	27	7	19·4		23	50·0		16	20·5
19	0	44·8		17	15 2	6	9	45·7	15	2	16·3
	10	40·5	28	3	10·9		19	41·4		12	12·0
	20	36·2		13	6·7	7	5	37·2		22	7·8
20	6	32·0		23	2·4		15	32·9	16	8	3·5
	16	27·7	29	8	58·1	8	1	28·7		17	59·2
21	2	23·4		18	53·9		11	24·4	17	3	55·0
	12	19·2	30	4	49·6		21	20·1		13	50·7
	22	14·9		14	45·3	9	7	15·9		23	46·5
	22 8	10·6	May 1	0	41·1		17	11·6	18	9	42·2

In case the rate of slackening in the motion of the great reddish spot should continue, the preceding end of the spot will be near the middle of the disk, when observations become feasible in August, about 40 or 50 minutes after the passages of the assumed First Meridian, and the interval will then slowly increase.

Ephemerides of the Satellites of Saturn, August to November 1882.
By A. Marth, Esq.

The following ephemerides of the five inner satellites are founded upon the same elements which have served in the preparation of the ephemerides (published in *Astron. Nachrichten*) for the last apparitions of *Saturn*, as the conjunctions in right ascension of the satellites with the planet's centre, observed by Prof. Asaph Hall, and published on page 308 of the *Monthly Notices*, indicate that the elements are sufficiently near the truth for the present purpose. The five satellites deviate so little from the plane of the ring, that it will be most suitable to treat their deviation as latitudes above this plane, the ascending node N and inclination J of which in reference to the plane of the Earth's equator being here assumed to be

1882, Aug.	2	$N = 126\overset{\circ}{\cdot}4424$	$J = 7\overset{\circ}{\cdot}0157$
Sept.	1	·4446	·0154
Oct.	1	·4467	·0152
	31	·4499	·0151
Nov.	20	126·4545	7·0148

the longitudes N of the node being reckoned from the point of the true equinox.

The assumed longitudes of the satellites in their orbits (*i.e.* their longitudes from the ascending node added to the right ascension N of the ascending node), referring to the time when the light arrives at the distance, the logarithm of which is 0·950, are the following:

		♂ Gr.	*Mimas.*	*Enceladus.*	*Tethys.*	*Dione.*	*Rhea.*
1882, Aug.	2		$193\overset{\circ}{\cdot}238$	$320\overset{\circ}{\cdot}442$	$350\overset{\circ}{\cdot}024$	$263\overset{\circ}{\cdot}960$	$149\overset{\circ}{\cdot}920$
Sept.	1		132·956	282·399	310·969	250·012	20·625
Oct.	1		72·675	244·355	271·913	236·063	251·329
	31		12·395	206·312	232·857	222·115	122·033
Nov.	30		312·116	168·269	193·801	208·167	352·737

The adopted daily sidereal motions of the five satellites and the corresponding times of their sidereal revolutions are:

		d	h	m	s
Mimas	$381\overset{\circ}{\cdot}99063$	0	22	37	6·08
Enceladus	262·73186	1	8	53	6·86
Tethys	190·69812	1	21	18	25·96
Dione	131·53503	2	17	41	9·33
Rhea	79·69012	4	12	25	11·87

The values for the latter four satellites were given at the end of the ephemerides for 1877 in the *Astr. Nachr.* No. 2155. The values for *Mimas* belong to Oct. 1, 1882, and depend on the hypothesis of accelerated motion introduced in the ephemeris for 1880, *Astr. Nachr.* No. 2328. The question must at present, and till proper observations are available, remain an open one.

In the following tables P is the position-angle of the minor axis of the ring; L+180° the planetocentric longitude of the Earth referred to the plane of the ring; Λ+180° that of the Sun or Λ−L the difference between the two. The apparent equatorial diameter of the ball and the diameter of the outer rim of the ring depend on Bessel's determinations 17″·053 and 39″·311 for the distance, the logarithm of which is 0·9796480. The assumed proportion of the polar axis of the ball to the equatorial diameter is 0 : 900.

In the tables for the satellites a and b are the semi-axes of the apparent orbits, their values depending on Bessel's determination of the major axis of the orbit of *Titan*, and $l-L$ are the longitudes of the satellites in their orbits reckoned from the points which are in superior conjunction with the planet's centre or are in opposition to the Earth in longitude.

The values of P, a, b and $l-L$ are to be interpolated for the times for which the apparent positions of the satellites are required, and the rectangular coordinates x and y, reckoned parallel to the axes of the ring and expressed in seconds of arc, or, if polar-coordinates are wanted, the position-angles p and distances s of the satellites in reference to the centre of the planet, are then found by

$$s \sin (p-P) = x = a \sin (l-L)$$
$$s \cos (p-P) = y = b \cos (l-L).$$

$\odot$ Gr. 1882.	P	L	Latitude of Earth above plane of ring.	Sun	A−L
Aug. 2	357°795	53°981	−23°616	−22°186	−6°486
7	357·756	54·294	23·659	22·234	6·610
12	357·723	54·565	23·693	22·282	6·692
17	357·695	54·793	23·718	22·330	6·730
22	357·673	54·976	23·734	22·378	6·723
27	357·656	55·113	23·741	22·425	6·670
Sept. 1	357·645	55·203	−23·740	−22·473	−6·570
6	357·640	55·245	23·731	22·520	6·422
11	357·641	55·239	23·713	22·566	6·225
16	357·648	55·184	23·687	22·613	5·979
21	357·661	55·082	23·654	22·660	5·686
26	357·680	54·934	23·612	22·706	5·347
Oct. 1	357·704	54·740	−23·563	−22·752	−4·963
6	357·733	54·504	23·507	22·798	4·536
11	357·768	54·228	23·445	22·843	4·068
16	357·807	53·915	23·377	22·888	3·563
21	357·850	53·570	23·304	22·933	3·025
26	357·897	53·196	23·226	22·978	2·460
31	357·946	52·800	23·145	23·023	1·872
Nov. 5	357·998	52·386	−23·062	−23·067	−1·265
10	358·051	51·960	22·978	23·112	0·646
15	358·105	51·529	22·893	23·156	−0·022
20	358·159	51·098	22·809	23·199	+0·601
25	358·212	50·674	22·728	23·243	1·218
30	358·264	50·263	22·651	23·286	+1·822

o^h Gr.	Diam. of Ball			Axis of Ring		Mimas.			
	Equat.	Phase. pr. l.	Polar.	Major.	Minor.	a_1	b_1	l_1-L	Diff.
1882.									°
Aug. 2	17.37	0.056	15.93	40.05	16.04	27.32	−10.94	138.26	1909.82
7	17.52	.058	16.07	40.40	16.21	27.56	11.06	248.08	.86
12	17.68	.060	16.21	40.75	16.38	27.80	11.17	357.94	.90
17	17.84	.061	16.36	41.12	16.54	28.05	11.28	107.84	1909.96
22	18.00	.062	16.51	41.50	16.70	28.31	11.39	217.80	1910.00
27	18.17	.061	16.66	41.88	16.86	28.57	11.50	327.80	.04
Sept. 1	18.34	.060	16.81	42.27	17.02	28.83	−11.61	77.84	1910.08
6	18.50	.058	16.96	42.65	17.16	29.09	11.71	187.92	.13
11	18.67	.055	17.12	43.03	17.30	29.35	11.80	298.05	.18
17	18.83	.051	17.26	43.40	17.44	29.61	11.89	48.23	.21
21	18.99	.047	17.41	43.77	17.56	29.86	11.98	158.44	.25
26	19.14	.042	17.55	44.12	17.67	30.09	12.05	268.69	.29
Oct. 1	19.28	.036	17.68	44.45	17.77	30.32	−12.12	18.98	1910.32
6	19.42	.031	17.80	44.76	17.85	30.53	12.18	129.30	.34
11	19.54	.025	17.91	45.04	17.92	30.73	12.22	239.64	.37
16	19.65	.020	18.01	45.30	17.97	30.90	12.26	350.01	.39
21	19.75	.015	18.10	45.52	18.01	31.05	12.28	100.40	.40
26	19.83	.010	18.17	45.70	18.02	31.18	12.29	210.80	.41
31	19.89	.005	18.22	45.85	18.02	31.28	12.29	321.21	.41
Nov. 5	19.94	.003	18.26	45.96	18.00	31.35	−12.28	71.62	1910.40
10	19.96	.001	18.29	46.02	17.96	31.39	12.25	182.02	.39
15	19.97	foll. l.	18.29	46.04	17.91	31.40	12.22	292.41	.37
20	19.96	.001	18.28	46.01	17.84	31.39	12.17	42.78	.35
25	19.93	.003	18.25	45.94	17.75	31.34	12.11	153.13	.32
30	19.88	0.005	18.20	45.83	17.65	31.26	−12.04	263.45	1910.29

o^h Gr.	Enceladus.				Tethys.			
	a_2	b_2	l_2-L	Diff.	a_2	b_2	l_2-L	Diff.
1882.			°	°			°	°
Aug. 2	35.05	−14.04	265.77	1313.47	43.38	−17.38	295.54	953.27
7	35.35	14.19	139.24	.51	43.76	17.56	168.81	.31
12	35.66	14.33	12.75	.56	44.15	17.74	42.12	.35
17	35.99	14.48	246.31	.60	44.55	17.92	275.47	.40
22	36.32	14.62	119.91	.65	44.96	18.10	148.87	.45
27	36.65	14.76	353.56	.69	45.37	18.27	22.32	.49
Sept. 1	36.99	−14.89	227.25	1313.74	45.79	−18.43	255.81	953.54
6	37.32	15.02	100.99	.78	46.20	18.59	129.35	.58
11	37.66	15.14	334.77	.83	46.61	18.75	2.93	.63
16	37.98	15.26	208.60	.87	47.02	18.89	236.56	.67
21	38.30	15.37	82.47	.91	47.71	19.02	110.23	.71
26	38.61	15.46	316.38	.95	47.79	19.14	343.94	.75

Enceladus.

Tethys.

0ʰ Gr. 1882.	a_2	b_2	l_2-L	Diff.	a_3	b_3	l_3-L	Diff.
Oct. 1	38″90	−15′55	190°33	1313·99	48″15	−19″25	217°69	953·79
6	39·17	15·62	64·32	1314·01	48·49	19·34	91·48	·83
11	39·42	15·68	298·33	·04	48·79	19·41	325·31	·85
16	39·64	1573	172·37	·07	49·07	19·47	199·16	·88
21	39·83	15·76	46·44	·08	49·31	19·51	73·04	·90
26	40·00	15·77	280·52	·10	49·51	19·53	306·94	·92
31	40·12	15·77	154·62	·10	49·67	19·52	180·86	·93
Nov. 5	40·22	−15·75	28·72	1314·10	49·78	−19·50	54·79	953·93
10	40·27	15·72	262·82	·10	49·85	19·46	288·72	·92
15	40·29	15·67	136·92	·08	49·87	19·40	162·64	·92
20	40·26	15·61	11·00	·07	49·84	19·32	36·56	·90
25	40·20	15·53	245·07	·04	49·77	19·23	270·46	·88
30	40·10	−15·44	119·11	1314·01	49·64	−19·12	144·34	953·85

Dione.

Rhea.

0ʰ Gr. 1882.	a_4	b_4	l_4-L	Diff.	a_5	b_5	l_5-L	Diff.
Aug. 2	55″56	−22″26	209°64	657·42	77″59	−31″08	95°73	398·17
7	56·05	22·49	147·06	·47	78·27	31·41	133·90	·22
12	56·54	22·72	84·53	·51	78·96	31·73	172·12	·26
17	57·06	22·95	22·04	·55	79·68	32·05	210·38	·31
22	57·58	23·18	319·59	·60	80·41	32·36	248·69	·35
27	58·11	23·40	257·19	·65	81·15	32·67	287·04	·40
Sept. 1	58·64	−23·61	194·84	657·69	81·89	−32·97	325·44	398·45
6	59·17	23·81	132·53	·74	82·63	33·26	3·89	·49
11	59·70	24·01	70·27	·79	83·37	33·53	42·38	·54
16	60·22	24·19	8·06	·83	84·10	33·79	80·92	·58
21	60·72	24·36	305·89	·87	84·80	34·02	119·50	·63
26	61·21	24·52	243·76	·92	85·48	34·24	158·13	·68
Oct. 1	61·67	−24·65	181·68	657·96	86·12	−34·43	196·81	398·71
6	62·10	24·77	119·64	657·99	86·72	34·59	235·52	·75
11	62·49	24·86	57·63	658·02	87·27	34·72	274·27	·79
16	62·85	24·94	355·65	·05	87·76	34·82	313·06	·81
21	63·15	24·98	293·70	·08	88·19	34·89	351·87	·84
26	63·41	25·01	231·78	·09	88·55	34·92	30·71	·86
31	63·61	25·01	169·87	·10	88·84	34·92	69·57	·87
Nov. 5	63·76	−24·98	107·97	658·11	89·04	−34·88	108·44	398·88
10	63·85	24·92	46·08	·11	89·16	34·81	147·32	·89
15	63·87	24·85	344·19	·10	89·20	34·70	186·21	·88
20	63·84	24·75	282·29	·09	89·15	34·56	225·09	·87
25	63·74	24·63	220·38	·08	89·01	34·39	263·96	·85
30	63·58	−24·49	158·46	658·05	88·79	−34·19	302·81	398·83

Approximate Greenwich Mean Times of conjunctions of the satellites with the centre of *Saturn*:

"*n*" inferior conj. with centre, or satellite exactly in the direction of the minor axis of the ring, north, moving from the following to the preceding side.

"*s*" superior conjunction, or satellite south, moving from the preceding to the following side.

1882.	h		1882.	h		1882.	h	
Aug. 3	1·4	Rh. n.	Aug. 14	14·7	Te. n.	Aug. 23	9 5	Rh. s.
	3·4	Di. s.		16·6	En. s.		14·4	En. n.
	6·8	Te. n.	15	9·1	En. n.		16·2	Di. n.
4	5·4	Te. s.		11·1	Di. n.		18·9	Mi. n.
	12·3	Di. n.		13·3	Te. s.	24	1·2	Te. n.
5	4·1	Te. n.	16	12·0	Te. n.		6·8	En. s.
	7·6	Rh. s.		14·8	Rh. n.		17·5	Mi. n.
	21·1	Di. s.		17·2	Mi. s.		23·3	En. n.
6	2·7	Te. s.		17·9	En. n.		23·9	Te. s.
7	1·4	Te. n.		20·0	Di. s.	25	1·1	Di. s.
	6·0	Di. n.	17	10·4	En. s.		15·7	En. s.
	13·9	Rh. n.		10·6	Te. s.		15·7	Rh. n.
	20·2	En. s.		15·8	Mi. s.		16·1	Mi. n.
8	0·1	Te. s.	18	4·8	Di. n.		22·5	Te. n.
	12·6	En. n.		9·3	Te. n.	26	8·1	En. n.
	14·9	Di. s.		14·5	Mi. s.		9·9	Di. n.
	22·7	Te. n.		19·3	En. s.		14·7	Mi. n.
9	20·1	Rh. s.		21·1	Rh. s.		21 2	Te. s.
	21·4	Te. s.	19	7 9	Te. s.	27	13·3	Mi. n.
	21·5	En. n.		11·7	En. n.		17·0	En. n.
	23·7	Di. n.		13·1	Mi. s.		18·7	Di. s.
10	13·9	En. s.		13·7	Di. s.		19·8	Te. n.
	20·0	Te. n.	20	6 6	Te. n.		22·0	Rh. s.
11	6·4	En. n.		11·7	Mi. s.	28	9·5	En. s.
	8·6	Di. s.		20·6	En. n.		11·9	Mi. n.
	18·7	Te. s.		22·5	Di. n.		18·5	Te. s.
	22·8	En. s.	21	3·3	Rh. n.	29	3·6	Di. n.
12	2·4	Rh. n.		5·3	Te. s.		10·6	Mi. n.
	15·3	En. n.		13·0	En. s.		17·1	Te. n.
	17·4	Te. n.		21·6	Mi. n.		18·4	En. s.
	17·4	Di. n.	22	3·9	Te. n.		21·9	Mi. s.
13	7·7	En. s.		7·4	Di. s.	30	4·2	Rh. n.
	16·0	Te. s.		20·2	Mi. n.		10 8	En. n.
14	2·3	Di. s.		21·9	En. s.		12·4	Di. s.
	8·6	Rh. s.	23	2 6	Te. s.		15·8	Te. s.

1882.	h		1882.	h		1882.	h	
Aug. 30	20·5	Mi. s.	Sept. 10	1·0	Te. n.	Sept. 19	11·5	Te. n.
31	14·5	Te. n.		9·9	En. n.		12·0	Rh. s.
	19·1	Mi. s.		11·2	Di. s.		15·4	Mi. s.
	19·7	En. n.		11·2	Rh. s.	20	1·0	Di. n.
	21·3	Di. n.		16·6	Mi. n.		10·1	Te. s.
Sept. 1	10·4	Rh. s.		23·6	Te. s.		14·0	Mi. s.
	12·1	En. s.	11	15·2	Mi. n.		16·5	En. s.
	13·1	Te. s.		18·7	En. n.	21	8·8	Te. n.
	17·7	Mi. s.		20·0	Di. n.		8·9	En. n.
2	6·1	Di. s.		22·3	Te. n.		9·9	Di. s.
	11·8	Te. n.	12	11·2	En. s.		12·7	Mi. s.
3	10·4	Te. s.		13·8	Mi. n.		18·2	Rh. n.
	13·4	En. n.		17·4	Rh. n.	22	7·4	Te. s.
	15·0	Di. n.		20·9	Te. s.		11·3	Mi. s.
	16·6	Rh. n.	13	4·8	Di. s.		17·8	En. n.
4	9·1	Te. n.		12·4	Mi. n.		18·7	Di. n.
	13·6	Mi. s.		19·6	Te. n.	23	6·1	Te. n.
	22·3	En. n.		20·1	En. s.		9·9	Mi. s.
	23·8	Di. s.	14	11·0	Mi. n.		10·2	En. s.
5	7·7	Te. s.		12·5	En. n.	24	0·4	Rh. s.
	12·2	Mi. s.		13·7	Di. n.		3·5	Di. s.
	14·8	En. s.		18·2	Te. s.		4·7	Te. s.
	22·8	Rh. s.		23·6	Rh. s.		19·1	En. s.
6	6·4	Te. n.	15	16·9	Te. n.		19·8	Mi. n.
	7·2	En. n.		21·0	Mi. s.	25	3·4	Te. n.
	8·6	Di. n.		21·4	En. n.		11·5	En. n.
	10·8	Mi. s.		22·5	Di. s.		12·4	Di. n.
7	5·0	Te. s.	16	13·8	En. s.		18·4	Mi. n.
	9·4	Mi. s.		15·5	Te. s.	26	2·0	Te. s.
	16·1	En. n.		19·6	Mi. s.		6·6	Rh. n.
	17·5	Di. s.	17	5·8	Rh. n.		17·0	Mi. n.
	20·7	Mi. n.		7·4	Di. n.		20·4	En. n.
8	3·7	Te. n.		14·2	Te. n.		21·2	Di. s.
	5·0	Rh. n.		18·2	Mi. s.	27	0·7	Te. n.
	8·5	En. s.		22·7	En. s.		12·9	En. s.
	19·3	Mi. n.	18	12·8	Te. s.		15·7	Mi. n.
9	2·3	Di. n.		15·1	En. n.		23·3	Te. s.
	2·3	Te. s.		16·2	Di. s.	28	6·0	Di. n.
	17·4	En. s.		16·8	Mi. s.		12·8	Rh. s.
	18·0	Mi. n.	19	7·6	En. s.		14·3	Mi. n.

1882.	h		1882.	h		1882.	h	
Sept. 28	21·7	En. s.	Oct. 9	19·6	Rh. n.	Oct. 18	18·4	Di. s.
	22·0	Te. n.		20·8	En. s.		18·4	En. n.
29	14·9	Di. s.	10	5·7	Te. n.		20·3	Rh. n.
	20·6	Te. s.		8·9	Mi. s.	19	10·9	En. s.
30	18·9	Rh. n.		13·2	En. n.		16·2	Te. s.
	19·3	Te. n.		13·5	Di. s.		19·1	Mi. s.
	23·7	Di. n.		20·2	Mi. n.	20	3·3	Di. n.
Oct. 1	15·5	En. s.	11	4·4	Te. s.		14·8	Te. s.
	17·9	Te. s.		5·6	En. s.		17·7	Mi. s.
2	7·9	En. n.		18·9	Mi. n.		19·8	En. s.
	8·5	Di. s.		22·1	En. n.	21	2·4	Rh. s.
	16·5	Te. n.		22·3	Di. n.		12·1	Di. s.
3	1·1	Rh. s.	12	1·8	Rh. s.		12·2	En. n.
	15·2	Te. s.		3·0	Te. n.		13·5	Te. n.
	16·8	En. n.		14·5	En. s.		16·3	Mi. s.
	17·3	Di. n.		17·5	Mi. n.	22	12·1	Te. s.
	18·6	Mi. s.	13	1·7	Te. s.		14·9	Mi. s.
4	9·2	En. s.		6·9	En. n.		20·9	Di. n.
	13·8	Te. n.		7·1	Di. s.		21·1	En. n.
	17·3	Mi. s.		16·1	Mi. n.	23	8·6	Rh. n.
5	2·2	Di. s.	14	0·3	Te. n.		10·7	Te. n
	7·3	Rh. n.		8·0	Rh. n.		13·5	En. s.
	12·5	Te. s.		14·7	Mi. n.		13·5	Mi. s.
	15·9	Mi. s.		15·8	En. n.	24	5·7	Di. s.
	18·1	En. s.		16·0	Di. n.		5·9	En. n.
6	10·6	En. n.		22·9	Te. s.		9·4	Te. s.
	11·0	Di. n.	15	8·3	En. s.		12·1	Mi. s.
	11·1	Te. n.		13·3	Mi. n.		22·4	En. s.
	14·5	Mi. s.		21·6	Te. n.	25	8·0	Te. n.
7	9·8	Te. s.	16	0·8	Di. s.		10·8	Mi. s.
	13·1	Mi. s.		11·9	Mi. n.		14·5	Di. n.
	13·5	Rh. s.		14·1	Rh. s.		14·8	Rh. s.
	19·4	En. n.		17·1	En. s.		14·8	En. n.
	19·8	Di. s.		20·2	Te. s.	26	6·7	Te. s.
8	8·4	Te. n.	17	9·6	En. n.		7·3	En. s.
	11·7	Mi. s.		9·6	Di. n.		9·4	Mi. s.
	11·9	En. s.		10·5	Mi. n.		23·4	Di. s.
9	4·7	Di. n.		18·9	Te. n.	27	5·3	Te. n.
	7·1	Te. s.	18	9·2	Mi. n.		20·9	Rh. n.
	10·3	Mi. s.		17·5	Te. s.	28	4·0	Te. s.

1882.	h
	8·2 Di. n.
9	2·6 Te. n.
	16·5 Mi. n.
	17·0 Di. s.
	17·4 En. n.
30	1·2 Te. s.
	3·1 Rh. s.
	9 9 En. s.
	15·1 Mi. n.
	23·9 Te. n.
31	1·8 Di. n.
	13·7 Mi. n.
	18·8 En. s.
	22·5 Te. s.
Nov. 1	9·2 Rh. n.
	10·7 Di. s.
	11·2 En. n.
	12·4 Mi. n.
	21·2 Te. n.
2	11·0 Mi. n.
	19·5 Di. n.
	19·8 Te. s.
	20·1 En. n.
3	9·6 Mi. n.
	12·5 En. s.
	15·4 Rh. s.
	18·5 Te. n.
4	4·3 Di. s.
	4·9 En. n.
	8·2 Mi. n.
	17·1 Te. s.
	19·5 Mi. s.
	21·4 En. s.
5	13·1 Di. n.
	13·8 En. n.
	15·7 Te. n.
	18·1 Mi. s.
	21·5 Rh. n.
6	6·2 En. s.
	14·4 Te. s.

1882.	h
Nov. 6	16·7 Mi. s.
	21·9 Di. s.
	22·7 En. n.
7	13·0 Te. n.
	15·1 En. s.
	15·3 Mi. s.
8	3·7 Rh. s.
	6·8 Di. n.
	7·6 En. n.
	11·7 Te. s.
	14·0 Mi. s.
9	10·3 Te. n.
	12·6 Mi. s.
	15·6 Di. s.
	16·4 En. n.
10	8·9 En. s.
	9·0 Te. s.
	9·8 Rh. n.
	11·2 Mi. s.
11	0·4 Di. n.
	7·6 Te. n.
	9·8 Mi. s.
	17·7 En. s.
12	6·2 Te. s.
	8·4 Mi. s.
	9·2 Di. s.
	10·2 En. n.
	16·0 Rh. s.
13	4·9 Te. n.
	7·0 Mi. s.
	18·1 Di. n.
	18·3 Mi. n.
	19·1 En. n.
14	3·5 Te. s.
	5·6 Mi. s.
	11·5 En. s.
	16·9 Mi. n.
	22·1 Rh. n.
15	2·2 Te. n.
	2·9 Di. s.

1882.	h
Nov. 15	3·9 En. n.
	15·5 Mi. n.
	20·4 En. s.
16	0·8 Te. s.
	11·7 Di. n.
	12·8 En. n.
	14·2 Mi. n.
	23·5 Te. n.
17	4·3 Rh. s.
	5·2 En. s.
	12·8 Mi. n.
	20·5 Di. s.
	21·7 En. n.
	22·1 Te. s.
18	11·4 Mi. n.
	14·1 En. s.
	20·8 Te. n.
19	5·3 Di. n.
	6·6 En. n.
	10·0 Mi. n.
	10·4 Rh. n.
	19·4 Te. s.
20	8·6 Mi. n.
	14·2 Di. s.
	15·4 En. n.
	18·0 Te. n.
21	7·2 Mi. n.
	7·9 En. s.
	16·6 Rh. s.
	16·7 Te. s.
	23·0 Di. n.
22	15·3 Te. n.
	16·7 En. s.
23	7·8 Di. s.
	9·2 En. n.
	14·0 Te. s.
	22·7 Rh. n.
24	12·6 Te. n.
	16·6 Di. n.
	18·1 En. n.

1882.	h		1882.	h		1882.	h
Nov. 25	10·5 En. s.		Nov. 27	10·2 Mi. s.		Nov. 29	7·4 Mi. s.
	11·3 Te. s.			10·3 Di. n.			13·1 En. s.
	13·0 Mi. s.			11·8 En. n.			18·7 Mi. n.
26	1·5 Di. s.		28	4·2 En. s.		30	3·9 Di. n.
	4·9 Rh. s.			7·2 Te. n.			4·5 Te. n.
	9·9 Te. n.			8·3 Mi. s.			5·6 En. n.
	11·6 Mi. s.			11·1 Rh. n.			17·2 Rh. s.
	19·4 En. s.			19·1 Di. s.			
27	8·5 Te. s.		29	5·8 Te. s.			

By means of this list of conjunctions, approximate values of the coordinates x and y, expressed in semi-diameters of the planet's equator, may be easily found for any other time t in the following little table, the argument of which is the interval τ between the time t and the time of the nearest (preceding or following) conjunction.

τ h	Mimas.		Enceladus.		Tethys.		Dione.		Rhea.	
	x	y	x	y	x	y	x	y	x	y
0	0·0	1·2	0·0	1·6	0·0	2·0	0·0	2·5	0·0	3·5
1	0·9	1·2	0·8	1·6	0·7	2·0	0·6	2·5	0·5	3·5
2	1·7	1·0	1·5	1·5	1·4	1·9	1·2	2·5	1·0	3·5
3	2·3	0·8	2·2	1·3	2·0	1·8	1·8	2·4	1·5	3·5
4	2·8	0·5	2·8	1·1	2·6	1·7	2·4	2·3	2·0	3·4
5	3·1	0·2	3·3	0·9	3·2	1·5	2·9	2·2	2·5	3·4
6	3·1		3·7	0·6	3·7	1·3	3·5	2·1	3·0	3·3
7	2·9		3·9	0·3	4·1	1·1	4·0	2·0	3·5	3·2
8	2·5		4·0	0·1	4·5	0·9	4·4	1·8	4·0	3·2
9					4·7	0·7	4·8	1·7	4·5	3·1
10					4·9	0·4	5·2	1·5	4·9	3·0
11					5·0	0·1	5·6	1·3	5·3	2·9
12							5·8	1·0	5·7	2·7
13							6·1	0·8	6·1	2·6
14							6·2	0·6	6·5	2·5
15							6·3	0·3	6·8	2·3
16							6·4	0·1	7·1	2·1
18									7·7	1·8
20									8·2	1·4
27									8·9	0·0

Differences of Right Ascension and Declination between Iapetus and the Centre of Saturn.

♂ Gr. 1882.	a−A s	δ−D ″	♂ Gr. 1882.	a−A s	δ−D ′
Aug. 2	−24·10	−187·0	Sept. 7	34·84	192·5
3	22·05	188·4	8	33·49	198·8
4	19·85	188·5	9	31·95	204·0
5	17·50	187·3	10	30·22	208·0
6	15·02	184·9	11	+28·30	+210·8
7	12·43	181·3	12	26·21	212·4
8	9·76	176·5	13	23·96	212·8
9	7·01	170·5	14	21·56	211·9
10	4·21	163·4	15	19·03	209·8
11	− 1·37	155·2	16	16·37	206·4
12	+ 1·50	−146·0	17	13·61	201·8
13	4·36	135·8	18	10·76	196·0
14	7·21	124·7	19	7·83	188·9
15	10·03	112·8	20	4·84	180·7
16	12·80	100·1	21	+ 1·81	+171·4
17	15·49	86·7	22	− 1·24	161·0
18	18·10	72·7	23	4·29	149·5
19	20·61	58·3	24	7·33	137·1
20	23·00	43·4	25	10·33	123·8
21	25·27	28·2	26	13·28	109·7
22	+27·39	− 12·8	27	16·15	94·8
23	29·35	+ 2·7	28	18·93	79·3
24	31·14	18·3	29	21·59	63·2
25	32·76	33·9	30	24·12	46·6
26	34·19	49·3	Oct. 1	−26·50	+ 29·7
27	35·42	64·4	2	28·72	+ 12·6
28	36·45	79·3	3	30·75	− 4·7
29	37·27	93·7	4	32·58	22·0
30	37·88	107·6	5	34·20	39·3
31	38·27	121·0	6	35·59	56·3
Sept. 1	+38·44	+133·7	7	36·74	72·9
2	38·39	145·7	8	37·64	89·1
3	38·12	156·9	9	38·28	104·8
4	37·62	167·3	10	38·66	119·7
5	36·91	176·7	11	−38·77	−133·9
6	35·98	185·1	12	38·61	147·1

0ʰ Gr. 1882.	α—A	δ—D	0ʰ Gr. 1882.	α—A	δ—D
Oct. 13	38·18	159·4	Nov. 6	23·46	59·6
14	37·47	170·5	7	26·11	42·5
15	36 50	180·5	8	28·59	25·1
16	35·27	189·2	9	30·88	− 7·5
17	33·78	196·6	10	+ 32·98	+ 10·2
18	32·05	202·6	11	34·87	27·6
19	30·09	207·2	12	36·55	44·9
20	27·92	210·3	13	38·00	62·0
21	−25·54	−212·0	14	39·21	78·7
22	22·98	212·1	15	40·18	94·9
23	20·25	210·8	16	40·90	110·6
24	17·37	208·0	17	41·36	125·6
25	14·37	203·7	18	41·58	139·8
26	11·27	197·9	19	41·54	153·1
27	8·08	190·8	20	+ 41·26	+ 165·4
28	4·82	182·4	21	40·72	176·9
29	− 1·53	172·7	22	39·94	187·3
30	+ 1·77	161·8	23	38·92	196·5
31	5·06	−149·8	24	37·66	204·5
Nov. 1	8·33	136·7	25	36·18	211·3
2	11·55	122·7	26	34·49	216·8
3	14·69	107·9	27	32·59	221·0
4	17·73	92·4	28	30·50	224·0
5	20·66	76·3	29	28·23	225·6
			Nov. 30	+ 25·79	+ 225·8

It would be well to test the correctness of this ephemeris by some early observations of *Iapetus*.

Ephemeris of the Satellite of Neptune, 1882–83.
By A. Marth, Esq.

P, angle of position of the minor axis of the satellite's apparent orbit, in the direction of superior conjunction.

a, b, major and minor semi-axis of the apparent orbit.

u−U, longitude of the satellite in its orbit reckoned from the point which is in superior conjunction with the planet, or in opposition to the Earth.

U+180°, planetocentric longitude of the Earth, reckoned in the satellite's orbit from the ascending node on the equator.

B, latitude of the Earth above the plane of the orbit.

o^h Gr.	P	a	b	u−U	Diff.	U	B
1882.							
Aug. 22	317·28	16·52	7·01	301·72	612·59	140·13	25·10
Sept. 1	317·26	16 61	7·04	194·31	·54	140·16	25·10
11	317·20	16·70	7 07	86·85	·48	140·25	25·07
21	317·11	16·78	7·09	339·33	·43	140·38	25·01
Oct. 1	316·99	16·85	7·10	231·76	·38	140·56	24·94
11	316·84	16 90	7·10	124·14	·35	140·78	24·85
21	316·67	16·94	7·09	16·49	·32	141·02	24·74
31	316·49	16·97	7·07	268·81	·30	141·29	24·61
Nov. 10	316·30	16·97	7·03	161·11	·29	141·56	24·48
20	316·12	16·96	6·99	53·40	·29	141·83	24·35
30	315·94	16·93	6·95	305·69	·30	142·09	24·23
Dec. 10	315·78	16·89	6·90	197·99	·33	142·33	24·11
20	315 64	16·83	6·84	90·32	·36	142·53	24·00
30	315·53	16·75	6·79	342·68	·41	142·69	23·91
1883.							
Jan. 9	315·46	16·67	6·74	235·09	·45	142·80	23·85
19	315·42	16·58	6·69	127·54	·40	142·86	23·81
29	315·41	16·48	6·65	20·04	·50	142·87	23·80
Feb. 8	315·44	16·39	6·62	272·60	·62	142·81	23·81
18	315·51	16·30	6·59	165·22	·67	142·71	23·85
28	315·61	16·21	6·57	57·89	612·72	142·55	23·91
Mar. 10	315·75	16·12	6·56	310·61		142·35	24·00

If the values of P, a, b, and u−U are interpolated for the times for which the apparent places of the satellite are required, the position-angle p and distances s are found by

$$s \sin (P-p) = a \sin (u-U),$$
$$s \cos (P-p) = b \cos (u-U).$$

The satellite moves in the direction of *decreasing* position-angles, and will be at its greatest elongations ("*nf*" in posit. P+90° and distance a, "*sf*" in posit. P−90°) and at its conjunctions ("*sup.*" in posit. P and distance b, "*inf.*" in posit. P−180°) at the following hours, Greenwich M. T.:—

"nf." elong.		"sup." conj.		"sp." elong.		"inf." conj.	
1882.	h		h		h		h
Aug. 21	11·6	Aug. 22	22·8	Aug. 24	10·1	Aug. 25	21·3
27	8·6	28	19·9	30	7·1	31	18·4
Sept. 2	5·7	Sept. 3	16·9	Sept. 5	4·2	Sept. 6	15·4
8	2·7	9	14·0	11	1·2	12	12·5
13	23·8	15	11 0	16	22·3	18	9·6

" nf." elong.		" sup." conj.		" sp." elong.		" inf." conj.	
1882.	h		h		h		h
Sept. 19	20·8	Sept. 21	8·1	Sept. 22	19·4	Sept. 24	6·6
25	17·9	27	5·2	28	16·4	30	37
Oct. 1	15·0	Oct. 3	2·3	Oct. 4	13·5	Oct. 6	0·8
7	12·1	8	23·3	10	10·6	11	21·9
13	9·2	14	20·4	16	7·7	17	19·0
19	6·3	20	17·5	22	4·8	23	16·1
25	3·4	26	14·6	28	1·9	29	13·2
31	0·5	Nov. 1	11·7	Nov. 2	23·0	Nov. 4	10·3
Nov. 5	21·6	7	8·9	8	20·1	10	7·4
11	18·7	13	6·0	14	17·2	16	4·5
17	15·8	19	3·1	20	14·3	22	1·6
23	12·9	25	0·2	26	11·5	27	22·7
29	10·0	30	21·3	Dec. 2	8·6	Dec. 3	19·8
Dec. 5	7·1	Dec. 6	18·4	8	5·7	9	16·9
11	4·2	12	15·5	14	2·8	15	14·0
17	1·3	18	12·6	19	23·9	21	11·1
22	22·4	24	9·7	25	21·0	27	8·2
28	19·5	30	6·8	31	18 1	Jan. 2	5·3
1883.							
Jan. 3	16·6	Jan. 5	3·9	Jan. 6	15·1	8	2·4
9	13·7	11	0·9	12	12·2	13	23·5
15	10·8	16	22·0	18	9·3	19	20·6
21	7·8	22	19·1	24	6·4	25	17·6
27	4·9	28	16·1	30	3·4	31	14·7
Feb. 2	1·9	Feb. 3	13·2	Feb. 5	0·5	Feb. 6	11·7
7	23·0	9	10·2	10	21·5	12	8·8
13	20·0	15	7·4	16	18·5	18	5·8
19	17·0	21	4·3	22	15·6	24	2·8
25	14·1	27	1·3	28	12 6	Mar. 1	23·8
Mar. 3	11·1	Mar. 4	22·3	Mar. 6	9·6	7	20 8

The Solar Eclipse, 1882, *May* 16.   By the Rev. S. J. Perry.

With the view of comparing the times of spectroscopic and of ordinary telescopic contact, I arranged to have the partial eclipse of May 16 observed by Mr. W. McKeon with the 8-inch

Equatoreal and a Browning automatic spectroscope of six prisms, and by Mr. J. Rooney with the 9½-inch Cassegrain and an ordinary negative eyepiece. The sky was perfect, and the dispersion used with the spectroscope was that of ten prisms of 60°, the pencil of light passing twice through each of the first five prisms of the instrument. The slit was kept tangent to the limb of the Sun at the calculated point of contact.

At G.M.T. 18ʰ 19ᵐ 21ˢ·85 the Moon was first seen on the upper limit of the chromosphere.

„ „ 18ʰ 19ᵐ 38ˢ·00 the Moon had half covered the chromospheric layer.

„ „ 18ʰ 19ᵐ 48ˢ·97 the Moon reached the base of the chromosphere, and the solar spectrum was darkened at the limb.

With the Cassegrain—

At G.M.T. 18ʰ 19ᵐ 36ˢ·00 the limb of the Sun near the point of contact suddenly became quite steady.

„ „ 18ʰ 19ᵐ 56ˢ·50 a very minute indentation was first observed. Definition excellent.

At the end of the eclipse—

G.M.T. 19ʰ 21ᵐ 9ˢ·90 the chromosphere was first seen at the point of last contact; possibly a second or so late.

„ 19ʰ 21ᵐ 20ˢ·50 the last contact was observed with the Cassegrain. Observation excellent.

The points of the solar limb cut by the Moon were watched very narrowly during the progress of the eclipse, and the chromosphere was always observed to be cut very sharply by the Moon's limb.

The bright chromospheric lines C and F near the lunar edge were seen projected far within the spectrum of the photosphere, and were both widened and curved, the displacement being very marked from the juxtaposition of the dark and bright lines.

The dark atmospheric lines were also observed to be curved in close proximity to the Moon.

Four faint bright lines were traceable between B and C in the solar spectrum.

After the eclipse, the whole chromosphere was measured. The mean height near the point of first contact was found to be 8″·54, with a slight prominence, whose maximum height was

22″·64, at 2° distance. The mean height of the chromosphere near the point of last contact was 10″·68, prominences being at 2° and 1° 30′ on either side, whose maximum heights were 33″·32 and 45″·28 respectively.

The director of the Lyons Observatory, M. Ch. André, has kindly sent me an account of his observations during the same eclipse; and I venture, with his permission, to extract a few lines from his letter, as they are of considerable interest at the present moment.

"Une observation, que nous venons de faire lors de l'éclipse partielle du 16 mai 1882, vient à l'appui de ces mêmes idées [explanation of the phenomena of the black drop by M. André]. Si l'on cherche à observer l'éclipse d'une tâche solaire par la lune en notant par exemple le contact d'un des bords de la tâche avec le bord de l'échancrure solaire, on observe constamment un ligament noir analogue à celui du passage de *Vénus* ou de *Mercure*. Après reflexion, ce fait, qui m'avait d'abord surpris, est très simple. Le phénomène en question est un effet du même ordre que celui du passage de *Vénus* : il en diffère en ce que

" 1° L'obscurité de la tâche est moindre que celle de *Vénus*.

" 2° Le corps obscur est fixe et le corps lumineux mobile.

" 3° Les deux corps qui arrivent au contact tournent ici leurs courbures en sens inverse.

"Les ligaments ne sont donc point, comme quelques astronomes le croient, des apparences spéciales aux passages des planètes sur le soleil. Ce sont bien certainement des phénomènes de diffraction dus à l'appareil optique qui sert aux observations ; on doit les traiter comme tels."

Another remark in the same letter also bears upon this subject. "L'observation du passage de *Mercure* à Ogden du 4 mai 1878 a complètement verifié toutes les conséquences que j'avais déduites de mes expériences."

Stonyhurst Observatory:
     1882, *June* 6.

Further Spectroscopic Observations of Comet a, 1882 (*Wells*), *made at the Royal Observatory, Greenwich.* ·

(*Communicated by the Astronomer Royal.*)

The spectrum of this interesting object has been observed on four occasions since the date of the first Note—viz. on May 13, May 20, May 31, and June 7. The most striking feature which it has exhibited has been the remarkable development of a bright line in the yellow, apparently coincident with the D lines of sodium.

This line was first remarked on May 31, when two measures of its position were obtained with the single-prism spectroscope. These gave its wave-length as 5902·5 and 5903·0 tenth-metres, or 10·5 and 11·0 tenth-metres from D towards the red. The measures were made with great difficulty, owing to the exceedingly awkward position of the spectroscope, and are hence very rough. The bright line seemed two or three times as bright as the continuous spectrum on which it appeared.

Another view of the comet was secured early in the morning of June 8 (at about June 7, $15^h 40^m$), when the bright line had become of the most extraordinary brilliancy, and so far outshone the continuous spectrum that the comet might almost be said to shine by monochromatic light. The telescopic appearance of the comet was in strict accordance with this view, as it showed a distinct planetary disk of an orange color, nearly as deep and vivid as that of *Mars*. The comet was seen in the finder till $15^h 50^m$, 4^m after sunrise. Only one measure could be obtained of the position of the line, and this measure was unfortunately rendered very rough from the difficulties of the comet's position, &c. The resulting wave-length was 5894·5 tenth-metres, assuming 5892·0 as the wave-length of the D lines. The displacement, which was towards the red, would indicate a motion of recession of 79 miles per second. The "Half-prism" spectroscope, with one half-prism, in the direct position, as used for measures of displacement of the lines in stellar spectra, was employed; magnifying power 14°, dispersion $18\frac{1}{2}°$ from A to H.

Besides this remarkable bright line, several trifling irregularities in the continuous spectrum were seen or suspected on different occasions.

On April 24, as stated in the former Note, two ill-defined maxima were remarked in the green and greenish-blue.

On May 13 one such ill-defined maximum was suspected near E, a little to the red of the sharp edge of the green band obtained from a Bunsen-flame.

On May 20 the spectrum presented much the same appearance, except that the fainter districts bounding the two maxima above mentioned rather attracted attention and three minima were observed instead of two maxima. These minima, or faint dark bands, seemed fairly sharp at the edges. Three measures were obtained of the most distinct, which was, however, an exceedingly difficult object. These gave its wave-length as 4818, 4811, and 4855 tenth-metres respectively. The middle dark band was measured as about 5500 tenth-metres. No measures could be obtained of the third dark band, which was near D but on the blue side of it.

The spectrum of the nucleus extended from about λ 4300 to λ 6150. The spectrum of the tail was also continuous, but visible only in the green. It might be mistaken for the green band of the Bunsen-flame, but seemed too regular and to extend too far towards the red.

On May 31 the spectrum presented a very similar appearance to that shown on May 20, but, being much brighter, details were better seen.

Two dark spaces were seen near F; the less refrangible one was measured, and its wave-length determined as 4862 tenth-

metres. It therefore probably is the F line. A bright streak
was seen further towards the red; this was measured as 5146
tenth-metres. A second bright streak was measured at 5328
tenth-metres. A dark band about 120 tenth-metres in breadth
was seen close to the bright line near D, but nearer the blue.

On May 22 and June 7 none of these minor irregularities in
the continuous spectrum were remarked. At best they were exceedingly difficult and ill-defined objects, and would easily be
overpowered by the strong twilight in which the comet was
observed on those two occasions. The observations were made
by Mr. Maunder throughout.

Royal Observatory, Greenwich:
1882, June 9.

Postscript.—The comet was seen by Mr. Maunder in full day-
light, just after perihelion passage, on June 10, 20^h, with the 12$\frac{3}{4}$-
inch Equatoreal, using negative eyepieces with powers of 60, 130,
220, and 310. With the lowest power, a light-red glass was
necessary, but with the powers 220 and 310 the light of the sky
was not too strong for the eye, and the comet could be easily seen
as a dull yellow stellar point of light, the disk being no larger
than that of a star. It was judged to be not so bright as *Capella*,
which had been seen previously, making every allowance for the
much brighter background on which the comet was seen. On
putting in the transit-micrometer (which has a much larger field)
to determine its place, the comet was lost, after having been seen
distinctly for a quarter of an hour, and it was not picked up
again afterwards. A white haze had gradually formed round
the Sun, and the sky became overcast about noon.

On the Spectrum of Comet Wells.   By M. N. C. Dunér.

(*Communicated by W. Huggins, Esq.*)

On the evening of June 3 in observing the spectrum of the
Comet Wells I remarked in the continuous spectrum of the
nucleus a considerably bright dot, which by reducing the breadth
of the slit turned out to be a narrow line. The following measurements of the wave-length show that the line was due to the
presence of sodium :—

λ
5897
5895
5894

The first set is much less accurate than the two others, being
determined with a less powerful ocular.

The bright line was not limited to the spectrum of the nucleus, but extended itself, though considerably fainter, far into the coma.

The observation was made with the Refractor of the Observatory at Lund (aperture 245 millimètres) and a universal spectroscope by Merz with one direct vision prism.

Lund : 1882, *June* 5.

Elements of Comet Wells, obtained graphically.
By F. C. Penrose, Esq.

T, very roughly computed			June 10·0	
$\pi - \Omega$	...	...	...	209° 0′
Ω	...	...	...	205 30
i	...	...	...	74 22
D or q	...	...	...	[8·77815]

Observations of Comet b, 1881.   By T. W. Backhouse, Esq.

The accompanying table gives the brightness of the head of this comet—*i.e.* the nucleus with its surrounding nebulosity ; the length and greatest width of the tail, as measured on Proctor's map from my drawings and notes of its position; and the direction of the tail—*i.e.* of the middle when it was of uniform brightness, or of the brightest line when one part was decidedly brighter than the rest of the width. The next column gives the part to which the observation of its direction refers. The last column but one gives the mode of observation: n. e. indicating the naked eye; specs., spectacles to correct my short sight; (3·5), a pair of field-glasses, power 3·5, aperture 1½ in.; (5) and (2·2), single field-glasses of these respective powers ; and (20) and (38), these powers of my Cooke's 4¼-in. Refractor. The last column gives the hindrances to a perfect view. There was more or less twilight in all the earlier observations, if not up to August 3 ; though at midnight on July 2, 3, and 4, and from July 22, it was almost imperceptible at the comet's position.

The comet was quite white. This circumstance I noted from June 29 to July 14.

Observations of Comet b, 1881.

Place and Time.	Brightness of Head.	Length of Tail.	Max. Width of Tail.	At Distance from Nucleus.	Central or Brightest Line of Tail: passes or points to	Central or Brightest Line of Tail: at distance from nucleus.	Modes of Observation.	Hindrances.
		°	°	°		°		
Tyndrum, Perthshire.								
June 29, 13h 18m	—	3½	3/5		About Polaris		(3·5)	Twilight
"	= α Urs. Maj.*; considerably fainter than Polaris †	Abt. 4 §			—		n. e.	"
13 29	—	Abt. 5½ §			—		"	"
Sunderland.								
July 1, 11 32	—	7½			3/5 (α, δ) Urs. Min.		"	"
11 50	—	13			1/3 (δ, ε) "		"	"
12 20	—	14½			δ Urs. Min. 1/4 (δ, ε) Urs. Min.	3 14½	"	"
12 40	Between ε* and ζ* Urs. Maj.; brighter than α*; brighter than γ Cass.*		1¾	10	—		"	"
Newhaven, Sussex.								
July 2, 11 5	= ε Urs. Maj.*; brighter than α Urs. Maj.	17½	2½	12	3/5 (α, δ) Urs. Min. 1/3 (δ, ε) "	for first 3° 4° to 17½°	"	" Not very clear
Abt. 12 30		20½	4¼	18	M. Cam. Slightly *f* P VI. 292 2/3 ? (P XII. 232, δ Urs. Min.	5 10½ 20		"

Place and Time.	Brightness of Head.	Length of Tail.	Max. Width of Tail.	At Distance from Nucleus.	Central or Brightest Line of Tail: passes or points to	Central or Brightest Line of Tail: at distance from nucleus.	Modes of Observation.	Hindrances.
Newhaven, Sussex.		°	°	°		°		
July 3, 10 5	—				M. Cam.	3½	„	Twilight
10 25	—	12½	1¼	8	3/5 (δ, ε) Urs. Min. 2/3 (δ, ε) „	lower part most of tail	„	„
11 20	—				Slightly *p* M Cam. M. Cam.	up to 1¾° 3½	(5)	„
12 10	—	17	3¾	12¼	P VI. 292 3/7 (α Urs. Min., P XII. 232)	9 16	n. e.	„
12 15	= α Urs. Min.*; = η Urs. Maj.*				—		„	„
4, 12 5	—	16	4½	11	P VI. 292, or perhaps a little *f* it 5/8 (α Urs. Min., P XII. 232)	8 15	„	„
13 0	Scarcely fainter than α Urs. Min.*; = ζ Urs. Maj.*				—		„	„
Darlington.								
July 7, 10 10	Far fainter than α Urs. Min.				—		„	„
10 50	Much fainter than ζ Urs. Maj.				—		„	„
10 50	—	Abt. 3			—		(2·2)	„
12 0	Between α and γ Urs. Maj.?	8½	1·4		—		n. e.	„
North Berwick.								
July 14, 11 50	Rather brighter than κ Draconis				—		„	„
11 55	—				The star at R.A. 155° N.D. 83°	1¾	(2·2)	„

Place and Time.	Brightness of Head.	Length of Tail.	Max. Width of Tail.	At Distance from Nucleus.	Central or Brightest Line of Tail: passes or points to	Central or Brightest Line of Tail: at distance from nucleus.	Modes of Observation.	Hindrances.
North Berwick.								
July 14, 11 55		$7\frac{1}{2}^{\circ}$	$\frac{8}{7}$	$6\frac{1}{2}^{\circ}$	$\frac{3}{5}$ (ε, ζ) Urs. Min.	7°	n. e.	Twilight
12 15	= ε Cass.	$8\frac{1}{2}$			—		"	" Perhaps moonlight
Sunderland.								
July 20, 11 4	= κ Drac.				—		"	Twilight
12 5	Rather brighter than κ Drac.; about half way from ε Cass. to ζ Urs. Min.* in brightness	11		11	$\frac{2}{5}$ (ε, ζ) Urs. Min. $\frac{3}{5}$ (ε, ζ) "	4 11	"	"
22, 12 0		$9\frac{1}{2}$			$\frac{1}{2}$ (ε, ζ) Urs. Min.		"	Very little twilight
27, 10 15	Considerably fainter than κ Drac.; scarcely brighter than δ Urs. Min.; rather brighter than ε				—		"	Twilight
12 15	Rather brighter than δ Urs. Min.; slightly brighter than λ Drac.; considerably fainter than κ; $\frac{2}{3}$ from κ Drac. to δ Urs. Min. in brightness	$9\frac{1}{2}$	2	$7\frac{1}{2}$	ζ Urs. Min.	$9\frac{1}{2}$	"	
29, 12 0	—	$8\frac{1}{3}$	1	Abt. $3\frac{1}{2}$	$\frac{1}{3}$ (ζ, ε) Urs. Min. ζ, or $\frac{1}{2}^{\circ}$ n of it	2 $8\frac{1}{3}$	"	Not very clear
31, 12 37	—	8	$1\frac{1}{4}$	$3\frac{2}{3}$	$\frac{3}{5}$ or $\frac{2}{3}$ (ε, ζ) Urs. Min. $\frac{11}{12}$ (ε, ζ)	2 8	specs.	
12 45	Slightly brighter than ζ or ε Urs. Min.; rather fainter than δ				—		n. e.	

Place and Time.	Brightness of Head.	Length of Tail.	Max. Width of Tail.	At Distance from Nucleus.	Central or Brightest Line of Tail: passes or points to	Central or Brightest Line of Tail: at distance from nucleus.	Modes of Observation.	Hindrances.
		°		°		°		
Sunderland.								
Aug. 1, 10 42	—				$\frac{2}{3}$ (ε, ζ) Urs. Min.	2	(3·5)	Not so clear as lastnight
12 10	—	7½			ζ Urs. Min.	7½	n. e.	
3, 11 20	—				$\frac{3}{4}$ (ε, ζ) Urs. Min.	1¾	(3·5)	Hazy
4, 12 3	—	6½	1¾		ζ Urs. Min.	6½	n. e.	
12 10	Scarcely = 4 Urs. Min., but considerably brighter than 5				—		"	
12 30	—				Much to *n.* of ζ Urs. Min.		(3·5)	
12 50	= about Comet *c* ‡				—		n. e.	Twilight
5, 12 25	—	6¼			ζ Urs. Min.		"	Not very clear
10, 10 45	Invisible						n. e. & specs.	Moonlight
10 45	—	1$\frac{1}{7}$			—		(3·5)	"
14, 10 40	Far fainter than the star at R.A. 205°, N.D. 78½°	2½			ζ Urs. Min.		"	"
18, 10 10	—	1					n. e.	Slight light from town
10 15	—	3¼	1	Alt. 2¾	Slightly *s.* of θ Urs. Min.	3¾	(3·5)	Less light if any
12 10	—	3¾			θ Urs. Min.	3¾	"	Moonlight

Place and Time	Brightness of Head	Length of Tail	Max. Width of Tail	At Distance from Nucleus	Central or Brightest Line of Tail		Modes of Observation	Hindrances
					passes or points to	at distance from nucleus		
Sunderland. Aug. 25, 10 30	Slightly brighter than P XIV. 273	2⅔°	Broad		—	°	(3·5)	
Sept. 7, 10 5	—	40'			—		(20)	Full moon
14, 11 15	Considerably fainter than 20 Urs. Min.	25'	Abt. 8'		About 5° s. of f		(20)	Moonlight
11 30	Invisible	None			—		(3·5)	„
11 30					—		n. e.	„
12 5	—	28'			—		(38)	„
16, 10 45	Invisible	40'	Abt. 11'		About 7° s. of f		(20)	
11 0					—		n. e.	
11 0	Invisible	30'			—		(3·5)	
19, 8 20	—				—		n. e.	
8 20	Considerably fainter than 20 Urs. Min.. or than the star at R.A. 244°, N.D. 73¾°	10'			—		(3·5)	
8 50	—	35'	22'		{ 5° or 10° s. of f 15° or 20° s. of f	{ 10' 35'	(20)	
Scarborough. Sept. 26, 10 10	Visible, together with a much fainter star near				—		n. e.	

* Means that the object, compared with Comet *b*, was more favourably situated; † much more so; ‡ less so.

§ Estimated; no stars to compare it with.

Remarks on the Tail.

June 29.—Curvature suspected. (3·5.)

July 1.—Probably curved, in lower part. (Naked eye.)

July 2, 11ʰ 5ᵐ.—Slightly curved. For the first 3° the middle of the tail points to $\frac{3}{4}$ (a, δ) *Ursæ Minoris*; but this is not the brightest line, for the *p* side is slightly the brightest part in the lower portion of the comet. (Naked eye.)

12ʰ 30ᵐ.—The tail has a slight appearance of bifurcation, because above this lower portion the *p* side is exceeded in brightness by the continuation of the *f* side,* which, however, ceases to be the side, since much fainter light appears on its left. The tail then becomes much more indefinite, curving considerably to the left. .It is doubtful from my notes whether it passed $\frac{2}{5}$ or $\frac{3}{5}$ from P XII. 232 to δ *Urs. Min.*, but the former appears to have been correct. (Naked eye.)

July 3, 11.20.—The *p* edge much better defined than the *f* edge. This superior definition begins to be perceptible rather more than 1° from the nucleus, and increases beyond M *Cam.* (5.)

12.10.—With the exception of the first degree or two from the head, the main part of the tail is quite straight; but the appearance of curvature is produced by the diffused *f* part of the tail becoming brighter relative to the rest as it recedes from the nucleus. This diffused part causes the *f* edge of the tail to be worse defined than the *p* throughout, except at the beginning and end. (Naked eye.)

July 4, 11.42.—At 6° from the nucleus the tail is as bright as the brightest parts of the Milky Way in *Cygnus,* which are in rather darker sky. (Naked eye.)

12.5.—The *p* side the brightest from M *Cam.* to a distance of 10° from the nucleus. Curvature of lower part of tail very slight indeed. The diffused *f* part of the tail brighter relatively than on the 3rd. The *p* and *f* sides of the tail are so different that, whereas the *p* side seems an emanation from the head, the *f* side seems rather an emanation from the *p* side. (Naked eye.)

July 14, 11.55.—Tail slightly curved.

July 20, 12.5.—Tail straight, but in the lower part the brightest line is nearer the *n* than the *s* edge. (Naked eye.)

July 27, 12.15.—Ditto. ζ *Urs. Min.* is at the end of the tail, the *s* edge ending a little *s* of it; the *n* edge does not reach so far, pointing to $\frac{3}{4}$ (ϵ, ζ). The whole tail faint, except close to the nucleus. (Naked eye.)

July 31, 12.37.—The end of the tail is a prolongation of the *s* side of the earlier part, but is slightly curved, concave to *s.* (Naked eye.)

Aug. 4, 12.3.—Straight. The *s* edge extends furthest. (Specs.)

---

* M *Cam.* is in the brightest line of this, and P VI. 292 slightly precedes it.

Place and Time.	Brightness of Head.	Length of Tail.	Max. Width of Tail.	At Distance from Nucleus.	Central or Brightest Line of Tail — passes or points to	— at distance from nucleus	Modes of Observation.	Hindrances.
Sunderland.				°		°		
Aug. 25, 10 30	Slightly brighter than P XIV.273	2⅓°	Broad		—		(3·5)	Full moon
Sept. 7, 10 5	—	40'			—		(20)	Moonlight
14, 11 15	Considerably fainter than 20 Urs. Min.	25'	Abt. 8'		About 5° s. of f		(20)	"
11 30		None			—		(3·5)	"
11 30	Invisible	28'			—		n. e.	"
12 5		40'	Abt. 11'		About 7° s. of f		(38)	
16, 10 45	Invisible	30'			—		(20)	
11 0					—		n. e.	
11 0	Invisible				—		(3·5)	
19, 8 20	Invisible	10'			5° or 10° s. of f	10'	n. e.	
8 20	Considerably fainter than 20 Urs. Min.. or than the star at R.A. 244°, N. D. 73¾°	35'	22'		15° or 20° s. of f	35'	(3·5) }	
8 50	—				—		(20)	
Scarborough.								
6, 10 10	Visible, together with a much fainter star near						n. e.	

... that the object, compared with Comet b, was more favourably situated; † much more so; ‡ less so. § Estimated; no stars to compare it with.

July 4.—With a prism of Chance's dense flint glass, and no slit or any other instrument, the most conspicuous part of the spectrum is the two less refrangible bright bands (or rather points, as they appear when so observed) in the spectrum of the head. The continuous spectrum of the nucleus is also plain ; and the third band, though it is much fainter than the least refrangible one. The middle band much the brightest. With two prisms the spectrum is still plainer ; but I can see no other band. As to the tail, I could not be sure that its spectrum was otherwise than simply continuous.

July 22, 11.55.—With the miniature spectroscope on the telescope, the three bright bands are still very conspicuous in the spectrum of both nucleus and coma, especially the latter; the fourth band very faintly visible in that of the nucleus only. The brightest band is visible to a distance of at least 4' or 5' preceding the nucleus, and the two other bright ones nearly as far; while *f* the nucleus—*i.e.* along the tail (though this direction is not along its axis)—these are visible to a distance of 4½', and the brightest band to 8'.

Aug. 4.—The three bright bands still conspicuous. The continuous spectrum relatively much fainter than before. The fourth band invisible.

Sept. 16.—The brightest band conspicuous, and the least refrangible one plain, but the third (the most refrangible) scarcely visible. There is also a faint continuous spectrum, but no trace of the conspicuous bright longitudinal line that formed the continuous spectrum of the nucleus in the early part of this comet's appearance.

This is the only comet in which I have certainly seen the fourth bright band, though I suspected one about the same place in the great comet of 1874 and in Comet *a*, 1882. I did not measure its position, but if the above estimate is correct, it must be rather less refrangible than the violet carbon band—say about wave-length 436; the bands I suspected in the comets of 1874 and 1882 seemed even less refracted : it is therefore improbable that I can have seen any of the bands photographed by Dr. Huggins.

Sunderland:
    1882, *June* 5.

Sextant Observations of Comet b, 1881.   By Capt. J. F. Parson.

(*Communicated by Captain H. Toynbee, R.N.*)

	G. M. T.			
	d	h	m	s
1881, June 24, in about 44½° N. 9° W.	23	13	48	00

	°	'
Angular distance from Pole Star	43	41
„ „ *Capella*	4	43

	G. M. T.
	d h m s

June 25, in about 41½° N. 12° W.   25 10   4 30

    Angular distance from Pole Star   36 3

       ,,   ,,   β *Ursæ Majoris* 42 13

June 29, in about 30° 20′ N. 20′ W.   29 11   6 00

    Angular distance from Pole Star   23 47

       ,,   ,,   α *Ursæ Majoris* 29 56

Tail increasing in length since last seen.

July 3, in about 21° N. 25° W.   3 10   3 00

    Angular distance from Pole Star   15 20 ⎱ *

       ,,   ,,   α *Ursæ Majoris* 24 50 ⎰

On the possible Existence of Perturbations in Cometic Orbits during the Formation of Nuclear Jets; with Suggestions for their Detection. By C. E. Burton, Esq.

The sudden and extensive changes which took place in the nuclear jets of Comet *b*, 1881, together with the great brilliancy of some of these streams in the neighbourhood of the nucleus, suggested and appeared to justify the conclusion that if the apparent emission of matter from the nucleus or its immediate neighbourhood really took place under the influence of transformed solar radiant energy, there must be corresponding deviations from the original path of the nucleus, measuring the force so brought into action. The visibility of these perturbations will evidently depend, as far as we are concerned, (1) on the distance of the comet from the Earth; (2) on the force by which the matter of the jet is thrown off; (3) on the ratio of the mass thrown off at any instant to that of the nucleus; (4) on the inclination of the direction of the jet to the orbit and to the line of sight, as well as on the angle between the line of sight and the orbit; (5) on the duration of the jet as compared with that of the series of observations. It seems probable that disturbances normal to the orbit would be easier to detect than retardation or acceleration of orbital velocity.

At the discovery of Comet Wells, and for some little time afterwards, it seemed probable that there would be opportunities of testing the suggestions contained in this note, and it is possible that such may yet occur.

The data for an investigation of the probable magnitude of the disturbing effects which form the subject of this note can hardly be said to exist at present, but it seems not unlikely that displacements amounting to several seconds of arc may take place in the course of 24 hours, and that some of the discordance between the computed and observed places are due mainly to the cause now suggested. It is of course difficult to make use of meridian observations, even at the same station and by the

* Distances only approximate: Moon too bright to get good observations.

same observer, when a matter of this kind is under investigation, the time interval between successive observations being far too great; and if observations at different stations separated by sufficiently short arcs are combined, the difficulty of making proper allowance for variations of personal and instrumental equation comparable in magnitude with the quantities sought becomes insuperable.

For this reason it would seem that the work required could best be accomplished by a single observer with the aid of a heliometer, the nucleus of a suitable comet being referred to at least two stars in the neighbourhood of its path at short intervals —*e.g.* half-hourly—for several hours together. The same stars should be employed during the entirety of a series of observations, so that the places of the nucleus concluded therefrom may be affected by a *constant* error, as far as the points of reference are concerned. It will obviously be advisable to make the intervals of observation as nearly equal as possible, to diminish the risk in after-interpolation if it should be decided to treat the observations by the method of differences. Any *per saltum* change in the path of the nucleus thus detected would—especially if coincident with the appearance of a jet—be open to suspicion as being due to the suggested cause, or at least as belonging to a different class from gravitational disturbances.

Loughlinstown, Co. Dublin :
    1882, *June* 6.

Note on Lunar Photographs taken after Enlargement of the Primary Image by an Eyepiece. By C. E. Burton, Esq.

Having found by experiments made* with rapid commercial gelatino-bromide dry plates that a dense image was produced by exposure to *direct* moonlight for 4 seconds, and that a dense image of the moon itself resulted in the same time, after the enlargement of the image formed by a small object-glass to the extent of 8 diameters, I requested and obtained the kindly-accorded permission of Dr. Wentworth Erck to make use of his 7½-in. Alvan Clark Equatoreal for further experiments. After some preliminary trials the difference between the focus of the most active photographic rays and the best visual focus was determined by the spectroscope in the usual manner, but after the insertion of a Kellner eyepiece and for the conjugate focus of the latter. The equivalent focal length of the Kellner was about 1 in. and the enlargement of the primary image was 6 diameters. Under these conditions it was found that to bring the yellow rays to a focus after having set to the best focus for the rays in the neighbourhood of F, the eye-tube had to be pushed in 0·10 in. or 0·11 in. Images of printing density were obtained on May 29 and May 31 in 10 and 12 seconds, though the Moon was low, and at the last date the east wind had brought reddish

* 1882, May 27, 11ʰ 30ᵐ Dublin M.T.

haze with it, increasing the time necessary for exposure. The slow motion in R. A. was used on May 31 to counteract the Moon's proper motion. The image was sharp but not steady, and the photographs lack sharpness, though they justify expectations of success under more favourable conditions. A faint image of *Mizar* (both components) was obtained in 10 seconds ± 1 sec., the separation of the two images being very satisfactory (0·10 in.) and = 3 times the diameter of the denser. Ferrous oxalate was employed for development.

Previous trials* had been made with a 9-in. reflector on an altazimuth mounting, but the difficulty of following was found to be too great under the then existing circumstances, though it is anticipated that it may be overcome during the short exposures now required.

Loughlinstown :
1882, *June* 5.

Transits of Red Spot across the Apparent Central Meridian of Jupiter, with the Corresponding Longitudes, according to the System of Longitudes of Mr. A. Marth (Monthly Notices, vol. xli. No. 7), the Times reduced to the Meridian of Greenwich. Observed at Morrison Observatory, Glasgow, Missouri. By Professor C. W. Pritchett.

Date.	Prec. End.	Middle.	Foll. End.	Wt.	Long. Prec. End.	Long. Middle.	Long Foll. End.
	h m	h m	h m		°	°	°
1881, July 4		22 16 4	22 47·9	4		5·9	24·9
11	22 32·6	23 4·6		3	347·9	7·2	
16	21 40·3	22 12·8	22 44·2	4	347·9	7·5	26·5
23	22 23·8	22 56·3	23 28·3	4	346·5	6·2	25·5
28		22 6·3	22 37·8	4		7·7	26·8
Aug. 9	21 24·6	21 57·0	22 28·5	4·5	347·0	6·6	28·4
21		22 25·1		3·5			28·3
28	22 7·8	22 40·8	23 11·3	4	351·0	11·0	29·4
Sept. 4	22 51·1	23 24·1	23 54·3	3	350·3	10·3	28·5
11	23 33·8	0 7 8		3	349·4	9·9	
19		20 47·3	21 17·3	3·5		12·6	30·8
20		16 39·3	17 9·3	3·5		13·2	31·4
Oct. 4	17 26·8	17 59·0	18 30·3	3	349·3	10·9	28·0
7		15 34·3				12·8	
9	16 38·7	17 10·7	17 40·8	3	352·7	12 8	30·2
14		16 18·4				13·1	
19	14 53·9	15 25·5	15 53·8	5	354·8	13·8	31·0
31	14 39·8	15 12·8		3	352·9	12·7	

* 1882, April 25, 8ʰ + , D.M.T. Dense impression with 15ˢ exposure to an image magnified about 7 diameters.

Date.		Prec. End.	Middle.	Foll. End.	Wt.	Long. Prec. End.	Long. Middle.	Long Foll. End.
		h m	h m	h m		°	°	°
1881, Nov.	5	13 50·3	14 21·8	14 48·8	3	355·8	14·8	31·2
	12	14 34·3	15 5·3	15 36·3	2·5	356·3	15·5	33·8
	14	16 12·0	16 40·3	17 10·3	3	356·5	13·6	31·7
	19	15 18·1	15 49·8		3	356·6	15·8	
	24	14 24·3	14 56·3	15 27·3	3·5	356·8	16·1	34·9
	26	16 2·5	16 33·6		3	357·2	16·0	
Dec.	1	15 9·8	15 41·3	16 9·8	3·5	357·8	16·9	34·1
	6	14 18·3	14 49·3	15 20·0	3·5	359·2	17·9	36·5
	15	16 44·3	17 14·3	17 43·3	3·5	1·5	19·7	37·1
	23	13 18·3	13 48·3	14 18 3	5	0·5	18·6	36·8
	25		15 27·3	15 56·8	5		19·6	36·9
1882, Jan.	6	14 50 3	15 18·8	15 48·3	3	1·4	18·7	36·6
	11		14 37·8		3			38·3
	13	15 38·3	16 6·3		2·5	2·8	19·7	
Feb.	2		13 10·5		4			39·3
	4		14 19·3	14 48·8	2·5		21·2	40·3
	9		13 33·3	14 1·3	3		24·8	41·7

Observations of the Phenomena of Jupiter's Satellites, made at Mr. Edward Crossley's Observatory, Bermerside, Halifax. By J. Gledhill, Esq.

Date. 1881.	Satellite and Phenomena.		G.M.T.	Naut. Alm.	Remarks.
			h m s	h m s	
Oct. 3	II. Ec. D.	Fading away?	10 52 0	10 55 7	Sky not good.
		Certainly fading	53 0		
		Still visible	56 0		
		Gone	56 30		
12	II. Tr. I.	First contact	9 16 0	9 15 0	Very bad definition.
		Inner contact	9 18 0		
14	I. Sh. I.	Just on disk	10 45 0	10 42 0	Bad sky.
	I. Tr. I.	Outer contact	11 25 0	11 26 0	„
		Inner „	11 28 0		„
Nov. 20	II. Tr. I.	First contact	9 40 0	9 42 0	Fair.
		Bisection	9 43 c		
		Inner contact	9 44 0		Good.
	II. Sh. I.	First seen	10 4 0	10 4 0	Windy; clouds.
		Inner contact	10 7 0		
22	II. Ec. R.	First seen	7 32 49	7 33 41	
		Full?	7 36 15		

Date.	Satellite and Phenomena.	G.M.T.			East Asc.			Remarks.
1881.		h	m	s	h	m	s	
Nov. 22	I. Tr. I. Outer contact	8	55	0	8	55	1	
	Bisection		58	1				
	Inner contact		59	0				
	I. Sh. I. First seen	9	9	44	9	10	0	
	Inner contact	9	12	0				
	I. Tr. E. Outer contact	11	8	0	11	7	0	
	I. Sh. E. Just off	11	23	0	11	22	0	Uncertain
25	III. Sh. I. Just within	11	11	0				Much motion
	III. Tr. E. Outer contact	11	26	0	11	26	0	Rough observation.
29	II. Oc. D. Outer contact	6	45	0	6	45	0	
	Inner „	6	49	0				
	II. Re. R. First seen	10	8	10	10	9	1	
	Full disk	10	10	30				
	I. Tr. I. First contact	10	40	0	10	39	0	
	Bisection		42	0				
	Inner contact		44	c				
	I. Sh. I. Inner contact	11	6	0				Bad sky.
Dec. 7	I. Oc. D. First contact	9	32	30	9	35	0	Much motion.
	Bisection		35	0				„
	Second contact		36	30				
14	I. Oc. D. Last seen	11	21	0	11	21	0	
22	II. Tr. E. Just off	10	40	30	10	43	0	
	I. Sh. I. Bisection	11	18	0				
	Just within	11	19	0				
1882. Jan. 24	I. Re. R. First seen	7	23	21	7	23	33	Good observation.
	Full?		29	0				
Feb. 15	II. Oc. D. First contact	9	29	30	9	29	0	Bad definition.
	Inner contact	9	32	30				„
Mar. 3	I. Tr. E. Inner contact	7	27	0	7	30	0	
	Outer „	7	31	0				
11	I. Re. R. First seen	7	56	38	7	56	46	Good sky.
	Full?	7	57	49				
18	I. Re. R. First seen	9	52	50	9	52	26	

Rosenvoida, Halifax:
    1882, April 25.

MONTHLY NOTICES

OF THE

ROYAL ASTRONOMICAL SOCIETY.

Vol. XLII.   Supplementary Number.   No. 9.

Addition to the Ephemeris for Physical Observations of Jupiter.
On the Motion of the White Spot. By A. Marth, Esq.

Next to the great reddish spot, of which observations have been made since the autumn of 1878, the small white spot, a little south of *Jupiter's* equator, which seems to have first attracted attention in the autumn of 1880, promises, by its long continued visibility, to be of special service in the study of the changes which are going on upon the surface of *Jupiter.* The variations in the rate of motion and in the brightness of the spot render careful and assiduously continued observations of its position and of its appearance essential for any further investigations. In order to enable readers to see how far the known observations of the spot can be represented on the assumption of regular motion, I give a list of all the observed passages of the spot across the middle of the disk, which have come to my knowledge, together with the corresponding longitudes in a system in which the daily rate of rotation is assumed to be $=878°·46$. These longitudes are connected with the system of longitudes adopted in the ephemeris by the equation

Corresponding long. $=$ Long. of $♃$'s Central Meridian $+$ Corr. for phase

$$+ 336°·0 + 8°·04 \ (t - t_o),$$

in which t is the time of the observed passage, and $t_o = $ Aug. 2, 1882, 0^h Gr. The period of rotation corresponding to the daily rate $878°·46$ is $9^h \ 50^m \ 7^s·42$, while that corresponding to the assumed rate of rotation of the First Meridian is $9^h \ 55^m \ 34^s·47$.
The history of the development of the spot before Oct. 16,

L L

1880, requires clearing up, and I must leave it, for the present, an open question whether the earlier observations refer really to the same object. The observations of Kortazzi are found in the *Bulletin de l'Acad. de St.-Pétersbourg,* tome 27, p. 142, those of Schmidt in No. 2353 of the *Astron. Nachrichten*; the others are taken from papers in the *Monthly Notices,* letters in the *English Mechanic,* or from private communications. The places of observing and the apertures of the instruments employed are :—

Barnard; Nashville, Tennessee; 5 in. refractor.

Corder; Writtle, near Chelmsford; 4·5 in. reflector.

Dennett; Southampton; 9·5 in. reflector.

Denning; Bristol; 10 in. reflector.

Kortazzi; Nicolajew; 6·5 in. refractor.

McCance; Putney; 10 in. reflector.

Schmidt; Athens; 6·5 in. dialyt.

A. S. Williams; West Brighton; 5·2 in. reflector and 2·7 in. refractor.

In the case of Schmidt's observations, two values of the resulting longitudes are given, the first corresponding to the time of the observed passage as directly observed, the second to the time after applying the correction which Schmidt himself has deduced, and about which I must content myself with referring to his paper. The abbreviations are: Ath. Athens, Gr. Greenwich, Nic. Nicolajew, b. bright, f. faint, v. very.

Observed Time of Passage.				Observer.	Corr. Long.	
1880, Aug. 27	11 18·7	Nic.		Kortazzi	17·7	
Sept. 5	11 34·0	„		„	14·2	
7	12 43·1	„			13·6	
16	13 0·5	„		„	11·4	
18	11 50	Gr.	v b	Denning	3·7	
21	10 58·1	Nic.		Kortazzi	9·8	
Oct. 16	10 51·4	Ath.		Schmidt	30·6	30·3
18	12 2·3	„		„	31·0	29·9
19	7 36·8	„		„	27·7	27·3
20	11 38	Gr.	v b	Denning	31·2	
21	8 46·6	Ath.		Schmidt	27·3	27·2
23	9 52·8	„		„	24·8	24·7
28	7 56·1	„		„	26·3	26·2
29	11 53·3	Gr.		Williams	27·4	
30	7 25	„		Corder	22·3	
30	7 28	„	v b	Denning	24·1	
30	9 4·6	Ath.		Schmidt	25·1	25·1

Observed Time of Passage.				Observer.	Corr. Long.	
1880, Nov.	2	14 15	Gr.	b	Denning	28·0
	3	9 47	,,	b	,,	23·0
	4	5 25	,,	b	,,	21·6
	4	7 3·0	Ath.		Schmidt	23·5	23·3
	6	8 12·0	,,		,,	22·6	22·6
	8	7 48	Gr.		Denning	22·8
	8	9 23·4	Ath.		Schmidt	23·2	22·8
	9	5 4·9	,,		..	23·9	22·8
	11	6 20·1	,,			26·8	26·5
	12	11 54·6	,,			29·3	25·9
	15	8 38·6	,,		,,	24·1	23·9
	17	8 14	Gr.		Denning	24·9
	17	8 15	..		Corder	25·5
	17	8 15	,,		Williams	25·5
	17	8 15·5	,,		Dennett	25.8
	17	9 52·0	Ath.		Schmidt	26·8	25·7
	19	9 20	Gr.		Corder	22·1
	19	9 23	,,	v b	Denning	23·9
	19	9 26·5	,,		Dennett	25·8
	19	9 28	..		McCance	27·0
	20	5 0	,,		Corder	21·9
	20	5 2	,,	v b	Denning	23·1
	21	10 35	,,		McCance	24·7
	21	10 41	,,	b	Denning	28·4
	22	6 15	..	..	McCance	30·6
	22	6 18	..		Corder	32·4
	22	16 4·2	,,		Barnard	23·9
	26	8 40	,,		McCance	26·6
	26	8 44	,,	v f	Denning	29·1
	27	4 19	,,	v b	,,	25·8
	27	4 21			Williams	27·1
	28	9 51	,,	v b	Denning	26·8
	29	5 27	..	v b	,,	24·1
Dec.	2	12 14·0	,,		Barnard	27·6
	3	7 54	,,	f	Denning	27·4
	6	14 36·2	,,		Barnard	27·8
	10	7 14	,,	v b	Denning	31·5
	12	8 27		v b	,,	32·8
	15	5 16	..	v b		31·3

Observed Time of Passage.					Observer.	Corr. Long.
1880, Dec. 21	8 48	Gr.			McCance	30·6
22	8 49	„			Denning	31·2
23	10 4	„	▽b		„	33·6
24	5 36·5	„			McCance	28·7
25	11 17				Williams	34·7
28	7 59	„	▽b		Denning	28·8
30	9 11	„			McCance	29·4
30	9 15	„	▽b		Denning	31·8
31	4 47				McCance	27·6
1881, Jan. 6	8 26·5	„			„	30·2
6	8 27	„	b		Denning	30·5
7	4 5		f		„	29·0
7	4 6	..			McCance	29·6
13	7 37	„	▽▽b		Denning	27·8
15	8 53	„			Williams	30·7
16	4 28	„	.▽b		Denning	27·4
Feb. 23	7 5	„	f		„	15·6
July 10	14 45	„	▽b			16·4
17	13 56	„	▽b			15·0
24	13 7	..	▽f			13·8
26	14 19		b			14·5
Aug. 2	13 36	„	b			17·1
4	14 47		b			17·2
8	17 8	„	f			16·9
17	17 31		▽b			16·9
18	13 12		b			17·2
27	13 33	„	▽▽b		„	16·3
Sept. 1	11 38	„	f		..	18·6
7	15 13				„	20·7
9	16 26	..	.		Williams	22·3
15	10 10:	„	f		Denning	23·9
15	10 10·5	„			Williams	24·3
16	15 33				„	19·5
19	12 9	„	▽▽b		Denning	16·8
21	13 30	„	▽▽b		„	17·2
25	15 46	..			Williams	14·3
26	11 26·5	„			..	14·6
27	16 46	..			„	8·0
·28	12 40	„	b		Denning	16·5

Observed Time of Passage.				Observer.	Corr. Long.
1881, Sept. 30	13 52	Gr.	vf	Denning	17˚5
Oct. 3	10 49	„	vvf	„	21·6
5	12 3	„	vb	„	23·9
11	15 36	,,		Williams	25·3
17	9 13	„	vb	Denning	29·1
25	13 58	„	vb	„	25·7
25	13 58·5	„		Williams	26·0
26	9 36	..		„	24·5
28	10 44	„		Denning	23·2
30	11 52·5	„		Williams	22·2
30	11 54	„	vb	Denning	23·1
Nov. 6	11 6	..		Williams	24·0
7	16 37			„	24·5
14	5 56	„	b	Denning	23·6
15	11 32	„	vb	„	27·2
22	10 44	„	f	„	28·0
22	10 47·5	„	vf	Williams	30·1
23	6 25	..	vf	Denning	28·5
23	6 26	„	vf	Williams	29·2
23	16 16	„	vf	Denning	29·1
27	8 45		b	„	28·2
27	8 55			Corder	34·3
28	4 27	„	vb	Denning	29·4
28	14 18	„	b	Williams	29·9
29	9 52	„	vvb	Denning	26·2
29	9 54·5	„	b	Williams	27·7
Dec. 1	11 ·55		b	„	28·4
2	6 44	„	vvb	Denning	27·2
3	12 14	„	vvb	„	27·0
7	4 41	„	vvb	„	24·7
7	4 45			Corder	27·1
7	14 33	„	vb	Denning	25·8
8	10 20	..		Corder	30·0
10	11 26	„	vb	Denning	27·3
12	12 30			Williams	23·4
13	8 12	„	vvb	Denning	24·5
13	8 12	„	vf	Williams	24·5
18	6 12	„	b	Denning	23·7
20	7 26	„ ..	b.	„	25·7

Observed Time of Passage.					Observer.	Corr. Long.
1881, Dec.	21	12 54	Gr.	b	Williams	24·3
	22	8 33	„	vvb	Denning	23·5
	23	4 11	„	vvb	„	22·2
	24	9 43	„	vb	„	23·2
	24	9 44	„	vf	Williams	23·8
	29	7 50	„		Corder	26·4
1882, Jan.	3	5 44	„	vf	Denning	21·6
	5	7 5	„	vvf	„	27·8
	6	12 34	„	vvf	„	26·9
	7	8 13	„	f		26·1
	9	9 27		b		28·0
	10	5 8	..	vb		28·4
	13	11 52	„	vvb	„	29·9
	16	8 42	„	vvb	„	29·1
	21	6 38	..	b	..	25·2
Feb.	1	8 22	„	f	„	30·2
	3	9 32	..	vb	Williams	29·5
	8	7 38	„	f	Denning	31·4
	10	8 50	„	b	Williams	31·9
	12	10 0	„	f	Denning	31·1
	13	5 43		f	„	32·4
	17	8 7		vf	..	33·5
	19	9 23	„	vvf	„	36·5
	19	9 23	„	vf	Williams	36·5
	21	10 33	„	f	Denning	35·6
	26	8 34	„	vvb	„	34·3
Mar.	3	6 30	„	vvb	„	29·9
	12	6 59	„	vvb	„	31·7
	14	8 5	„	b	„	28·4
	14	8 12	„	vvb	Williams	32·6
	16	9 19	„	vvb	Denning	29·9
	16	9 21	„	b	Williams	31·1
	30	7 41	„	vf	Denning	25·5
Apr.	6	7 1	„	vvb	„	28·3
	8	8 12	..	b		28·0
July	29	15 49	„	vb		29·8
Aug.	5	15 4	„	vb		30·8
	7	16 9	„	vb	„	27·1

In case the white spot should remain visible and keep approximately its mean rate of motion, it may be looked for near the central meridian at the times which follow or precede the predicted passages of the First Meridian, given on pp. 391–95, by the interval indicated in the following table:—

1882.				1883.							h	
Aug. 26	Oct.	10	Nov. 24	Jan. 7	Feb.	21	Apr.	7	6 *after* F. Mer.			
30		14	28	12		26		11	5	,,		,,
Sept. 4		19	Dec. 3	16	Mar.	2		16	4	,,		,,
9		23	7	21		7		20	3	.,		,,
13		28	12	25		11		25	2	,,		,,
18	Nov. 1		16	30		16		29	1	,,		,,
22		6	21	Feb. 3		20	May	4	0	,,		,,
27		10	25	8		25		9	1 *before*			,,
Oct. 1		15	30	12		29		13	2	,,		,,
6		19	Jan. 3	17	Apr.	3		18	3	,,		,,
10		24	8	21		7		22	4	,,		,,
15		28	12	26		12		27	5	,,		,,

Cannot some of the many possessors of sufficiently powerful telescopes be induced to take part in the interesting observations of *Jupiter's* spots?

Ephemerides of the Satellites of Saturn, December 1882, *to March* 1883.   By A. Marth, Esq.

The following ephemerides of the five inner satellites are a continuation of those published in the last number of the *Monthly Notices*, so that it will be sufficient to refer to the explanations there given.

The ascending node N and inclination J of the plane of the ring in reference to the plane of the Earth's equator are here assumed to be

1882, Nov.	30	N = 126°·4545	J = 7°·0148
Dec.	30	·4590	·0144
1883, Jan.	29	·4621	·0140
Feb.	28	·4639	·0136
Mar	30	126·4661	7·0134

The assumed longitudes of the satellites in their orbits are (*v.* page 396)—

☌ Gr.	Mimas.	Enceladus.	Tethys.	Dione.	Rhea.
1882, Nov. 30	312°116	168°269	193°801	208°167	352°737
Dec. 30	251·839	130·225	154·746	194·219	223·442
1883, Jan. 29	191·562	92·182	115·691	180·270	94·147
Feb. 28	131·287	54·139	76·635	166·322	324·851
Mar. 30	71·012	16·095	37·579	152·373	195·555

☌ Gr.	P	L	Latitude of Earth above plane of ring.	Sun	A−L
1882.					
Nov. 30	358°264	50°263	−22°651	−23°286	+ 1°822
Dec. 5	358·313	49·870	22·578	23·329	2·409
10	358·359	49·500	22·512	23·372	2·972
15	358·402	49·160	22·453	23·415	3·506
20	358·441	48·853	22·402	23·458	4·007
25	358·475	48·584	22·361	23·500	4·471
30	358·503	48·355	22·330	23·542	4·894
1883.					
Jan. 4	358·526	48·169	−22·310	−23·583	+ 5·274
9	358·544	48·030	22·302	23·624	5·609
14	358·556	47·940	22·305	23·666	5·894
19	358·561	47·898	22·320	23·707	6·130
24	358·560	47·905	22·347	23·747	6·318
29	358·552	47·961	22·385	23·788	6·457
Feb. 3	358·539	48·066	−22·435	−23·828	+ 6·547
8	358·520	48·220	22·495	23·868	6·589
13	358·494	48·421	22·566	23·908	6·584
18	358·463	48·668	22·646	23·948	6·533
23	358·426	48·959	22·735	23·987	6·438
28	358·384	49·292	22·832	24·026	6·301
Mar. 5	358·337	49·666	−22·936	−24·065	+ 6·123
10	358·285	50·079	23·047	24·104	5·907
15	358·228	50·528	23·163	24·142	5·654
20	358·167	51·010	23·284	24·180	5·368
25	358·103	51·523	23·409	24·218	5·051
30	358·035	52·067	−23·536	−24·256	+ 4·706

0ʰ Gr.	Equat.	Diam. of Ball Phase. foll. l.	Polar.	Axis of Ring Major.	Minor.	Mimas. a_1	b_1	l_1-L	Diff.
1882.									
Nov. 30	19·88	0·005	18·20	45·83	17·65	31·26	−12·04	263·45	1910·29
Dec. 5	19·81	·007	18·14	45·67	17·54	31·16	11·96	13·74	·25
10	19·73	·013	18·06	45·48	17·41	31·04	11·88	123·99	·20
15	19·63	·019	17·97	45·25	17·28	30·87	11·79	234·19	·15
20	19·51	·024	17·86	44·99	17·14	30·69	11·69	344·34	·10
25	19·39	·030	17·74	44·69	17·00	30·49	11·60	94·44	·05
30	19·25	·035	17·62	44·38	16·86	30·27	11·50	204·49	1910·00
1883.									
Jan. 4	19·10	·040	17·48	44·04	16·72	30·04	−11·40	314·49	1909·94
9	18·95	·045	17·34	43·68	16·58	29·79	11·31	64·43	·88
14	18·79	·050	17·19	43·31	16·44	29·54	11·21	174·31	·83
19	18·62	·053	17·04	42·93	16·30	29·28	11·12	284·14	·77
24	18·46	·056	16·89	42·54	16·18	29·02	11·03	33·91	·72
29	18·29	·058	16·74	42·16	16·05	28·76	10·95	143·63	·67
Feb. 3	18·12	·059	16·58	41·77	15·94	28·49	−10·87	253·30	1909·62
8	17·95	·059	16·43	41·39	15·84	28·23	10·80	2·92	·57
13	17·80	·059	16·29	41·01	15·74	27·98	10·74	112·49	·53
18	17·63	·057	16·14	40·65	15·65	27·73	10·68	222·02	·48
23	17·48	·055	16·00	40·29	15·57	27·48	10·62	331·50	·44
28	17·33	·052	15·87	39·95	15·50	27·25	10·57	80·94	·41
Mar. 5	17·19	·049	15·74	39·62	15·44	27·03	−10·53	190·35	1909·38
10	17·05	·045	15·62	39·31	15·39	26·82	10·50	299·73	·35
15	16·93	·041	15·51	39·02	15·35	26·62	10·47	49·08	·32
20	16·81	·037	15·40	38·75	15·32	26·43	10·45	158·40	·30
25	16·70	·034	15·30	38·49	15·29	26·26	10·43	267·70	1909·28
30	16·60	0·028	15·21	38·26	15·28	26·10	−10·42	16·98	

0ʰ Gr.	Enceladus. a_2	b_2	l_2-L	Diff.	Tethys. a_3	b_3	l_3-L	Diff.
1882.								
Nov. 30	40·10	−15·44	119·11	1314·01	49·64	−19·12	144·34	953·85
Dec. 5	39·97	15·35	353·12	1313·97	49·48	19·00	18·19	·82
10	39·80	15·24	227·09	·94	49·27	18·86	252·01	·79
15	39·60	15·12	91·03	·89	49·02	18·72	124·80	·74
20	39·37	15·00	334·92	·85	48·73	18·57	359·54	·71
25	39·11	14·88	208·77	·80	48·42	18·42	233·25	·66
30	38·83	14·75	82·57	·74	48·07	18·26	106·91	·60

Enceladus.

☽ Gr. 1883.	a_2	b_2	l_2-L	Diff.
Jan. 4	38″54	−14″63	316°31	1313·70
9	38·22	14·51	190·01	·64
14	37·90	14·39	63·65	·58
19	37·57	14·27	297·23	·53
24	37·23	14·16	170·76	·48
29	36·89	14·05	44·24	·43
Feb. 3	36·55	−13·95	277·67	1313·38
8	36·22	13·86	151·05	·33
13	35·89	13·77	24·38	·29
18	35·57	13·70	257·67	·25
23	35·26	13·63	130·92	·20
28	34·96	13·57	4·12	·17
Mar. 5	34·67	−13·51	237·29	1313·14
10	34·40	13·47	110·43	·10
15	34·15	13·43	343·53	·07
20	33·91	13·40	216·60	·05
25	33·68	13·38	89·65	1313·03
30	33·48	−13·37	322·68	

Tethys.

a_2	b_2	l_2-L	Diff.
47″70	−18″11	340°51	953·54
47·32	17·96	214·05	·50
46·91	17·81	87·55	·45
46·50	17·66	321·00	·40
46·09	17·52	194·40	·35
45·67	17·39	67·75	·30
45·25	−17·27	301·05	953·25
44·83	17·15	174·30	·19
44·43	17·05	47·49	·15
44·03	16·95	280·64	·11
43·65	16·87	153·75	·07
43·28	16·79	26·82	·03
42·92	−16·73	259·85	953·00
42·59	16·67	132·85	952·96
42·27	16·63	5·81	·93
41·97	16·59	238·74	·91
41·70	16·57	111·65	952·88
41·44	−16·55	344·53	

Dione.

☽ Gr. 1882.	a_4	b_4	l_4-L	Diff.
Nov. 30	63″58	−24″49	158°46	658·05
Dec. 5	63·37	24·33	96·51	658·01
10	63·10	24·16	34·52	657·98
15	62·78	23·98	332·50	·94
20	62·41	23·79	270·44	·91
25	62·01	23·59	208·35	·86
30	61·57	23·39	146·21	·81
1883. Jan. 4	61·10	−23·19	84·02	657·77
9	60·60	23·00	21·79	·71
14	60·09	22·81	319·50	·66
19	59·56	22·62	257·16	·61
24	59·03	22·44	194·77	·55
29	58·49	22·27	132·32	·51

Rhea.

a_2	b_2	l_2-L	Diff.
88″79	−34″19	302°81	398·83
88·49	33·98	341·64	·80
88·12	33·74	20·44	·77
87·67	33·48	59·21	·74
87·16	33·22	97·95	·70
86·59	32·95	136·65	·65
85·98	32·67	175·30	·60
85·32	−32·39	213·90	398·56
84·63	32·12	252·46	·51
83·91	31·85	290·97	·46
83·18	31·59	329·43	·41
82·43	31·34	7·84	·35
81·68	31·11	46·19	·31

♂ Gr.	a_s	b_s	$l_s - L$	Diff.	a_s	b_s	$l_s - L$	Diff.
		Dione.				Rhea.		
1883.				°				°
Feb. 3	57·95	−22·12	69·83	657·46	80·93	−30·89	84·50	398·26
8	57·42	21·97	7·29	·41	80·19	30·68	122·76	·21
13	56·90	21·84	304·70	·36	79·46	30·50	160·97	·17
18	56·39	21·71	242·06	·32	78·75	30·32	199·14	·12
23	55·90	21·60	179·38	·29	78·07	30·17	237·26	·08
28	55·43	21·51	116·67	·24	77·41	30·04	275·34	·04
Mar. 5	54·98	−21·42	53·91	657·21	76·77	−29·92	313·38	398·00
10	54·55	21·35	351·12	·17	76·17	29·82	351·38	397·97
15	54·14	21·30	288·29	·14	75·60	29·74	29·35	·94
20	53·76	21·25	225·43	·11	75·07	29·68	67·29	·91
25	53·41	21·22	162·54	657·09	74·58	29·63	105·20	397·88
30	53·08	−21·20	99·63		74·13	−29·60	143·08	

Approximate Greenwich Mean Times of conjunctions of the satellites with the centre of *Saturn*:
" n " inferior conj., north; " s " superior conj., south.

1882.	h			1882.	h			1882.	h		
Nov. 30	17·2	Rh.	s.	Dec. 5	10·4	Mi.	n.	Dec. 9	17·7	Di.	s.
Dec. 1	3·1	Te.	s.		15·2	Di.	n.		17·8	Rh.	s.
	12·7	Di.	s.		17·1	En.	n.	10	12·1	En.	s.
	14·4	En.	n.		20·4	Te.	n.		13·6	Te.	s.
	16·0	Mi.	n.	6	9·1	Mi.	n.		14·8	Mi.	s.
2	1·8	Te.	n.		9·5	En.	s.	11	2·5	Di.	n.
	6·9	En.	s.		19·0	Te.	s.		4·6	En.	n.
	14·6	Mi.	n.	7	0·1	Di.	s.		12·2	Te.	n.
	21·6	Di.	n.		1·9	En.	n.		13·4	Mi.	s.
	23·4	Rh.	n.		7·7	Mi.	n.	12	0·0	Rh.	n.
3	0·4	Te.	s.		11·7	Rh.	n.		10·9	Te.	s.
	13·2	Mi.	n.		17·6	Te.	n.		11·4	Di.	s.
	15·7	En.	s.	8	6·3	Mi.	n.		12·1	Mi.	s.
	23·1	Te.	n.		8·9	Di.	n.		13·4	En.	n.
4	6·4	Di.	s.		10·8	En.	n.	13	5·9	En.	s.
	8·2	En.	n.		16·3	Te.	s.		9·5	Te.	n.
	11·8	Mi.	n.	9	3·3	En.	s.		10·7	Mi.	s.
	21·7	Te.	s.		4·9	Mi.	n.		20·2	Di.	n.
5	0·6	En.	s.		14·9	Te.	n.	14	6·2	Rh.	s.
	5·5	Rh.	s.		16·2	Mi.	s.		8·2	Te.	s.

1882. h	1882. h	1883. h
Dec. 14 9·3 Mi. s.	Dec. 27 13·2 Te. s.	Jan. 5 12·8 Mi. n·
14·8 En. s.	13·9 Mi. s.	12·9 En. s.
15 5·0 Di. s.	15·1 En. n.	20·0 Rh. s.
6·8 Te. n.	19·2 Rh. s.	23·7 Te. s.
7·2 En. n.	28 7·6 En. s.	6 2·3 Di. s.
7·9 Mi. s.	11·9 Te. n.	5·3 En. n.
16 5·5 Te. s.	12·5 Mi. s.	11·4 Mi. n.
6·5 Mi. s.	21·3 Di. s.	22·4 Te. n.
12·3 Rh. n.	29 10·5 Te. s.	7 10·0 Mi. n.
13·8 Di. n.	11·1 Mi. s.	11·2 Di. n.
16·1 En. n.	16·5 En. s.	14·2 En. n.
17 4·1 Te. n.	30 1·4 Rh. n.	21·1 Te. s.
5·1 Mi. s.	6·2 Di. n.	8 2·2 Rh. n.
8·5 En. s.	8·9 En. n.	6·6 En. s.
22·7 Di. s.	9·2 Te. n.	8·6 Mi. n.
18 2·8 Te. s.	9·8 Mi. s.	19·7 Te. n.
18·5 Rh. s.	31 1·3 En. s.	20·0 Di. s.
19 1·4 Te. n.	7·8 Te. s.	9 7·3 Mi. n.
7·5 Di. n.	8·4 Mi. s.	15·5 En. s.
9·8 En. n.	1·50 Di. s.	18·4 Te. s.
20 0·1 Te. s.	1·78 En. n.	10 4·9 Di. n.
16·3 Di. s.	1883.	5·9 Mi. n.
22·7 Te. n.	Jan. 1 6·5 Te. n.	8·0 En. n.
21 0·7 Rh. n.	7·0 Mi. s.	8·4 Rh. s.
21·3 Te. s.	7·6 Rh. s.	17·0 Te. n.
22 1·2 Di. n.	10·2 En. s.	11 4·5 Mi. n.
20·0 Te. n.	23·8 Di. n.	13·7 Di. s.
23 6·9 Rh. s.	2 2·7 En. n.	15·7 Te. s.
10·0 Di. s.	5·1 Te. s.	15·8 Mi. s.
18·6 Te. s.	5·6 Mi. s.	16·8 En. n.
24 17·3 Te. n.	3 3·8 Te. n.	12 9·3 En. s.
18·8 Di. n.	4·2 Mi. s.	14·3 Te. n.
25 13·0 Rh. n.	8·7 Di. s.	14·4 Mi. s.
13·8 En. s.	11·5 En. n.	14·6 Rh. n.
15·9 Te. s.	13·8 Rh. n.	22·5 Di. n.
26 3·7 Di. s.	4 2·4 Te. s.	13 1·7 En. n.
6·2 En. n.	4·0 En. s.	13·0 Te. s.
14·6 Te. n.	14·2 Mi. n.	13·0 Mi. s.
15·3 Mi. s.	17·5 Di. n.	18·2 En. s.
27 12·5 Di. n.	5 1·1 Te. n.	14 7·4 Di. s.

1883.	h			1883.	h			1883.	h		
Jan.	14	10·6	En. n.	Jan.	25	19·5	Te. n.	Feb.	4	4·8	Rh. n.
		11·6	Te. n.		26	2·2	En. s.			5·3	Mi. s.
		11·7	Mi. s.			3·9	Rh. n.			6·1	Te. n.
		20·8	Rh. s.			6·4	Mi. n.			16·4	En. s.
	15	3·1	En. s.			15·0	Di. n.		5	4·0	Mi. s.
		10·3	Te. s.			18·2	Te. s.			4·7	Te. s.
		10·3	Mi. s.		27	5·0	Mi. n.			5·0	Di. s.
		16·2	Di. n.			11·1	En. s.			8·9	En. n.
	16	8·9	Mi. s.			16·8	Te. n.		6	1·3	En. s.
		8·9	Te. n.			23·8	Di. s.			3·4	Te. n.
		12·0	En. s.		28	3·5	En. n.			11·0	Rh. s.
	17	1·1	Di. s.			3·7	Mi. n.			13·8	Di. n.
		3·0	Rh. n.			10·1	Rh. s.			13·9	Mi. n.
		7·6	Te. s.			15·0	Mi. s.		7	2·1	Te. s.
	18	6·2	Te. n.			15·5	Te. s.			10·2	En. s.
		9·9	Di. n.		29	8·7	Di. n.			12·5	Mi. n.
	19	4·9	Te. s.			12·4	En. n.			22·7	Di. s.
		9·2	Rh. s.			13·6	Mi. s.		8	0·7	Te. n.
		18·8	Di. s.			14·1	Te. n.			2·6	En. n.
	20	3·6	Te. n.		30	4·8	En. s.			11·1	Mi. n.
		14·6	En. s.			12·2	Mi. s.			17·2	Rh. n.
	21	2·2	Te. s.			12·8	Te. s.			19·1	En. s.
		3·6	Di. n.			16·3	Rh. n.			23·4	Te. s.
		7·1	En. n.			17·5	Di. s.		9	7·5	Di. n.
		15·4	Rh. n.		31	10·8	Mi. s.			9·7	Mi. n.
	22	0·9	Te. n.			11·4	Te. n.			11·5	En. n.
		12·5	Di. s.			13·7	En. s.			22·0	Te. n.
		23·5	Te. s.	Feb.	1	2·4	Di. n.		10	4·0	En. s.
	23	8·4	En. s.			6·2	En. n.			8·4	Mi. n.
		10·6	Mi. n.			9·5	Mi. s.			16·4	Di. s.
		21·3	Di. n.			10·1	Te. s.			20·7	Te. s.
		21·6	Rh. s.			22·5	Rh. s.			23·5	Rh. s.
		22·2	Te. n.		2	8·1	Mi. s.		11	7·0	Mi. n.
	24	0·8	En. n.			8·8	Te. n.			12·9	En. s.
		9·2	Mi. n.			11·3	Di. s.			19·4	Te. n.
		17·3	En. s.			15·1	En. n.		12	1·2	Di. n.
		20·8	Te. s.		3	6·7	Mi. s.			5·3	En. n.
	25	6·2	Di. s.			7·4	Te. s.			5·6	Mi. n.
		7·8	Mi. n.			7·5	En. s.			18·0	Te. s.
		9·7	En. n.			20·1	Di. n.		13	5·7	Rh. n.

1883.	h		1883.	h		1883.	h	
Feb. 13	10·1	Di. s.	Feb, 27	20·6	Te. s.	Mar. 14	15·1	Rh. s.
	16·7	Te. n.	28	11·6	Di. n.		23·3	Te. s.
14	6·7	En. s.		19·3	Te. n.	15	13·1	Di. s.
	15·3	Te. s.	Mar. 1	1·5	Rh. s.		21·9	Te. n.
	18·9	Di. n.		17·9	Te. s.	16	20·6	Te. s.
15	12·0	Rh. s.		20·4	Di. s.		21·4	Rh. n.
	14·0	Te. n.	2	16·6	Te. n.		22·0	Di. n.
16	3·8	Di. s.	3	5·3	Di. n.	17	19·3	Te. n.
	12·7	Te. s.		7·8	Rh. n.	18	6·8	Di. s.
17	11·3	Te. n.		15·3	Te. s.		17·9	Te. s.
	12·7	Di. n.	4	13·9	Te. n.	19	3·7	Rh. s.
	18·2	Rh. n.		14·2	Di. s.		15·7	Di. n.
18	10·0	Te. s.	5	12·6	Te. s.		1·66	Te. n.
	21·5	Di. s.		14·1	Rh. s.	20	0·6	Di. s.
19	8·6	Te. n.		23·0	Di. n.		15·3	Te. s.
20	0·5	Rh. s.	6	11·3	Te. n.	21	10·0	Rh. n.
	6·4	Di. n.	7	7·9	Di. s.		13·9	Te. n.
	7·3	Te. s.		9·9	Te. s.	22	9·4	Di. n.
21	6·0	Te. s.		20·3	Rh. s.		12·6	Te. s.
	15·2	Di. n.	8	8·6	Te. n.	23	11·3	Te. n.
22	4·6	Te. s.		16·8	Di. n.		16·3	Rh. s.
	6·7	Rh. n.	9	7·3	Te. s.		18·3	Di. n.
23	0·1	Di. n.	10	1·6	Di. s.	24	9·9	Te. s.
	3·3	Te. n.		2·6	Rh. s.	25	3·2	Di. n.
24	2·0	Te. s.		5·9	Te. n.		8·6	Te. n.
	9·0	Di. s.	11	4·6	Te. s.		22·5	Rh. n.
	13·0	Rh. s.		10·5	Di. n.	26	7·3	Te. s.
25	0·6	Te. n.	12	3·3	Te. n.		12·1	Di. s.
	17·8	Di. n.		8·9	Rh. n.	27	5·9	Te. n.
	23·3	Te. s.		19·4	Di. s.		20·9	Di. n.
26	19·3	Rh. n.	13	1·9	Te. s.	28	4·6	Te. s.
	22·0	Te. n.	14	0·6	Te. n.		4·8	Rh. s.
27	2·7	Di. s.		4·2	Di. n.			

By means of this list of conjunctions and of the little table given on page 404, approximate places of the satellites may be easily found for any hour, as the table gives the coordinates of the satellites at intervals of an hour, reckoned from the time of the nearest (preceding or following) conjunction with the centre. Observers who are desirous to follow the motions of the satellites will do well to lay down the data of the table graphically on a sufficiently large scale, so that by marking the corresponding times for the night of observing they may get information about the places of the satellites at a glance. The places change in the direction of decreasing position-angles.

The following ephemerides of *Titan* and *Iapetus* give the positions of the two satellites in reference to the circle of declination.

Differences of Right Ascension and Declination between Titan and Iapetus, and the Centre of Saturn.

		Titan.		Iapetus.	
♂ Gr.		α−A s	δ−D $''$	α−A s	δ−D
1882, Nov.	30	+ 14·25	+ 19·6	+ 25·79	+ 225·8″
Dec.	1	+ 12·12	+ 46·4	23·20	224·6
	2	+ 8·15	+ 66·1	20·48	222·1
	3	+ 2·90	+ 75·3	17·64	218·2
	4	− 2·80	+ 72·3	14·69	213·1
	5	− 8·02	+ 57·4	11·66	206·6
	6	− 11·89	+ 32·8	8·56	198·9
	7	− 13·76	+ 2·9	5·42	190·0
	8	− 13·37	− 27·5	+ 2·25	+ 180·0
	9	− 10·82	− 53·4	− 0·93	+ 168·9
	10	− 6·60	− 71·0	4·09	156·7
	11	− 1·39	− 77·8	7·22	143·6
	12	+ 4·02	− 73·3	10·30	129·6
	13	+ 8·84	− 58·4	13·30	114·8
	14	+ 12·39	− 35·3	16·21	99·3
	15	+ 14·19	− 7·2	19·01	83·3
	16	+ 13·93	+ 21·8	21·68	66·7
	17	+ 11·62	+ 47·5	24·21	49·7
	18	+ 7·55	+ 65·8	26·56	32·5
	19	+ 2·31	+ 73·8	− 28·73	+ 15·1
	20	− 3·29	+ 69·8	− 30·71	− 2·3
	21	− 8·32	+ 54·2	32·48	19·7
	22	− 11·95	+ 29·7	34·03	37·0
	23	− 13·57	+ 0·2	35·34	53·9
	24	− 12·96	− 29·2	36·41	70·4
	25	− 10·28	− 53·8	37·23	86·4
	26	− 6·01	− 70·1	37·79	101·7
	27	− 0·86	− 75·9	38·09	116·3
	28	+ 4·42	− 70·6	38·13	130·1

		Titan.		Iapetus.	
☾ Gr.		α−A	δ−D	α−A	δ−D
1882, Dec.	29	+ 9·04	−55·3	37·91	142·9
	30	+ 12·38	−32·2	37·42	154·7
	31	+ 13·95	− 4·7	36·67	165·4
1883, Jan.	1	+ 13·51	+23·4	−35·67	−174·9
	2	+ 11·09	+48·0	34·43	183·2
	3	+ 6·98	+65·1	32·95	190·2
	4	+ 1·81	+72·0	31·24	195·8
	5	− 3·64	+67·2	29·33	200·0
	6	− 8·47	+51·3	27·22	202·9
	7	−11·86	+26·9	24·93	204·3
	8	−13·27	− 1·9	22·47	204·3
	9	−12·51	−30·3	19·87	203·0
	10	− 9·75	−53·8	17·14	200·2
	11	− 5·50	−69·0	14·30	196·1
	12	− 0·44	−73·8	11·38	190·6
	13	+ 4·67	−68·0	8·39	183·9
	14	+ 9·10	−52·5	5·35	176·0
	15	+ 12·23	−29·7	− 2·29	−166·9
	16	+ 13·62	− 2·7	+ 0·77	−156·8
	17	+ 13·06	+24·5	3·82	145·7
	18	+ 10·57	+48·0	6·83	133·6
	19	+ 6·49	+64·1	9·78	120·7
	20	+ 1·41	+70·1	12·66	107·1
	21	− 3·87	+64·8	15·44	92·9
	22	− 8·49	+48·8	18·12	78·1
	23	−11·68	+24·7	20·67	62·9
	24	−12·93	− 3·5	23·08	47·4
	25	−12·06	−31·0	25·34	31·6
	26	− 9·28	−53·5	+27·43	− 15·7
	27	− 5·09	−67·8	+29·35	+ 0·3
	28	− 0·15	−71·9	31·08	16·2
	29	+ 4·79	−65·7	+32·62	+31·9
	30	+ 9·04	−50·2	33·95	47·4
	31	+ 12·00	−27·7	35·08	62·6
Feb.	1	+ 13·26	− 1·3	36·00	77·3
	2	+ 12·61	+25·1	36·70	91·5
	3	+ 10·12	+47·8	37·18	105·1
	4	+ 6·10	+63·2	37·45	118·1

		Titan.		Iapetus.	
0ʰ Gr.		α—A s	δ—D	α—A s	δ—D
1883, Feb.	5	+ 1·15	+ 68″·5	37·50	130″·4
	6	− 3·97	+ 62·8	37·34	141·8
	7	− 8·42	+ 46·8	36·96	152·4
	8	− 11·44	+ 23·0	36·37	162·1
	9	− 12·57	− 4·6	35·58	170·9
	10	− 11·65	− 31·3	34·60	178·7
	11	− 8·89	− 53·1	33·42	185·4
	12	− 4·79	− 66·7	32·06	191·0
	13	+ 0·01	− 70·4	30·52	195·6
	14	+ 4·80	− 64·0	28·82	199·1
	15	+ 8·89	− 48·5	26·96	201·4
	16	+ 11·73	− 26·3	24·96	202·6
	17	+ 12·90	− 0·4	22·83	202·7
	18	+ 12·22	+ 25·4	20·57	201·6
	19	+ 9·76	+ 47·5	18·21	199·4
	20	+ 5·82	+ 62·3	15·75	196·1
	21	+ 0·99	+ 67·3	13·21	191·7
	22	− 3·98	+ 61·4	10·61	186·2
	23	− 8·27	+ 45·4	7·96	179·7
	24	− 11·18	+ 21·9	5·27	172·2
	25	− 12·24	− 5·2	+ 2·56	+ 163·7
	26	− 11·31	− 31·4	− 0·16	+ 154·3
	27	− 8·61	− 52·7	2·86	144·0
	28	− 4·60	− 65·8	5·54	133·0
Mar.	1	+ 0·07	− 69·3	8·17	121·2
	2	+ 4·72	− 62·8	10·75	108·7
	3	+ 8·70	− 47·5	13·26	95·6
	4	+ 11·44	− 25·5	15·68	82·0
	5	+ 12·57	0·0	18·00	67·9
	6	+ 11·90	+ 25·5	20·20	53·5
	7	+ 9·49	+ 47·2	22·27	38·8
	8	+ 5·65	+ 61·6	24·20	23·9
	9	+ 0·94	+ 66·4	− 25·97	+ 8·9
	10	− 3·91	+ 60·4	− 27·58	− 6·2
	11	− 8·09	+ 44·5	29·01	21·1
	12	− 10·93	+ 21·3	30·25	35·8
	13	− 11·96	− 5·5	31·30	50·3
	14	− 11·10	− 31·4	32·14	64·5

0^h Gr.	Titan		Iapetus	
	$a-A$ s	$\delta-D$	$a-A$ s	$\delta-D$
1883, Mar. 15	− 8·42	−52·3	32·78	78·2
16	− 4·52	−65·3	33·21	91·3
17	+ 0·04	−68·7	33·42	103·9
18	+ 4·59	−62·2	33·42	115·6
19	+ 8·47	−47·0	33·19	126·7
20	+11·18	−25·2	32·76	136·9
21	+12·29	+ 0·1	32·11	146·1
22	+11·65	+25·3	31·25	154·4
23	+ 9·31	+46·8	30·19	161·7
24	+ 5·56	+61·2	28·94	167·9
25	+ 0·95	+66·0	27·50	173·0
26	− 3·79	+60·0	25·88	176·9
27	− 7·90	+44·2	24·10	179·7
28	−10·70	+21·1	22·17	181·3
29	−11·74	− 5·5	20·10	181·7
30	−10·88	−31·3	−17·90	−181·0

Observations of Comet a 1882. By E. J. Stone, Esq.

The following is the series of observations of Comet a, 1882, made with the Transit-Circle of the Radcliffe Observatory, Oxford, when passing *sub polo* :—

Ref.	Day.	G.M.T.	Observed R.A.	Observed N.P.D. (uncorrected for Parallax)	Obs.
	1882	h m s	h m s	° ′ ″	
(a)	May 12	8 57 20·13	0 14 22·90	15 32 53·4	R.
(b)	13	9 18 33·31	0 39 36·12	15 54 2·8	W.
(c)	15	9 57 21·31	1 26 23·60	17 8 33·7	R.
(d)	16	10 14 15·71	1 47 17·34	18 0 13·6	W.
(e)	17	10 29 20·28	2 6 20·93	19 0 10·4	R.
(f)	18	10 42 34·30	2 23 33·69	20 7 31·6	W.
(g)	19	10 54 4·86	2 39 2·69	21 21 18·6	R.
(h)	20	11 3 59·82	2 52 55·84	22 40 44·1	W.
(i)	21	11 12 28	3 5 (21·75)	24 5 (18)	R.
(j)	22	11 19 38·44	3 16 30·14	25 33 (54)	R.
(k)	24	11 30 41·92	3 35 28·55	28 42 28·8	R.
(l)	25	11 35 0	—	30 21 48·0	F.B.
(m)	26	11 38 14·07	3 50 55·05	32 4 1·2	R.
(n)	27	11 40 54·82	3 57 32·80	33 49 7·1	W.
(o)	29	11 44 34·11	4 9 5·80	37 27 44·1	R.
(p)	31	11 46 21·43	4 18 46·53	41 18 26·7	R.

Observers' Notes :—

(*a*), (*b*) Very faint, but observations fair. (*c*) Very faint at times; observation fair on the whole. (*d*) Nucleus sometimes showed as a bright point, but generally not well defined, and would scarcely stand any illumination of field. Observation, though difficult, very fair. (*e*) Observation good. (*f*) Observation considered very good. Nucleus very sharp at times. (*g*) Difficult, but considered fairly good. Nucleus faint at times. (*h*) Faint. Observation good. (*i*) Observation only approximate. Sky cloudy. (*j*) R.A. good. N.P.D. very rough from a single bisection when extremely faint.

General Notes (*a*) to (*j*).—In the telescope, the light of the head on the night of May 18, the nucleus being better defined than on any other night when the observations were made by me, was certainly not brighter than an eighth magnitude star (W.)

Brightness = Eight in star-magnitude (R.) May 21 and 22, cloudy.

(*k*) Difficult observation, but considered fairly good. Clouds passing. 7½ star-magnitude.

(*l*) The comet was as bright as a 7 or 7⅓ magnitude star, but cloud prevailed nearly the whole time of the transit: only one bisection made.

(*m*) Observation pretty good. As bright (in telescope) as a 7–6 mag. star.

(*n*) Observation very satisfactory. Nucleus a bright point equal to 6⅓ star-magnitude.

(*o*) Observation good. Brightness in star-magnitude = 6–5. *Note.*—May 29, 13ʰ. The comet and tail are both visible with the naked eye.

(*p*) Very good observation. Brightness in telescope = 4½ mag. *Note* at 10½ʰ.—Compared comet with stars near for magnitude, and found it (to the naked eye) identical in brightness with δ Persei = 3rd magnitude. Observers—W. = Mr. WICKHAM. R. = Mr. ROBINSON. F.B. = Mr. F. BELLAMY.

Observation of Comet a 1882. By L. G. Puckle, Esq.

(*Communicated by Capt. H. Toynbee, R.N.*)

On Friday, June 23, at 7ʰ 20ᵐ P.M., the ship being in latitude 2° 30′ N., and longitude 104° 33′ E., we observed a large comet a little to the southward of the planet *Venus*.

The following angles were observed with a sextant :—

					°	′	″
Altitude of *Venus*	...	...	...		13	30	0
Altitude of nucleus of comet		...			12	0	0
Angle between nucleus and *Venus*	...				6	30	0

The nucleus was well defined and bright, like a star of the second magnitude, the tail spreading out like a fan to the extent of about two and a half degrees of altitude, as visible with the naked eye. The tail stretched upwards from the horizon, with a slight curve in it towards the southward, at the upper end.

The comet remained visible until it sank behind the clouds on the horizon. Careful search was made for it the next night, but without success. We did not see it again, even with the aid of a telescope.

On our arrival at Hong Kong on June 30 we made inquiries about it, but found that nobody had seen it or heard anything about it until we ourselves reported it.

The Solar Eclipse of 1882, *May* 17, *observed at Vizagapatam.*
By A. V. Nursinga Row, Esq.

The partial eclipse of the Sun of 1882, May 17, was observed at the Observatory, Daba Gardens, Vizagapatam (Latitude $17° 42' 9''$ N., Longitude $5^h 33^m 32^s·3$ E., according to the Trigonometrical Survey). There were no clouds, but the sky was hazy, and the Sun was surrounded by a halo during the whole time of the eclipse. The beginning of the eclipse was observed by observer Verabadroodoo at local sidereal time $4^h 25^m 17^s$; the end was observed by Mr. A. V. Juggarow at local sidereal time $6^h 57^m 21^s$.

On the Solution of Kepler's Problem.   By Prof. Ch. V. Zenger.

The various modes of calculating the true anomaly by means of the mean anomaly and eccentricity of the planetary orbit considered by Mädler, Leverrier, and J. W. L. Glaisher, require a considerable amount of calculation, partly owing to the slow convergence of the series employed.

Writing

$$E - e \sin E = F = m (T - t),$$

where E denotes the true anomaly, e the eccentricity, m the mean anomaly, and T the time of perihelion, in the following manner—

$$E - F = e \sin E,$$

it becomes obvious, from the small amount of the eccentricity, that $E - F$ will be always small, and the series will rapidly converge.

Now

$$E - F = \sin(E - F) + \frac{1}{6}\sin^3(E - F) + \frac{3}{40}\sin^5(E - F) + \frac{15}{336}\sin^7(E - F) + \&c.;$$

and we thus have

$$\frac{\sin (E - F)}{\sin E} = \frac{e}{1 + \frac{1}{6}\sin^2(E - F) + \frac{3}{40}\sin^4(E - F) + \frac{15}{336}\sin^6(E - F) + \ldots}$$

and therefore

$$\cot F - \cot E = \frac{e \csc F}{1 + \frac{1}{6}\sin^2(E-F) + \frac{3}{40}\sin^4(E-F) + \frac{15}{336}\sin^6(E-F)}.$$

For the first approximation $E-F$ may be replaced by its maximum value, viz.

$$E - F = ew \sin 90° = ew,$$

where ew is the angular value of the eccentricity; the first approximation thus is

$$\cot E = \cot F - \frac{e \csc F}{1 + \frac{1}{6}\sin^2 ew + \frac{3}{40}\sin^4 ew + \frac{15}{336}\sin^6 ew}.$$

The value thus obtained gives the angular difference

$$E - F = e \sin E.$$

We have also

$$dE = \frac{dF}{1 - e \cos E},$$

and correcting by the known variation of E, we get immediately with the first approximation, a value correct to the seconds, as the following parallel calculations of the true anomaly of *Mars* by Mädler's and the present method show, without the application of the *regula falsi.*

Mädler's method.—Suppose the true anomaly of *Mars* required for

$$\text{April 24, } 13^h\ 25^m\ 15^s\cdot0,$$

$$\frac{T = \text{Jan.} \quad 8. \quad 9^h\ 44^m\ \ 0^s\cdot0}{T - t = \quad 107, \quad 3^h\ 41^m\ 15^s\cdot0},$$

$$m = \frac{360°}{686\cdot97964} = 0°\ 31'\ 26''\cdot516,$$

$$m(T-t) = F = 56°\ 9'\ 7''\cdot4, \quad e = 0\cdot0932168,$$

$$ew = 19227''\cdot24 = 5°\ 20'\ 27''\cdot04.$$

Mädler says the value of E must be comprised between $59°\ 9'\cdot12$ and $61°\ 29'\cdot55$, and puts $E = 60°$ for the first approximation,

$$\log \sin 60° = 9\cdot9375306,$$

$$\log ew = \frac{4\cdot2839195}{4\cdot2214501} = \log 16651'''\cdot53,$$

$$E - ew \sin E = 60° - 4°\ 37'\ 31''\cdot33 = \quad 56°\ 22'\ 28''\cdot67,$$

$$\frac{F = \quad 56°\ 9'\ \ 7''\cdot40}{dF = -\ \ 0°\ 46'\ 38''\cdot73}.$$

For comparison, I now give the first approximation obtained by my own method.

$$\cot E = 0.670645 - \cfrac{0.112243}{1 + \cfrac{\sin^2(5°\ 20'\cdot45)}{6} + \cfrac{3\sin^4(5°\ 20'\ 45)}{40}},$$

$$\cot E = 0.670645 - \frac{0.112243}{1.01448} = 0.670645 - 0.110390,$$

$$\cot E = 0.560255,\quad E = 60°\ 45'\ 36'',$$

$$\log \sin 60°\ 45'\ 36'' = 9.94079,\quad F' = ew = 4°\ 39'\ 35''\cdot6,$$

$$\log ew \qquad = \underline{4.28392}$$
$$4.22471 = \log 16775''\cdot6,$$

$$E - F' = 60°\ 45'\ 36''\cdot0 - 4°\ 39'\ 35''\cdot6 = 56°\ 6'\ 0''\cdot4,$$

$$\underline{F = 56°\ 9'\ 7''\cdot4}$$
$$\Delta F = 0°\ 3'\ 7''\cdot0$$

$$\Delta F : \Delta F' = 46'\cdot645 : 3'\cdot12 = 15 : 1.$$

The first approximation by my own method is fifteen times better than Mädler's result.

But using the differential relation

$$dE = \frac{187''}{1 - 0.09322 \cos 60°\ 45'\cdot6} = 195''\cdot92,$$

the corrected true anomaly will be

$$E + \delta E = E_1 = 60°\ 45'\ 36''\cdot0 + 3'\ 15''\cdot92,$$
$$E_1 = 60°\ 48'\ 51''\cdot92.$$

Mädler has to go through three approximative calculations to find finally the true anomaly

$$E = 60°\ 48'\ 53''\cdot78,$$

giving an error

$$\Delta F = 0°\ 0'\ 0''\cdot04.$$

It is obvious how rapidly the proposed method, using only one approximation, gives to $1''\cdot80$ the true value of E.

The next approximation would bring the value of E to

$$60°\ 48'\ 53''\cdot72.$$

Prague:
   1882, *July* 17.

Postscript.

The largest planetary eccentricity not exceeding 0.4, it will in all cases be sufficient to take the two first terms of powers of $\sin(E - F)$, which may be replaced for the first approximation by

$$\theta_0 = 1 + \frac{1}{6}\sin^2(ew) + \frac{3}{40}\sin^4(ew).$$

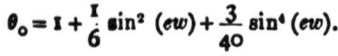

It will greatly shorten computation if θ_0 be given by a small table for the eight principal planets.

The following table gives for those planets the terms $\log \theta_0$ and $\log \frac{e}{\theta_0}$:

I. $\log \theta_0$ and $\log \left(\frac{e}{\theta_0}\right)$.

	ew	θ_0	$\log \theta_0$	$\log \frac{e}{\theta_0}$	$\log e$
☿	0·2056048	11° 46′ 55″·89	1·082021	0·0342357	9·2787978	9·3130335
♀	0·0068433	0 23 31·53	1·000008	0·0000037	7·8352619	7·8352656
♁	0·0167701	0 57 39·08	1·000047	0·0002041	8·2243316	8·2245357
♂	0·0932611	6 43 41·21	1·002288	0·0009882	8·9687025	8·9697007
♃	0·0482388	2 45 49·97	1·000388	0·0001685	8·6832281	8·6833966
♄	0·0559956	2 55 49·92	1·000436	0·0001893	8·7479645	8·7481538
♅	0·0465775	2 40 7·30	1·000361	0·0001568	8·6680194	8·6681762
♆	0·0087195	0 29 48·53	1·000013	0·0000054	7·9404862	7·9404916

The first approximation may be immediately obtained by the equation

$$\log (\cot F - \cot E_0) = \log\left(\frac{e}{\theta_0}\right) + \log \operatorname{cosec} F = \log\left(\frac{e}{\theta_0}\right) - \log \sin F.$$

Let the number to the 6th decimal be N, and we get

$$\cot E_0 = -N + \cot F.$$

Computing now with the value E_0, that of F_0, $\delta F = F - F_0$, becomes a known value, and by means of Table II. the correction of E is directly obtained by the equation

$$\log \delta E = \log \delta F + \log (1 - e \cos E)^{-1} = \log \delta F + \log (1 + e \cos E).$$

II. $\log (1 - e \cos E)^{-1}$.

E=	10°	20°	30°	40°	50°	60°	70°	80°	90°
☿	0·09826	0·09323	0·08197	0·07443	0·06157	0·04711	0·03166	0·01579	0·00000
♀	0·00293	0·00280	0·00257	0·00228	0·00192	0·00148	0·00101	0·00052	0·00000
♁	0·00723	0·00690	0·00635	0·00562	0·00471	0·00366	0·00250	0·00126	0·00000
♂	0·05317	0·05659	0·04639	0·04079	0·03396	0·02619	0·01816	0·00891	0·00000
♃	0·02112	0·02014	0·01852	0·01634	0·01368	0·01060	0·00723	0·00365	0·00000
♄	0·02463	0·02347	0·02158	0·01904	0·01589	0·01232	0·00840	0·00424	0·00000
♅	0·02040	0·01944	0·01787	0·01628	0·01320	0·01023	0·00748	0·00353	0·00000
♆	0·00375	0·00371	0·00329	0·00292	0·00362	0·00190	0·00129	0·00066	0·00000

Réponse à M. Marth. Par M. Loewy.

(Communicated by Mr. Stone.)

Dans la réunion du mois de Mai dernier, M. Marth a fait à la Société Royale Astronomique une nouvelle communication relative à la détermination des flexions, et dans cette note il prétend avoir rencontré dans mes écrits quelques assertions extraordinaires, et il s'applique à les combattre avec énergie.

J'ai lu attentivement et à plusieurs reprises le Mémoire de M. Marth, et, par surcroît de précaution, j'ai prié quelques-uns de mes amis de le lire et de le comparer avec ce que j'ai publié moi-même, pour voir s'ils seraient plus capables que moi de retrouver quelque chose qui ait pu donner lieu à M. Marth de m'attribuer les assertions singulières qu'il combat si victorieusement; et si, d'un autre côté, M. Marth répond à une seule des objections que j'ai faites contre les nombreux inconvénients résultant de l'emploi de son appareil.

En vérité, M. Marth ne répond à aucune de mes objections, et il se contente de démontrer, avec beaucoup de sagacité, du reste, que le centre optique ne se déplace pas lorsque l'objectif oscille légèrement autour de ce centre; c'est là un théorème d'optique élémentaire bien connu qui d'ailleurs n'avait rien à faire dans la question qui nous occupe actuellement.

Dans mon Mémoire, j'ai indiqué (pages 9 à 18, et 56 à 62), en exposant la théorie de mon appareil, tous les mouvements que peut subir la lentille auxiliaire pendant la rotation de la lunette, et j'ai mis en équation seulement ceux qui sont de nature à produire un déplacement du centre optique en excluant tous les autres, et j'ai naturellement compris dans cette exclusion tous les mouvements oscillatoires autour du centre optique.

Je me suis dispensé de donner à propos de chacun de ces mouvements une démonstration quelconque, car je supposais chez mes lecteurs non seulement quelques notions d'optique élémentaire, mais même des connaissances en optique supérieure. Peut-être M. Marth a-t-il, bien involontairement sans doute, omis de lire les pages notées ci-dessus, ou peut-être, cédant à l'entraînement d'une imagination un peu prévenue, a-t-il cru trouver les assertions singulières qu'il combat si victorieusement, mais dont, malheureusement pour lui, on ne trouve aucune trace dans la note que j'ai eu l'honneur de soumettre à la Société Royale Astronomique. Et d'autre part, comme je viens de le dire, j'ai précisément indiqué dans le Mémoire sur la Flexion le contraire des assertions que me prête M. Marth.

Quoiqu'il en soit, je vais, pour terminer cette discussion, exposer de nouveau avec quelques détails les déplacements que peut subir pendant la rotation, quand il s'agit par exemple de la flexion en hauteur, une lentille introduite dans le cube central de la lunette.

Il peut se produire :

1° Un déplacement de l'objectif dans son ensemble et perpendiculairement à l'axe optique par suite du mode d'attache de l'appareil au cube.

2° Un glissement également perpendiculaire à l'axe optique de l'un des verres par rapport à l'autre.

3° Une sorte de déplacement ou d'écartement angulaire des verres de l'objectif.

4° Des déformations dues aux mouvements 2 et 3 ou à des flexions différentes subies par les verres pendant la rotation.

Les mouvements 1 et 2 provoquent naturellement un déplacement direct de la ligne de visée.

Le mouvement 3, quoique ne provoquant pas le déplacement matériel du centre optique, altère les images dans le réticule pendant la rotation de la lunette. Le mouvement 2 peut provoquer également une déformation.

Ces déplacements ou ces altérations sont parfois assez notables et peuvent faire naître des erreurs systématiques ; il y a toujours lieu d'en tenir compte, surtout quand il s'agit de déterminer la variation de la ligne de visée avec la plus haute précision : c'est-à-dire à quelques centièmes près de seconde d'arc.

C'est un fait maintenant bien connu que dans beaucoup de lunettes, et surtout dans la plupart des petits instruments, la véritable flexion n'existe pas, et qu'en réalité les variations qu'on y rencontre souvent sont produites ou par de légers déplacements de la monture portant l'objectif ou par les mouvements 2 et 3 dans la monture.

Pendant plusieurs années ces causes ont provoqué une variation de la ligne de visée à la lunette de Gambey, à l'Observatoire de Paris. Le même fait existe au petit cercle méridien de Brunner installé à l'Observatoire de Montsouris (*Mémoires du Bureau des Longitudes*, tome i.). Lors de nos opérations de longitude entre Paris et Brégentz, nous avons, M. Oppolzer et moi, constaté un effet analogue avec l'instrument sortant des ateliers de MM. Troughton et Simms. Plus tard dans les déterminations de différences de longitude entre Paris et Berlin, nous avons trouvé une variation très-notable dans la ligne de visée, à la lunette de Rigaud, et pour la faire disparaître il nous a fallu démonter et remonter à nouveau l'objectif de ce petit instrument (*Mémoires du Bureau des Longitudes*, tome ii., qui va paraître prochainement). Et la valeur numérique des variations de la ligne de visée dans ces divers instruments est comprise entre 1″ et 3″.

Il faudrait plusieurs pages pour faire la liste de tous les instruments dans lesquels on a pu constater des déplacements de la ligne de visée, par suite des mouvements 1 ou 2 de l'objectif, et d'ailleurs c'est un fait bien connu de tous les astronomes qui ont étudié avec soin leurs lunettes.

En dehors de ces déplacements directs des images dans le réticule, on se trouve parfois en présence, comme nous l'avons indiqué plus haut, de déformations pouvant causer de légères

erreurs systématiques : il est donc inadmissible pour évaluer ces
effets dans la détermination de la variation de la ligne de visée
d'employer un appareil qui peut provoquer des erreurs analogues
plus considérables encore.

Déjà en 1859, M. Przemowski en faisant à l'observatoire de
Varsovie des observations méridiennes avec une lunette de
4 pouces, construite par Fraunhofer, a été frappé de la déformation qui se produisait dans les images pendant la rotation, et
reconnut que cette déformation qui donnait lieu à des erreurs
sensibles dans les observations tenait au décentrage : c'est alors
qu'il fit construire un appareil spécial pour mieux évaluer la
nature et l'étendue de la déformation correspondant à chacun
des mouvements 2 ou 3. J'ai eu moi-même, en ces derniers
temps, l'occasion de constater le même fait, en examinant les
qualités optiques de plusieurs instruments destinés à l'observation du prochain passage de Vénus.

En variant la position de la lunette, les anneaux de diffraction
se déformaient sensiblement, et de circulaires devenaient ellip-
tiques et quelquefois absolument irréguliers. Ces déformations
étaient même si considérables que nous étions tout d'abord tentés
de les attribuer à un défaut de construction, mais une autre
position de la lunette les augmentait ou parfois les faisait disparaître, de sorte qu'on ne pouvait imputer ces altérations des
images qu'à un déplacement de l'un des verres de l'objectif, à
un effet de décentrage ou de flexion.

Des mouvements encore plus dangereux peuvent se manifester
lorsqu'on introduit dans le cube de la lunette un miroir.

Mais, en résumé, qu'il s'agisse d'un déplacement direct de la
ligne de visée ou d'une déformation des images, on ne peut
a priori affirmer qu'elles n'existeront pas, quelles que soient
l'habileté du constructeur et la perfection de l'instrument. C'est
l'expérience seule qui plus tard peut les faire trouver.

Tous ces faits si connus des astronomes au courant des études
de haute précision, m'ont engagé à réunir sur un seul disque de
verre toutes les pièces nécessaires à mon appareil, et pour le
rendre invariable dans sa monture et absolument indéformable
pendant la rotation de la lunette, j'ai déterminé par des expériences directes l'épaisseur qu'il fallait donner à ce bloc de verre
pour pouvoir le coller dans sa monture, le serrer même à force
dans son barillet sans l'exposer à éprouver d'altération ou de
déformation sensible. Cette épaisseur égale à 0·25 du diamètre
permet d'obtenir ce résultat qu'on ne saurait atteindre avec des
valeurs inférieures.

En rendant ainsi toutes les déformations impossibles j'ai pu
par une expérimentation rigoureuse déterminer tous les mouvements de translation et de rotation du seul disque de verre qui
constitue mon appareil.

On voit par tout ce qui précède la multiplicité des déplacements auxquels peut donner lieu l'appareil de M. Marth
composé de cinq pièces optiques disposées de telle façon que

l'image doit traverser deux objectifs ou deux fois le même objectif
et un miroir. Les déformations des images ou leurs déplacements deviennent alors si nombreux et si considérables qu'il n'y
a aucune méthode possible pour les éviter ou les évaluer. C'est
ce qui explique pourquoi cet appareil est depuis vingt ans
demeuré sans application possible.

Je n'ai plus rien à ajouter sur les nouvelles explications que
fournit M. Marth dans la note du mois de Mai; il y donne lui-
même la réfutation la plus complète de son appareil; il nie *a
priori* les mouvements même les plus probables, et donne pour
ceux qu'il veut bien admettre des méthodes d'évaluation tout à
fait défectueuses.

Je ne doute pas que M. Marth ne présente son appareil à la
Société Royale Astronomique, mais ce dont je doute énormé-
ment, c'est qu'il puisse apporter à l'appui une série d'expérimen-
tations donnant des résultats précis, et concordant avec les
valeurs trouvées par les procédés astronomiques connus.

Je ne crois pas devoir continuer la discussion sur cette
matière. Aujourd'hui, il sera facile aux astronomes de trouver
dans les notes et Mémoires publiés des deux côtés tous les
éléments nécessaires pour asseoir leur jugement avec facilité et
sécurité. Je ne répondrai même plus à M. Marth s'il lui vient
encore à l'esprit de m'accuser de quelque nouvelle hérésie
relative à une question de mécanique ou de géométrie élémen-
taires.

INDEX.

LIST OF PRESENTS

RECEIVED DURING THE SESSION OF 1881–82,

AND OF

BOOKS PURCHASED WITH THE TURNOR FUND

DURING THE SAME PERIOD,

FORMING:

APPENDIX XXXIII.

To the Catalogue of the Library of the Royal Astronomical Society.

Abbadie, A. d', Recherches sur la verticale, 8vo.
Bruxelles, 1881 — The Auth

Abney, Capt. W. de W., Photography with emulsions: a treatise on the theory and practical working of the collodion and gelatine emulsion processes, 8vo.
London, 1882

Actuaries, Institute of, Journal, Nos. 123–126, 8vo.
London, 1881–2 — The Insti

Airy, Sir G. B., Account of observations of the Transit of Venus, 1874, Dec. 8, made under the authority of the British Government, and of the reduction of the observations, 4to. *London*, 1881 — The sioners of Admiral

———————— On a systematic interruption in the order of numerical values of vulgar fractions, when arranged in a series of consecutive magnitudes, 8vo. *London*, 1881 — The A

———————— Logarithms of the values of all vulgar fractions with numerator and denominator not exceeding 100, arranged in order of magnitude, 8vo. *London*, 1881

———————— A new method of clearing the lunar distance, 8vo. *London*, 1881

American Academy of Arts and Sciences, Proceedings, new series, vol. viii.–vol. xi. 1, 8vo. — The A
Boston and Cambridge, 1881–82

g

·ety.   American Philosophical Society, Transactions, new series, vol.
        xv. 3, 4to.   *Philadelphia*, 1881

————————————————— Proceedings, Nos. 107–109,
        8vo.   *Philadelphia*, 1880–82

tors.   ———— Journal of Mathematics, vol. iii. 4, and iv. 1, 4to.
                                                Cambridge, 1880–81

————————————— Science, series III. vol. xxii. No. 127—
        vol. xxiii. 138, 8vo.   *New Haven*, 1881–82,

y. Amsterdam, Koninklijke Akademie van Wetenschappen,
        Verhandelingen, Deel xxi. 4to.   *Amsterdam*, 1881

————————————————————— Verslagen en mede-
        deelingen, Afdeeling Natuurkunde, tweede reeks, Deel
        xvi. 8vo.   *Amsterdam*, 1881

————————————————————————— Afdeeling
        Letterkunde, tweede reeks, Deel x. 8vo. *Amsterdam*, 1881

————————————————————— Jaarboek voor 1880, 8vo.
                                                Amsterdam, 1881

————————————————— Catalogus van de
        boekerij, Deel iii. 2, 8vo.   *Amsterdam*, 1881

————————————————— Processen - Verbaal,
        1880–81, 8vo.   *Amsterdam*, 1881

————————————————— Tria carmina latina,
        8vo.   *Amsterdam*, 1881

Editor.   Analyst, The, Nos. 64–75, 8vo.   *London*, 1881–82

Author.   Åstrand, J. J., Om en ny methode för lösning af trinomiske
          ligninger af n^{te} grad. 8vo.   *Kristiania*

Editor.   Astronomical Register, The, Nos. 223–234, 8vo.
                                                London, 1881–82.

          Astronomie (l') Revue mensuelle d'astronomie populaire, de
          météorologie et de physique du globe, publiée par
          Camille Flammarion, Nos. 1–6, 8vo.   *Paris*, 1882

·   Astronomische Nachrichten, Nos. 2378–2433, 4to.
                                                Kiel, 1881–82

          Athenæum, The, Nos. 2798–2849, 4to.   *London*, 1881–82

Author.   Backlund, O., Elemente und Ephemeride des Encke'schen
          Cometen für das Jahr 1881, 4to.   *St. Petersburg*, 1881

          ———————— Zur theorie des Encke'schen Cometen, 4to.
                                                St. Petersburg, 1881

Baird, Capt. A. W., and E. Roberts, Tide-tables for the Indian ports for the year 1882, 8vo.   *London,* 1881   E. Roberts, Esq.

Ball, L. de, Declination von 200 Sternen +49° bis 51°, nach Beobachtungen im ersten Vertical am Passageninstrument der Herzogl. Sternwarte zu Gotha, 4to. *Kiel,* 1882   The Author.

Batavia, Magnetical and Meteorological Observatory, Observations, vol. v. 4to.   *Batavia,* 1881   The Observatory

———————————————————————, Regenwaarnemingen in Nederlandsch-Indië, door Dr. P. A. Bergsma, Jaargang II., III., 8vo.   *Batavia,* 1881–82   ———

——— Natuurkundige Vereeniging in Nederlandsch Indië, natuurkundig Tijdschrift, Deel xl., 8vo.   The Society.
Batavia, 1881

Becker, L., Untersuchungen über die allgemeinen Störungen der Feronia, 4to.   *Bonn,* 1882   The Author.

Belfast Natural History and Philosophical Society, Proceedings for the session 1880–81, 8vo.   *Belfast,* 1882   The Society.

Berlin, Königlich-preussische Akademie der Wissenschaften Monatsbericht, 1881, Feb.–Dec.   *Berlin,* 1881–82   The Academy

———Königlich-preussisches geodätisches Institut, Publicationen:

 Das Mittelwasser der Ostsee by Swinemünde, bearbeitet von W. Seibt.   The Institute

 Die Ausdehnungscoefficienten der Küstenvermessung, von A. Westphal.   ———

 Astronomisch-geodätische Arbeiten in den Jahren 1879–1880, 4to.   *Berlin,* 1881   ———

 Das Rheinische Dreiecksnetz. III. Die Netzausgleichung, 4to.   *Berlin,* 1882   ———

 Zur enstehungsgeschichte der Europäischen Gradmessung, 4to.   *Berlin,* 1882,   ———

 Präcisions-Nivellement der Elbe, zweite Mittheil ung ausgeführt von Wilhelm Seibt, 4to.   *Berlin,* 1881   ———

 Das Hessische Dreiecksnetz, 4to.   *Berlin,* 1882   ———

——— Physikalische Gesellschaft, Fortschritte der Physik im Jahre 1876, 8vo.   *Berlin,* 1880–81   The Society.

ᵉ tory, a.

Berliner Astronomisches Jahrbuch für 1883, 8vo.

Berlin, 1881

———————————————— Circular, Nos. 161-183, 8vo. *Berlin,* 1881-82

my. Bologna, Accademia delle Scienze dell' Istituto, Memorie, serie III. tomo x. 3, 4; serie IV. tomo i. 4to.

Bologna, 1879-80

————————————————————— Indici generali dei X Tomi della terza serie, 4to. *Bologna,* 1880

.M.'s rnment ndia.

Bombay, Government Observatory, magnetical and meteorological observations in the years 1871 to 1878, under the superintendence of Charles and F. Chambers, fol.

Bombay, 1881

———————————————— Report on the administration of the meteorological department of Western India for the year 1880-81, fol. *Bombay,* 1881

ety. Bordeaux, Société des Sciences Physiques et Naturelles, Mémoires, série II. tome iv. 2, 3, 8vo.

Paris et Bordeaux, 1881

useum. Bristol Museum and Library, Report of the Proceedings at the 11th annual meeting, 1882, 8vo. *Bristol,* 1882

ᴉe tion.

British Association for the Advancement of Science, Report of the 51st meeting, held at York, 1881, 8vo.

London, 1882

ᴣ tory.

Bruxelles, Observatoire Royal, Annuaire, 1882, 16mo.

Bruxelles, 1881

Buda Pest, Magyar Tudományos Akadémia :

cademy. ———————————————— Értekezések a mathematikai Tudományok Köréböl, vii. 3, 6-18, 8vo.

Buda Pest, 1879-80

——————————————————————— természettudományok Köréböl, ix. 20-25; x. 1-18, 8vo.

Buda Pest, 1879-80

———————————————— Mathematikai es természettudományok Közlemenyek, xvi. 8vo. *Buda Pest,* 1881

———————————————— Literarische Berichte aus Ungarn, iv. 1-4, 8vo. *Buda Pest,* 1880

Buda Pest, Magyar Tudományos Akadémia, Ungarische   The Academy.
Revue, 1881,1-3, 8vo.   *Buda Pest,* 1881

Bulletin des Sciences Mathématiques et Astronomiques,   The Editors.
série II. tome v. 1, 4-12, 8vo.   *Paris,* 1881-82

Bureau des Longitudes, Annuaire pour l'an 1881-82, 16mo.   Bureau des
Paris, 1881-82   Longitudes.

Burnham, S. W., Quadruple stars, 8vo.   *London,* 1881   The Author.

Calcutta, Asiatic Society of Bengal, Journal, Part I. vol.   The Society.
xlix. extra no. to vol. l. 5 ; and Part II. vol. l. 3-4.
Calcutta, 1881-82

————————————————— Proceedings, 1881, 4—   ————
1882, 1, 8vo.   *Calcutta,* 1881-82

Canada, Geological and Natural History Survey, Report of   The Survey.
progress for 1879-80, 8vo.   *Montreal,* 1881

Canadian Journal and Proceedings of the Canadian Institute,   The Canadian
new series, vol. i. 2, 8vo.   *Toronto,* 1881   Institute.

————— Meteorological Office, Report of the meteorological   Canadian
service of the Dominion of Canada for the year 1879,   Meteorolo-
8vo.   *Ottawa,* 1881   gical Office.

Cape of Good Hope, Royal Observatory, the Cape Catalogue   The
of Stars, deduced from observations made in the years   Observatory.
1834-40, and reduced to the epoch 1840, under the
superintendence of E. J. Stone, 8vo.

Cape Town, 1878

————————————————————— Catalogue of   The Lords
12,441 stars for the epoch 1880. E. J. Stone. 4to.   Commis-
(3 copies.)   *London,* 1881   sioners of the
                                Admiralty.

Cape Town, South African Philosophical Society, Trans-   The Society.
actions, vols. i. and ii. 8vo.
Cape Town and Cambridge, 1878-81

Capron, J. R., A plea for the rainband, 8vo.   *London,* 1881   The Author.

Carpmael, C., Report of the meteorological service of the   The Canadian
Dominion of Canada, 8vo.   *Ottawa,* 1882   Meteorolo-
                                gical Office.

Carruthers, Rev. G. T., An attempt to prove Newton's law   The Author.
of attraction for a resisting medium, 8vo.   *Roorkee,* 1881

Chasles, Michel, Catalogue de la Bibliothèque scientifique,   The
historique et littéraire, 8vo.   *Paris,* 1881   Publisher.

Society. Cherbourg, Société Nationale des Sciences Naturelles et
Mathématiques, tome xxii. 8vo. *Paris et Cherbourg*, 1879

——————————————————————— Catalogue de la
Bibliothèque, 1re partie, 2me édition, 8vo.
Cherbourg, 1881

Director Chicago Astronomical Society and Dearborn Observatory,
the Report for the years 1880 and 1881, 8vo. *Chicago*, 1880–81

rary. ——— Public Library, ninth annual report, 8vo.
Chicago, 1881

Cincinnati Observatory Publications, No. 6. Micrometrical
measurements of double stars, 1879–80, 8vo.
Cincinnati, 1882

her. Clifford, W. K., Mathematical fragments, being facsimiles of
unfinished papers relating to the theory of Graphs, fol.
London, 1881

ry. Coimbra, Universidade, Ephemerides astronomicas calculadas
para o meridiano do Observatorio, para o anno 1883, 8vo.
Coimbra, 1881

Author. Cole, J. E., The Earth's orbit and distance from the Sun, 16mo.
New York, 1882

u des Connaissance des Temps ou des mouvements célestes à l'usage
tudes. des astronomes et des navigateurs, 1881–83, 8vo.
Paris, 1879–81

rs. T. Cooke & Sons, T., Catalogue of astronomical and scientific
& Sons. instruments, 4to.   *York*, 1881

e Cordoba, Observatorio Nacional Argentino, Resultados, vol.
atory. ii.: Observaciones del año 1872, 4to. *Buenos Aires*, 1881

——————————————————————— Anales de la
Oficina meteorologica Argentina, tomo ii.: Climas de
Bahia Blanca, y Corrientes, 4to.   *Buenos Aires*, 1881

Author. Crosland, N., The astronomy of the future, 8vo.   *London*

emy. Dijon, Académie des Sciences, Arts et Belles-Lettres, Mémoires, série III., tome vi., 1880, 8vo.
Dijon et Paris, 1881

Author. Doberck, W., The inventor of the telescope, 8vo.   *London*

——————————— On γ *Virginis* considered as a revolving
double star, 4to.   *Dublin*

Doberck, W., On the binary stars σ *Coronæ*, τ *Ophiuchi*, γ *Leonis*, ζ *Aquarii*, 36 *Andromedæ*, and ι *Leonis*, 4to. | The Aut
Dublin, 1875

———————— On ω *Leonis* considered as a revolving double star, 4to. *Dublin*, 1876

———————— On the first comet of 1845, and on μ² *Boötis*, considered as a revolving double star, 4to. *Dublin*, 1875

———————— On the binary stars 44 *Boötis*, η *Cassiopeiæ* and μ *Draconis*, 4to. *Dublin*.

Dublin, Royal Irish Academy, Transactions (Science), vol. xxviii. 1-6, 4to. *Dublin*, 1880-81 | The Academ

———————————————— (Polite Literature and Antiquities), vol. xxvii. 4, 4to. *Dublin*, 1880

———————————————— Proceedings (Science), vol. iii. series II. 5-6, 8vo. *Dublin*, 1880-81

———————————————— (Polite Literature and Antiquities), vol. ii. series II. 2, 8vo.
Dublin, 1881

Dun Echt Observatory Circulars, Nos. 20-52. | The Ear Crawfo
Dun Echt, 1881-82

Dunsink Observatory, Astronomical Observations and Researches, Part IV., 4to. *Dublin*, 1882 | The Observa

Eastman, J. R., and C. W. Pritchett, The Longitude of Morrison Observatory, Glasgow, Mo., 4to. | The U Nava Observa
Washington, 1881

Edinburgh, Royal Society, Transactions, vol. xxx. 1, for the session 1880-81, 4to. *Edinburgh*, 1881 | The Soci

———————————————— Proceedings for the session 1880-81, 8vo. *Edinburgh*, 1881

Ellis, W., Longitude by telegraph, 8vo. *London*, 1882 | The Au

English Mechanic, the, Nos. 848-898, folio. *London*, 1881-82 | The Edi

Europäische Gradmessung, sechste allgemeine Conferenz, Verhandlungen, 4to. *Berlin*, 1881 | The Confere

Faye, H., Cours d'Astronomie de l'École Polytechnique, 1ʳᵉ partie, 8vo. *Paris*, 1881 | The Au

Ferrari, G. S., La luce zodiacale, 8vo. *Roma*, 1881

uthor. Fleming, S., The adoption of a prime meridian to be common to all nations, 8vo. *London*, 1881

Fonvielle, W., Une visite à la grande comète de 1881, 16mo, *Paris*, 1881

tute. Franklin Institute, Journal, series III. vol. lxxxii. 1—vol. lxxxiii. 6, 8vo. *Philadelphia*, 1881–82

uthor. Freeman, Rev. A., Table showing the time and place of the transit of any star across the prime vertical circle in latitude 52°12'10", 8vo. *Cambridge*, 1882

ety. Genève, Société de Physique et d'Histoire Naturelle, Mémoires, tome xxvii. 2me partie, 4to. *Genève*, 1881

Genova, Giornale della Società di Letture e Conversazioni Scientifiche, v. n., 8vo. *Genova*, 1881–82

Geological Society, Quarterly Journal, Nos. 146–149, 8vo. *London*, 1881–82

Glasgow Philosophical Society, Proceedings, vol. xiii. 1, 1880–81, 8vo. *Glasgow*, 1881

uthor. Gordon, Major E. S., On fixing positions by the more simple astronomical observations, for the use of R. A. officers engaged in exploration, 8vo. *London*, 1881

Society. Göttingen, Königliche Gesellschaft der Wissenschaften, Abhandlungen, Band xxvii., xxviii., 4to. *Göttingen*, 1881–82

———————————————— Gelehrte Anzeigen, 1881, 1, 2, 8vo. *Göttingen*, 1881

———————————————— Nachrichten, 1881, 8vo. *Göttingen*, 1881

Royal rvatory. Greenwich, Royal Observatory, Astronomical and magnetical and meteorological observations made in the year 1879 under the direction of Sir George Biddell Airy, K.C.B., 4to. *London*, 1881

——— ———————————— Results of Astronomical Observations, 1879, 4to. *London*, 1881

———————————————— Results of Magnetical and Meteorological Observations, 1879, 4to. *London*, 1881

———————————————— Description of the Greenwich Time-signal system, 4to. *London*, 1881

Greenwich, Royal Observatory, Rates of chronometers on The Ro
trial for purchase by the Board of Admiralty, 1881, 4to. Observa

London, 1881

——————————————— Spectroscopic and Photographic Results, 1880, 4to. *London,* 1881

——————————————— Report of the Astronomer
Royal to the Board of Visitors, 1882, 4to. *London,* 1882

Guccia, J., Sur une classe de surfaces réprésentables point The Au
par point, sur un plan, 8vo. *Paris,* 1880

Gyldén, H., Om banan af en punkt, som rör sig i en sferoids
eqvatorsplan under inverkan af den Newtonska attrak-
tionskraften, 8vo. *Stockholm,* 1880

——————— Undersökningar af theorien för Himlakropparnas
rörelser, I., 8vo. *Stockholm,* 1881

——————— Ueber die Theorie der Bewegungen der Himmelskörper, 4to. *Kiel*

Haarlem, Musée Teyler, Archives, Serie II., 1, 28vo. Musée Te
Haarlem, 1881

Halle, Kaiserlich-Leopoldinisch-Carolinische deutsche Aka- The Acad
demie der Naturforscher, Band xli. 1, 2, 4to.
Halle, 1879-80

——————————————Leopoldina, Heft 17, 1881,
4to. *Halle,* 1881

Hamilton, G., A law of elliptic motion, deduced from the laws G. Hamil
of gravitation and compound rotation, 8vo. Esq.
Liverpool, 1873

——————— Progressive changes of form in rotating sphe-
roids (Sun, planets, &c.) from cooling and contraction o
their mass, 8vo. *Liverpool,* 1874

——————————————— Remarks on, by
J. Clerk Maxwell, 8vo.

Harkness, Professor W., On the relative accuracy of different The Au
methods of determining the solar parallax, 8vo.
New Haven, 1881

Harvard College, Astronomical Observatory, 36th Annual The Ob
Report of the Director, E. C. Pickering, 8vo. vatory
Cambridge, U.S., 1882

e
tory,
wa.

Hasselberg, B., Beiträge zur Spectroscopie der Metalloide, 8vo. *St.-Pétersbourg*, 1881

———— Zur Spectroscopie des Wasserstoffs, 8vo.
St.-Pétersbourg, 1881

W.
ord.

Helmholtz, H., Wissenschaftliche Abhandlungen, Band i. 1, 8vo
Leipzig, 1881

Society.

Helsingfors, Finska Vetenskaps-Societeten, Öfversigt af Förhandlingar xxii. 1879–80, 8vo. *Helsingfors*, 1880

—————————————————— Bidrag till Kännedom af Finlands Natur och Folk. Häftet 13, 14, 8vo
Helsingfors, 1880

———————————————— Observations météorologiques, vol. vii. année 1879, 8vo.
Helsingfors, 1882

Author.

Holden, E. S., Observations on the light of telescopes used as night-glasses, 8vo. *New Haven*, 1881

e U.S.
aval
tory.

———— Investigation of the objective and micrometers of the 26-inch Equatoreal constructed by Alvan Clark and Sons, 4to. *Washington*, 1881

British
logical
titute.

Horological Journal, The, Nos. 275–286, 8vo.
London, 1881–82

Author.

Howe, A. H., A theoretical inquiry into the physical cause of epidemic diseases, 8vo. *London*, 1865

.M.'s
ent in
a.

India, Great Trigonometrical Survey, Account of the operations, prepared under the directions of Major-General J. T. Walker, vol. vi., 4to. *Dehra Dun*, 1880

—————————————— General Report on the operations during 1879–80. Prepared under the superintendence of Major-General J. T. Walker, fol. *Calcutta*, 1881

eteoro-
cal
ce.

International Polar Commission, Communications, Part III., 8vo. *St. Petersburg*, 1882

Author.

Janssen, J., Note sur la photographie de la Comète *b* 1881, obtenue à l'observatoire de Meudon, 8vo. *Paris*, 1882

Editor.

Journal of Science, Third Series, Nos. 91–102, 8vo.
London, 1881–82

The
'versity.

Kasan, Université Impériale, Bulletin et Mémoires (en russe), 1880, 8vo. *Kasan*, 1879–80

Kjöbenhavn, Kongelige Danske videnskabernes Selskabs   The Soci
Skrifter, 6te-Række, naturvidensk og mathematisk Afdeeling, Band i. 3, 4, 5, ii. 1, 2, 4to.   *Kjöbenhavn*, 1881

———————————————————— Oversigt, 1880, 2–1882, 1,
8vo.   *Kjöbenhavn*, 1880–82

Knowledge, an illustrated magazine of science, Nos. 1, 2, 4,   The Edi
4to.   *London*, 1881

Konkoly, N. de, Beobachtungen angestellt am Astro-   Dr. d
physikalischen Observatorium in O Gyalla, Band iii.,   Konko
4to.   *Halle*, 1881

Langley, S. P., The Bolometer and radiant energy, 8vo.   The Au
                                            Cambridge, 1881

——————— Sur la distribution de l'énergie dans le spectre
solaire normal, 4to.   *Paris*

Laplace, P. S., Œuvres complètes; publiées sous les au-   The Acad
spices de l'Académie des Sciences, tome iv., 4to.   .of Scien
                                            Paris.
                                            Paris, 1880

Leeds Philosophical and Literary Society, Annual Report   The Soci
for 1880, 8vo.   *Leeds*, 1881

Leipzig, Astronomische Gesellschaft, Publicationen No. XVI.,   Prof.
Syzygien-Tafeln für den Mond, von Prof. Th. von   Oppo
Oppolzer, 4to.   *Leipzig*, 1881

——————————————————— Vierteljahrsschrift,   The Soci
Jahrgang xvi. 1–xvii. 2, 8vo.   *Leipzig*, 1881–82

——— Königlich Sächsische Gesellschaft, Abhandlungen der
mathematisch-physicalischen Classe, VI., 8vo.
                                            Leipzig, 1880

———————————————————— Berichte über die
Verhandlungen der mathematisch-physischen Classe,
1880, i. ii., 8vo.   *Leipzig*, 1880–81

——— Fürstlich Jablonowskische Gesellschaft, Jahresbericht, 1880, 8vo.   *Leipzig*, 1880

Lindhagen, A., Nicolai Coppernici de hypothesibus motuum   M. A
coelestium a se constitutis commentariolus, 8vo.   Lindhag
                                            Stockholm, 1881

Lisboa, Academia Real das Sciencias, Memorias, classe de   The
sciencias mathematicas, physicas et naturaes, nova serie,
tomo v. 2, vi. 1, 4to.   .   *Lisboa*, 1877–81

Society. Lisboa, Academia Real das Sciencias, Jornal, xxviii., xxix., 8vo.   *Lisboa*, 1880

———————————————————————— Sessão publica em 9 de Junho de 1880, 8vo.   *Lisboa*, 1880

——— Sociedade de Geographia, Boletim, Serie II. 4–10, 8vo.
                                                                                                 Lisboa, 1881

ibrary. Liverpool Free Public Library, 29th Annual Report, 8vo.
                                                                                            Liverpool, 1882

'ety. ——— Literary and Philosophical Society, Proceedings, vol. xxxiii., xxxiv., 8vo.   *Liverpool*, 1879–80

u°des ⸴udes. Loewy, Observatoires astronomiques en province, année 1880, 8vo.   *Paris*, 1881

uthor. ——— Éphémérides des étoiles de culmination lunaire et de longitude, pour 1882, 4to.   *Paris*

uthors. ——— Et Périgaud, Étude des flexions du grand cercle meridien (flexion en distance polaire, flexion latérale, flexion de l'axe instrumental), et de la forme des tourillons, à l'aide de l'appareil imaginé par M. Loewy, 4to.   *Paris*, 1881

———————————————————————— Détermination de la flexion horizontale, de la flexion latérale, et de la flexion de l'axe instrumental du cercle méridien de Bischoffsheim, à l'aide du nouvel appareil, 4to.   *Paris*, 1881

uthor. Loomis, E., Contributions to Meteorology, 15th and 16th papers, 8vo.   *New Haven*, 1881–82

. L. ce, Esq. Martin, B., *Venus* in the Sun, being an explication of the rationale of that great phenomenon; *and* E. Halley, Dissertation on the method of determining the parallax of the Sun by the Transit of *Venus*, June 6, 1761, 4to.
                                                                                              London, 1761

tory, ngen. Mayer, Tobias, Grössere Mondkarte, nebst Detailzeichnungen, zum ersten herausgegeben von der Königlichen Stern- warte zu Göttingen (13 sheets, containing 40 photo- graphs).   *Göttingen*, 1881

's ent ictoria. Melbourne Observatory, Monthly record of results of observa- tions in meteorology, terrestrial magnetism, &c., taken under the superintendence of R. L. J. Ellery, Oct. 1880– Feb. 1881, 8vo.   *Melbourne*

Melbourne Observatory, Sixteenth Report of the Board of Visitors, together with the annual Report of the Government Astronomer, fol.   *Melbourne,* 1881   H.M.'s Government in Victoria.

Meteorological Office, Quarterly Weather Report, 1876, 1–1876, Appendix III., 4to.   *London*   The Meteorological Office

———————— Hourly readings from the self-recording instruments at the seven Observatories in connection with the office, April–November, 1880, fol.   *London*   ————

———————— Report of the International Meteorological Committee, Meeting at Berne, 1880, 8vo.   *London,* 1881   ————

————————— Report of the Meteorological Council of the Royal Society for the year ending March 31, 1881, 8vo.   *London,* 1882   ————

Meteorological Society, Quarterly Journal, New Series, Nos. 38–41, 8vo.   *London,* 1881–82   The Society.

———————— Index to the publications of the English meteorological societies, 1839 to 1881, 8vo.   *London,* 1881   ————

———————— The meteorological record, with remarks on the weather, by William Marriott, 8vo.   *London,* 1881–82   ————

———————— Hints to meteorological observers, with instructions for taking observations, and tables for their reduction, by William Marriott, 8vo. *London,* 1881   ————

———————— The snow-storms of January 17 to 21, 1881, 8vo.   *London,* 1881   ————

Meyer, W., Recherches sur *Saturne,* ses anneaux et ses satellites, 4to.   *Genève,* 1881   The Author.

———————— sur l'enrégistrement des battements de secondes d'une pendule au moyen du microphone, 8vo.   *Genève,* 1881   ————

Milano, Reale Osservatorio di Brera, Pubblicazioni No. 19, fol.   *Milano,* 1881   The Observatory.

Moncalieri, Bolletino mensuale pubblicato per cura dell' Osservatorio Centrale del real Collegio Carlo Alberto, ser. II. vol. i.—ii. 1, 4to.   *Torino,* 1881–82   Italian Meteorological Association.

Academy. Montpellier, Académie des Sciences et Lettres, Mémoires de la section des sciences, tome x., année 1880, 4to.
Montpellier, 1881

The
· Moscou, Observatoire de, Annales, publiées par Th. Bredichin, vol. vii. 2, 4to.   *Moscou*, 1881

Society. Moçambique, Sociedade de Geographia, Boletim, Ser. I. No. 1–6, 8vo.   *Moçambique*, 1881

Academy. München, Königlich-bayerische Akademie der Wissenschaften, Abhandlungen der math.-physicalischen Classe, Band xiv. 1, 4to.   *München*, 1881

——————————————————————— Sitzungsberichte der mathematisch-physicalischen Classe, 1881, iii.—1882, ii. 8vo.   *München*, 1881–82

.W.
berck. Naturen, Et illustreret maanedsskrift for populær naturvidenskab, udgivet af C. Krafft, 6te. Aarg. 2, 4to.
Kristiania, 1882

Editor. Naturforscher, der, Jahrgang xiv. No. 25—Jahrgang xv. No. 23, 4to.   *Berlin*, 1881–82

Lords
.s.
of the
iralty. Nautical Almanac and Astronomical Ephemeris for the year 1885, 8vo.   *London*, 1881

Society. Neuchatel, Société des Sciences Naturelles, Bulletin, tome xii. 2, 8vo.   *Neuchatel*, 1881

.Naval
atory. Newcomb, S., Observations of the transit of *Venus*, December 8–9, 1874, made and reduced under the direction of the Commission created by Congress, Part I. 4to.
Washington, 1880

American
emeris
· ——— Catalogue of 1098 standard clock and zodiacal stars, 4to.   *Washington*

Editor. ——— Populäre Astronomie, deutsche vermehrte Ausgabe, bearbeitet durch R. Engelmann, 8vo.   *Leipzig*, 1881

Society. New South Wales, Royal Society, Journal and Proceedings for 1880, vol. xiv. 8vo.   *Sydney*, 1881

Library. New York, Astor Library, 33rd Annual Report of the trustees, 8vo.   *Albany*, 1882

The
rwegian
ernment. Norske Nordhavs-Expedition, 1876–78, iii. Zoologi, Gephyrea, ved D. C. Danelssen og Johan Koren, 4to.
Christiania, 1881

Nursingrow, A. V., Results of meteorological observations at   The Author.
Daba Gardens, Vizagapatam, 1880, 8vo.   *Madras,* 1881

Observatory, The, a monthly review of astronomy, edited by   The Editors.
W. H. M. Christie and E. W. Maunder, Nos. 51-62, 8vo.
London, 1881-82

Olsen, O. T., The fisherman's nautical almanac, 1882, 8vo.   The Author.
Hull, 1881

Oppolzer, Th. von, Ist das Newton'sche Attractionsgesetz zur   ——
Erklärung der Bewegungen der Himmelskörper aus-
reichend ? Hat man Veranlassung, dasselbe nur als
Näherungsausdruck zu bezeichnen ? 4to.

Oudemans, J. A. C., Determination à Utrecht de l'azimut   Geodesic Com-
d'Amersfoort, 4to.   •   *La Haye,* 1881   mission of the
Netherlands.

Oxford, Radcliffe Library and Museum, Catalogue of books   The Radcliffe
added during the year 1881, 4to.   *Oxford,* 1882   Trustees.

Palermo, R. Osservatorio, Pubblicazioni, anni 1880-81, 4to.   The
Palermo, 1882   Observatory.

———————————— Bulletino Meteorologico, anno xv.,   ——
vol. xv. 1879, 4to.   *Palermo,* 1881

Paris, Académie des Sciences de l'Institut de France,   The Academy.
Comptes Rendus hebdomadaires, tome xcii. 22-tome xciv.
22, 4to.   *Paris,* 1881-82

—— École Polytechnique, Journal, tome xxxi. cahier 50,   École Poly-
4to.   *Paris,* 1881   technique.

—— Observatoire de, Annales, Observations, 1870, 1874-78,   The
4to.   *Paris,* 1881   Observatory.

———————————— Rapport annuel pour 1880-82, par   ——
M. le Contre-Amiral Mouchez, 4to.   *Paris,* 1881-82

—— Société Mathématique de France, Bulletin, tome ix.   The Society.
4—tome x. 2, 8vo.   *Paris,* 1881-82

———————— Philomathique, Bulletin, série VII., tome iv.   ——
2—tome vi. 2, 8vo.   *Paris,* 1881-82

Parkhurst, H. M., Astronomical Tables, comprising logarithms   The Author.
from 3 to 100 decimal places, and other useful tables.
Revised Edition, 16mo.   *New York,* 1881

Pearson, Rev. James, An elementary treatise on the tides   ——
(two copies), 8vo.   *London,* 1881

Author. Perrotin, J., Visite à divers Observatoires de l'Europe, 8vo.
Paris, 1881

Petrosemolo, G., Dimostrazione del metodo di Lyon per la riduzione delle distanze lunari, e formole per calcolare la tavola occorrente, 8vo. *Livorno,* 1881

Society. Physical Society of London, Proceedings, vol. iv. part v. 8vo. *London,* 1880–81

Photographic Society of Great Britain, Journal and Transactions, New Series, vol. v. 9—vi. 8, 8vo.
London, 1881–82

Author. Pickering, E. C., Photometric measurements of the variable stars β *Persei* and DM 81° 25', made at the Harvard College Observatory, 8vo. *Cambridge, U.S.,* 1881

———————— Large Telescopes, 8vo.
Cambridge, U.S., 1881

ttee. Poids et Mesures, Comité International des, Procès-Verbaux des Séances de 1881, 8vo. *Paris,* 1882

Potsdam, Astrophysikalisches Observatorium, Publicationen, Band ii., 4to. *Potsdam,* 1881

Prag, k. k. Sternwarte, astronomische, magnetische und meteorologische Beobachtungen, im Jahre 1881, 4to.
Prag, 1882

Author. Prince, C. L., Observations upon the temperature, pressure, and rainfall of the past winter, fol. 1882.

atory. Pulkowa, Nicolai Haupt-Sternwarte, librorum in bibliotheca speculae Pulcovensis contentorum catalogus systematicus. Pars II., ab E. Lindemanno elaborata, 8vo.
Petropoli, 1880

———————— Jahresbericht am 20 Mai, 1881, dem Comité abgestaltet vom Director O. Struve, 8vo.
St. Petersburg, 1881

itors. Revue Scientifique, série III., année i., semestre I. No. 24— année ii., semestre I. No. 22, 4to.
Paris, 1881–82

tory. Rio de Janeiro, Observatoire Impériale, Annales, publiées par Emm. Liais, extrait du premier volume (two copies), 4to.
Rio de Janeiro, 1881

Rio de Janeiro, Observatoire Impériale, Bulletin astronomique et météorologique, Nos. 1–6, 4to. *Rio de Janeiro,* 1881 — The Observatory.

Roma, Società Italiana delle Scienze, detta dei XL.; Memorie di matematica e di fisica, serie III. tomo iii. 4to. *Roma,* 1879 — The Society.

—— R. Accademia dei Lincei, Atti, anno 278, serie III., Transunti, vol. v. 13—vol. vi. 10, 4to. *Roma,* 1881–82 — The Academy

—— Accademia Pontificia de' nuovi Lincei, Atti, anno xxxiii. 7—anno xxxiv. 5, 4to. *Roma,* 1880–81 ——

—— Memorie della Società degli Spettroscopisti Italiani, vol. x. 4—vol. xi. 4, 4to. *Roma,* 1881–82 — The Society

—— Pontificia Università Gregoriana, vol. xix. 9—vol. xxi. 2, 4to. *Roma,* 1881–82 — Sig. G. Ferrari.

Royal Society of London, Philosophical Transactions, vol. 172, ii. 4to. *London* — The Society

————————— Proceedings, Nos. 213–219, 8vo. *London,* 1881–82 ——

—— Geographical Society, Journal, vol. l. 8vo. *London,* 1881 ——

————————— Index to vols. i.–x 8vo. *London,* 1881 ——

————————— Proceedings, and Monthly Record of Geography, new monthly series, vol. iii. 7—vol. iv. 6, 8vo. *London,* 1881–82 ——

————————— Classified Catalogue of the Library, 8vo. *London,* 1871 ——

Royal Asiatic Society of Great Britain and Ireland, Journal, new series, vol. xiii. 2—vol. xiv. 2, 8vo. *London,* 1881–82 — The Society

————————— Bombay branch, Journal, Nos. 37–39, 8vo. *Bombay,* 1880–81 ——

—— Institution of Great Britain, Proceedings, vol. ix. 4, 8vo. *London,* 1881 — The Institution.

—— United Service Institution, Journal, Nos. 110–115, 8vo. *London,* 1881–82 ——

Rugby School Natural History Society, Report for the year 1881, 8vo. *Rugby,* 1882 — The Society.

h

Author. Russell, H. C., Papers read before the Astronomical Section of the Royal Society of New South Wales, 1878–79:—

Some new double stars and southern binaries.

Recent changes in the surface of *Jupiter*.

The Wentworth hurricane.

The "Gem" cluster in *Argo*.

Thunder and hail storms in New South Wales.

Note upon a sliding-scale for correcting barometer readings to 32° Fahr. and mean sea-level.

————— Transit of *Mercury*, Nov. 8, 1881, 8vo.

Sydney, 1882

emy. St.-Pétersbourg, Académie Impériale des Sciences, Bulletin, tome xxvii. 3—tome xxviii. 1, 4to.   *St.-Pétersbourg,* 1881–82

ry. San Fernando, Istituto y Observatorio di Marina, Anales, Seccion II. : Observaciones meteorológicas, año 1880, folio.   *San Fernando,* 1881

Author. Sang, E., On the comet's [Comet *b*, 1881] close approach to the star B.A.C. 1881 (*Camelopardalus,* H. 21), 8vo.

Edinburgh, 1881

Schiaparelli, G. V., Osservazioni astronomiche e fisiche sull' asse di rotazione e sulla topographia della pianeta Marte, Memoria III., 4to.   *Roma,* 1881

Schwedoff, Th., Les configurations de la grande comète de 1882 (*a*) prédites d'après la théorie des ondes cosmiques, fol.   *Odessa,* 1882

————— Sur l'origine de la grèle, 8vo.   *Odessa,* 1882

Editor. Science, vol. ii., Nos. 70 and 71, 4to.

London and New York, 1881

n Scien- ————— Observer, Nos. 33–35, 8vo.   *Boston,* 1881
Society.
e Editor. Scientific Roll, the, and magazine of systematized notes, edited by A. Ramsay, Part I., Nos. 4–6, 8vo.

London, 1881–82

Author. Seeliger, H., Untersuchungen über die Bewegungsverhältnisse in dem dreifachen Sternsystem ζ *Cancri,* 4to.   *Wien,* 1881

lenogra- Selenographical Journal, Nos. 40–52, 8vo (two copies).
hical
iety.   *London,* 1881–82

Sidereal Messenger, the, edited by W. W. Payne, vol. i. Nos.   The Edi
2, 3, 8vo.   *Northfield, Minn.,* 1882

Sirius : Zeitschrift für populäre Astronomie, Band xiv. 6–
Band xv. 6, 8vo.   *Leipzig,* 1881–82

Slafter, Rev. E. F., History and cause of the incorrect lati-   The Au
tudes as recorded in the journals of the early writers,
navigators and explorers relating to the Atlantic coast of
North America, 1535–1740, 8vo.   *Boston,* 1882

Smyth, C. P., Gaseous spectra in vacuum tubes under small
dispersion and at low electric temperature, 4to.
                                             Edinburgh, 1881

———————— On the constitution of the lines forming the low
temperature spectrum of oxygen, 4to.   *Edinburgh,* 1881

Society of Arts, Journal, Nos. 1491–1542, 8vo.   The Soci
                                             London, 1881–82

Stockholm, Kongliga Svenska Vetenskaps - Akademiens The A
Handlingar, Ny Följd, Bandet xiv–xvii., 4to.
                                             Stockholm, 1876–79

————————————————————— Öfversigt   af
Förhandlingar, Bandet xxxiv.–xxxvii., 8vo.
                                             Stockholm, 1877–81

————————————————————— Bihang, Ban-
det iv. v., 8vo.   *Stockholm,* 1877–80

———————————————————————— Lefnadsteck-
ningar, Bandet ii. 1, 8vo.   *Stockholm,* 1878

Stockwell, J. N., The theory of the Moon's motion deduced   The Aut
from the law of universal gravitation, 4to.
                                             Philadelphia, 1881

Struve, O., Ueber den Doppelsterne Σ 60=η *Cassiopeiæ*, 8vo.   The
                                   St.-Pétersbourg, 1881   Observa
                                                              Pulkov
Sydney, Government Observatory, Results of astronomical   H.M.
observations made in the years 1877 and 1878, 8vo.   Gov
                                             Sydney, 1881   in N.

————————————————————— Results of meteorological
observations during 1878–79, under the direction of
H. C. Russell, 8vo.   *Sydney,* 1880–81

' Sydney, Government Observatory, Results of rain and river
      observations made in New South Wales during 1880–81,
      8vo.   *Sydney.* 1881–82

· Tasmania, Royal Society of, Monthly Notices, papers and
      proceedings, and report for 1879, 8vo. *Hobart Town.* 1880

   Tebbutt, J., On the progress and present state of astronomical
      science in New South Wales, 8vo.

   ——————— On a new and remarkable star in the constellation *Ara*, 8vo.   *Sydney,* 1877

   ——————— On the longitude of the Sydney Observatory,
   8vo.   *Sydney,* 1880

   ——————— Note on the opposition magnitudes of *Uranus*
   and *Jupiter*, 8vo.   *Sydney,* 1880

   ——————— On the orbit-elements of Comet I, 1880, 8vo.
                                                Sydney, 1880

   ——————— On the star Lacaille 2145 8vo.   *Sydney,* 1881

   ——————— On the variable star *R. Carinæ*, 8vo.
                                                Sydney, 1881

   ——————— Comet II. 1881, 8vo.   *Sydney*

   Terby, F., Observations des Comètes *b* et *c*, faites à Louvain,
      8vo.   *Bruxelles,* 1881

   Tiflis, physikalisches Observatorium, meteorologische Beo-
      bachtungen, 1880, 8vo.   *Tiflis,* 1881

   ——————————————————— magnetische Beobachtungen, 1880, 8vo.   *Tiflis,* 1881

   Toulouse, Observatoire astronomique, magnétique et météoro-
      logique, Annales, tome i., 1873–78, 4to. *Paris,* 1880

   ——————— Académie des Sciences, Inscriptions et Belles-
   Lettres ; Mémoires, série VIII. tome ii. 2—tome iii. 1,
   8vo.   *Toulouse,* 1880–81

   Torino, Reale Accademia delle Scienze, Memorie, serie II.
      tomo xxxii., xxxiii., 4to.   *Torino,* 1880–81

   ——————————————————— Atti, vol. xvi. 5—
   vol. xvii. 4, 8vo.   *Torino,* 1881–82

   ——————————————————— Bolletino dell Osservatorio della Regia Università di Torino, anno xv. 1880,
   fol.   *Torino,* 1881

Tromholt, S., Sur les périodes de l'aurore boréale, d'après les observations faites à Godthaab en Groenland, fol.   The Author.
Copenhague, 1882   The College.

University College, London, Calendar for the session 1881–82, 8vo.   *London*, 1880

Venus, passage de, Conférence Internationale, procès-verbaux, fol.   *Paris*, 1881   The French Government.

Villarceau, Y., Théorie de la flexion plane des solides, et conséquences relatives, tant à la construction des lunettes astronomiques, qu'à la réglementation de ces appareils, pour les affranchir des déviations de l'axe optique produites par la flexion, 4to.   *Paris*, 1881   The Author.

———— Remarques à l'occasion du Mémoire de MM. Loewy et Périgaud sur la flexion des lunettes, 4to.   The Author.
Paris, 1881

Washburn Observatory of the University of Wisconsin, Contributions, No. 1., 4to.   *Madison*, 1881   The Observatory.

Washington, Office of the American Ephemeris and Nautical Almanac, astronomical papers, vol. I. pt. v., On Gauss's method of computing secular perturbations, by G. W. Hill, 4to.   *Washington*, 1881   Office of American Ephemeris.

———— Smithsonian Institution, Contributions to knowledge, vol. xxiii., 4to.   *Washington*, 1881   The Institution.

———————————— Miscellaneous Collections, vols. xviii.–xxi., 8vo.   *Washington*, 1880–81   ————

———— United States Naval Observatory, Astronomical and meteorological observations made during the year 1876, 2 vols. 4to.   *Washington*, 1880   The Observatory.

Weiss, E., Beobachtungen des Venusdurchganges vom 8 Dec. 1874 in Jassy, 8vo.   *Wien*, 1875   The Author.

———— Planeten und Cometen-Beobachtungen, Sept. 1862 –Dec. 1864, 8vo.   *Wien*   ————

———— Sternschnuppen-Beobachtungen, ausgeführte in den Jahren 1867–70, zusammengestellt von E. W., 8vo.   *Wien*, 1873   ————

Weyer, G. D. E., Über die Berechnung des wahrscheinlichsten Chronometerganges aus einer Reihe von Standbeobachtungen, und über Gewichtsbestimmungen aus Standunterschieden der Chronometer, 4to.   1881   ————

Author.   Weyer, G. D. E., Über die kürzeste Berechnungsart der Monddistanzen im nautischen Gebrauch, 4to.   1881

———————— Die Wiedererscheinung der Methode und Tafel von Elford als sogenannte Neger-Tafel, 4to.   1881

emy.   Wien, Kaiserliche Akademie der Wissenschaften, Sitzungsberichte der mathematisch-naturwissenschaftlichen Classe, Abth. I. Band lxxxii. 3, 4, 5—Band lxxxiii. 1–4, 8vo.

Wien, 1881

————————————————————————— Circular,

Nos. xl.—xlvi., 8vo.   *Wien,* 1881

Klein.   Wochenschrift für Astronomie, Meteorologie und Geographie, neue Folge, Jahrg. 25, No. 22, 8vo.   *Cöln,* 1882

Author.   Wolf, R., Astronomische Mittheilungen, No. liv., 8vo.

Zürich, 1881

———————— Quelques resultats déduits de la statistique solaire, I., 4to.   *Roma*

ray,   Wray, W., Catalogue of telescopes, &c., 8vo (2 copies).

London, 1881

e   Yale College, Report of the Board of Managers of the
rvatory.   Winchester Observatory, 1880–81, 8vo. *New Haven,* 1881

Author.   Young, C. A., The Sun, 8vo.   *New York,* 1881

Society.   Zoological Society of London, Transactions, vol. xi. 5, 6, and general index to vols. i.–x., 4to.   *London,* 1881–82

————————————————— Proceedings, 1881, 1–4, 8vo.

London, 1881–82

nstitute.   Zürich, Meteorologisches Central-Anstalt, schweizerische meteorologische Beobachtungen, v. n. 4to.   *Zürich,* 1881

Society.   ———— Naturforschende Gesellschaft, Vierteljahrsschrift, Jahrgang xxiv. xxv. 8vo.   *Zurich,* 1879–80

. Wolf.   ———— Schweizerische geodätische Commission. Das schweizerische Dreiecksnetz. Band I., Die Winkelmessungen und Stationsausgleichungen, 4to.   *Zurich,* 1881

*MSS., Drawings, Photographs, &c. presented to
the Society.*

ecutors   Bernaerts, G., Observations du Soleil, 1874–79, 6 vols. fol.
e late   and 7 vols. 4to (MSS.).

Bernaerts, G., Drawings of *Jupiter* and *Mars* 3 vols. 4to. (MSS.)　　The Execu of the M. Bern

Mars, drawings of, made at the Royal Observatory, Green-wich, 1879 and 1881–82.　　The Rc Observa

Sang, E., Diagrams of the Transit of *Venus*, 1882, with descriptions (MS.).　　The Au

Photograph of the nebula in *Orion*.　　Dr. H.

Photographs of drawings of *Jupiter*.　　Lord

————— of the solar eclipse of 1882, May 17.　　H. Pra

————of Sun-spots.

————— the lunar eclipse of 1881, Dec. 5.　　J. Capron,

————— of Mr. Tebbutt's Observatory, Windsor, New South Wales.　　J. Tebb Esq.

————— of the Washburn Observatory.　　Prof. E Holde

————— Vienna Observatory.　　Prof. E.

————— Strassburg Observatory.

Two photographs of the Meridian Photometer.　　] Pic

Photograph of tomb of J. Harrison.　　Mr.

Photographic portrait of Mr. Birt.　　Rev. Ri

Model of Mr. Birt's arrangement for determining the tints on the lunar surface.

Leaden token, used as a ticket of admission by the Members of the Spitalfields Mathematical Society.　　W. God Esq.

Books purchased with the Turnor Fund.

Astronomische Nachrichten, General Register (Bde i.—lxxx.), 4 vols., 4to.　　*Hamburg und Leipzig,* 1851–75

Bessel, F. W., Recensionen, herausgegeben von Rudolf Engelmann, 8vo.　　*Leipzig,* 1878

Copernicus, 1881–82, 4to.　　*Dublin*

Flammarion, C., Curiosités du Ciel, 8vo.　　*Paris*

Higgins, W. H., The names of the stars and constellations, compiled from the Latin, Greek, and Arabic, with their derivations and meanings, 8vo.　　*London,* 1882

Merriman, M., Elements of the method of least squares, 8vo.

London, 1877

Nature, 1881–82, 4to.   *London*

Philosophical Magazine, 1881–82, 8vo.   *London*

St. Petersburg, Commentarii Academiæ Scientiarium Imperialis Petropolitanæ, tomi i.–viii., 4to.

Petropoli, 1740–52

Thomson, Sir W., and P. G. Tait, Treatise on natural philosophy, vol. i., part i., new edition, 8vo.   *Cambridge,* 1879

Vénus, Recueil de Mémoires, rapports et documents rélatifs au passage de, 1874, 2 vols., 4to.   *Paris,* 1876–77

Zeitschrift für Instrumentenkunde, 1881–82, 4to.   *Berlin*

Lightning Source UK Ltd.
Milton Keynes UK
UKHW021926180219
337529UK00011B/954/P